A First Course in Differential Equations, Modeling, and Simulation

A First Course in Differential Equations, Modeling, and Simulation shows how differential equations arise from applying basic physical principles and experimental observations to engineering systems. Avoiding overly theoretical explanations, the textbook also discusses classical and Laplace transform methods for obtaining the analytical solution of differential equations. In addition, the authors explain how to solve sets of differential equations where analytical solutions cannot easily be obtained. Incorporating valuable suggestions from mathematicians and mathematics professors, the third edition:

- Reworks the chapter "Response of First and Second Order Systems" to include the system response to step changes, impulses, rectangular pulses, and sinusoid forcing functions as well as the response of coupled first- and second-order ordinary differential equations (ODEs); it also introduces Bode plots to analyze the frequency response of second-order ODEs and the principle of oscillation modes in coupled second-order ODEs
- Adds a new section on springs and dampers in series or parallel
- Includes new content on Simulink® and modeling
- Contains new exercises that can be used as projects and answers to many of the end-of-chapter problems
- Features new end-of-chapter problems and updates throughout

This textbook provides students with a practical understanding of how to apply differential equations in modern engineering and science.

A solutions manual and files of all figures in the text are available to adopting professors.

A First Course in Differential Equations, Modeling, and Simulation

3rd Edition

Carlos A. Smith, Scott W. Campbell,
and Ryan G. Toomey

CRC Press
Taylor & Francis Group
Boca Raton London New York

CRC Press is an imprint of the
Taylor & Francis Group, an **informa** business

Designed cover image: Shutterstock

Third edition published 2026
by CRC Press
2385 NW Executive Center Drive, Suite 320, Boca Raton FL 33431

and by CRC Press
4 Park Square, Milton Park, Abingdon, Oxon, OX14 4RN

CRC Press is an imprint of Taylor & Francis Group, LLC

First edition published by CRC Press 2016
Second edition published by CRC Press 2021

ISBN: 978-1-041-02421-7 (hbk)
ISBN: 978-1-041-02422-4 (pbk)
ISBN: 978-1-003-61916-1 (ebk)

DOI: 10.1201/9781003619161

Typeset in Times
by Apex CoVantage, LLC

Access the Support Material: https://www.routledge.com/9781041024217

Carlos A. Smith

This book is dedicated to the Lord, my God. For his daily blessings have made this work possible.

To the love of my life and of the entire family, my wife Cristina – "Mimi."

To the Livingstons – Tim, Cristina, Sophia, and Christopher.

To the Smiths – Carlos, Jennifer, Alex, Nicolas, and Kara K.

To the Smiths in Miami – Rene, Terina, Eddy, Mikey, Miriam, Stephanie, Sabrina (John Muia), Lucas (Karla), Kelly, Finn, and Phoebe.

To my school "De La Salle," to my teachers for having taught me the principles in life, "God, Country, and Family," to my classmates, and to previous graduates, for their friendship and being great role models.

To my former students!!

Special Dedication.

*CAS also wants to dedicate this edition to a couple of brothers, Lieutenant Kevin Livingston (*rip*), and Brian Livingston, JD. Kevin was a mechanical engineer graduate from the Naval Academy in Annapolis. Kevin was an outstanding young man that made us all in his family immensely proud. Kevin trained as a helicopter pilot but ended up serving in the carrier Enterprise during the Gulf War.*

Brian Livingston, Kevin's brother, is another outstanding individual but this time, not so young. CAS owes a great debt to Brian for his encouragement in writing this third edition. His suggestions about how to write it, related to end of chapters problems, commas, paragraphs, and the use of . . . were inspiring during the writing stages.

There is a third brother, S. Timothy Livingston, another outstanding young man who is not mentioned in this special dedication. Timothy was incredibly lucky, *for marrying my daughter and he is mentioned just above in the "normal dedications."*

And to my dearest homeland . . . Cuba.

Scott W. Campbell

Dedicated to:

My late parents, Ruth and Robert, who gave me the opportunities they didn't have, and to my wife and best friend, Gwendolyn, for helping me to make use of them.

My late father-in-law, Richard Jones, for inspiration both in teaching and in engineering, and to my daughter Alicia Morrow, for encouraging me to follow my muse.

Boris Vuksanovich and Roger Smith, my junior high and high school math teachers, for making mathematics challenging, doable and, most of all, fun.

I'd also like to add my thanks to Carlos A. Smith, for being the driving force behind both this book and the course that inspired it.

Ryan G. Toomey

I dedicate this to my parents Jill and Kenton, my wife Norma, and my son Markus. I also thank Carlos A. Smith for allowing me to contribute to the third edition of this textbook. Finally, I dedicate this to all the students, past and future, who inspire my passion for teaching.

Contents

Preface....xiii
Authors....xv

Chapter 1 Introduction....1

1.1 An Introductory Example....1
1.2 Differential Equations....5
1.2.1 Initial and Boundary Conditions....6
1.2.2 Ordinary and Partial DEs....8
1.3 Modeling....9
1.4 Forcing Functions....10
1.5 Book Objectives....13
1.6 Summary....14

Chapter 2 Objects in a Gravitational Field....16

2.1 An Example....16
2.2 Antidifferentiation – Technique for solving first-order ordinary differential equations....17
2.3 Back to Section 2.1....18
2.4 Another Example....19
2.5 Separation of Variables – Technique for Solving First-Order Ordinary Differential Equations....20
2.6 Back to Section 2.4....21
2.7 Equations, Unknowns, and Degrees of Freedom....22
2.8 Partial Fractions Expansion....25
2.8.1 Unrepeated Real Roots – Algebra....27
2.8.2 Repeated Real Roots – Algebra....28
2.8.3 Unrepeated Real Roots – Limit Concept....30
2.8.4 Repeated Real Roots – Limit Concept....31
2.8.5 Complex Roots – Limit Concept....32
2.8.6 First-Order Expansion....32
2.9 Summary....34

Chapter 3 Classical Solutions of Ordinary Linear Differential Equations....40

3.1 Examples of Differential Equations....40
3.1.1 Bioengineering – Chapter 8....40
3.1.2 Mechanical Translational – Chapter 6....41
3.1.3 Fluid System – Chapter 8....41
3.1.4 Thermal System – Chapter 9....42
3.1.5 Electrical Circuit – Chapter 10....42
3.2 Definition of a Linear Differential Equation....43
3.3 Integrating Factor Method....44
3.3.1 Development of the Integrating Factor Method....46

3.4 Solution of Homogeneous Differential Equations 47
3.4.1 Characteristic Equation 47
3.4.2 Roots of the Characteristic Equation 51
3.4.3 Qualitative System Response 52
3.5 Solution of Nonhomogeneous Differential Equations 55
3.5.1 Homogeneous Solution (Natural Response) and Nonhomogeneous Solution (Forced Response) 56
3.5.2 Undetermined Coefficients 57
3.5.3 Multiple Forcing Functions 68
3.6 Variation of Parameters 70
3.6.1 Second-Order Systems 70
3.6.2 Higher-Order Systems 77
3.7 Handling Nonlinearities and Variable Coefficients 80
3.7.1 Taylor Series 80
3.7.2 Linearization of Nonlinear Differential Equations 81
3.8 Transient and Final Responses 87
3.9 Summary 90

Chapter 4 Laplace Transforms 101

4.1 Definition of the Laplace Transform 101
4.2 Properties and Theorems of the Laplace Transform 104
4.2.1 Linearity Property 104
4.2.2 Real Differentiation Theorem 105
4.2.3 Real Integration Theorem 106
4.2.4 Real Translation Theorem 106
4.2.5 Final Value Theorem 107
4.2.6 Complex Differentiation Theorem 107
4.2.7 Complex Translation Theorem 108
4.2.8 Initial Value Theorem 108
4.3 Solution of Differential Equations Using Laplace Transform 110
4.3.1 Inversion by Partial Fractions Expansion 112
4.3.2 Handling Time Delays 130
4.4 Transfer Functions 132
4.5 Algebraic Manipulations Using Laplace Transforms 134
4.6 Deviation Variables 136
4.7 Summary 140

Chapter 5 Response of First- and Second-Order Systems 148

5.1 First-Order Systems 148
5.1.1 Response to a Step Function 150
5.1.2 Response to an Impulse and a Rectangular Pulse 153
5.1.3 Response to a Sinusoidal Signal 157
5.1.4 Transfer Functions for First-Order ODEs 160
5.2 Second-Order Systems 160
5.2.1 Response to a Step Function 162
5.2.2 Response to an Impulse and a Rectangular Pulse 166
5.2.3 Response to a Sinusoidal Signal 168
5.2.4 Transfer Function 171

5.3 Coupled First-Order ODEs ... 171
5.4 Coupled Undamped Second-Order ODEs ... 173
5.4.1 Weakly Coupled Oscillators and Beats ... 178
5.5 Examples ... 180
5.6 Some Concluding Remarks (Before the Summary) ... 186
5.7 Summary ... 187

Chapter 6 Mechanical Systems: Translational ... 196

6.1 Mechanical Law, Systems Components, and Forces ... 196
6.1.1 Mechanical Law ... 196
6.1.2 System Components ... 196
6.1.3 Forces ... 199
6.2 Types of Systems ... 200
6.2.1 Undamped System ... 201
6.2.2 Damped System ... 204
6.3 D'Alembert's Principle and Free-Body Diagrams ... 207
6.4 Examples ... 209
6.5 Springs and Dampers in Series or Parallel ... 218
6.5.1 Springs in Parallel ... 218
6.5.2 Springs in Series ... 219
6.5.3 Dampers in Parallel ... 220
6.5.4 Dampers in Series ... 220
6.5.5 Springs and Dampers in Series ... 221
6.6 Vertical Systems ... 222
6.7 Summary ... 225

Chapter 7 Mechanical Systems: Rotational ... 238

7.1 Mechanical Law, Moment of Inertia, and Torque ... 238
7.1.1 Mass Moment of Inertia ... 239
7.1.2 Torque ... 241
7.2 Torsion Springs ... 244
7.3 Rotational Damping ... 246
7.4 Gears ... 249
7.5 Systems with Rotational and Translational Elements ... 252
7.6 Summary ... 254

Chapter 8 Mass Balances ... 268

8.1 Conservation of Mass ... 268
8.1.1 Multicomponent Systems ... 270
8.1.2 Types of Processes ... 272
8.2 Flow Rates and Concentrations ... 273
8.3 Elements and Experimental Facts ... 275
8.3.1 Flow Element ... 275
8.3.2 Liquid Service ... 275
8.3.3 Gas Service ... 277
8.4 Examples ... 277
8.5 Expressions for Mass Transport and Chemical Reactions ... 283
8.5.1 Mass Transport ... 283

8.5.2 Chemical Reactions 285
8.5.3 Batch Processes 287
8.6 Additional Examples 290
8.7 Application to Bioengineering Processes 298
8.7.1 Compartmental Modeling 298
8.7.2 Biological Reactions 301
8.7.3 Fermentation 307
8.8 Final Comments 311
8.9 Summary 311
References 311

Chapter 9 Thermal Systems 326

9.1 Conservation of Energy 326
9.2 Modes of Heat Transfer 326
9.3 Conduction 327
9.4 Convection 328
9.5 Conduction and Convection in Series 329
9.6 Accumulated or Stored Energy 331
9.7 Some Examples 332
9.8 Heat Transfer in a Flow System 339
9.9 Thermal Effects in a Reactive System 343
9.10 Boundary Value Problems in Heat Transfer 345
9.11 Summary 358

Chapter 10 Electrical Systems 366

10.1 Some Definitions and Conventions 366
10.2 Electrical Laws, Components, and Initial Conditions 367
10.2.1 Electrical Laws 367
10.2.2 Electrical Components 369
10.2.3 Initial Conditions 372
10.3 Examples of Electrical Circuits 373
10.3.1 Undamped Circuit (Natural Frequency and Resonance) 383
10.4 Additional Examples 386
10.5 Energy and Power 397
10.6 RC Circuits as Filters 407
10.6.1 High-Pass Filter 409
10.6.2 Low-Pass Filter 411
10.7 Summary 414

Chapter 11 Numerical Simulation 427

11.1 Numerical Solution of Differential Equations 427
11.2 Euler's Method for First-Order Ordinary Differential Equations 428
11.3 Euler's Method for Second-Order Ordinary Differential Equations 429
11.4 Step Size 431
11.5 More Sophisticated Methods 432
11.6 Representation of Differential Equations by Block Diagrams 432

11.6.1 Basic Blocks 432
11.6.2 Guidelines for Constructing Block Diagrams 435
11.6.3 Some Additional Examples 436
11.6.4 Some Additional Source Blocks 439
11.7 Additional Examples 441
11.8 Some Additional Topics on Simulink, and Modelling 448
11.8.1 Writing Equations 448
11.8.2 Trial and Error (Convergence) 448
11.8.3 Another Helpful Simulation Tip 451
11.9 Summary 454
Reference 454

Answers to Selected Problems 467
Index 509

Preface

The third edition continues our desire to be student-oriented and that of making differential equations meaningful from an engineering/science point of view. We seek to make the presentation readable, simple, and practical to motivate the student to learn the subject. Equally, we hope to provide students with the necessary tools to facilitate future courses and engineering practice.

Using simple physics, Chapters 1 and 2 introduce dynamic modeling, the definition of differential equations, two simple methods for obtaining their analytical solution, and a method to follow when modeling. These chapters are used to introduce the subject and to motivate the readers for important things to come.

Chapter 3 presents the classical methods for obtaining the analytical solution of differential equations. The chapter also discusses in detail the meaning and engineering importance of the roots of characteristic equations. The Laplace transform method is the subject of Chapter 4. The meaning of transfer function, as well as the power of Laplace transforms for obtaining the analytical solution of coupled differential equations, is presented in detail. Chapter 5 presents in detail the response of first- and second-order differential equations.

Chapters 6 through 10 present the modeling (which is the set of equations that describe the behavior/response of systems) of translational and rotational mechanical systems, fluid systems, biochemical systems, thermal systems, and electrical systems. These chapters follow the same structure in that the basic applicable physical law opens each section, followed by the applicable element equations and physical facts, and finally, by several examples. In an effort to integrate the book, in several examples in these chapters, we refer to previous material related to the response and stability of systems without obtaining the analytical solution. Chapter 11 presents simulation (which is obtaining the behavior/response of the model using numerical methods/computer), with examples developed in previous chapters.

The material can be covered in an order and degree of completion that is tailored to the pedagogical philosophy of the instructor. Certainly, the material can be covered in the order it is presented where the students first develop the full set of mathematical skills before applying them to physical systems. Chapter 11 can easily be covered at any moment in the course. Support materials are available to adopting professors on the Routledge website (https://www.routledge.com/9781041024217).

A few problems at the end of the modeling chapters in this third edition suggest the student obtain both the analytical solution (taught in Chapters 2, 3, and 4) and the simulation solution (Chapter 11) and compare the results. Although not all the problems suggest this exercise, it can be suggested for most problems. Therefore, in this case, it would be beneficial to cover Chapter 11 right after Chapter 5 before entering Chapter 6 and on. Good luck and good learning!

Carlos A. Smith
Scott W. Campbell
Ryan G. Toomey
Tampa, Florida

Authors

Carlos A. Smith is Professor Emeritus of chemical engineering at the University of South Florida. He has been on the faculty for 52 years, serving in different capacities. Professor Smith has lectured in Europe and many countries in Latin America. He is the coauthor of three editions of a textbook on process control and the author of another book on the same subject. These books have been translated into Spanish and Portuguese.

Scott W. Campbell served on the faculty of the Department of Chemical, Biological & Materials Engineering at the University of South Florida from 1986 until he retired in 2022. After a year away, he has returned to teach part-time, including the course that uses this textbook. He has authored or co-authored over 60 technical peer-reviewed articles, mostly in the area of thermodynamics, and has received numerous teaching awards at the department, college, university, and state levels.

Ryan G. Toomey is a professor of chemical engineering at the University of South Florida. Following receipt of his BS (University of California Berkeley, 1996) and PhD (University of Minnesota, 2002) in chemical engineering, he was a postdoctoral associate with the Institute for Microsystems Technology at the University of Freiburg (Germany). His research activities focus primarily on using responsive, surface-tethered polymer networks to mediate interfacial interactions. His group is concerned with structure–property–function relationships of stimuli-sensitive polymers in confined geometries and how volume-phase transitions in thin films can be harnessed to direct and control adsorption and desorption phenomena at surfaces. He is the recipient of a Camille and Henry Dreyfus New Faculty Award (2005), a National Science Foundation CAREER Award (2007), and the Outstanding American Institute of Chemical Engineers student chapter advisor award (2023).

1 Introduction

This chapter looks to motivate you to read and study this book. We hope to open your mind to a fascinating and practical engineering tool and do so by introducing the topics of the book using a simple, commonsense example. The example shows the use of basic differentials learned in calculus courses and introduces differential equations and their analytical and numerical solutions (simulation) as well as the subject of mathematical modeling.

1.1 AN INTRODUCTORY EXAMPLE

Consider the following tank where two streams enter the tank, mix, and a single stream exits through a valve.

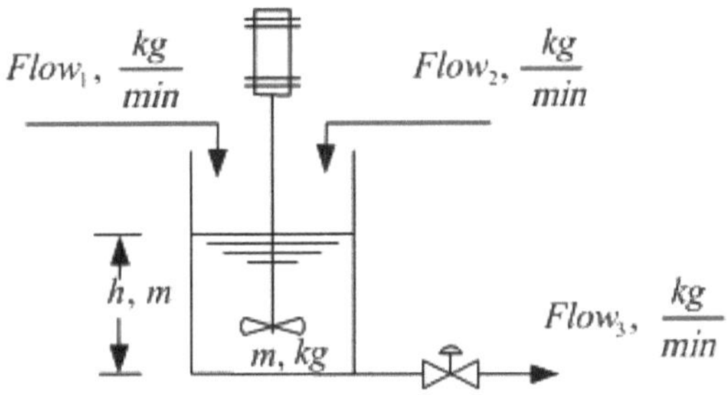

FIGURE 1.1 Mixing tank

Let us suppose that the process is at a state when nothing is changing; that is, the same amount of flow exits out as comes in, and of course, the liquid level is constant. We refer to this condition as the "steady state," or, say, that the system is "at rest." At this steady operation, we can write

$$Flow_1 + Flow_2 = Flow_3 \tag{1.1}$$

or

$$Flow_1 + Flow_2 - Flow_3 = 0 \tag{1.2}$$

Note that we can also word the Equation 1.2 as

$$\sum Flows\ in - \sum Flows\ out = 0 \tag{1.3}$$

Let us now use the following notation:

$$w_1 = Flow_1; w_2 = Flow_2; w_3 = Flow_3$$

and the units of each term are kg/min. w's are used in this book to express mass flow rates (mass/time). So we can write

$$w_1 + w_2 - w_3 = 0 \tag{1.4}$$

DOI: 10.1201/9781003619161-1

Equations 1.1, 1.2, 1.3, and 1.4 are *algebraic equations*.

Let us suppose that $w_1 = 200\ \text{kg/min}$ and $w_2 = 100\ \text{kg/min}$; Equation 1.4 indicates that

$$200 + 100 - w_3 = 0$$

and

$$w_3 = 300\ \text{kg/min}$$

Ifs w_2 changes to $200\ \text{kg/min}$, Equation 1.4 indicates that now $w_3 = 400\ \text{kg/min}$.

The outlet flow of liquid through the valve depends, among other things, on the liquid level in the tank, h as indicated in the figure, which engineers often refer to as the head of liquid. This dependence, or relation, may be expressed as

$$w_3 = C_V \sqrt{h} \tag{1.5}$$

where C_V is called the valve coefficient, and it is what the engineer calculates when sizing the valve. For this example, assume $C_V = 166.7\ \text{kg}/\text{m}^{1/2} \cdot \text{min}$. Substituting Equation 1.5 into 1.4 gives

$$w_1 + w_2 - 166.7\sqrt{h} = 0 \tag{1.6}$$

For the steady operation, Equation 1.6 provides the necessary liquid level to deliver the outlet flow:

$$200 + 100 - 166.7\sqrt{h} = 0$$
$$h = 3.24\ \text{m}$$

If w_2 becomes $200\ \text{kg/min}$, the new necessary liquid level to deliver $400\ \text{kg/min}$ is

$$200 + 200 - 166.7\sqrt{h} = 0$$
$$h = 5.76\ \text{m}$$

So Equation 1.4 provides the new outlet flow, and Equation 1.6 provides the liquid level in the tank necessary to deliver the new outlet flow. These equations describe the liquid in the tank; they are the *mathematical model* of the tank.

When flow w_2 changes from $100\ \text{kg/min}$ to $200\ \text{kg/min}$ the level in the tank must change from $3.24\ \text{m}$ to $5.76\ \text{m}$. Although the change in inlet flow may be very close to instantaneous, the change in level is not instantaneous. It takes some amount of time to go from the initial level to the new level. The model given by Equation 1.6 does not provide, or is not related to, the time it takes to accomplish the change. This model only provides the level required to produce the new output flow. If the time to reach the new level is important, then another model is needed.

The new model needs to describe *how fast the level in the tank changes* when any of the inlet flows, or both, changes. To see how this equation develops, consider Equation 1.3 again:

$$\sum \textit{Flows in} - \sum \textit{Flows out} = 0 \tag{1.3}$$

or, noting that the flows are specifically mass flows,

$$\sum \begin{matrix}\textit{Rate of mass} \\ \textit{entering system}\end{matrix} - \sum \begin{matrix}\textit{Rate of mass} \\ \textit{exiting system}\end{matrix} = 0 \tag{1.7}$$

The expression *Rate of mass* refers to flows in units of *mass/time* (kg/ min in this example). Equation 1.7 only refers to the streams entering and exiting the system (the tank in this case); it does not account for the mass inside the system. Thus, it does not describe what happens to the mass or liquid level in the tank when the entering and exiting streams are not equal. To account for the mass inside the process and develop the desired model, we rewrite Equation 1.7 as

$$\sum \begin{matrix} \textit{Rate of mass} \\ \textit{entering system} \end{matrix} - \sum \begin{matrix} \textit{Rate of mass} \\ \textit{exiting system} \end{matrix} = \begin{matrix} \textit{Rate of change of mass} \\ \textit{accumulated in system} \end{matrix} \tag{1.8}$$

Applying this equation to the tank,

$$w_1 + w_2 - w_3 = \frac{dm}{dt} \tag{1.9}$$

where m is the mass inside the tank (accumulated) as indicated in Figure 1.1.

As we have learned in calculus, the derivative $\frac{dy}{dx}$ means the *change of* y *with respect to* x. So, the term $\frac{dm}{dt}$ means the change *of the mass in the tank* (m) *with respect to time* (t), or in other words, *how fast the mass in the tank changes* or the *rate of change of the mass in the tank*. A positive derivative indicates that the mass in the tank is increasing because there is more mass entering the tank than exiting; a negative derivative indicates that the mass in the tank is decreasing, or depleting, because there is more mass exiting the tank than entering.

The mass of liquid accumulated in the tank is given by the volume of liquid times its density (mass/volume). The volume of liquid is given by the level of liquid times the cross-sectional area of the tank. All of this is mathematically written as

$$m = \rho V = \rho\, Ah \tag{1.10}$$

where ρ = density of fluid – assumed constant in this example at $1000 \frac{\text{kg}}{\text{m}^3}$

V = volume of liquid in tank, m^3

A = cross-sectional area of tank – assumed constant at 0.146 m^2, and

h = height (level) of liquid in the tank, as shown in Figure 1.1.

From Equation 1.10,

$$\frac{dm}{dt} = \rho A \frac{dh}{dt} \tag{1.11}$$

Substituting Equations 1.5 and 1.11 into Equation 1.9,

$$w_1 + w_2 - C_V \sqrt{h} = \rho A \frac{dh}{dt} \tag{1.12}$$

or

$$\rho A \frac{dh}{dt} + C_V \sqrt{h} = w_1 + w_2 \tag{1.13a}$$

or

$$146\frac{dh}{dt}+166.7\sqrt{h}=w_1+w_2 \tag{1.13b}$$

Equation 1.13 is a *differential equation*, and its solution describes the liquid level in the tank as a function of the inlet flows; Equation 1.12 is the same differential equation written a bit differently. *The solution of these equations gives the height* h *as a function of time when* w_1 *and/or* w_2 *change.* Figure 1.2 shows the level response when w_2 changes from 100 to 200 kg/min at $t = 5$ min. The input flow w_2 changes as a step (instantaneously), but it takes some time for the level to go from its original to its final value. Note that the solution/response provides the final height of liquid after a *very long time*, that is, as time gets large, which is the same information provided by Equation 1.6. Thus, *the solution of the differential equation describes how the level in the tank changes as a function of time, including its final steady value.* That is, the solution of the differential equation describes how fast and how much the liquid level changes. The differential equation provides more information than the algebraic equation.

We will learn how to develop the differential equation (the model) and how to obtain its solution in this book. Do not worry, it is fun and very practical.

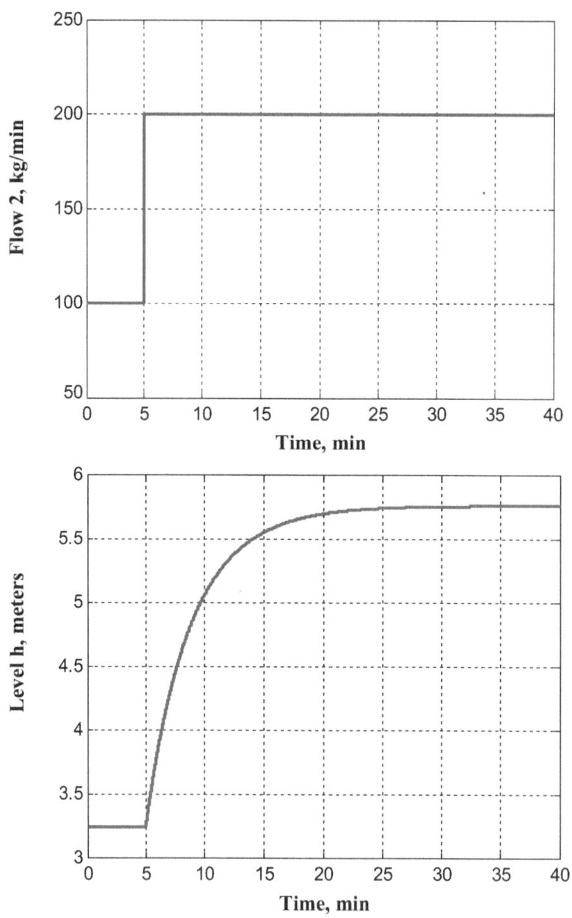

FIGURE 1.2 Response of level in the tank.

1.2 DIFFERENTIAL EQUATIONS

Up to this moment in your study of engineering and science you have become quite familiar with *algebraic equations*, for example,

$$x^3+3x^2+4x=3t^2+4t \tag{1.14}$$

From this equation, we can solve for x at any t. We could repeat this for different values of t and make a graph of x versus t.

Many natural and man-made phenomena and systems cannot be described by algebraic equations, or the description provided by these equations is not complete (as in the example in Section 1.1). In these cases, *differential equations (DEs)* may provide the required description; an example of a DE is

$$\frac{d^3x(t)}{dt^3}+3\frac{d^2x(t)}{dt^2}+4\frac{dx(t)}{dt}+x(t)=F(t) \tag{1.15}$$

Note that the variable x that is differentiated must be a function of t otherwise the differential would be zero. That is the reason for writing $x(t)$. What often happens is that, because obviously x is a function of t, we often drop the (t) term and simply write

$$\frac{d^3x}{dt^3}+3\frac{d^2x}{dt^2}+4\frac{dx}{dt}+x=F(t) \tag{1.16}$$

A differential equation is an equation containing one or more derivatives of an unknown function and perhaps the function itself. In Equation 1.15 $x(t)$, or x in Equation 1.16, is the unknown function; in Equation 1.13, h is the unknown function.

The unknown function is called the *dependent variable*; the variable by which this dependent variable is differentiated by what is called the *independent variable*, t *in Equations 1.15, 1.16, and 1.13*; note that all appearances of the dependent variable are on the left side of the equal sign. The function on the right-hand side of the equal sign, $F(t)$ in Equations 1.15 and 1.16, or w_1 and w_2 in Equation 1.13, is called the *forcing function* because once it changes, it "forces" the dependent variable to change.

It is common in mathematics to use primes (′) or dots (•) to represent derivatives, that is,

$$\frac{dx}{dt}=x'=\dot{x};\quad \frac{d^2x}{dt^2}=x''=\ddot{x}, \text{etc}$$

In these notes, we may use the prime notation for derivatives. Therefore, we could have written Equation 1.16 as

$$x'''+3x''+4x'+x=F(t)$$

and for Equation 1.13,

$$\rho Ah'+C_V\sqrt{h}=w_1+w_2$$

The *order* of a DE is the highest derivative in the equation. For example, Equation 1.16 is a third-order DE, and Equation 1.13 is a first-order DE.

To summarize,

Order (3) — Dependent variable (x) — Forcing function

$$\frac{d^3x}{dt^3}+3\frac{d^2x}{dt^2}+4\frac{dx}{dt}+x=F(t)$$

Independent variable (t)

1.2.1 Initial and Boundary Conditions

A differential equation by itself is not enough to provide the complete solution. Consider the following equation from physics and mathematics:

$$\frac{dx}{dt}=v \tag{1.17}$$

where x is position, t is time, and v is velocity. Assuming v is constant at 10 m/s,

$$\int dx=\int v\,dt$$

$$x=vt+C=10t+C \tag{1.18}$$

and a possible graph of this solution is shown in Figure 1.3.

Both lines, and many others, in Figure 1.3b satisfy Equation 1.18. Equation *1.18 represents a family of solutions,* depending on the value of *C, which satisfies the differential equation.* To know which curve, and value of C, is the correct one for the physical situation at hand, a value of x at some time t must be known. Usually, based on the physics of the situation, x at $t = 0$ *is* specified, or known, and indicated as $x(0)$. That is,

$$x\big|_{t=0}=x(0) \tag{1.19}$$

where the vertical line after the x denotes "evaluated at." This specification is called an *initial condition.* Thus, the complete mathematical model requires the differential equation *and* the initial condition, or

$$\frac{dx}{dt}=v \tag{1.20a}$$

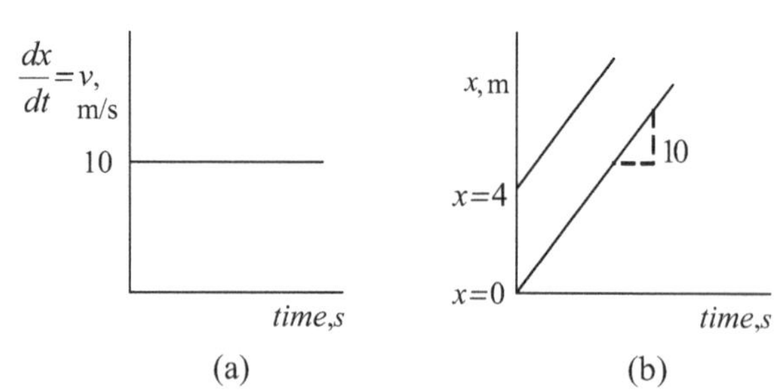

FIGURE 1.3 Graph of Equation 1.18

and

$$x\big|_{t=0} = x(0) \tag{1.20b}$$

We call this an *initial value problem*, and it can now be solved by applying the initial condition in Equation 1.18:

$$x(0) = 10(0) + C \quad \Rightarrow \quad C = x(0)$$

or

$$x = x(0) + 10t$$

In obtaining Equation 1.18, we made use of indeterminate integrals. We can also use the limits of integration to obtain the final solution:

$$\begin{aligned} \int_{x(0)}^{x} dx = \int_0^t v\,dt = \int_0^t 10\,dt \Rightarrow x\big|_{x(0)}^{x} &= 10t\big|_0^t \\ x - x(0) &= 10(t-0) \\ x &= x(0) + 10t \end{aligned} \tag{1.21}$$

For an *nth-order* DE n *initial conditions* are needed to complete the model. That is, for Equation 1.16, the following initial conditions $x''(0)$, $x'(0)$, and $x(0)$ are necessary – an initial condition for each of the first *n – 1* derivatives and one for the unknown function itself.

For example, consider the electrical circuit shown in Figure 1.4. Chapter 10 shows that the DE that describes the voltage drop v_C across the capacitor when the switch is closed is Equation 10.27b:

$$3.75\times10^{-6}\frac{d^2v_C}{dt^2} + 3.0\times10^{-4}\frac{dv_C}{dt} + v_C = v_S \tag{10.27b}$$

To obtain the solution of this second-order differential equation two initial conditions are required. If originally the capacitor is discharged, we state this as $v_C(0) = 0$ V (V stands for volts); that is, at $t = 0$, there is no voltage drop across the capacitor; this is the first initial condition. If originally the circuit is at "rest" or in a "steady" condition this is stated as $\frac{dv_C}{dt}\big|_{t=0} = 0$ V/s (V/s stands

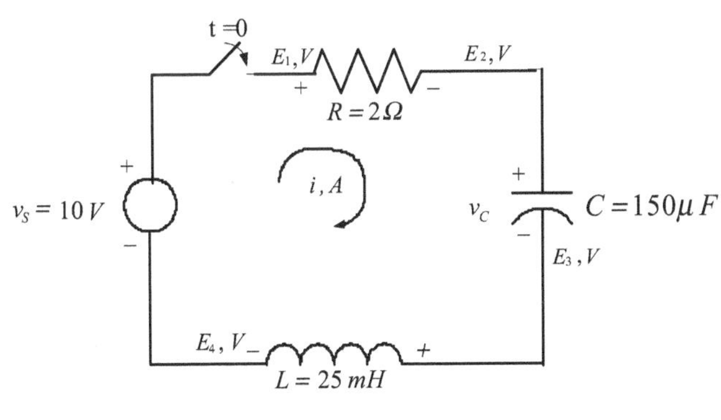

FIGURE 1.4 Resistance, inductor, capacitor circuit

for volts/second); this is the second initial condition. This statement of the first derivative of v_C equal to zero makes sense. Remember that writing $\frac{dv_C}{dt}|_{t=0}$ is "shorthand" for saying *the rate of change of* v_C *at* $t=0$. Because originally v_C is steady that is, nothing is changing, *the rate of change of* v_C *at* $t=0$ *equals 0 V/s*, or $\frac{dv_C}{dt}|_{t=0}=0$ V/s.

Thus, if in any system the dependent variable is at a steady value (at rest) before the forcing function changes, we state that at that moment $\frac{d(\text{dependent variable})}{dt}\Big|_{t=0}=0.$

It is simple to realize that if this is the case, the higher derivatives of the dependent variables are also 0. Think about it, use the example of displacement x, velocity $\frac{dx}{dt}$, and acceleration $\frac{d^2x}{dt^2}$ to help you understand.

It is important to realize how the physical system conditions are used to write the initial conditions. Remember, in developing the equations describing physical systems, the mathematics *must* reflect the physical conditions, and all the terms must have units that make sense. We will remind you of this fact in future chapters.

The independent variable of the differential equations we have shown in this first chapter has been time t. This will be the case in almost all the equations discussed in this book because it is the most common. There are sometimes, however, when other independent variables are used. For example, in several fields of engineering, DEs describing the dependent variable as a function of distance from some point develop. For instance, the DE could be

$$5\frac{dT}{dl}+3T=2q$$

where T is temperature along a length l of a solid pipe and q is the energy entering the pipe. In this case, the dependent variable T is a function of l. To obtain the complete solution, we need to know the value of T at some l. The term *initial condition* that we have used to describe this value implies the value of the dependent variable at some time t. So, to avoid this implication, when the independent variable is not time, we use the term *boundary condition* to refer to this value needed to obtain the solution.

1.2.2 Ordinary and Partial DEs

Before continuing with new material, we would like to state once more that the dependent variable must be a function of the independent variable, otherwise, the differential is zero. To stress this fact many books would write Equation 1.13 as

$$\rho A\frac{dh(t)}{dt}+C_V\sqrt{h(t)}=w_1(t)+w_2(t) \tag{1.22}$$

Writing $h(t)$ the equation explicitly shows that the dependent variable h is a function of the independent variable, t, and by writing $w_1(t)$ and $w_1(t)$, the equation also explicitly also shows that the forcing functions are a function of time. Often, however, as we said earlier, writers often skip that step; that is, they write Equation 1.13 instead of Equation 1.22. In this book, we will not explicitly show the functionality; that is, we will most often write Equation 1.13 and not Equation 1.22.

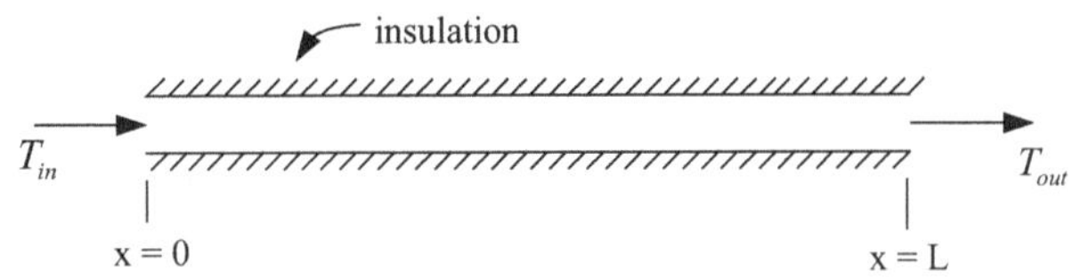

FIGURE 1.5 Well-insulated pipe.

The dependent variable in the DEs discussed so far is a function of only one independent variable; for example, in the case of the mixing tank in Section 1–1 the level h is only a function of time t. These DEs are called *ordinary differential equations* or *ODEs*.

Sometimes the dependent variable may be a function of more than one independent variable. Consider Figure 1.5, which shows a well-insulated long pipe in which a liquid flows.

At the initial steady condition, the temperature of the liquid is the same all along the pipe, let us call this temperature $T_{initial}$, and, of course, at that condition $T_{in} = T_{initial}$.

Suppose now that at some time, $t = 0$, the inlet temperature T_{in} increases by 30°C. The liquid temperature inside the pipe starts increasing at some rate (not instantaneously) to its final value but, not at the same time all along the pipe. That is, the temperature next to the entrance starts changing *before* the temperature at $0.1L$ starts changing, and this temperature at $0.1L$ starts changing *before* the temperature at $0.2L$ starts changing, and so on. That means that the temperature along the pipe, call it T, is a function of (depends on) time t and distance x down the pipe; we say that $T(t, x)$. In this case, the differential equation modeling the temperature in the pipe will contain partial derivatives of $T(t, x)$, such as $\frac{\partial T(t,x)}{\partial x}$ and $\frac{\partial T(t,x)}{\partial t}$; these equations are called *partial differential equations* or *PDEs*. For a first-order PDE, the initial condition is of the form $T(t=0) = T_{initial}$ for all x, and $T(x=0) = T_{initial}$ for t.

This book focuses only on ODEs! We raise the issue here because sometimes the ODEs we develop and solve contain simplifying assumptions, without which would require PDEs. You will learn more about the assumptions, and when they are applicable, in your later engineering/science courses.

1.3 MODELING

As we have seen in the previous sections, *modeling* in the context of this book refers to mathematical modeling. Specifically, a mathematical model is an equation, or set of equations, that describes a system. Engineering and science students have been writing mathematical models since physics courses. At that time, the models were simple and most often were composed of a single equation. The reader may remember Newton's equation, $F = ma$, which provides (models) the force generated by a device of mass m when moving with an acceleration a. As we progress through engineering and science, the models become a bit more complex and composed of several equations. The fact is engineering, and science students have been developing mathematical models almost "from day one!"

Note the steps taken in developing the model, Equation 1.13; these are the typical steps in developing any model. First, we started using a basic physical law; in the example shown this was Equation 1.8, which, although we did not name it, is the law of conservation of mass (we will visit it again in Chapter 8). We then continued with other necessary equations such as Equations 1.5 and 1.10, which we referred to as process components or phenomena occurring. Thus, in developing a model, we start from a basic physical law, also sometimes referred to as governing equation, followed by equations describing phenomena taking place, physical elements, etc. We will stress this procedure in every modeling chapter.

Note also that the resulting model, Equation 1.13, was obtained substituting Equations 1.5 and 1.10 into Equation 1.9. Equation 1.13 contains only one unknown, the liquid level h; the flow rates of streams 1 and 2 are not considered unknowns because it is up to the user to decide how to change them. Sometimes, in more extensive models, these substitutions may result in a rather large and complex single equation. It is better in these cases not to substitute the equations and consider the entire set of equations as the model. For example, in the mixing tank process, we could have said that the model for the tank was composed of Equation 1.9 – always the basic law as the first equation – and Equations 1.5 and 1.10; in this case, there are three equations with three unknowns, w_3, m, and h.

We have referred to Equation 1.6 as the model of the tank shown in Figure 1.1. However, Equation 1.12 is also another model of the tank. The difference is that the solution of Equation 1.12 provides the response of the level as a function of time, while Equation 1.6 only provides the level at the end of time once the steady conditions are reached again. We refer to models described by DEs when time is the independent variable as *dynamic models*; we refer to the algebraic equations, when time is not involved in the models, as *steady-state models*.

1.4 FORCING FUNCTIONS

Let us look in more detail at differential equations and at what we have referred to as forcing functions; we use the example of the tank in Section 1.1. The resulting model is Equation 1.13, or

$$\rho A\frac{dh}{dt}+C_V\sqrt{h}=w_1+w_2 \tag{1.13}$$

Now suppose that originally there is no flow into the tank; that is, $w_1=w_2=0\ \text{kg/min}$. In this case,

$$\rho A\frac{dh}{dt}+166.7\sqrt{h}=0$$

Because the system is at a steady condition, $\frac{dh}{dt}=0$ m/min (m/min stands for meters/minute) and the equation indicates $h=0\ \text{m}$. Obviously, with $h=0\ \text{m}$, w_3 is also 0 kg/min (there is no flow rate out, which makes sense). If nothing happens, the level h will stay at the value of 0 meter. So, for h (the dependent variable in the example) to change, something must happen – or force it. In this system, the input flows w_1 and/or w_2 are what could change, and this is why we refer to them as forcing functions.

Let us extend this discussion a bit more and return to the original system when $w_1=200\ \text{kg/min}$, $w_2=100\ \text{kg/min}$, and $h=3.24$ m. As long as w_1 and w_2 remain at these values, the liquid level h will also remain constant at 3.4 m. Once more, for a change in level (the dependent variable) to occur, w_1 and/or w_2 (the forcing functions) must vary.

Note that at a steady condition the differential equation, Equation 1.13, becomes equal to the algebraic equation, Equation 1.6, because $\frac{dh}{dt}=0$. Thus, under steady conditions DEs, when time is the independent variable, become algebraic equations; only during changes is that this is not the case.

We have just presented that the forcing functions are the terms that when change, they force the dependent variable to change. These forcing functions, or changes, could enter the system at any time. A way to show when the change occurs is using the term $u(t-a)$, which is a shorthand notation for

$$u(t-a) \equiv \begin{cases} 0 & for\ t < a \\ 1 & for\ t \geq a \end{cases}$$

Note that the value of the $u(t-a)$ term is either 0 or 1. It starts at 0 and changes to 1 at $t = a$. When $t - a$ is equal to or greater than 0, the term is 1. Thus, the mathematical notation $u(t-a)$ acts very much like a switch. For $t < a$, the switch is open, and its value is 0; for $t \geq a$, the switch is closed, and its value is 1.

As an example, consider that at some time, $t = a$, $w_1(t)$ changes from 200 kg/min to 250 kg/min instantaneously; this is shown in Figure 1.6.

As the figure shows, the change is in the form of a step. In this case, the expression

$$w_1(t) = 200 + 50\, u(t-a)$$

expresses the change of $w_1(t)$ from 200 to 250 kg/min at time equal to a. For $t < a$, $u(t-a) = 0$ and $w_1(t) = 200 + 50\,(0) = 200$ kg/min. For $t \geq a$, $u(t-a) = 1$ and $w_1(t) = 200 + 50\,(1) = 250$ kg/min.

The following examples should help further understanding the use of $u(t-a)$.

Example 1.1

Consider the function $f(t)$ shown in Figure 1.7. Develop the mathematical expression for $f(t)$.

There are three changes in this figure at times 4, 6, and 8; each change has a magnitude of 3. Because there are three changes, there should be three switches $u(t-a)$ at a = 4, 6, and 8. The expression is

$$f(t) = 1 - 3u(t-4) + 3u(t-6) - 3u(t-8)$$

At $t < 4$ all u's are zero, and $f(t) = 1$ as the figure shows. At $t \geq 4$ but less than 6, $u(t-4) = 1$, $u(t-6) = u(t-8) = 0$, and $f(t) = -2$ as the figure shows. At $t \geq 6$ but less than 8, $u(t-4)$

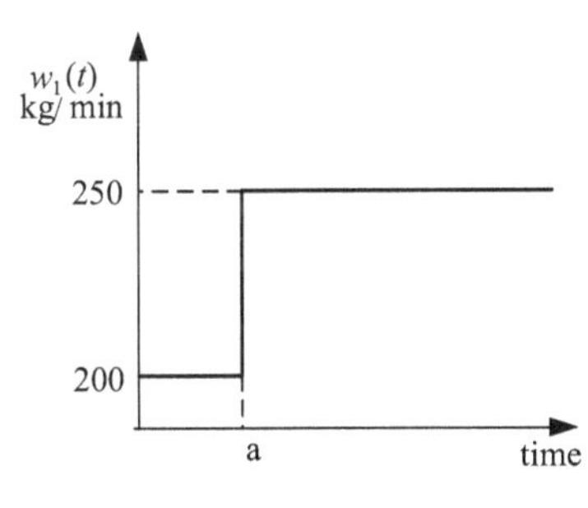

FIGURE 1.6 Step change in flow of stream 1

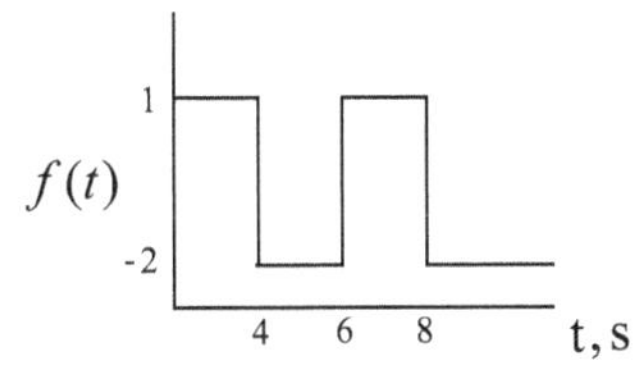

FIGURE 1.7 Forcing function for Example 1.1

$= u(t-6) = 1$, $u(t-8) = 0$, and $f(t) = 1$ as the figure shows. Based on this presentation, the reader can understand why $f(t) = -2$ after $t \geq 8$.

Example 1.2

Consider the function $f(t)$ shown in Figure 1.8. Develop the mathematical expression for $f(t)$.

This is a ramp change starting at 4 minutes. The expression is

$$f(t) = 5 + f_1(t)u(t-4)$$

That is, for $t < 4$ $u(t-4) = 0$, and no matter what $f_1(t)$ is, it does not affect $f(t)$ and $f(t) = 5$. Right at $t = 4$ $u(t-4)$ becomes 1, and $f_1(t)$ starts to be added to 5, affecting $f(t)$. Because it is a ramp affecting $f(t)$, the expression for $f_1(t)$ must be that of a ramp such that right at $t = 4$ it has a value of zero (0) but, from then on, increases. This expression is (flash back to algebra!)

$$f_1(t) = \frac{7}{4}(t-4)$$

Therefore,

$$f(t) = 5 + \frac{7}{4}(t-4)u(t-4)$$

Before concluding with this section, it is worthwhile to stress again that the value of the expression $u(t-a)$ is only 0 or 1 depending on its argument $(t-a)$. When $t-a < 1$, the expression has a value of zero (0); when $t-a > 1$, the expression has a value of one (1). Therefore, $u(t-a)$ indicates when the function by which it is multiplied by starts to have an effect. In the last example, the ramp starts having an effect at $t = 4$.

Example 1.3

Suppose that a power supply to an electrical circuit provides 10 *V* (*V* stands for volts); we refer to this power as v_{supply} It is planned to add to this power, at time $t = 300$ s, another voltage as a sine wave with an amplitude of 10 *V* and frequency of 60 *radians/s*. What is the mathematical expression describing this planned addition?

Let us refer to the added voltage as v_{added} and it can be described as $v_{\text{added}} = 10 \sin 60t$. We could write

$$v_{\text{supply}} = 10 + v_{\text{added}}\, u(t) = 10 + 10 \sin 60t\, u(t)$$

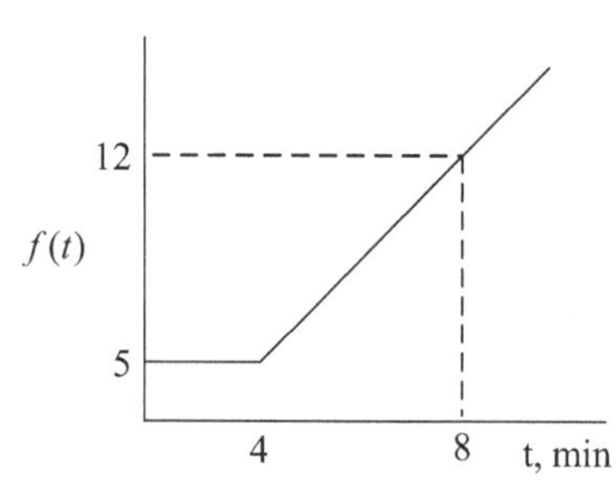

FIGURE 1.8 Forcing function for Example 1.2

This says that the voltage for $t < 0$ is 10 V and that at $t = 0$ it becomes $10 + 10 \sin 60t$. However, the plan is to start adding this new voltage starting at $t = 300$ s. For this specification, we write

$$v_{\text{supply}} = 10 + 10 \sin 60t\, u(t - 300)$$

Note now that up to just before $t = 300$ s, $u(t - 300) = 0$ and the sine expression will have no effect on v_{supply}. However, right at $t = 300$ s $u(t - 300)$ becomes one (1), and the supply voltage becomes

$$v_{\text{supply}} = 10 + 10 \sin 60(300)$$

Thus, there is a drastic change in voltage, from 10 to 11.32 V, and then it continues moving as a sine wave. If a smooth transition from 10 V to a sine wave is desired, we then express the change as

$$v_{\text{supply}} = 10 + \sin 60(t - 300)u(t - 300)$$

Do you understand? Think about it.

Example 1.4

Suppose a torque T of 20 N (N stands for Newtons) is applied to a rotating machine. It is planned to stop the torque at time $t = 3$ min. Express this change mathematically.

This statement indicates a change in torque from 20 to 0 N, a step change of –20 N, at $t = 3$ min.

$$T = 20 - 20u(t - 3)$$

1.5 BOOK OBJECTIVES

This book has the following three well-defined objectives:

1. Analytical solution of DEs
2. Mathematical modeling
3. Computer solution of models – simulation

We have stated that Equation 1.13 describes the liquid level in the tank. As it is, the equation by itself does not tell us what is the level at some time after one of the input flows, or both, have changed. That is, we would like to have an expression of h as a function of t, or $h = f(t)$. To obtain such a function, we must "solve" the DE. The *analytical solution* of a DE is one that finds the function that satisfies the DE, including its initial or boundary condition, as we showed in Section 1.2 and resulted in Equation 1.21. Obtaining the analytical solution of differential equations is the sole subject of the usual "Differential Equations course" taught by the "Mathematics Department" and taken by engineers and scientists. This book presents various methods for obtaining the analytical solution of ODEs.

Mathematical modeling consists of developing the set of equations that describes a system; this set of equations is termed the *model*. The book focuses on models that contain one or more ODEs; models may also contain algebraic equations.

Developing a completely exact model is often rather difficult. Sometimes, the theory of the phenomenon is not completely known; sometimes, the experimental facts are not available; or sometimes, there is no need to include all the details – often we simply look for trends. Always, however, *the user of the model must understand its limitations, be skeptical of the results, and analyze them to decide if they make sense. Analysis of the results is imperative for the user to build confidence in the model.*

Often the analytical solution of DEs is rather difficult or impossible to obtain because the equations are very nonlinear, or several coupled equations constitute the model. In these instances, the use of computer solutions using numerical methods, or specifically designed software packages for this purpose, may be the only possible way to reach a solution. We call *simulation* the solution of mathematical models using computers. The book discusses a couple of numerical methods and a software package to solve the models.

1.6 SUMMARY

This chapter presented an example where algebraic equations do not provide the necessary information required for analysis or design; in this case, DEs did so. The chapter also defined some terms related to DEs.

The example showed how to develop the model of a system. Section 1.2 discussed the usual procedure in developing a model, and it is quite worthwhile to stress it again in this summary. The first step in modeling is the use of a *law* (physical, chemical, biological, sociological, financial/economical, etc.) that applies to the system being modeled. The second step usually consists of using *equations describing phenomena involved, physical elements, and/or experimental or empirical fact(s)* that may be needed in the law. An example of these two steps is shown in the tank example presented in Section 1.1. The modeling started with the application of the law of conservation of mass, followed by a flow element equation expressing the outlet flow through the valve, Equation 1.5, and another equation relating the mass accumulated in the tank to the liquid level, Equation 1.10. Obviously, other equations may also be used in models but, the use of a law is basic. We will stress this procedure in all our modeling efforts.

An important comment about modeling is that the user must understand well the limitations of the model (often resulting from assumptions taken during its development). Engineers and scientists use modeling for designing, specifying, and debottlenecking systems. Oftentimes, very expensive decisions are made based on the solutions obtained from these modules. More importantly, safety considerations/decisions may be taken based on the solutions. Thus, it is imperative for the engineer/scientist to be skeptical of the results and become as convinced as possible that the results are realistic (make sense) and are possibly correct.

Problems

Problem 1.1

Write the mathematical expression for the function $f(t)$ for each of the following cases:

Problem 1.2

Consider the tank of Section 1.1 and its steady state condition of $w_1 = 200\ \text{kg/min}$, $w_2 = 100\ \text{kg/min}$, and $h = 3.24$ m. Suppose that at time equal to 10 min, w_1 changes to 400 kg/min, and after a long time, the response of the liquid reaches a new steady level.

a. How would you mathematically express the forcing function $w_1(t)$?
b. Using the model developed in Section 1.1, what is the new steady liquid level after the forcing function is applied?

Problem 1.3

Figure P1.2 shows the thrust curve of a model rocket engine. Develop the equation that provides the thrust as a function of time.

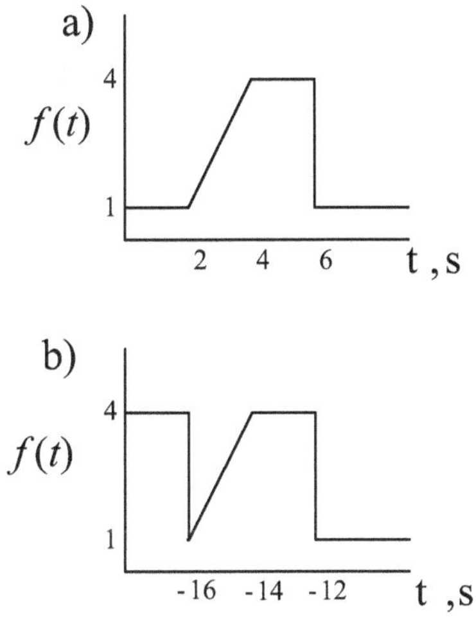

FIGURE P1.1 Graphs for Problem 1.1

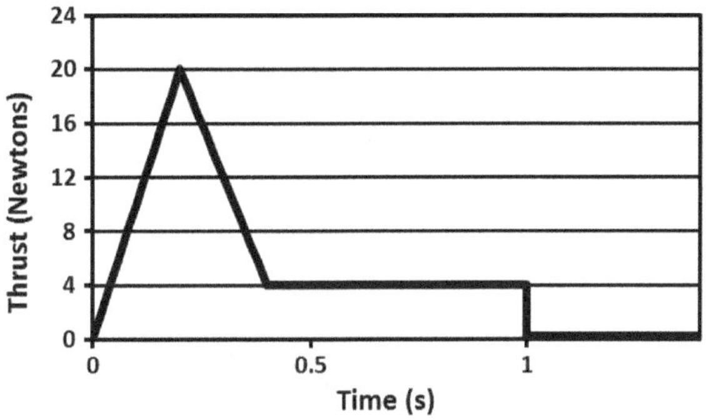

FIGURE P1.2 Thrust curve for a model rocket engine.

2 Objects in a Gravitational Field

This chapter is essentially an extension of the previous chapter in that it shows how basic physics helps with developing mathematical models and how basic calculus helps in obtaining the analytical solutions of models. The chapter shows a method to keep track of the equations and unknowns while developing models; we strongly recommend the reader use it. The chapter also presents a couple of methods to solve differential equations. An object in a gravitational field (gravity acting on objects) is the example used in the chapter.

As we mentioned in the previous chapter, *any modeling starts by using a basic physical law followed by equations describing physical elements and/or experimental facts of phenomena related to the physics of the system.* In the case of objects in a gravitational field, the physical law that helps is Newton's second law:

$$\vec{F} = m\vec{a}$$

where the arrows above the force F and the acceleration a indicate that these quantities are vectors (meaning that direction and magnitude must be specified). Because in this chapter we will only be considering forces acting on the object to be perpendicular to the ground, in the "y direction," such as gravity, we write the law again as

$$\sum F_y = ma_y \tag{2.1}$$

The summation sign is used in F_y when there is more than one force acting on the object. In modeling the systems in this chapter, we have to choose a direction, up or down, as positive. In all cases, unless otherwise specified, the positive direction is the "up direction."

2.1 AN EXAMPLE

An object of 20 kg is held in the air 30 m above the ground; Figure 2.1 shows the setup. Develop models that describe the velocity and position of the object as a function of time once it is released.

As an object falls the two most common forces acting on it are gravity and air resistance. In this first example, we neglect air resistance. Thus, only the force of gravity acts on the object, and using Newton's second law, Equation 2.1,

$$F_y = ma_y$$

Using the definition of acceleration from physics and mathematics, $a_y = \dfrac{dv_y}{dt}$, the equation becomes

$$m\frac{dv_y}{dt} = F_y \tag{2.2}$$

The phenomenon in this case is the force of gravity acting on the object,

$$F_y = F_g = -\text{mg} \tag{2.3}$$

DOI: 10.1201/9781003619161-2

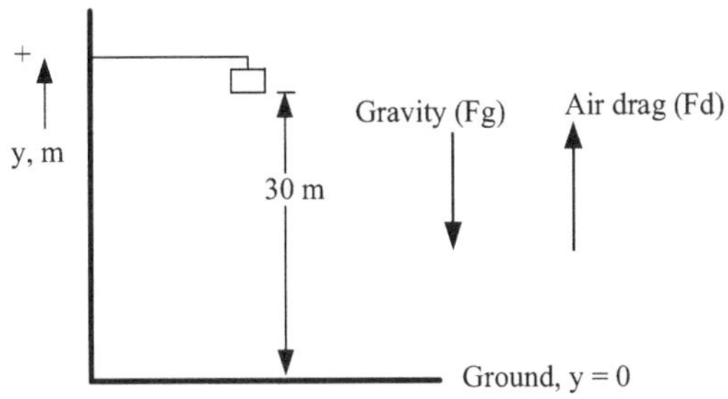

FIGURE 2.1 Object held above ground

where g is the acceleration due to gravity (at sea level 9.8 m/s^2 or 32.2 ft/s^2); the minus sign indicates that the force of gravity is down, which is opposite to the positive direction (up). From Equations 2.2 and 2.3,

$$m\frac{dv_y}{dt} = -\text{mg}$$

$$\frac{dv_y}{dt} = -\text{g} \tag{2.4}$$

Using the definition of velocity learned in physics and mathematics,

$$\frac{dy}{dt} = v_y \tag{2.5}$$

To complete the model, we need to specify the initial conditions. There are two differential equations in this example: Equation 2.4 with v_y and Equation 2.5 with y as dependent variables. Thus, we need to obtain $v_y(0)$ and $y(0)$. Because the object is released from rest, that is, at that moment it is not moving,

$$\left.\frac{dy}{dt}\right|_{t=0} = v_y(0) = 0 \text{ m/s} \tag{2.4a}$$

and because the object is originally 30 m above the ground,

$$y(0) = 30 \text{ m} \tag{2.5a}$$

ground is $y = 0$ m.

Equations 2.4 and 2.4a provide the model and initial condition of the velocity as a function of time; the analytical solution of this model provides v_y as a function of time t. Equations 2.5 and 2.5a provide the model and initial condition of the position as a function of time; the analytical solution of this model provides y as a function of t. We next present the first technique for obtaining the analytical solution of differential equations, that of antidifferentiation.

2.2 ANTIDIFFERENTIATION – TECHNIQUE FOR SOLVING FIRST-ORDER ORDINARY DIFFERENTIAL EQUATIONS

If the differential equation is of the form,

$$\frac{dx}{dz} = f(z)$$

that is, *if everything on the right side is only a function of the independent variable, z in this case, or a constant, the equation can be solved analytically using antidifferentiation:*

$$dx = f(z)dz$$

Using *indefinite integrals,*

$$\int dx = \int f(z)dz \Rightarrow x = \int f(z)dz + C$$

The constant C is obtained using the initial or boundary condition. Using *definite integrals,*

$$\int_{x(0)}^{x} dx = \int_{z=0}^{z} f(z)dz$$

where the initial or boundary condition is used as a limit of integration.

For example, consider

$$\frac{dx}{dz} = 3z^2 + 2, \text{ with } x(0) = 4$$

Using indefinite integrals,

$$\int dx = \int (3z^2 + 2)dz \Rightarrow x = z^3 + 2z + C$$

applying the boundary condition,

$$4 = 0^3 + 2(0) + C \Rightarrow C = 4$$

Finally,

$$x = z^3 + 2z + 4$$

It is now left to the reader (don't you dislike this comment?) to show how using definite integral results in the same solution.

2.3 BACK TO SECTION 2.1

Let us return to the example shown in Figure 2.1. Using the antidifferentiation technique, we integrate Equation 2.4:

$$\begin{aligned} \int_0^{v_y} dv_y &= -g\int_0^t dt \\ v_y - 0 &= -g(t-0) \\ v_y &= -gt \end{aligned} \tag{2.6}$$

Using Equation 2.6, we can integrate Equation 2.5:

$$\int_{30}^{y} dy = \int_0^t v_y dt = -g\int_0^t t dt$$

$$y - 30 = -\frac{1}{2}g(t^2 - 0)$$

$$y = 30 - \frac{1}{2}gt^2 \tag{2.7}$$

Equations 2.6 and 2.7 are the expressions for the velocity and position of the object at any time after it has been released. We can now use them to answer several questions such as how long will it take the object to hit ground, t_{final}, and what will its velocity be, $v_{y\,final}$. Equation 2.7 answers the time question; at ground level $y = 0$:

$$0 = 30 - \frac{1}{2}gt_{final}^2 \Rightarrow t_{final} = \sqrt{\frac{60}{9.8}} = 2.47 \text{ s}$$

Equation 2.6 answers the second question about the velocity when the body hits the ground:

$$v_{y\,final} = -gt_{final} = -9.8(2.47) = -24.25 \text{ m/s}$$

Note that the final velocity is a negative value, indicating that it is downward (positive is up); velocity is a vector quantity requiring direction and magnitude. The final speed (a scalar) is 24.25 m/s.

Note from Equation 2.7 that as time increases, the position y decreases; however, the equation only has physical significance up to $y = 0$ (ground; unless a hole has been dug!). Engineering equations *must have physical significance*; otherwise, they are not of much use to engineers.

As mentioned in Chapter 1, and we take this opportunity to reiterate again, the veracity of the solution depends on the assumptions taken in developing the model. The main assumption in the model developed was neglecting the effect of air resistance.

2.4 ANOTHER EXAMPLE

The models developed in Section 2.1 assumed no air resistance opposing the falling object. Develop now the same models but this time considering the *drag force* due to the air resistance.

Newton's second law is again the starting point:

$$\Sigma F_y = m\frac{dv_y}{dt} \tag{2.8}$$

This time we have two forces, the force due to gravity and the force due to the air resistance:

$$\Sigma F_y = F_g + F_d \tag{2.8a}$$

where F_d is the drag force due to air resistance. The gravity force as we know pulls the object down, but the air drag force F_d slows down the object drop by pushing upward on it (see Figure 2.1). The expression for the force of gravity provides an equation:

$$F_g = -\text{mg} \tag{2.9}$$

We now need an expression for F_d, which is another phenomenon occurring. A "crude" expression/model is to assume that this force is proportional to the velocity, or

$$F_d \sim v_y \Rightarrow F_d = -Pv_y \tag{2.10}$$

where ~ indicates "is proportional to" and P is an empirically obtained proportionality constant. The minus sign is needed because the velocity is down – a negative value – but the drag force due to the air resistance is upward – in the positive direction. Thus, the minus sign multiplied by the negative velocity yields a positive force as it should.

Equations 2.8, 2.8a, 2.9, and 2.10 constitute the model that provides the velocity of the object under the assumption that the forces applied to it are the force of gravity and the air resistance force given by Equation 2.10. For the model of the position of the object, we use again the definition of velocity,

$$\frac{dy}{dt} = v_y \tag{2.11}$$

The model for the velocity is composed of four equations, and using simple algebra, we obtain a single equation. Substituting Equations 2.9, and 2.10 into Equation 2.8 gives

$$m\frac{dv_y}{dt} = -\text{mg} - Pv_y \tag{2.12a}$$

or

$$\frac{dv_y}{dt} = -g - rv_y \tag{2.12b}$$

where $r = P/m$. Equation 2.12b can also be written as

$$\frac{dv_y}{dt} + rv_y = -g \tag{2.12c}$$

If there is no air resistance, $P = 0$, $r = 0$. Equation 2.12b becomes equal to Equation 2.4. The initial conditions are still the same as given by Equations 2.4a and 2.5a

Equation 2.12b cannot be solved using antidifferentiation because the dependent variable v_y appears on the right side of the equation:

$$\frac{dv_y}{dt} + rv_y = -g \Rightarrow \frac{dv_y}{dt} = -g - rv_y$$

We now offer another method that may be used when antidifferentiation cannot be used.

2.5 SEPARATION OF VARIABLES – TECHNIQUE FOR SOLVING FIRST-ORDER ORDINARY DIFFERENTIAL EQUATIONS

If an ordinary differential equation $\frac{dx}{dz} = f(x,z)$ can be put in the form $\frac{dx}{dz} = X(x)Z(z)$ factoring the function $f(x, z)$ into the product of two functions each of a single variable – it is separable and can be solved by

$$\int \frac{dx}{X(x)} = \int Z(z)dz$$

For example,

$$\frac{dx}{dz} = 2xz + 3x \text{ with } x(0) = 4$$

$$\frac{dx}{dz} = x(2z+3)$$

then

$$\int_4^x \frac{dx}{x} = \int_0^z (2z+3)dz$$

$$\ln x - \ln 4 = (z^2 + 3z$$

$$x = 4e^{(z^2+3z)}$$

2.6 BACK TO SECTION 2.4

Using the separation of variables, we can integrate Equation 2.12b,

$$\frac{dv_y}{dt} = -g - rv_y = -1\left(g + rv_y\right)$$

and

$$\int_0^{v_y} \frac{dv_y}{(g+rv_y)} = -1\int_0^t dt$$

$$\frac{1}{r}\ln(g+rv_y) - \frac{1}{r}\ln(g) = -t \quad (2.13a)$$

and after some algebraic manipulations (it may be good practice for the reader to check this next result),

$$v_y = \frac{g}{r}e^{-rt} - \frac{g}{r} = \frac{g}{r}\left[e^{-rt} - 1\right] \quad (2.13b)$$

The position y can be obtained by antidifferentiation. From Equations 2.11, 2.5a, and 2.13b,

$$\int_{30}^{y} dy = \int_0^t v_y dt = \int_0^t \frac{g}{r}\left[e^{-rt} - 1\right]dt$$

After integration and some simple rearrangement (again, practice) we get

$$y = 30 - \frac{g}{r}t + \frac{g}{r^2}\left[1 - e^{-rt}\right] \quad (2.14)$$

At this point, we stress once again that the starting point was a physical law, Newton's second law, followed by a couple of equations describing the phenomena due to the forces of gravity and air resistance.

As mentioned in Section 2.4, the model of the drag force due to air resistance ($F_d = -Pv_y$) is rather "crude." Experiments suggest that a better model may be for the drag force to vary with the square of the velocity. Think about the solution using this new model; it is the topic of a problem at the end of the chapter.

Before concluding the section, let's discuss the meaning of *terminal velocity*, and to explain it, consider Equation 2.12a:

$$m\frac{dv_y}{dt} = -\text{mg} - Pv_y \tag{2.12a}$$

As discussed, the first term on the right-hand side is the force due to gravity, which is constant and down; the second term is air drag force and opposes the motion of the object; thus, it is up. When the object is released, the velocity is slow but as it continues falling, the velocity increases, and therefore, the drag force also increases. If enough distance is available (that means before getting to the ground), the up-drag force becomes equal to the down gravity force, and the velocity reaches a constant value, $\frac{dv_y}{dt} = 0$. This constant velocity is the terminal velocity, which can easily be calculated from Equation 2.12a:

$$v_{y\,\text{terminal}} = -\frac{m}{P}\text{g}$$

Substituting $v_{y\,\text{terminal}}$ into Equation 2.13b, it can be seen that the time to reach terminal velocity is infinite. One might ask, however, technically speaking, how long it would take to reach it. In Chapter 5 we develop an engineering rule of thumb that answers this question.

2.7 EQUATIONS, UNKNOWNS, AND DEGREES OF FREEDOM

The models developed in the previous section are rather simple in that they are composed of very few equations. As the reader proceeds in his/her field of study the models become more complex because principally they will be composed of more equations. Thus, it behooves us to use, *from the very beginning*, a procedure to develop these models in an organized fashion, and we recommend the following procedure.

We use the example of Section 2.4 to show the procedure. Applying Newton's second law gives

$$\sum F_y = ma_y$$

and using the definition of acceleration,

$$m\frac{dv_y}{dt} = \sum F_y \tag{2.15}$$

Note that in this equation, we know the value of m but not that of v_y or $\sum F_y$, thus, there is one equation with two unknowns. As we well know, there must be the same number of equations as unknowns before reaching a solution. To *explicitly* show that there is one equation and two unknowns, *we write*

$$m\frac{dv_y}{dt} = \sum F_y \tag{2.15}$$

1 equation, 2 unknowns $\left[v_y, \sum F_y\right]$.

NOTE →

So we need one more equation. The analytical solution of the differential equation provides the velocity v_y; thus, we need an equation for $\sum F_y$

$$\sum F_y = F_g + F_d \tag{2.16}$$

2 equations, 4 unknowns $[F_g, F_d]$

NOTE AGAIN

This new equation adds two new unknowns; increase by 1 the number of equations, by 2 the numbers of unknowns, and write in the new unknowns.

The expression for the force of gravity provides another equation,

$$F_g = -\text{mg} \tag{2.17}$$

3 equations, 4 unknowns.

NOTE AGAIN

This equation does not add any unknown because we know both m and g; we just increase by 1 the number of equations. We now need an expression for F_d, and as discussed before, this is

$$F_d = -Pv_y \tag{2.18}$$

4 equations, 4 unknowns.

NOTE ONCE MORE

The velocity v_y is an unknown, but it was already accounted for in Equation 2.15 and thus, there is no need to do so again.

For the model of the position of the object, we use the definition of velocity,

$$\frac{dy}{dt} = v_y \tag{2.19}$$

5 equations, 5 unknowns $[y]$

AND ONCE MORE

This time the position y of the object is a new unknown; we add one more equation and unknown and add y to the list of unknowns.

Equations 2.15 through 2.19 are the set of equations that describe the system. There are 5 equations with 5 unknowns; thus, the system of equations can be solved. Note that in Equation 2.18, we have 4 equations and 4 unknowns, and based on that, we could have stopped there. But, at that moment, we did not have the position y of the object.

The writing of equations and unknowns every time an equation is written seems somewhat foolish for beginners, and many students may also feel that this is "theory" and that there is no need to do it. The example just used is simple enough to remember which quantities are unknown. But, believe us, as you proceed with your studies, you will encounter larger systems, and if you have one that requires 45 equations to describe 45 unknowns or even one that requires 10 equations to describe 10 unknowns, you better have an organized procedure to successfully complete the model!

Another benefit of this method is that writing the unknowns indicates what equation you need to write to complete the model. This indeed is very helpful!

We will continue using this procedure throughout the book and strongly encourage the reader to do the same.

Degrees of freedom (*DoF*) is defined as the difference between the number of unknowns and the number of independent equations; that is,

$$\textit{Degrees of Freedom} = \textit{Number of unknowns} - \textit{Number of independent equations} \tag{2.20}$$

Thus, to solve any set of equations, $DoF = 0$! The reader may ask why the term *Degrees of Freedom*. The reason is that if all possible equations that apply to the system have been written that is, if there is no more equation available, and the difference in Equation 2.20 is not zero, that is $DoF \neq 0$, then the engineer/designer has the "freedom" to select/specify the value(s) of any of the unknown(s) necessary to make $DoF = 0$. The system as it is, is not correctly specified (systems could be over or underspecified). Obviously, the selection of what unknown(s) and its (their) value(s) must be done with common sense and engineering judgment.

Let us use the same example once more to illustrate this term of *DoF* in more detail. This time we change the problem statement to "*An object is held in the air 30 m above the ground and released. Develop the models that describe the velocity and position of the object as a function of time considering the drag force due to the air resistance.*" Note that the mass of the object is not specified this time.

As last time, we start with the basic law and the definition of acceleration:

$$m\frac{dv_y}{dt} = \sum F_y \tag{2.21}$$

1 equation, 3 unknowns $\left[\sum F_y, m, v_y\right]$.

Comparing this equation with Equation 2.15, we now have one more unknown, m. The next equation is the expression for the force, Equation 2.16:

$$\sum F_y = F_g + F_d \tag{2.16}$$

2 equations, 5 unknowns $[F_d, F_g]$.

We continue with the other equations shown:

$$F_g = -\text{mg} \tag{2.17}$$

3 equations, 5 unknowns;

$$F_d = -Pv_y \tag{2.18}$$

4 equations, 5 unknowns.

For the model of the position of the object, we use again the definition of velocity:

$$\frac{dy}{dt} = v_y \tag{2.19}$$

5 equations, 6 unknowns $[y]$.

At this point, we still have 1 degree of freedom. After thinking about it, we realize that we have not missed writing/considering any other equation and have not made any mistake counting equations and unknowns. Thus, as it is, we cannot solve this set of equations. The fact that we have more unknowns than equations indicates that we have not specified enough quantities. In this case, the engineer must specify one of the unknowns to reach a solution, and of course, the solution obtained is the correct one for the specification chosen. So the engineer starts considering the possible specifications. A couple of them come to mind: (a) Specify the mass of the object, and (b) specify the time it should take the object to hit the ground. Which is the best, or correct, specification depends on the objective of the system/process. Choosing the mass of the object,

$$m = specification \tag{2.22}$$

6 equations, 6 unknowns.

There is now zero (0) *DoF* and we can solve the model. Think how you would use the other specification mentioned; it is a problem at the end of the chapter. In Chapters 6–10, we will encounter many other examples and realize how the procedure just explained helps in the modeling effort.

The example just presented was the case of an *underspecified* system. Often, examples of systems that are *over-specified* exist. Commonly, these cases occur when more than needed operating issues, or equipment sizes, have been specified.

2.8 PARTIAL FRACTIONS EXPANSION

Often, rational terms (ratios of polynomials, $\frac{f(x)}{g(x)}$), such as

$$\frac{1}{\left(a_2x^2 + a_1x + a_0\right)} \quad \text{and} \quad \frac{b_1x + b_0}{x^2(a + x)}$$

or the like, develop in reaching a solution. From this point on, the mathematical manipulations necessary to reach the solution may not be simple, if possible. This may be the case in straight integration when applying antidifferentiation and the separation of variables (e.g., consider if it is necessary to integrate the second term just shown) and in obtaining the inverse Laplace transform, as we will study in Chapter 4. The necessary mathematical manipulations may be possible if these terms are split into simpler ones. British physicist Oliver Heaviside (1850–1925) introduced the mathematical technique of partial fractions expansion that is used to obtain these simpler terms. In this section, we present the technique. Again, *this section shows you how to take a term that may be difficult to apply some needed mathematics and break it into simpler terms to make it easier to apply the needed mathematics.*

Consider the term

$$y(x) = \frac{1}{(a_1x + a_0)\left(a_2x^2 + a_3x + a_4\right)}$$

First Step

The first step in obtaining the partial fractions expansion (splitting the term into simpler terms) is to factor the denominator as follows:

$$(a_1x + a_0)\left(a_2x^2 + a_3x + a_4\right) = a_1\left(x + \frac{a_0}{a_1}\right)a_2\left(x^2 + \frac{a_3}{a_2}x + \frac{a_4}{a_2}\right) = a_1a_2(x - r_1)(x - r_2)(x - r_3)$$

where r_1 is the root of the first term (actually $r_1 = -\frac{a_0}{a_1}$) and r_2 and r_3 are the roots of the quadratic term, that is, the values of x so that $a_2(x-r_2)(x-r_3)$ satisfy the equation,

$$a_2x^2 + a_3x + a_4 = 0$$

Note that before factoring the term, it was divided by the coefficient of the highest power of the variable, and then the roots were calculated; that is,

$$(a_1x + a_0)(a_2x^2 + a_3x + a_4) = a_1\left(x - \frac{a_0}{a_1}\right)a_2\left(x^2 + \frac{a_3}{a_2}x + \frac{a_4}{a_2}\right) = a_1(x - r_1)a_2(x - r_2)(x - r_3)$$

The roots are always the same whether the polynomial is divided by a_2 or not. However,

$$a_2x^2 + a_3x + a_4 = a_2(x-r_2)(x-r_3) \text{ but } a_2x^2 + a_3x + a_4 \neq (x-r_2)(x-r_3)$$

is equal to (arrow under the first "=") — is not equal to (arrow under "≠")

Please be careful.

For a quadratic or second-degree polynomial, the roots can be calculated by the standard quadratic formula (referring to the earlier quadratic equation):

$$r_2r_3 = \frac{-a_3 \pm \sqrt{a_3^2 - 4a_2a_4}}{2a_2}$$

For higher-degree polynomials, the reader is referred to any numerical methods text for a root-finding procedure. Most electronic calculators can find the roots of third- and higher-degree polynomials. Computer programs such as Mathcad® and MATLAB® provide functions for finding the roots of polynomials of any degree.

To use a numerical example, consider $2s^2 + 16s + 14 = 0$. The roots are $r_1 = -7$ and $r_2 = -1$, but, as just explained, $2s^2 + 16s + 14$ is not equal to $(s+7)(s+1)$:

$$2s^2 + 16s + 14 \neq (s+7)(s+1)$$

What it is correct is

$$2s^2 + 16s + 14 = 2\left(s^2 + 8s + 7\right) = 2(s+7)(s+1)$$

So, please be careful during the expansion.

Second Step

Once the denominator is factored into first-degree terms, the second step consists of expanding the original term into a sum of the simpler terms as follows:

$$y(x) = \frac{1}{(a_1x + a_0)\left(a_2x^2 + a_3x + a_4\right)} = \frac{1}{a_1a_2(x + \frac{a_0}{a_1})\left(x^2 + \frac{a_3}{a_2}x + \frac{a_4}{a_2}\right)} = \frac{1/a_1a_2}{(x-r_1)(x-r_2)(x-r_3)}$$

$$y(x)=\frac{1/a_1a_2}{(x-r_1)(x-r_2)(x-r_3)}=\frac{A_1}{x-r_1}+\frac{A_2}{x-r_2}+\frac{A_3}{x-r_3} \tag{2.23}$$

Third Step

The third step consists of obtaining A_1, A_2, and A_3. The way to do this depends on the roots; roots can be real unrepeated, real repeated, or complex conjugates. There are two methods to obtain the A terms (1) algebra – straight substitution – and (2) the limit concept of calculus. We next present the two methods as they apply to the different types of roots.

We will be using Equation 2.23 to show how to obtain the A terms. This equation seems to be somewhat complex because so many terms – a*'s,* r*'s,* x*'s, and* A*'s* – are shown. It is worthwhile (if not necessary) before beginning, to remind you that the a*'s and* r*'s* are constants; if we were to use their numerical values, the equation will not look so complex. x is the variable, and the A's are the terms we are looking for.

2.8.1 Unrepeated Real Roots – Algebra

To obtain the A coefficients in Equation 2.23, set the right-hand side under the common denominator as

$$y(x)=\frac{A_1(x-r_2)(x-r_3)+A_2(x-r_1)(x-r_3)+A_3(x-r_1)(x-r_2)}{(x-r_1)(x-r_2)(x-r_3)} \tag{2.24}$$

and finally,

$$\begin{aligned}y(x)&=\frac{1/a_1a_2}{(x-r_1)(x-r_2)(x-r_3)}\\&=\frac{(A_1+A_2+A_3)x^2-\left[A_1(r_2+r_3)+A_2(r_1+r_3)+A_3(r_1+r_2)\right]x+A_1r_2r_3+A_2r_1r_3+A_3r_1r_2}{(x-r_1)(x-r_2)(x-r_3)}\end{aligned}$$

Equating coefficients of equal terms in the numerator,

$$A_1+A_2+A_3=0;\ -A_1(r_2+r_3)-A_2(r_1+r_3)-A_3(r_1+r_2)=0;\ A_1r_2r_3+A_2r_1r_3+A_3r_1r_2=1/a_1a_2$$

resulting in three equations with three unknowns (the A coefficients). From these equations, using simple (?) algebra,

$$A_1=\frac{1/a_1a_2}{(r_1-r_2)(r_1-r_3)};\qquad A_2=\frac{1/a_1a_2}{(r_2-r_1)(r_2-r_3)};\qquad A_3=\frac{1/a_1a_2}{(r_3-r_1)(r_3-r_2)}$$

Therefore,

$$y(x)=\frac{1/a_1a_2}{(x-r_1)(x-r_2)(x-r_3)}=\frac{\dfrac{1/a_1a_2}{(r_1-r_2)(r_1-r_3)}}{x-r_1}+\frac{\dfrac{1/a_1a_2}{(r_2-r_1)(r_2-r_3)}}{x-r_2}+\frac{\dfrac{1/a_1a_2}{(r_3-r_1)(r_3-r_2)}}{x-r_3}$$

Note what we now have. Instead of one term (the original one), we have three terms but much simpler ones (instead of a third-order original equation, we now have three first-order terms). This, as we shall see, facilitates any further mathematical manipulation. Remember, the terms in the numerators are the A coefficients that are constant depending on the numerical values of the a's and r's.

2.8.2 Repeated Real Roots – Algebra

Let us assume that $r_1 = r_2$, yielding *repeated roots*; the expansion is carried out as follows:

$$y(x) = \frac{1}{(a_1x + a_0)(a_2x^2 + a_3x + a_4)} = \frac{1/a_1a_2}{(x - r_1)^2(x - r_3)} \tag{2.25}$$

In this case with repeated root, the expansion is as shown:

$$y(x) = \frac{1/a_1a_2}{(x - r_1)^2(x - r_3)} = \frac{A_1}{(x - r_1)^2} + \frac{A_2}{x - r_1} + \frac{A_3}{x - r_3}$$

Note how the expansion is done, specifically the repeated term. In general, if root r_1 is repeated m times, the expansion is carried out as follows:

$$y(x) = \frac{A_1}{(x - r_1)^m} + \frac{A_2}{(x - r_1)^{m-1}} + \ldots + \frac{A_m}{x - r_1} + \ldots \tag{2.26}$$

At this point, we proceed in the same manner with algebraic substitution as before:

$$y(x) = \frac{1/a_1a_2}{(x - r_1)^2(x - r_3)} = \frac{A_1}{(x - r_1)^2} + \frac{A_2}{x - r_1} + \frac{A_3}{x - r_3} = \frac{A_1(x - r_3) + A_2(x - r_2)(x - r_3) + A_3(x - r_1)^2}{(x - r_1)^2(x - r_3)}$$

$$y(x) = \frac{1/a_1a_2}{(x - r_1)^2(x - r_3)} = \frac{(A_2 + A_3)x^2 - (A_2(r_1 + r_3) + 2r_1)x + (A_2r_1r_3 + A_3r_1^2 - A_1r_3)}{(x - r_1)^2(x - r_3)}$$

Equating coefficients of equal terms in the numerator yield the A coefficients:

$$A_1 = -\frac{1}{a_1a_2(r_3 - r_1)} \quad A_2 = -\frac{1}{a_1a_2(r_3 - r_1)^2} \quad A_3 = \frac{1}{a_1a_2(r_3 - r_1)^2}$$

Therefore,

$$y(x) = \frac{1/a_1a_2}{(x - r_1)^2(x - r_3)} = \frac{-\dfrac{1}{a_1a_2(r_3 - r_1)}}{(x - r_1)^2} + \frac{-\dfrac{1}{a_1a_2(r_3 - r_1)^2}}{x - r_1} + \frac{\dfrac{1}{a_1a_2(r_3 - r_1)^2}}{x - r_3}$$

Once more, the original mathematical expression has been changed into simpler terms that simplify the next mathematical calculations.

Let us now present a couple of examples.

Example 2.1

Consider the reaction $A \rightarrow P$. The following model describes the concentration C_p of the product component P in a batch reaction (Chapter 8).

$$\frac{dC_P}{dt} + 2C_P^2 = C_A$$

C_A is the concentration of the reactant component; the concentration terms are in mole/m³. The reactor is initially empty and at $t = 0$ reactant A is added with a concentration of 2 mole/m³, that is, $C_A = 2u(t)$, and obviously, at that moment, there is no product yet; that is, $C_P(0) = 0$ mole/m³. Obtain the amount of time for C_P to reach 0.6 mole/m³; the time unit is hour.

From the model,

$$\frac{dC_P}{dt} = C_A(t) - 2C_P^2$$

and using the separation of variables,

$$\int_0^{0.6} \frac{dC_P}{C_A(t) - 2C_P^2} = \int_0^{0.6} \frac{dC_P}{2 - 2C_P^2} = \int_0^t dt$$

or

$$t = \frac{1}{2}\int_0^{0.6} \frac{dC_P}{1 - C_P^2}$$

Evaluating the integral is not easy. Using partial fraction expansion for the integrand expression,

$$\frac{1}{1 - C_P^2} = \frac{1}{(1 + C_P)(1 - C_P)} = \frac{A_1}{1 + C_P} + \frac{A_2}{1 - C_P}$$

and following the method previously explained, we obtain the A coefficients:

$$\frac{1}{1 - C_P^2} = \frac{A_1(1 - C_P) + A_2(1 + C_P)}{(1 + C_P)(1 - C_P)} = \frac{(A_2 - A_1)C_P + (A_1 + A_2)}{(1 + C_P)(1 - C_P)}$$

from which $A_2 - A_1 = 0$ and $A_1 + A_2 = 1 \Rightarrow A_1 = 0.5$ and $A_2 = 0.5$.

Then the integral is easily evaluated as

$$t = \frac{1}{2}\int_0^{0.6} \frac{dC_P}{1 - C_P^2} = \frac{1}{2}\int_0^{0.6} \frac{1}{(1 + C_P)(1 - C_P)} dC_P = \frac{1}{2}\left[\int_0^{0.6} \frac{0.5}{1 + C_P} dC_P + \int_0^{0.6} \frac{0.5}{1 - C_P} dC_P\right]$$

$$t = \frac{1}{4}\left[\ln(1 + C_P)\Big|_0^{0.6} - \ln(1 - C_P)\Big|_0^{0.6}\right] = \frac{1}{4}\ln\left(\frac{1.6}{0.4}\right)$$

$$t = 0.347 \text{ h}$$

To connect a discussion in Chapter 1 to this example, the forcing function in this example was simply written as C_A. However, the problem states that "at $t = 0$, reactant A is added with a concentration of 2 mole/m³." Using the notation, we learned in Chapter 1 about step functions, we could have written the forcing function as $C_A u(t)$. Note that because the integration starts at $t = 0$ then $u(t) = 1$ and $C_A u(t) = C_A$. Does this expression of $C_A u(t)$ make any difference in the solution and results? No, not really. It only explicitly says that there is no C_A for $t < 0$.

Example 2.2

Suppose the following equation describes the concentration X of a component in a fermentation reaction (Chapter 8). Obtain the time required to go from the initial concentration of 0.6 mmole/L to a concentration of 1.0 mmole/L; the time unit is hour.

$$\frac{dX}{dt}=\frac{X^2(-0.5+X)}{X+1} \quad \text{with } X(0)=0.6\ \text{mmole/L}$$

Using the separation of variables, we get

$$\frac{X+1}{X^2(-0.5+X)}dX=dt$$

The integration of the first term is rather difficult. Thus, before integration, we need to factor the term into a simpler one. Using partial fraction expansion,

$$\frac{X+1}{X^2(-0.5+X)}=\frac{A_1}{X^2}+\frac{A_2}{X}+\frac{A_3}{-0.5+X}$$

$$\frac{X+1}{X^2(-0.5+X)}=\frac{A_1}{X^2}+\frac{A_2}{X}+\frac{A_3}{-0.5+X}=\frac{A_1(-0.5+X)+A_2X(-0.5+X)+A_3X^2}{X^2(-0.5+X)}$$

and equating coefficients of equal terms yields $A_1=-2$; $A_2=-6$; $A_3=6$.

Knowing these values of the A coefficients, we can integrate the expressions:

$$\int_{0.6}^{1.0}\frac{X+1}{X^2(-0.5+X)}dX=-2\int_{0.6}^{1.0}\frac{1}{X^2}dX-6\int_{0.6}^{1.0}\frac{1}{X}dX+6\int_{0.6}^{1.0}\frac{1}{-0.5+X}dX=\int_0^t dt$$

$$t=5.25\ \text{h}$$

It may be a good idea to make sure you obtain this answer (practice some integration).

2.8.3 Unrepeated Real Roots – Limit Concept

We have so far looked at the use of algebraic substitution for obtaining the A coefficients. We now present another method that significantly reduces the amount of algebra.

Recalling Equation 2.23, let us calculate A_1 first:

$$y(x)=\frac{1/a_1a_2}{(x-r_1)(x-r_2)(x-r_3)}=\frac{A_1}{x-r_1}+\frac{A_2}{x-r_2}+\frac{A_3}{x-r_3} \tag{2.23}$$

and multiplying both sides by $x-r_1$,

$$\frac{(1/a_1a_2)(x-r_1)}{(x-r_1)(x-r_2)(x-r_3)}=\frac{(1/a_1a_2)}{(x-r_2)(x-r_3)}=A_1+\frac{A_2(x-r_1)}{x-r_2}+\frac{A_3(x-r_1)}{x-r_3}$$

It would be great if now we could solve for A_1, which could be done by getting rid of the other two coefficients, A_1 and A_2. Note that this is easily done by taking the limit $x\to r_1$:

$$A_1=\lim_{x\to r_1}\left[\frac{(1/a_1a_2)}{(x-r_2)(x-r_3)}\right]=\frac{1}{a_1a_2(r_1-r_2)(r_1-r_3)}$$

By taking the limit $x \to r_1$, the terms containing A_1 and A_2 on the right-hand side drop out because their numerator becomes zero.

To obtain A_2, we follow a similar procedure, that is, multiply both sides of the equation by its denominator $(x - r_2)$:

$$\frac{(1/a_1a_2)(x-r_2)}{(x-r_1)(x-r_2)(x-r_3)} = \frac{(1/a_1a_2)}{(x-r_1)(x-r_3)} = \frac{A_1(x-r_2)}{x-r_1} + A_2 + \frac{A_3(x-r_2)}{x-r_3}$$

Taking the limit $x \to r_2$,

$$A_2 = \lim_{x \to r_2}\left[\frac{(1/a_1a_2)}{(x-r_1)(x-r_3)}\right] = \frac{1}{a_1a_2(r_2-r_1)(r_2-r_3)}$$

By taking the limit $x \to r_2$ the terms containing A_1 and A_3 drop out because again their numerators become zero.

We follow a similar procedure for obtaining A_3. Essentially, for the case of *unrepeated real roots*, the procedure we have followed can be described by the formula:

$$A_k = \lim_{s \to r_k}\left[(x-r_k)y(x)\right] \tag{2.27}$$

2.8.4 Repeated Real Roots – Limit Concept

Recall Equation 2.25 showing two repeated roots:

$$y(x) = \frac{1}{(a_1x+a_0)(a_2x^2+a_3x+a_4)} = \frac{1/a_1a_2}{(x-r_1)^2(x-r_3)}$$

As presented previously, for repeated roots, the expansion is,

$$y(x) = \frac{1/a_1a_2}{(x-r_1)^2(x-r_3)} = \frac{A_1}{(x-r_1)^2} + \frac{A_2}{x-r_1} + \frac{A_3}{x-r_3}$$

The coefficient A_1 is calculated as before, using Equation 2.27:

$$A_1 = \lim_{x \to r_1}\frac{d}{dx}\left[(x-r_1)^2 y(x)\right] = \frac{1}{a_1a_2(r_1-r_3)}$$

Coefficient A_3 is also obtained using Equation 2.27:

$$A_3 = \lim_{x \to r_1}\frac{1}{1!}\frac{d}{dx}\left[(x-r_3)^2 y(x)\right] = \frac{1}{a_1a_2(r_3-r_1)^2}$$

The same procedure cannot be used to obtain the A_2 coefficient (try it!). In this case, A_2 is calculated using the following:

$$A_2 = \lim_{x \to r_1}\frac{1}{1!}\frac{d}{dx}\left[(x-r_1)^2 y(x)\right] = -\frac{1}{a_1a_2(r_1-r_3)^2}$$

All three terms are now calculated with minimum algebraic manipulations.

In general, if root r_1 is repeated m times, the expansion is carried out as follows:

$$y(x) = \frac{A_1}{(x-r_k)^m} + \frac{A_2}{(x-r_k)^{m-1}} + \ldots + \frac{A_m}{x-r_k} + \ldots \tag{2.28}$$

and the coefficients are calculated using

$$A_1 = \lim_{x \to r_k} \left[(x-r_k)^m y(x) \right] \tag{2.29}$$

$$A_k = \lim_{x \to r_k} \frac{1}{(k-1)!} \frac{d^{k-1}}{dx^{k-1}} \left[(x-r_k)^m y(x) \right] \tag{2.30}$$

for $k = 2, \ldots, m$.

2.8.5 Complex Roots – Limit Concept

As previously shown, sometimes the roots are complex; that is, for a second-order equation, the roots may be $r_1 = \alpha + i\beta$ and $r_2 = \alpha - i\beta$, where α is the real component and β the imaginary component. In this case, there are two possible ways to apply partial fraction expansion. We next present both ways.

2.8.6 First-Order Expansion

In this case, we proceed as usual:

$$y(x) = \frac{1/a_1 a_2}{(x-r_1)(x-r_2)(x-r_3)} = \frac{A_1}{x-r_1} + \frac{A_2}{x-\alpha-i\beta} + \frac{A_3}{x-\alpha+i\beta}$$

The A_1 coefficient is obtained as before:

$$A_1 = \lim_{x \to r_1} \left[\frac{1/a_1 a_2}{(x-\alpha-i\beta)(x-\alpha+i\beta)} \right] = \frac{1}{a_1 a_2 \left(r_1^2 - r_3^2 \right)}$$

The A_2 and A_3 coefficients are also obtained as before, except that a bit more algebra is required because of the complex numbers – be careful this is where mistakes occur! That is,

$$A_2 = \lim_{x \to r_2} \left[\frac{1/a_1 a_2}{(x-r_1)(x-r_2)} \right] = \lim_{x \to \alpha+i\beta} \left[\frac{1/a_1 a_2}{(\alpha+i\beta-r_1)(\alpha+i\beta-\alpha+i\beta)} \right] = \frac{1/a_1 a_2}{(\alpha+i\beta-r_1)(2i\beta)}$$

$$A_2 = \frac{1/a_1 a_2}{\left[-2\beta^2 + 2i\beta(\alpha-r_1) \right]}$$

This denominator for A_2 is a complex expression, which is not useful for further mathematical manipulations. To eliminate the complex expression from the denominator, multiply the numerator and denominator of the coefficient by the complex conjugate of the denominator.

$$A_2 = \frac{1/a_1 a_2}{\left[-2\beta^2 + 2i\beta(\alpha-r_1) \right]} \cdot \frac{\left[-2\beta^2 - 2i\beta(\alpha-r_1) \right]}{\left[-2\beta^2 - 2i\beta(\alpha-r_1) \right]} = \frac{-2\beta^2 - 2i\beta(\alpha-r_1)}{a_1 a_2 \left[4\beta^4 + 4\beta^2 (\alpha-r_1)^2 \right]} = \sigma + i\gamma \tag{2.31}$$

where

$$\sigma = \frac{-2\beta^2}{a_1 a_2\left[4\beta^4 + 4\beta^2(\alpha - r_1)^2\right]} \qquad \gamma = \frac{-2\beta(\alpha - r_1)}{a_1 a_2\left[4\beta^4 + 4\beta^2(\alpha - r_1)^2\right]}$$

Following the exact procedure for A_3,

$$A_3 = \lim_{x \to \alpha - i\beta}\left[(x - \alpha + i\beta)\, y(x)\right] = \sigma - i\gamma \tag{2.32}$$

In these cases of complex roots, the A coefficients will always be complex conjugates. Then,

$$y(x) = \frac{1/a_1 a_2}{(x - r_1)(x - r_2)(x - r_3)} = \frac{A_1}{x - r_1} + \frac{\sigma + i\gamma}{x - \alpha - i\beta} + \frac{\sigma - i\gamma}{x - \alpha + i\beta} \tag{2.33}$$

Example 2.3

Simplify the following expression:

$$y(t) = \frac{2x + 11}{x^2 + 8x + 25}$$

Splitting the denominator into simpler terms and following the method just presented,

$$y(x) = \frac{2x + 11}{x^2 + 8x + 25} = \frac{2x + 11}{(x + 4 - i3)(x + 4 + i3)} = \frac{A_1}{x + 4 - i3} + \frac{A_2}{x + 4 + i3}$$

$$A_1 = \lim_{x \to -4 + i3}\left[(x + 4 - i3)y(x)\right] = \frac{2(-4 + i3) + 11}{i6} = \frac{(3 + i6)}{i6}$$

And $A_2 = \frac{(3 + i6)}{i6}$

Yielding

$$y(x) = \frac{3 + i6}{(i6)(x + 4 - i3)} + \frac{3 - i6}{(i6)(x + 4 + i3)}$$

This result shows an imaginary number in the denominator. To remove this imaginary number, we simply multiply the numerator and the denominator by this number,

$$y(x) = \frac{(3 + i6)(i6)}{(i6)(x + 4 - i3)(i6)} + \frac{(3 - i6)(i6)}{(i6)(x + 4 + i3)(i6)}$$

$$= \frac{(-36 + i18)}{(-36)(x + 4 - i3)} + \frac{(-36 - i18)}{(-36)(x + 4 + i3)}$$

So, $A_1 = \frac{-36 + i18}{-36} = 1 - i0.5$ and $A_2 = \frac{-36 - i18}{-36} = 1 + i0.5$

Finally, $y(x) = \frac{1 - i0.5}{x + 4 - i3} + \frac{1 + i0.5}{x + 4 + i3}$

Chapter 4 will show the meaning of, and how to handle, the complex numbers in the denominator.

Partial fraction expansion allows having any-order denominator $g(x)$ with a numerator $f(x)$ less than the degree of the denominator. If the numerator $f(x)$ happens to be of greater order than the denominator $g(x)$, the correct procedure is to first divide them and then expand the remainder. For example,

$$y(x)=\frac{f(x)}{g(x)}=\frac{x^3+8x^2+22x+19}{x^2+6x+9}=x+2+\frac{x+1}{x^2+6x+9}$$

This completes the presentation of the development of partial fractions. We have seen how a "complex mathematical function" is split into "simpler mathematical functions." Thus, simplifying the use of further mathematical manipulations as integration. Partial fractions are extensively used in Laplace transforms as presented in Chapter 4.

2.9 SUMMARY

Hopefully, this chapter was easy to read and provided more motivation to read and study the book. The chapter clearly showed that modeling starts from a basic law and that equations describing phenomena occurring in the system are used to complement the development. The procedure of counting equations and unknowns, while developing a model, was presented, and it was shown why it is helpful to do so. The concept of degrees of freedom was introduced, and we discussed its meaning. The chapter also showed two ways to solve first-order differential equations, antidifferentiation, and the separation of variables and the use of partial fraction expansion for the simplification of rational expressions.

Problems

Problem 2.1

Solve the following initial value problems for $y(t)$ by antidifferentiation or by separation of variables.

a. $e^{-2t}\frac{dy}{dt}=y^{-1}(1-e^{-2t});\ y(0)=0$

b. $\frac{dy}{dt}=y\cos(t)+y;\ y(0)=2$

c. $\frac{dy}{dt}=\frac{t+2}{y};\ y(0)=2$

d. $\frac{d}{dt}\left[t\frac{dy}{dt}\right]=2t;\ y(1)=1,\ y'(1)=3$

e. $\frac{d^2y}{dt^2}=32e^{-4t};\ y(0)=1,\ y'(0)=0$

f. $\frac{1}{t^2}\frac{dy}{dt}=y;\ y(0)=1$

g. $\frac{dy}{dt}=-y^2e^{2t};\ y(0)=1$

h. $\frac{dy}{dt}-(2t+1)y=0;\ y(0)=2$

i. $\frac{dy}{dt}+4ty^2=0;\ y(0)=1$

j. $\frac{d^2y}{dt^2} = \cos(t/2);\ y(0) = 0,\ y'(0) = 1$

k. $\frac{d^2y}{dt^2} = 4e^{-2t} - 8\,sin(2t);\ y'(0) = 0,\ y(0) = 0$

l. $\frac{d}{dt}\left[e^{-2t}\frac{dy}{dt}\right] = e^{-2t};\ y(0) = 0,\ y'(0) = 0$

Problem 2.2

Use separation of variables to solve each of the following differential equations; classify each equation as linear or nonlinear.

a. $\frac{du}{dy} = y^2 - 1$

b. $\frac{dT}{dt} = -0.0002(T-5)$

c. $\frac{dy(x)}{dx} = e^{2y}$

d. $\frac{du}{dt} = u^2 \cos \pi t;\ u(0) = -1/2$

e. $\frac{dy}{dt} = \left(\frac{y^2+4y-5}{y}\right)t$

f. $\frac{du}{dt} = u^3 + 6u^2 + 11u + 6$

g. $\frac{dy}{dt} + (t+3)y + 4t + 12 = 0;\ y(0) = 0$

h. $\frac{dy}{dt} + 2y = y\,cos(t);\ y(0) = 1$

i. $\frac{dy}{dt} + e^{-t}(y+1)(y-1) = -e^{-t};\ y(0) = -1$

Problem 2.3

Example 8.11 develops the following model describing the production of a high-cost product by an enzymatic reaction,

$$-\frac{10C^S}{12+C^S} = \frac{dC^S}{dt}$$

C^S is the concentration of the substrate in the reactor, with $C^S(0) = 13.33\ ^{\text{mmoles}}\!/_{\text{L}}$. Obtain the analytical solution providing time t as a function of C^S.

Problem 2.4

Consider the following first-order linear differential equation:

$$\tau\frac{dy(t)}{dt} + y(t) = Kx(t)$$

where $y(t)$ is the dependent variable, t is the independent variable, $x(t)$ is the forcing function, and τ and K are constants; $y(0)$ is the initial condition. Assume that $x(t)$ changes from an initial value of $x(0)$ to a value of $x(0)+D$ in a step manner at time equal to zero; that is, $x(t)=x(0)+Du(t)$. Show that the solution of this equation is given by

$$y(t)=y(0)+KD\left(1-e^{-\frac{t}{\tau}}\right)$$

Hint: Note that $y(0)=Kx(0)$.

Problem 2.5

The model that describes the velocity of the block shown in Figure P6.3 is

$$2\frac{dv}{dt}+4v=16u(t)$$

Assuming that the initial condition is $v(0)=0$ obtain the analytical equation that describes the velocity of the block.

Problem 2.6

The model that describes the currents through resistances R_1 and R_3 in Problem 10.22 are given by

$$i_1=1.43\times10^{-4}\ v_S+0.286\ i_2$$

and

$$2\times10^{-3}\frac{di_2}{dt}+21428i_2=0.286\ v_S$$

The initial conditions are $i_{R2}(0)=2.66\times10^{-4}\ A$ and $i_1(0)=2.93\times10^{-3}\ A$. For $v_S=20+20u(t)V$, solve the model to obtain expressions for $i_1(t)$ and $i_2(t)$.

Problem 2.7

The logistic expression is often used to describe the growth rate of microorganism (biomass) in fermentation reactions and environmental engineering processes. Assume that the following model (Example 8.13) describes the growth rate of the biomass in a glucose to gluconic acid fermentation reaction:

$$0.3768\,X\left(1-\frac{X}{18.91}\right)=\frac{dX}{dt}$$

where X is the concentration of biomass with an initial condition $X(0)=9.52$ gm/L.

Obtain the analytical solution and graph X vs. t for the first 10 units of time.

Problem 2.8

The following equation describes the pressure $P(t)$ in a tank containing a gas when it is filled (Problem 8.2),

$$6.915C_V\sqrt{1000-P(t)}=1.08\frac{dP(t)}{dt}$$

The gas enters the tank through a valve that must be sized by choosing the numerical value of C_V. The initial pressure in the tank is 100 kPa. The filling procedure calls for raising the gas pressure from its initial value of 100 kPa to a final value of 500 kPa in 4 minutes. Obtain the C_V that would satisfy this requirement.

Problem 2.9

The following reaction describes the concentration of reactant A in a reaction taking place in a batch reactor (Chapter 8).

$$\frac{dC^A}{dt}=-85.28\left(C^A\right)^2\left(0.132+0.5C^A\right)$$

the time unit is hour. The initial concentration is $C^A(0)=1.842\ \text{moles}/\text{liter}$. How much time does it take to reach $C^A(t)=0.1842\ \text{moles}/\text{liter}$?

Problem 2.10

In Example 8.2, the following model is developed,

$$12.5\frac{dx_3^{NaOH}}{dt}+x_3^{NaOH}=0.73x_1^{NaOH}$$

Using the initial condition $x_3^{NaOH}(0)=0.55$ and $x_1^{NaOH}=0.75-0.08\,u(t)$, show that the solution is

$$x_3^{NaOH}=0.55-0.0609\left(1-e^{-\frac{t}{12.5}}\right)$$

Problem 2.11

Using partial fraction expansion, simplify the following expressions.

a. $y(x)=\dfrac{4x+2}{x^2+4x+8}$

b. $y(x)=\dfrac{2x+10}{x^2+10x+41}$

Problem 2.12

A model rocket with a mass of 0.05 kg is launched vertically upward from an initial state of rest on the ground. When the engine is fired at $t=0$ its thrust provides a constant upward force of $F_T=2N$ (N stands for Newton) on the rocket. After 2 seconds, the fuel is consumed, and the thrust falls to zero. The drag force (caused by air resistance) on the rocket is $F_d=-kvN$, where v is the velocity and k is a drag coefficient equal to 0.01 kg/s.

a. Develop the model (differential equation and initial condition) that provides the velocity of the rocket as a function of time t while the engine is firing. Obtain the analytical solution (solve the differential equation) and determine the velocity of the rocket at the instant the fuel is consumed.

b. Develop the model (differential equation and initial condition) that provides the velocity of the rocket as a function of time after the fuel is consumed. Obtain the analytical solution (solve the differential equation) and determine how many seconds after the fuel is consumed the rocket will make upward progress.

Problem 2.13

A 500-kg dragster is moving at a velocity of 50 m/s. At time $t = 0$, a parachute is deployed. The drag force on the dragster by the parachute is $F_d = -kvN$, where v is the velocity and k is a drag coefficient equal to 100 kg/s. In a test to determine the stopping power of the chute, the driver makes sure there is plenty of pavement and does not use any braking. Although there are other sources of friction that would slow a coasting dragster, you may assume them negligible compared to the drag caused by the parachute.

a. Develop the mathematical model (differential equations) that provides the velocity and position of the dragster. Include the initial conditions.
b. Determine the time needed for the dragster's velocity to drop to 1 m/s.
c. Determine the distance that the dragster would have covered during that time.

Problem 2.14

Suppose a ball is thrown vertically at 35 m/s. Assuming no air resistance, how long does it take the ball to reach the maximum height? What is this maximum height? How long does it take the ball to come back to the ground after reaching the maximum height?

Problem 2.15

An object of 20 kg is held in the air 30 m above the ground and released. Assuming that the drag force due to air resistance is given by $F_d = 1.0|v_y|v_y$, develop the model for the velocity and position as a function of time. Can you think why it may be better to express the drag force due to air resistance as it is given instead of $F_d = 1.0v_y^2$? Also, what are the units of the constant 1.0?

Problem 2.16

An object is held in the air 30 m above the ground and released. Assume the drag force due to air resistance is given by $F_d = 1.0v_y$. What is the mass of the object necessary so that it reaches ground in 3 seconds? What is its velocity when it hits ground?

Problem 2.17

Consider the mixing tank of Section 1.1. At the initial steady state, the conditions are, $\overline{w}_1 = 200$ kg/min; $\overline{w}_2 = 100$ kg/min; $\overline{h} = 3.24$ m where the overbars indicate steady state conditions. Assume that at $t = 0$ both inlet flows are shut down but the exit valve is left open, calculate how long it takes to completely drain the tank.

Problem 2.18

A dragster crosses the finish line at a speed of 120 m/s, and should stop at a distance of 200 m. The brakes alone can generate a braking force of 16,200 N (assumed constant). It is proposed to supplement the braking force with air resistance by deploying a parachute when the dragster crosses the finish line. The mass of the driver and dragster is 900 kg. We may assume that the drag force on the dragster (which is designed for low drag) is negligible compared to the drag force on the chute (which is designed for high drag). This chute drag force F_{chute} is given by

$$F_{chute} = \frac{1}{2}\rho C_D A v^2$$

where ρ is the density of air (1.2 kg/m^3), A is the area of the parachute normal to the direction of travel ($A = \pi d^2/4$, where d is the chute diameter), C_D is equal to 1.5 for a dome-shaped chute, and v is the velocity of the dragster. The objective here is to determine the diameter d of the parachute that will stop the dragster in 200 m.

a. First determine whether the chute is even needed. Assume no parachute and neglect drag on the dragster. Using the given braking force, what distance would be required to bring the dragster to a stop?
b. If the answer to (a) was greater than 200 m, the chute is required. Determine the parachute diameter required to obtain the desired stopping distance.

Problem 2.19

A recent TV ad is the inspiration for this problem. A person with a mass of 100 kg jumps from a very tall building, and for the first 10 seconds is a free fall with hardly any air drag force. At the 10-second mark, a parachute opens, and it generates an upward force equal to Pv N, where the drag force coefficient $P = 200$ N/m/s. The initial velocity when the person propels from the building is $v(0) = 0$ m/s.

a. Obtain the model of the velocity and altitude of the person before the parachute opens.
b. Calculate the velocity and distance covered at the moment the parachute opens
c. Obtain the model describing the velocity and altitude after the parachute opens. How much time it takes for velocity to equal -5 m/s.

Problem 2.20

A model rocket car with mass $M = 0.04$ kg has an engine that provides a constant thrust force T of 0.2 N. The drag forces (in Newtons) on the car are modeled by $F_d = -kv$, where v is the velocity and k is a drag coefficient equal to 0.02 kg/s. The car is initially at rest when, at $t = 0$, the engine is fired for 4 seconds, moving the car *horizontally*. After 4 seconds, the car coasts until it eventually stops. For what *two* times will the car have a velocity of 5 m/s?

3 Classical Solutions of Ordinary Linear Differential Equations

This chapter presents the classical solutions of ordinary linear differential equations; Chapter 4 presents the Laplace Transform method. Chapter 2 presented the methods of antidifferentiation and separation of variables for first-order ordinary differential equations; we thought it was instructional at that time to show the reader the solution of the models that were being developed.

Specifically, the chapter starts by presenting several differential equations that are developed in the modeling chapters. We do this to motivate you to study the chapter and to show you how your future work as an engineer/scientist is related to the chapter. The Integrating Factor method for the solution of first-order linear differential equations is next presented. The methods for obtaining the analytical solution of homogeneous and nonhomogeneous (to be defined) of higher-order differential equations, as well as a method for handling nonlinearities and variable coefficients (to be also defined) follow. The chapter also includes a detailed discussion of the qualitative response of any system.

3.1 EXAMPLES OF DIFFERENTIAL EQUATIONS

Chapters 1 and 2 have already presented some differential equations. However, to motivate the reader to the importance of this and the next two chapters, we first present a few other models that are developed in the modeling Chapters 6 through 10. In each case, it is necessary for engineering purposes (design, development, etc.) to obtain the analytical equation that expresses the response of the dependent variable to the forcing function of interest.

3.1.1 BIOENGINEERING – CHAPTER 8

Example 8.9 shows that under some conditions a model that describes the concentration of glucose in the blood is

$$\frac{dC^G}{dt} + k_2 C^G = k_2 G_0 + k_1 C^C$$

where C^G is the concentration of glucose in mg/dL; G_0 is a constant at 90 mg/dL; k_1, k_2 are mass transfer coefficients; and C^C is the concentration of carbohydrates and, as shown in the example, equals $121.7\,e^{-0.0453t}$ mg/dL.

Using the values provided in the example, the model becomes

$$\frac{dC^G}{dt} + 0.0224C^G = 2.016 + 5.513e^{-0.0453t} \tag{3.1}$$

with initial condition $C^G(0) = 90\,\text{mg/dL}$.

DOI: 10.1201/9781003619161-3

3.1.2 Mechanical Translational – Chapter 6

Section 6.3 shows that the displacement (x in meters) of the cart shown in Figure 3.1 is described by the following model:

$$m\frac{d^2x}{dt^2}+P\frac{dx}{dt}+kx=f_A(t)$$

P is the dashpot (damper) constant in $\mathrm{N\cdot s/m}$; k is the spring constant in N / m; and m is the mass of the cart in kg.

Using the values provided in the section, the model becomes

$$10\frac{d^2x}{dt^2}+20\frac{dx}{dt}+100x=f_A(t) \tag{3.2}$$

with forcing function (applied force to cart) $f_A(t)=10u(t)\,\mathrm{N}$ (N stands for Newtons), and initial conditions $\left.\frac{dx}{dt}\right|_{t=0}=0\,{}^{\mathrm{m}}\!/_{\mathrm{s}}$ and $x(0)=0\,\mathrm{m}$.

3.1.3 Fluid System – Chapter 8

Example 8.1 shows that the level (h in meters) in the tank shown in Figure 3.2 is described by the following model:

$$\rho A\frac{dh}{dt}+1.5\rho\sqrt{11.32+9.8h}=w_1$$

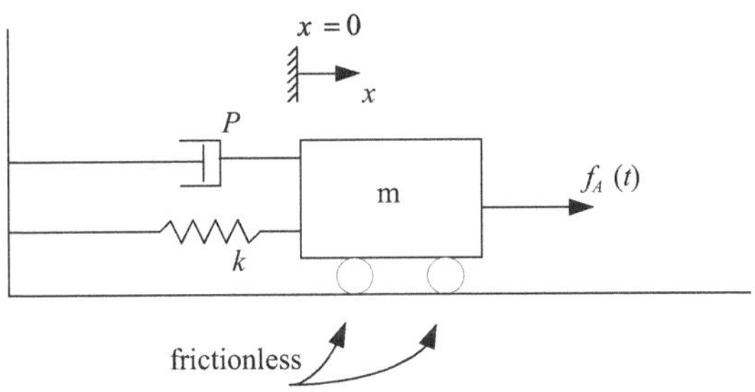

FIGURE 3.1 Mass–spring–dashpot system.

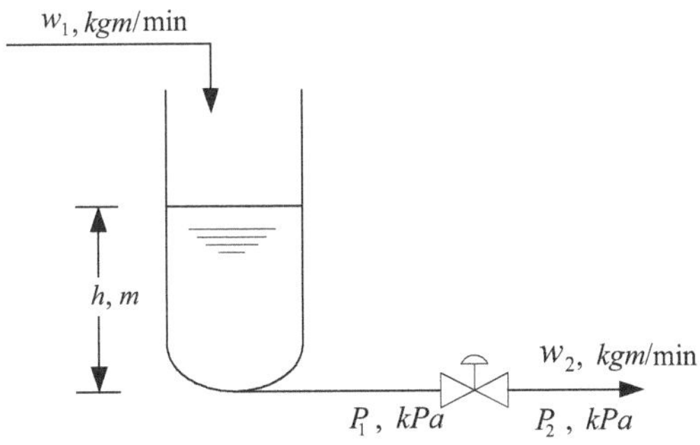

FIGURE 3.2 Liquid level.

where ρ is the density of the liquid in kg/m³; and A is the cross-sectional area of the tank in m².
Using the values provided in the example, the model becomes,

$$1000\frac{dh}{dt}+1500\sqrt{11.32+9.8h}=w_1 \tag{3.3}$$

with forcing function (flow into the tank) $w_1=1000+2000u(t)\ \text{kg/min}$, and initial condition $h(0)=3.38\ \text{m}$.

3.1.4 Thermal System – Chapter 9

Chapter 9 shows that the temperature (T in °C) of the plate of the iron shown in Figure 3.3 is described by the following model:

$$mC\frac{dT}{dt}+hAT=\dot{q}_{in}+hAT_A$$

where m is the mass of the plate in kg; C is the heat capacity of the plate in $\text{J/kg}\cdot{}^\circ\text{C}$; h is the heat transfer coefficient in $\text{J/s}\cdot\text{m}^2\cdot{}^\circ\text{C}$; A is the plate surface area in m^2; $\dot{q}_{in}$ is the energy into the iron in J/s; and T_A is the ambient air temperature in $^\circ\text{C}$.

Using the values provided in the example, the model becomes,

$$630\frac{dT}{dt}+0.8T=\dot{q}_{in}+20 \tag{3.4}$$

with forcing function (energy input to the iron) $\dot{q}_{in}=150\,u(t)$ W (W stands for Watts), and initial condition $T(0) = 25$ °C.

3.1.5 Electrical Circuit – Chapter 10

Example 10.3 shows that the voltage across the capacitor (v_C in Volts) in the circuit shown in Figure 3.4 is described by the following model:

$$LC\frac{d^2v_C}{dt^2}+RC\frac{dv_C}{dt}+v_C=v_S$$

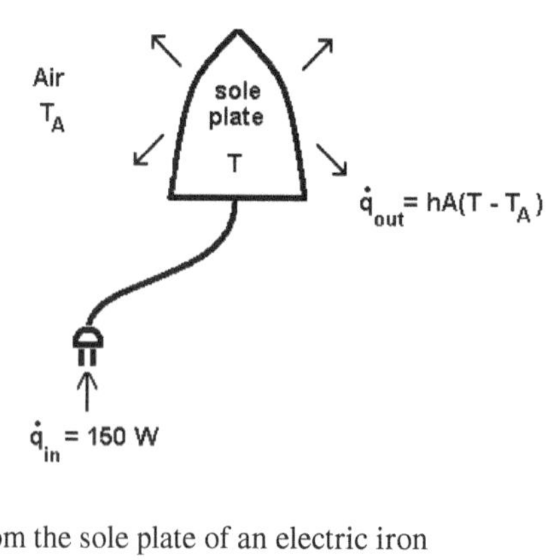

FIGURE 3.3 Heat transfer from the sole plate of an electric iron

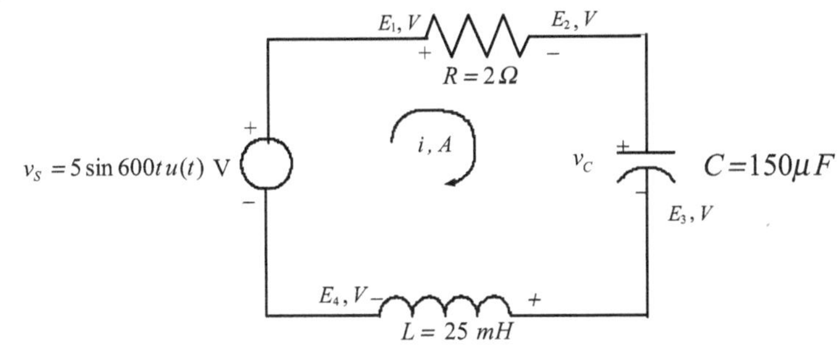

FIGURE 3.4 Electrical circuit.

where v_C is the voltage drop across the capacitor in V, v_S is the supplied voltage, and the other terms are shown in the figure. Using the numerical values shown,

$$3.75\times10^{-6}\frac{d^2v_C}{dt^2}+3\times10^{-4}\frac{dv_C}{dt}+v_C=v_S \tag{3.5}$$

with forcing function (supplied voltage) $v_S = 5\sin 600\,tu(t)$ V (V stands for Volts) and initial conditions $\left.\frac{dv_C}{dt}\right|_{t=0} = 0\,V/s$ and $v_C(0) = 0$ V.

3.2 DEFINITION OF A LINEAR DIFFERENTIAL EQUATION

A linear differential equation is one that can be put in the form

$$a_n(x)\frac{d^n y}{dx^n}+a_{n-1}(x)\frac{d^{n-1}y}{dx^{n-1}}+\cdots+a_0(x)y=r(x) \tag{3.6}$$

where $a_n, a_{n-1}, \cdots, a_0$ and r are either functions of *only* the independent variable x or constants. They do not have to be linear functions of the independent variable. As mentioned in Chapter 1, $r(x)$ is called the *forcing function* because when it changes, it *forces* the dependent variable y to change.

Consider the following differential equation:

$$b_2(x)y''+b_1(x)y'+b_o(x)y=R\,r(x)+S\,s(x) \tag{3.7}$$

This equation has two forcing functions and obviously could have any number of them. For a linear system, the addition of the solution of

$$b_2(x)y''+b_1(x)y'+b_o(x)y=Rr(x) \tag{3.8a}$$

plus the solution of

$$b_2(x)y''+b_1(x)y'+b_o(x)y=S\,s(x) \tag{3.8b}$$

is equal to a solution of Equation 3.7. That is, *the algebraic addition of the individual responses to several forcing functions to obtain the total response is exclusively a property of linear systems and it is called the principle of superposition.* This property is so fundamental that some authors define linear systems as those that satisfy the principle of superposition. Adding R times, the solution of

$$b_2(x)y''+b_1(x)y'+b_o(x)y=r(x)$$

to S times the solution of

$$b_2(x)y''+b_1(x)y'+b_o(x)y=s(x)$$

gives the same result.

If the right-hand side of a differential equation is equal to zero, such as

$$a_2(x)y''+a_1(x)y'+a_o(x)y=0 \tag{3.9}$$

it is called a *homogeneous* equation; otherwise, it is a *nonhomogeneous* equation.

The analytical solution of linear differential equations with coefficients that are function of the independent variable, such as $a_2(x)$, $a_1(x)$, . . ., and $a_o(x)$ in Equation 3.6, is rather difficult. Except for the integrating factor method, presented in Section 3.3, in this chapter, we study the analytical solutions to linear differential equations with *constant coefficients*, or

$$a_2\, y''+a_1\, y'+a_o\, y=r(x) \tag{3.10}$$

Because in dynamic systems the independent variable is time t, instead of x, these constant coefficients are also commonly referred to as *time-invariant coefficients.*

Example 3.1

Are these differential equations linear or nonlinear?

a. $m\, C_p \dfrac{dT}{dt}+hAT=\dot{Q}_{in}+hAT_{out} \Rightarrow$ linear

b. $\dfrac{dv}{dt}=-m\, g \Rightarrow \dfrac{dv}{dt}+(0)v=-m\, g \Rightarrow$ linear

c. $\dfrac{dh}{dt}=\dfrac{f_i-A_oC_D\sqrt{2gh}}{A} \Rightarrow A\dfrac{dh}{dt}+A_oC_D\sqrt{2gh}=f_i \Rightarrow$ nonlinear because of the $\sqrt{h}$ term

d. $\dfrac{dP}{dt}-k\, P=\alpha\, e^{-t} \Rightarrow$ linear

e. $\dfrac{dy}{dx}-2x^2y=2x^2 \Rightarrow$ linear

f. $\dfrac{dy}{dx}-2y^2=2x^2 \Rightarrow$ nonlinear because of the y^2 term

g. $1000\dfrac{dh}{dt}+1500\sqrt{11.32+9.8\,h}=\rho\, f_1 \Rightarrow$ nonlinear because of the second term on the left-hand side of the equal sign.

3.3 INTEGRATING FACTOR METHOD

The integrating factor method provides a solution to any first-order linear differential equation. Consider the following equation:

$$a_1(x)\frac{dy}{dx}+a_o(x)y=r(x)$$

The solution of the equation that can be put into the form

$$\frac{dy}{dx}+P(x)y=Q(x) \tag{3.11}$$

where $P(x)=\dfrac{a_0(x)}{a_1(x)}$ and $Q(x)=\dfrac{r(x)}{a_1(x)}$ is given by

$$y=e^{-\int P(x)dx}\left[\int Q(x)e^{\int P(x)dx}dx+C\right] \tag{3.12}$$

The use of the initial or boundary condition yields the integration constant C.

Example 3.2

From Equation 2.4,

$$\frac{dv}{dt}=-g \text{ with } v(0)=v_i$$

$$\frac{dv}{dt}+(0)v=-g\Rightarrow\frac{dv}{dt}+P(t)v=Q(t)$$

where $P(t)=0$ and $Q(t)=-g$
Then,

$$e^{-\int P(t)dt}=e^{-\int 0dt}=e^{-0}=1 \text{ and } e^{\int P(t)dt}=1$$

and

$$v=e^{-\int P(t)dt}\left[\int Q(t)e^{\int P(t)dt}dt+C\right]=1\left[\int -gdt+C\right]=-gt+C$$

Applying the initial condition, $C=v_i$,

$$v=v_i-gt$$

Example 3.3

From Equation 2.12c,

$$\frac{dv}{dt}+rv=-g \text{ with } v(0)=v_i$$

$$\frac{dv}{dt}+rv=-g\Rightarrow\frac{dv}{dt}+P(t)v=Q(t), \text{ where } P(t)=r \text{ and } Q(t)=-g$$

Then,

$$e^{-\int P(t)dt}=e^{-\int rdt}=e^{-rt} \text{ and } e^{\int P(t)dt}=e^{rt}$$

and

$$v = e^{-\int P(t)dt}\left[\int Q(t)e^{\int P(t)dt}dt + C\right] = e^{-rt}\left[\int -g e^{rt}dt + C\right] = e^{-rt}\left[-\frac{g}{r}e^{rt} + C\right] = -\frac{g}{r} + Ce^{-rt}$$

Applying the initial condition results in $C = v_i + \frac{g}{r}$:

$$v = (v_i + \frac{g}{r})e^{-rt} - \frac{g}{r}$$

The previous two differential equations could have been solved as well using separation of variables. This next example cannot be solved by separation of variables.

Example 3.4

$$m\,C_p \frac{dT}{dt} + hAT = \dot{q}_{in} \text{ with } T(0) = T_i$$

where $\dot{q}_{in} = \gamma e^{-\beta t}$ (an exponentially decaying energy source). Then,

$$\frac{dT}{dt} + \frac{hA}{m\,C_p}T = \frac{\gamma}{m\,C_p}e^{-\beta t} \Rightarrow \frac{dT}{dt} + a\,T = b\,e^{-\beta t}$$

$$a = \frac{hA}{mC_p}$$

$$b = \frac{\gamma}{m\,C_p}, \text{ where } P(t) = a \text{ and } Q(t) = b\,e^{-\beta t}\text{. So,}$$

$$e^{-\int P(t)dt} = e^{-\int a\,dt} = e^{-at} \text{ and } e^{\int P(t)dt} = e^{\int a\,dt} = e^{at}$$

Then

$$T = e^{-at}\left[\int b e^{-\beta t}e^{at}\,dt + C\right] = e^{-at}\left[\int b e^{(a-\beta)t}dt + C\right]$$

$$T = e^{-at}\left[\frac{b}{a-\beta}e^{(a-\beta)t} + C\right]$$

Applying the initial condition, $C = T_i - \frac{b}{a-\beta}$,

$$T = e^{-at}\left[T_i + \frac{b}{a-\beta}\left(e^{(a-\beta)t} - 1\right)\right]$$

3.3.1 Development of the Integrating Factor Method

It is worthwhile showing the reader the development of Equation 3.12. Let

$$\frac{dy}{dt} + P(t)y = Q(t) \tag{3.13}$$

Multiply both sides by an *integrating factor* $e^{\int P(t)dt}$,

$$\frac{dy}{dt}e^{\int P(t)dt} + P(t)ye^{\int P(t)dt} = Q(t)e^{\int P(t)dt} \tag{3.14}$$

Consider

$$\frac{d}{dt}\left(ye^{\int P(t)dt}\right) = e^{\int P(t)dt}\frac{dy}{dt} + y\frac{de^{\int P(t)dt}}{dt}$$

$$\frac{d}{dt}\left(ye^{\int P(t)dt}\right) = e^{\int P(t)dt}\frac{dy}{dt} + ye^{\int P(t)dt}\frac{d\int P(t)dt}{dt}$$

$$\frac{d}{dt}\left(ye^{\int P(t)dt}\right) = \frac{dy}{dt}e^{\int P(t)dt} + P(t)ye^{\int P(t)dt}$$

which is the same as the left-hand side of Equation 3.14. Thus,

$$\frac{d}{dt}\left(ye^{\int P(t)dt}\right) = Q(t)e^{\int P(t)dt}$$

$$\int d\left(ye^{\int P(t)dt}\right) = \int\left(Q(t)e^{\int P(t)dt}\right)dt$$

$$ye^{\int P(t)dt} = \int\left(Q(t)e^{\int P(t)dt}\right)dt + C$$

$$y = e^{-\int P(t)dt}\left[\int Q(t)e^{\int P(t)dt}dt + C\right]$$

which is Equation 3.12.

3.4 SOLUTION OF HOMOGENEOUS DIFFERENTIAL EQUATIONS

3.4.1 Characteristic Equation

The characteristic equation method provides the solution for an *nth order constant-coefficient (time-invariant) linear homogeneous ordinary* differential equation. To provide a background to this method, consider the following differential equation:

$$a_1\frac{dy}{dt} + a_o y = 0 \text{ with } y(0) = y_0 \tag{3.15}$$

Using the separation of variables method from Chapter 2, the solution is

$$y = Ce^{rt} = Ce^{-\frac{a_0}{a_1}t} \tag{3.16}$$

As presented in Chapters 1 and 2, Equation 3.16 provides a family of solutions depending on the value of C. However, the important solution, also called the complete solution or general solution, is the one that satisfies the initial condition. In this case,

$$y(0) = y_0 = Ce^{-\frac{a_0}{a_1}(0)} = c$$

or

$$y = y_0 e^{-\frac{a_0}{a_1}t} \tag{3.17}$$

Note that Equation 3.17 indicates that the solution is of the type $y = e^{rt}$, where r is some constant. This indication provides the background for the characteristic equation method. *The fundamental step of the characteristic equation method is assuming that the solution is* $y = e^{rt}$. Although Equation 3.15 is simple, let us use this new method to obtain its solution again; we then expand to higher-order differential equations.

Using the assumption $y = e^{rt}$, take its derivative, $\frac{dy}{dt} = re^{rt}$, and substitute it along with y into the differential equation,

$$a_1 re^{rt} + a_0 e^{rt} = 0$$

and dividing both sides by e^{rt} gives

$$a_1 r + a_0 = 0$$

This polynomial equation in r is called the *characteristic equation*; being of first order, it yields one root,

$$r = -\frac{a_0}{a_1}$$

Based on the assumption, the solution is

$$y = e^{-\frac{a_0}{a_1}t}$$

This expression satisfies the differential equation; we say that it is a *solution* of the differential equation. Multiplying y by any constant, $y = Ce^{-\frac{a_0}{a_1}t}$, is still a solution. Using the initial condition, we evaluate the constant to obtain the complete solution,

$$y(0) = y_0 = Ce^{-\frac{a_0}{a_1}(0)} = C$$

then

$$y = y_o e^{-\frac{a_o}{a_1}t}$$

which, of course, is Equation 3.17.

Let us now consider a second-order differential equation:

$$a_2 \frac{d^2 y}{dt^2} + a_1 \frac{dy}{dt} + a_0 y = 0 \text{ with } y(0) = y_0 \text{ and } \left.\frac{dy}{dt}\right|_{t=0} = y'_0$$

Using the assumption $y = e^{rt}$, taking the necessary derivatives $\frac{dy}{dt} = re^{rt}$ and $\frac{d^2 y}{dt^2} = r^2 e^{rt}$ – and substituting them into the differential equation,

$$a_2r^2e^{rt} + a_1re^{rt} + a_0e^{rt} = 0$$

or

$$a_2r^2 + a_1r + a_0 = 0$$

This is the characteristic equation in this case. Being of second order, it has two roots, r_1 and r_2. We say that there are two solutions,

$$y_1 = e^{r_1t} \text{ and } y_2 = e^{r_2t}$$

that satisfy the differential equation because each satisfies the equation. We can multiply each solution by a constant,

$$y_1 = C_1e^{r_1t} \text{ and } y_2 = C_2e^{r_2t}$$

and they are still solutions. Furthermore, we can also add these solutions and generate still another family of solutions:

$$y = C_1e^{r_1t} + C_2e^{r_2t}$$

Any of the preceding expressions for y (including y_1 and y_2) satisfy the differential equation.

We mentioned the following in Chapter 1 but believe it is important enough to stress it once again. We call "family of solutions" because no matter the values of C_1 and C_2, any of them satisfy the differential equation. However, only one specific value of C_1 and C_2 is correct for the physical system described by the equation. These correct values are obtained using the initial or boundary conditions.

Let us now summarize the method. Consider

$$a_n\frac{d^ny}{dt^n} + a_{n-1}\frac{d^{n-1}y}{dt^{n-1}} + \ldots + a_1\frac{dy}{dt} + a_0y = 0 \tag{3.18}$$

where a_n, a_{n-1}, a_1, and a_0 are *constant coefficients*. The *fundamental* step of this method is to assume a solution in the form $y = e^{rt}$. Using this assumption, we obtain each derivative in the differential equation:

$$\frac{d^ny}{dt^n} = r^n e^{rt};\cdots;\frac{d^2y}{dt^2} = r^2 e^{rt};\ \frac{dy}{dt} = re^{rt}$$

Substituting these derivatives and y into Equation 3.18,

$$a_n r^n e^{rt} + a_{n-1} r^{n-1} e^{rt} + \ldots + a_1 re^{rt} + a_0 e^{rt} = 0$$

and dividing both sides by e^{rt} gives

$$a_n r^n + a_{n-1} r^{n-1} + \ldots + a_1 r + a_0 = 0 \tag{3.19}$$

This equation is called the *characteristic equation*; it yields n roots, $r_n, r_{n-1}, \ldots, r_1$. Because there is more than one root, we express the complete solution as

$$y = \sum_{i=1}^{n} C_i\, e^{r_i t} \tag{3.20}$$

or, in this example,

$$y = C_n\, e^{r_n t} + C_{n-1}\, e^{r_{n-1} t} + \ldots + C_1\, e^{r_1 t}$$

The constants $C_n, C_{n-1}, \ldots,$ and C_1 are evaluated using the initial or boundary conditions.

This method is simple; the most difficult step is obtaining the roots of the characteristic equation (in case there are several of them, what are calculators or computers for?).

Example 3.5

Obtain the solution for the following differential equation,

$$\frac{d^2y}{dt^2} + \frac{dy}{dt} - 12\, y = 0 \text{ with } y'(0) = 0 \text{ and } y(0) = 3$$

Assuming the solution $y = e^{rt}$ and following the previous procedure, obtaining the necessary derivatives and substituting them into the differential equation yields the characteristic equation $\frac{dy}{dt} = r\, e^{rt}$ and $\frac{d^2y}{dt^2} = r^2\, e^{rt}$

$$r^2\, e^{rt} + r\, e^{rt} - 12\, e^{rt} = 0$$

and

$$r^2 + r - 12 = 0$$

Using the quadratic equation to find the roots, $r_1, r_2 = 3, -4$.

Finally,

$$y = \sum_{i=1}^{n} C_i\, e^{r_i t} = C_1\, e^{3t} + C_2\, e^{-4t}$$

Using the initial conditions yields

$$y(0) = 3 = C_1(1) + C_2(1) \Rightarrow 3 = C_1 + C_2$$
$$y' = 3C_1\, e^{3t} - 4C_2\, e^{-4t}$$
$$y'(0) = 0 = 3C_1(1) - 4C_2(1) \Rightarrow 0 = 3C_1 - 4C_2$$

From these equations, $C_1 = \frac{12}{7}$ and $C_2 = \frac{9}{7}$. Thus,

$$y = \frac{12}{7} e^{3t} + \frac{9}{7} e^{-4t}$$

3.4.2 Roots of the Characteristic Equation

Example 3.5 shows the case when the roots of the characteristic equation are real. However, the roots of a polynomial can be of other types. Let us consider these roots in more detail. To do so, consider a second-order differential equation; all findings equally apply to any nth order differential equation,

$$a_2 \frac{d^2 y}{dt^2} + a_1 \frac{dy}{dt} + a_0\, y = 0 \tag{3.21}$$

from which the following characteristic equation develops:

$$a_2\, r^2 + a_1\, r + a_0 = 0$$

and from the quadratic equation we obtain the roots

$$r_1, r_2 = \frac{-a_1 \pm \sqrt{a_1^{\,2} - 4 a_2\, a_0}}{2 a_2} \tag{3.22}$$

There are three possible cases depending on the value of the term under the square root:

1. $a_1^2 - 4 a_2\, a_0 > 0$ yielding two real roots r_1 and r_2
2. $a_1^2 - 4 a_2\, a_0 = 0$ yielding a single repeated real root
3. $a_1^2 - 4 a_2\, a_0 < 0$ yielding two complex roots at $\alpha \pm i\,\beta$

Case 1: Real Roots

For real roots, proceed as in Example 3.5, yielding

$$y = C_1\, e^{r_1 t} + C_2\, e^{r_2 t} \tag{3.23}$$

Case 2: Repeated Real Roots

For a single repeated root, both roots are *at* $r = -(a_1/2a_2)$. *The first term of the solution is* $y_1 = C_1 e^{rt}$. *The second term of the solution* is the same (except with different coefficient) *multiplied* by the independent variable, $y_2 = C_2 t e^{rt}$, then

$$y = C_1\, e^{rt} + C_2\, t\, e^{rt} \tag{3.24}$$

See Example 3.6.

Case 3: Complex Roots

For two complex roots, the roots are at $r_1 = \alpha + i\beta$ and $r_2 = \alpha - i\beta$, where

$$\alpha = -\frac{a_1}{2a_2} \text{ and } \beta = \frac{\sqrt{\left|a_1^2 - 4a_2 a_0\right|}}{2a_2}$$

where α is the real part of the root, and β is the imaginary part. Then,

$$y = C_1' e^{(\alpha+i\beta)t} + C_2' e^{(\alpha-i\beta)t} = C_1' e^{\alpha t} e^{i\beta t} + C_2' e^{\alpha t} e^{-i\beta t}$$
$$y = e^{\alpha t}\left[C_1' e^{i\beta t} + C_2' e^{-i\beta t}\right]$$

NOTE. Obtaining a good qualitative indication from this response is difficult because of the imaginary exponential powers. A better expression that avoids complex numbers is obtained using Euler's identity $e^{\pm i\beta t} = \cos\beta t \pm i\sin\beta t$. This expression is (see box for development),

$$y = e^{\alpha t}\left[C_1 \cos\beta t + C_2 \sin\beta t\right] \quad (3.25)$$

See Example 3.7.

Development of Equation 3.25

$$y = e^{\alpha t}\left[C_1' e^{i\beta t} + C_2' e^{-i\beta t}\right]$$

Using Euler's identity,

$$C_1' e^{i\beta t} + C_2' e^{-i\beta t} = C_1' \cos\beta t + iC_1' \sin\beta t + C_2' \cos\beta t - iC_2' \sin\beta t$$
$$C_1' e^{i\beta t} + C_2' e^{-i\beta t} = (C_1' + C_2')\cos\beta t + i(C_1' - C_2')\sin\beta t$$
$$C_1' e^{i\beta t} + C_2' e^{-i\beta t} = C_1 \cos\beta t + iC_2 \sin\beta t$$

and

$$y = e^{\alpha t}\left[C_1 \cos\beta t + iC_2 \sin\beta t\right]$$

The expression for y is now a bit more understandable compared to the original expression, but it still contains an imaginary term. Any constant multiple of a solution of a homogeneous linear equation is again a solution of that equation. So, following this analysis, we multiply the last term in the bracket by $1/i$ (this is a constant) $\frac{1}{\sqrt{-1}}$ and obtain

$$y = e^{\alpha t}\left[C_1 \cos\beta t + C_2 \sin\beta t\right]$$

constants C_1 and C_2 are obtained using the initial conditions. See Example 3.7.

3.4.3 Qualitative System Response

The characteristic equation method is particularly powerful because the *roots of the equation completely describe the qualitative response of the system.* Most of the important information about the system response can be obtained from these roots.

The relevant questions about the response from an engineering/science point of view are the following:

- Is the response stable? That is, *will the response remain bounded when forced by a bounded input?*
- Is the response monotonic or oscillatory?
- If monotonic and stable, how long will it take for the transients to die out?
- If oscillatory, what is the period of oscillation, and how long will it take for the oscillations to die out?

The first two questions are discussed in this chapter; the last two questions are discussed in Chapter 5 when we present the response of first- and second-order differential equations in more detail.

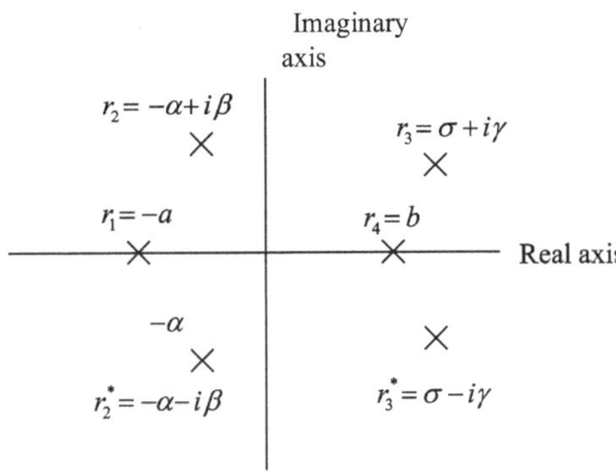

FIGURE 3.5 Roots of characteristic equation.

The important term of a *stable response* was introduced in the previous questions. We stress the definition just used: *A response is stable if it remains bounded when forced by a bounded input*; a stable response reaches a final value (this is what we call "bounded"). An *unstable response* is one that when forced by a bounded input (forcing function) continues moving up or down without stopping and reaching a final value or reaching an extreme and probably physical dangerous value. Please note that the bounded input must be one that reaches a final value.

The first two cases, Equations 3.23 and 3.24, result in exponential (monotonic) responses, and the third case, Equation 3.25, in an oscillatory response. The roots can be either real or complex; consider Figure 3.5.

Locations 1 and 4 correspond to Cases 1 and 2 when the roots are real, and locations 2 and 3 correspond to Case 3 when the roots are complex (the * denotes the complex conjugate). Now let us compare the solutions for all four locations, as given by Equations 3.23, and 3.25

Roots	Form of the solution
r_1	$y = C_1 e^{-at}$
r_2, r_2^*	$y = e^{-\alpha t}[C_1 \cos\beta t + C_2 \sin\beta t]$
r_3, r_3^*	$y = e^{\sigma t}[C_1 \cos\gamma t + C_2 \sin\gamma t]$
r_4	$y = C_1 e^{bt}$

Note that as the independent variable t increases, the solutions for the roots in locations 1 and 2 indicate that the response decays exponentially, due to the negative exponent, with oscillations superimposed in location 2. However, the solutions for the roots in locations 3 and 4 indicate that the responses increase without bounds, due to the positive exponent, with oscillations superimposed in location 3. Thus, the roots in locations 1 and 2 provide *stable responses*, and the roots in locations 3 and 4 provide *unstable responses*. The difference is in the location of the real part of the root. *For roots with negative real parts the response is stable; for roots with positive real parts, the response is unstable.* Furthermore, *for real roots the response is monotonic (locations 1 and 4), and for complex roots the response is oscillatory (locations 2 and 3).* We can also express these last statements as shown:

$$Root = \alpha \pm i\,\beta$$

Response is stable or unstable solely depending on the sign of α.
If negative, the response is stable;
If positive, the response is unstable.

Response is monotonic or oscillatory solely depending on the numerical value of β.
If $\beta = 0$, the roots are real and the response is monotonic, otherwise, the roots are complex or imaginary, and the response is oscillatory.

The fact that a system response may be oscillatory or not does not have anything to do with its stability. The system may be stable or unstable (only depending on the sign of α) and may be oscillatory or monotonic (only depending on the numerical value of β). The reader may want to think about the response for $r_5 = 0$ and for $r_6, r_6^* = 0 \pm i\beta$.

Note also that because for homogeneous differential equations there is no forcing function, *the qualitative behavior of the system does not depend on the type of forcing function* f(t) *only on the characteristics of the system itself.*

So we can learn much about the qualitative response of systems by just looking at the location of the roots of the characteristic equation. Obviously, this analysis of the location of the roots of the characteristic equation applies to any order differential equation.

Example 3.6

Obtain the solution of

$$y'' + 6y' + 9y = 0 \text{ with } y(0) = 1 \text{ and } y'(0) = 1$$

Assuming $y = e^{rt}$, then

$$\frac{dy}{dt} = re^{rt} \text{ and } \frac{d^2y}{dt^2} = r^2 e^{rt}$$

Substituting $y, \dfrac{dy}{dt}$, and $\dfrac{d^2y}{dt^2}$ into the differential equation and dividing both sides of the resulting equation by e^{rt} yields the characteristic equation

$$r^2 + 6r + 9 = 0$$

with roots $r_1 = r_2 = -3$. Both roots (repeated) are in the real negative axis, indicating that the response should be monotonic and stable.

From Equation 3.24,

$$y = C_1 e^{-3t} + C_2 t e^{-3t}$$

applying initial conditions $C_1 = 1$ and $C_2 = 4$.

Therefore,

$$y = e^{-3t} + 4t e^{-3t}$$

Please make sure you know how to obtain the values of C_1 and C_2 using the initial conditions. Example 3.5 shows an example.

The only terms in the solution are exponentials, indicating a monotonic response (there are no sine or cosine terms that would indicate oscillations). Also, because the power of the exponential terms is negative, as the independent variable t becomes very large, the value of the exponential term decays down to zero, and the response reaches a final value in this case $y|_{t=\infty} = y(\infty) = 0$.

Example 3.7

Obtain the solution of

$$y''+8y'+20y=0 \text{ with } y(0)=1 \text{ and } y'(0)=0$$

Assuming $y=e^{rt}$, then

$$\frac{dy}{dt}=re^{rt} \text{ and } \frac{d^2y}{dt^2}=r^2e^{rt}$$

Substituting y, $\frac{dy}{dt}$, and $\frac{d^2y}{dt^2}$ into the differential equation and dividing both sides of the resulting equation by e^{rt} yields the characteristic equation,

$$r^2+8r+20=0$$

Applying the quadratic solution to obtain the roots of the characteristic equation, $r_1, r_2=-4\pm i\,2$. Both roots have negative real parts indicating a stable response; they are also complex conjugates indicating an oscillatory response.

From Equation 3.25,

$$y=e^{-4t}\left[C_1\cos 2t+C_2\sin 2t\right]$$

Applying initial conditions, $C_1=1$ and $C_2=2$.

$$\text{Therefore, } y=e^{-4t}\left[\cos 2t+2\sin 2t\right]$$

This final solution indicates that the system is indeed stable because as the independent variable t becomes very large, the exponential term decays to zero and the response reaches a final value. The sine and cosine terms indicate an oscillatory response before reaching its final value.

3.5 SOLUTION OF NONHOMOGENEOUS DIFFERENTIAL EQUATIONS

The method for obtaining the *complete solution, y,* of a nonhomogeneous differential equation calls for dividing the solution into two parts, the *complementary solution*, y_C, and the *particular solution*, y_P, or

$$y=y_C+y_P \tag{3.26}$$

The form of the particular solution y_P *solely depends on the form of the forcing function*, and this is why sometimes it is also called the *forced response*. Because the particular solution only depends on the forcing function, it does not have anything to do with the system itself. The complementary solution is the one related to the system, including the initial conditions.

Consider

$$a_2\frac{d^2y}{dt^2}+a_1\frac{dy}{dt}+a_0y=f(t) \tag{3.27}$$

Substituting Equation 3.26 into Equation 3.27:

$$a_2\frac{d^2(y_C+y_P)}{dt^2}+a_1\frac{d(y_C+y_P)}{dt}+a_0(y_C+y_P)=f(t)$$

$$a_2\frac{d^2y_C}{dt^2}+a_1\frac{dy_C}{dt}+a_0\,y_C+a_2\frac{d^2y_P}{dt^2}+a_1\frac{dy_P}{dt}+a_0\,y_P=f(t)$$

Because y_P is the solution that depends on the forcing function of Equation 3.27,

$$a_2\frac{d^2y_P}{dt^2}+a_1\frac{dy_P}{dt}+a_0\,y_P=f(t)$$

and therefore,

$$a_2\frac{d^2y_C}{dt^2}+a_1\frac{dy_C}{dt}+a_0\,y_C=0$$

Note that the solution of the complementary part is the same as the solution of the *corresponding homogeneous equation*, and we'll refer to that solution from now on as y_H, or

$$y=y_H+y_P \tag{3.28}$$

Thus, the complete solution is the summation of a "solution of the corresponding homogeneous equation" plus a "particular solution of the nonhomogeneous equation." The *principle of superposition* guarantees that the summation provides the complete solution. As mentioned earlier, and as we shall see next, *the form of the particular solution depends only on the form of the forcing function,* and it is not related at all to the initial conditions.

It may be somewhat more obvious now, than before, to note that the solution of homogeneous differential equations is independent of the type of forcing function affecting the system. Thus, it is *only dependent on the system itself,* and it is why sometimes it is called the *natural response.* From Equation 3.28,

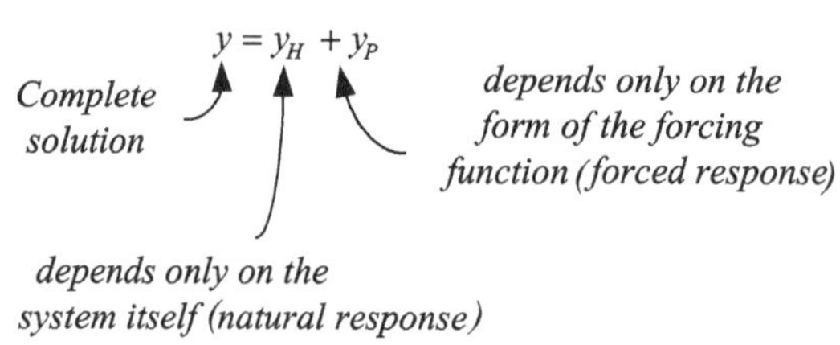

3.5.1 Homogeneous Solution (Natural Response) and Nonhomogeneous Solution (Forced Response)

It is worthwhile to delve a bit more on the statement that the solution of the corresponding homogeneous equation, y_H, depends only into the system itself and that of the particular solution, y_P, depends only on the form of the forcing function.

This statement is easily understood by considering the example of the electrical system shown in Figure 3.4. Section 3.1.5 states that the model of the circuit is given by

$$LC\frac{d^2v_C}{dt^2}+RC\frac{dv_C}{dt}+v_C=v_S$$

This is a second-order constant-coefficient (time-invariant) linear nonhomogeneous ordinary differential equation; thus, the solution is

$$v_C=v_{C_H}+v_{C_P}$$

The corresponding homogeneous solution is obtained by applying the characteristic equation method to

$$LC\frac{d^2v_{C_H}}{dt^2}+RC\frac{dv_{C_H}}{dt}+v_{C_H}=0$$

In more detail,

$$LC\frac{d^2v_{C_H}}{dt^2}+RC\frac{dv_{C_H}}{dt}+v_{C_H}=0$$

$$\updownarrow \qquad \updownarrow \qquad \updownarrow$$

$$a_2\frac{d^2v_{C_H}}{dt^2}+a_1\frac{dv_{C_H}}{dt}+a_0v_{C_H}=0$$

$$roots=\frac{-a_1\pm\sqrt{(a_1)^2-4a_2a_0}}{2a_2}=\frac{-RC\pm\sqrt{(RC)^2-4LC}}{2LC}$$

Therefore, it is obvious that the roots of v_{C_H}, and consequently the response behavior, only depend on the model of the system (circuit) and the values of each component in it. That is, it "depends on the system itself"; it does not have anything to do with the forcing function being applied.

By comparison, the form of the particular solution v_{C_P} only depends on the form of the forcing function; it does not have anything to do with the system.

As we show in later chapters, in designing a system the engineer selects the components and their values to obtain a desired response (desired performance) for a specific forcing function. Therefore, in designing the system, the engineer specifies the corresponding homogeneous equation.

3.5.2 Undetermined Coefficients

The undetermined coefficients method provides the *particular solution* of an *n*th order constant-coefficient (time invariant) linear nonhomogeneous ordinary differential equation. The method consists of the following:

- Based on the form of the forcing function, select a form of the particular solution (also referred to as a "trial solution") – see Table 3.1.
- If the forcing function involves a sine or cosine, the particular solution should contain *both* a sine and a cosine (see Examples 3.11 and 3.15).
- If any part of the particular solution is a solution of the corresponding homogeneous equation, multiply that particular solution by the independent variable. Repeat, if necessary; that is, once multiplied by the independent variable, if the result is still a solution of the corresponding homogeneous equation, multiply again by the independent variable (see Example 3.10).

TABLE 3.1
Solutions for the Particular Solution

$a_2\,y''+a_1\,y'+a_0\,y=f(t)$

	Form of the Forcing Function, *f(t)*	Form of the Particular Solution
1 →	$a_nt^n+a_{n-1}t^{n-1}+\cdots+a_1t+a_0$	$A_nt^n+A_{n-1}t^{n-1}+\cdots+A_1t+A_0$
2 →	$(a_nt^n+a_{n-1}t^{n-1}+\cdots+a_1t+a_0)e^{qt}$	$(A_nt^n+A_{n-1}t^{n-1}+\cdots+A_1t+A_0)e^{qt}$
3 →	$(a_nt^n+a_{n-1}t^{n-1}+\cdots+a_1t+a_0)\cos pt$ $+(b_nt^n+b_{n-1}t^{n-1}+\cdots+b_1t+b_0)\sin pt$	$(A_nt^n+A_{n-1}t^{n-1}+\cdots+A_1t+A_0)\cos pt$ $+(B_nt^n+B_{n-1}t^{n-1}+\cdots+B_1t+B_0)\sin pt$
4 →	$ae^{qt}\cos pt+be^{qt}\sin pt$	$Ae^{qt}\cos pt+Be^{qt}\sin pt$

Let us explain this table before presenting examples. (*Note: The table and procedure seem a bit complex, but it is not. Just give it a chance; it is not difficult, and that is why we are showing so many examples*). The left column presents four possible forms of the forcing function *f*(*t*), and the corresponding row in the right column provides the suggested form of the particular solution. Each possible form is numbered. Consider the following cases of forcing functions.

Case 1. $f(t) = 7t^3 + 5t + 3$. This forcing function is a polynomial in the independent variable. It corresponds to the first row of the table. Because the square term t^2 is missing, we could write $f(t) = 7t^3 + 0t^2 + 5t + 3$, and the suggestion given by that row in the table is for the form of the particular solution to be $y_P = A_3 t^3 + A_2 t^2 + A_1 t + A_0$.

Case 2. $f(t) = 8$. This forcing function is still a polynomial but this time of zero order. Thus, the suggestion given by the first row is still a polynomial but this time of zero order or, $y_P = A_0$.

Case 3. $f(t) = \left(7t^2 + 5t + 3\right)e^{-9t}$. This forcing function is a polynomial in the independent variable multiplied by an exponential term. It corresponds to the second row of the table. The suggestion given by that row is for the form of the particular solution to be $y_P = \left(A_2 t^2 + A_1 t + A_0\right)e^{-9t}$.

Case 4. $f(t) = \left(7t^2 + 5t + 3\right)\cos 5t$. This forcing function is a polynomial in the independent variable multiplied by a cosine term. It corresponds to the third row of the table. The suggestion given by that row in the table is for the form of the particular solution to be $y_P = \left(A_2 t^2 + A_1 t + A_0\right)\cos 5t + \left(B_2 t^2 + B_1 t + B_0\right)\sin 5t$. Recall the second bullet prior to Table 3.1: "If the forcing function involves a sine or cosine, the particular solution should contain *both* a sine and a cosine."

Case 5. $f(t) = 3e^{-9t}\sin 4t$. This forcing function corresponds to the fourth row of the table. The suggestion is $y_P = Ae^{-9t}\sin 4t + Be^{-9t}\cos 4t$. The second bullet before Table 3.1 still applies.

The following examples show in detail the use of the table and the presentations given by the bullets prior to the table. Particularly, read the last paragraphs in Examples 3.8, 3.9, and 3.11.

Example 3.8

Obtain the solution of

$$y'' + y' - 12y = 12t - 72t^2 \text{ with } y'(0) = 0 \text{ and } y(0) = 3$$

From Equation 3.28,

$$y = y_H + y_P \tag{3.28}$$

Remember, the complete solution is the summation of the corresponding homogeneous solution, y_H, and the particular solution, y_P. The corresponding homogeneous equation is

$$y_H'' + y_H' - 12y_H = 0$$

Using the characteristic equation method, assume $y_H = e^{rt}$, and after substituting it and its derivatives into the homogeneous equation, it yields the characteristic equation,

$$r^2 + r - 12 = 0$$

with roots $r_1 = 3$ and $r_2 = -4$. One root, r_1, is located on the positive real axis, and the other root, r_2, is on the negative real axis; thus, these roots indicate an unstable response. The fact that both roots are real (no imaginary parts) indicates that the response is monotonic. From Equation 3.20,

$$y_H = C_1 e^{3t} + C_2 e^{-4t} \tag{3.29}$$

Now the particular solution. The forcing function, $12t - 72t^2$, is a polynomial in the independent variable *t*, so Table 3.1 suggests as particular solution (first row in the table),

$$y_P = A_2 t^2 + A_1 t + A_0$$

from where the derivatives are

$$y_P' = 2\,A_2\,t + A_1 \text{ and } y_P'' = 2\,A_2$$

To obtain the *A* coefficients, we substitute y_P, y_P'', and y_P'' into the particular equation:

$$\begin{aligned}
y_P'' + y_P' - 12 y_P &= f(t) \\
y_P'' + y_P' - 12 y_P &= 12t + 72t^2 \\
2A_2 + 2A_2 t + A_1 - 12(A_2 t^2 + A_1 t + A_0) &= 12t - 72t^2 \\
(2A_2 + A_1 - 12A_0) + (2A_2 - 12A_1)t - 12A_2 t^2 &= 12t - 72t^2
\end{aligned}$$

This last equation shows two polynomials, one on each side of the equal sign. Equating the coefficients of equal terms in these polynomials yields A_2, A_1, and A_0.

$$\begin{aligned}
12A_2 &= 72 \Rightarrow A_2 = 6 \\
2A_2 - 12A_1 &= 12 \\
2(6) - 12A_1 &= 12 \Rightarrow A_1 = 0 \\
2A_2 + A_1 - 12A_0 &= 0 \\
2(6) + 0 - 12A_1 &= 0 \Rightarrow A_0 = 1
\end{aligned}$$

Thus,

$$y_P = 1 + 6t^2 \tag{3.30}$$

Substituting Equations 3.29 and 3.30 into Equation 3.28,

$$y = C_1 e^{3t} + C_2 e^{-4t} + 6t^2 + 1$$

and now, applying the initial conditions gives $C_1 = \frac{8}{7}$, and $C_2 = \frac{6}{7}$. Therefore,

$$y = \underbrace{\frac{8}{7}e^{3t} + \frac{6}{7}e^{-4t}}_{\text{Due to the system itself and initial conditions (natural response)}} + \underbrace{6t^2 + 1}_{\text{Due to the forcing function (forced response)}}$$

We conclude with a note about the particular solution when the forcing function is a polynomial. One should use a particular solution that is of the same order as the forcing function polynomial and include all lower terms, whether they are in the forcing function or not. For instance, we would use $y_P = A_2 t^2 + A_1 t + A_0$ for each of the following forcing functions $12t - 72t^2, 12 - 72t^2, 12 - 12t - 72t^2$, and $-72t^2$. Make sure you understand this presentation as how to use the table.

Example 3.9

Obtain the solution of

$$y'' + y' - 12y = e^{2t} \text{ with } y'(0) = 0 \text{ and } y(0) = 3$$

The corresponding homogeneous equation is the same as in the previous example, and thus, its solution is given by Equation 3.29,

$$y_H = C_1 e^{3t} + C_2 e^{-4t} \tag{3.29}$$

This time the forcing function is an exponential, so Table 3.1 suggests as particular solution (second row in the table),

$$y_P = A_0 e^{2t}$$

from which the derivatives are

$$y_P' = 2A_0 e^{2t} \quad and \quad y_P'' = 4A_0 e^{2t}$$

Substituting y_P, y_P', and y_P'' into the differential equation:

$$\begin{aligned} y_P'' + y_P' - 12y_P &= e^{2t} \\ 4A_0 e^{2t} + 2A_0 e^{2t} - 12A_0 e^{2t} &= e^{2t} \\ 4A_0 + 2A_0 - 12A_0 &= 1 \Rightarrow A_0 = -\frac{1}{6} \end{aligned}$$

Thus,

$$y_P = -\frac{1}{6} e^{2t} \tag{3.31}$$

Substituting Equations 3.29 and 3.31 into Equation 3.28 and applying the initial conditions gives

$$y = \underbrace{1.857 e^{3t} + 1.309 e^{-4t}}_{\text{Due to the system itself and initial conditions (natural response)}} \underbrace{- 0.167 e^{2t}}_{\text{Due to the forcing function (forced response)}}$$

In this example the forcing function is a constant (1) multiplying an exponential (e^{2t}). Essentially, we could say that a polynomial in t (independent variable) of zero order is multiplying the exponential. This is why the second row in the table suggests that the particular solution should be a polynomial of zero order (A_0) multiplying the exponential. If the forcing function had been te^{2t} (a first-order polynomial multiplying the exponential), the table would have suggested a first-order polynomial multiplying the exponential, or $y_P = (A_1 t + A_0)e^{2t}$. Note that this discussion is like the one in the previous example.

Example 3.10

Obtain the solution of

$$y'' + y' - 12y = e^{3t} \text{ with } y'(0) = 0 \text{ and } y(0) = 3$$

The corresponding homogeneous equation is the same as in the previous two examples, and thus, its solution is given by Equation 3.29:

$$y_H = C_1 e^{3t} + C_2 e^{-4t} \tag{3.29}$$

The particular solution suggested by Table 3.1, $y_P = A_0 e^{3t}$, does not work because it is part of the corresponding homogeneous solution (try using it and see why it does not work). So the suggestion presented in the third bullet before Table 3.1 is to multiply y_P by the independent variable *t*, or

$$y_P = A_0 t e^{3t}$$

from which

$$y_P' = A_0 e^{3t} + 3A_0 t e^{3t} \text{ and } y_P'' = 6A_0 e^{3t} + 9A_0 t e^{3t}$$

Substituting $y_{p,}$ y_P', and y_P'' into the differential equation as in the previous examples gives

$$y_P = \frac{1}{7} t e^{3t} \tag{3.32}$$

Adding Equations 3.29 and 3.32,

$$y = C_1 e^{3t} + C_2 e^{-4t} + \frac{1}{7} t e^{3t}$$

and using the initial conditions gives $C_1 = 1.694$ and $C_2 = 1.306$. Therefore,

$$y = 1.694 e^{3t} + 1.306 e^{-4t} + 0.143 t e^{3t}$$

Before presenting more examples, it is worthwhile to point out once more that all three previous examples have the same corresponding homogeneous equation, and therefore, the form of y_H is the same. The forcing functions were all different, and therefore, the particular solutions were also different.

Example 3.11

The following differential equation describes the displacement of an undamped mass–spring system (Chapter 6 discusses mechanical systems),

$$m y'' + k y = F \sin \omega t$$

Obtain its analytical solution when

a. $\omega \neq \sqrt{k/m}$.
b. $\omega = \sqrt{k/m}$.

Being a nonhomogeneous differential equation, from Equation 3.28,

$$y = y_H + y_P \tag{3.28}$$

We start by finding the solution to the corresponding homogeneous equation,

$$m\,y_H'' + k\,y_H = 0$$

Assuming $y_H = e^{rt}$, obtaining its second derivative, substituting it along with y_H into the earlier differential equation, and dividing both sides of the resulting equation by e^{rt} yields the characteristic equation,

$$mr^2 + k = 0$$

with roots $r_1 = i\sqrt{k/m}$ and $r_2 = -i\sqrt{k/m}$. Using Equation 3.25,

$$y_H = C_1 \cos\left(\sqrt{k/m}\,t\right) + C_2 \sin\left(\sqrt{k/m}\,t\right)$$

This solution indicates a continuous oscillatory response. The frequency of this corresponding homogeneous response or natural response is $\sqrt{k/m}$ radians/time. We call this the *natural frequency*, or *resonant frequency*, and denote it by ω_n.

a. For the particular solution, let us first assume that $\omega \neq \sqrt{k/m}$. In this case, Table 3.1 suggests as particular solution (third and fourth rows in the table),

$$y_P = A \cos\omega t + B \sin\omega t$$

from which

$$y_P' = -A\omega \sin\omega t + B\omega \cos\omega t \text{ and } y_P'' = -A\omega^2 \cos\omega t - B\omega^2 \sin\omega t$$

Substituting y_P and y_P'' into

$$\begin{aligned} m y_P'' + k\,y_P &= F \sin\omega t \\ -A\omega^2 m \cos\omega t - B\omega^2 m \sin\omega t + kA \cos\omega t + kB \sin\omega t &= F \sin\omega t \\ (kA - A\omega^2 m)\cos\omega t + (kB - B\omega^2 m)\sin\omega t &= F \sin\omega t \end{aligned}$$

And equating the coefficients of equal terms in this equation,

$$kA - A\omega^2 m = 0 \Rightarrow A = 0$$

$$kB - B\omega^2 m = F \Rightarrow B = \frac{F}{k - \omega^2 m}$$

Therefore, $y_P = \dfrac{F}{k-\omega^2 m}\sin\omega t$.

Finally, from Equation 3.28,

$$y = C_1\cos\left(\sqrt{k/m}\,t\right) + C_2\sin\left(\sqrt{k/m}\,t\right) + \frac{F}{k-\omega^2 m}\sin\omega t$$

The response is just a continuous oscillating curve. These types of systems are referred to as *undamped systems* and are further presented in Chapters 5 and 6. Note that as given at the beginning of the example, the differential equation $-my''+ky = F\sin\omega t$ – is of the second order without a first derivative.

b. Let us now assume that $\omega = \sqrt{k/m}$; that is, the frequency of the forcing function is equal to the natural or resonant frequency, $\omega = \omega_n$. In this case, the suggestion from Table 3.1 as a particular solution is in itself a solution of the corresponding homogeneous part and will not work. The suggestion in this case is to multiply the proposed expression from Table 3.1 by the independent variable. The particular solution is then

$$y_P = t\left[A\cos\left(\sqrt{k/m}\,t\right) + B\sin\left(\sqrt{k/m}\,t\right)\right]$$

from which

$$y_P' = A\cos\left(\sqrt{k/m}\,t\right) + B\sin\left(\sqrt{k/m}\,t\right) + t\left[-\sqrt{k/m}A\sin\left(\sqrt{k/m}\,t\right) + \sqrt{k/m}\,B\cos\left(\sqrt{k/m}\,t\right)\right]$$

and

$$\begin{aligned} y_P'' = {} & -2A\sqrt{k/m}\sin\left(\sqrt{k/m}t\right) + 2B\sqrt{k/m}\cos\left(\sqrt{k/m}t\right) \\ & + t\left[-A(k/m)\cos\left(\sqrt{k/m}t\right) - B(k/m)\sin\left(\sqrt{k/m}t\right)\right] \end{aligned}$$

Substituting y_P and y_P'' into

$$my_P'' + ky_P = F\sin\omega t$$

yields

$$-2Am\sqrt{k/m}\sin\left(\sqrt{k/m}\,t\right) + 2Bm\sqrt{k/m}\cos\left(\sqrt{k/m}\,t\right) = F\sin\left(\sqrt{k/m}\,t\right)$$

And equating the coefficients of equal terms gives $A = -\dfrac{F}{2m\sqrt{k/m}}$ and $B = 0$. Therefore, the particular solution is,

$$y_P = -\frac{F}{2m\sqrt{k/m}}t\cos\left(\sqrt{k/m}\,t\right)$$

and the complete solution is

$$y = C_1\cos\left(\sqrt{k/m}\,t\right) + C_2\sin\left(\sqrt{k/m}\,t\right) - \frac{F}{2m\sqrt{k/m}}t\cos\left(\sqrt{k/m}\,t\right)$$

Note that when $\omega = \omega_n$, y becomes unbounded as t increases. That is, the response oscillates with increasing amplitude; this type of response is referred to as *resonance*.

Sections 5.2.1.3 and 6.2.1 discuss in further details undamped systems and the terms of natural or resonant frequency and resonance.

In this example the forcing function is a constant (F) multiplying a sine term ($\sin \omega t$). Essentially, we could say that a polynomial in t (independent variable) of zero order is multiplying the sine term. This is why the third and fourth rows in the table suggest that the particular solution should be a polynomial of zero order (A_0) multiplying the sine *plus* another polynomial of zero order multiplying a cosine term. Remember, the second bullet prior to the table states: "If the forcing function involves a sine or cosine, the particular solution should contain *both* a sine and a cosine." If the forcing function had been $t \sin \omega t$ (a first-order polynomial multiplying the sine), the table would have suggested a first-order polynomial multiplying the sine plus another first-order polynomial multiplying a cosine term, or $y_P = (A_1 t + A_0)\sin \omega\, t + (B_1 t + B_0)\cos \omega\, t$.

Example 3.12

Obtain the solution of

$$y'' + 7y' + 12y = f(t) \text{ with } y'(0) = 0 \text{ and } y(0) = 3$$

for the following two forcing functions:

a. $f(t) = 6u(t)$
b. $f(t) = 12t - 72t^2$

Being a nonhomogeneous differential equation, from Equation 3.28,

$$y = y_H + y_P \tag{3.28}$$

The corresponding homogeneous equation is,

$$y_H'' + 7y_H' + 12y_H = 0$$

Using the characteristic equation method, assume $y_H = e^{rt}$, and after substituting into the corresponding homogeneous equation and following the usual procedure yields the characteristic equation:

$$r^2 + 7r + 12 = 0$$
$$(r+4)(r+3) = 0 \Rightarrow r_1 = -3;\ r_2 = -4$$

Both roots are on the negative real axis, indicating a stable monotonic response.

From Equation 3.20,

$$y_H = C_1 e^{-3t} + C_2 e^{-4t} \tag{3.33}$$

Up to this point, the solution for parts (a) and (b) is the same because the corresponding homogeneous solution is independent of the type of forcing function; we now consider each forcing function.

a. When $f(t) = 6u(t)$ Table 3.1 suggests the following particular solution (a polynomial of zero-order),

$$y_P = A_0$$

from which

$$y_P' = 0 \text{ and } y_P'' = 0$$

Substituting y_P, y_P'', and y_P'' into

$$y_P'' + 7y_P' + 12y_P = 6,$$
$$0 + 7(0) + 12A_0 = 6 \Rightarrow 12A_0 = 6 \Rightarrow A_0 = 0.5$$

Thus,

$$y_P = 0.5 \tag{3.34}$$

Substituting the solutions y_H and y_P into Equation 3.28,

$$y = C_1 e^{-3x} + C_2 e^{-4x} + 0.5$$

and applying the initial conditions gives

$$y = 10e^{-3t} - 7.5e^{-4t} + 0.5 \tag{3.35}$$

This solution shows a stable and monotonic response.

b. When $f(t) = 12t - 72t^2$ Table 3.1 suggests the following particular solution (a polynomial of second order),

$$y_P = A_2 t^2 + A_1 t + A_0$$

from which

$$y_P' = 2A_2 t + A_1 \text{ and } y_P'' = 2A_2$$

Making the usual substitutions of y_P, y_P', and y_P'' into the differential equation, and solving for A_0, A_1, and A_2 as presented before gives $A_0 = -3.66$, $A_1 = 8$, $A_2 = -6$. Thus,

$$y_P = -6t^2 + 8t - 3.66 \tag{3.36}$$

Substituting Equations 3.33 and 3.36 in Equation 3.28,

$$y = C_1 e^{-3t} + C_2 e^{-4t} - 6t^2 + 8t - 3.66$$

and applying the initial conditions gives $C_1 = 18.64$ and $C_2 = -11.98$:

$$y = 18.64e^{-3t} - 11.98e^{-4t} - 6t^2 + 8t - 3.66 \tag{3.37}$$

The roots indicated a stable monotonic response. The response given by the final result is monotonic as indicated by the roots of the characteristic equation, but it does not reach a final value as expected from the indication given by the roots. Can the reader think why this is the case (easy answer!)?

Example 3.13

The model and initial conditions, and forcing function for the mechanical system shown in Figure 3.1 are

$$10\frac{d^2x}{dt^2}+20\frac{dx}{dt}+100x=f_A(t) \tag{3.2}$$

$$\left.\frac{dx}{dt}\right|_{t=0}=0\,\mathrm{m/s} \text{ and } x(0)=0\,\mathrm{m}, \text{ and } f_A(t)=10u(t)\,\mathrm{N}$$

Obtain the analytical solution of the model.

The corresponding homogeneous equation is

$$10\frac{d^2x_H}{dt^2}+20\frac{dx_H}{dt}+100x_H=0 \tag{3.38}$$

and the characteristic equation,

$$10r^2+20r+100=0$$

with roots $r_1=-1+i3$ and $r_2=-1-i3$. Therefore, the corresponding homogeneous solution is

$$x_H=e^{-t}\left[C_1\cos 3t+C_2\sin 3t\right] \tag{3.39}$$

Because of the form of the forcing function, the particular solution is

$$x_P=A_0$$

and $\frac{d^2x_P}{dt^2}=\frac{dx_P}{dt}=0$. Substituting x_P, $\frac{dx_P}{dt}$, and $\frac{d^2x_P}{dt^2}$ into

$$10\frac{d^2x_P}{dt^2}+20\frac{dx_P}{dt}+100x_P=f_A(t)$$

$$100A_0=10 \text{ and } A_0=0.10$$

The particular solution is then

$$x_P=0.1 \tag{3.40}$$

The total solution is obtained by adding Equations 3.39 and 3.40,

$$x=e^{-t}\left[C_1\cos 3t+C_2\sin 3t\right]+0.1$$

and applying the initial conditions $C_1=-0.1$ and $C_2=-0.033$. The analytical solution is then

$$x=0.1-e^{-t}\left[0.1\cos 3t+0.033\sin 3t\right] \tag{3.41}$$

Example 3.14

The temperature (T in °C) of the sole plate of an iron shown in Figure 3.3 is described by the following model:

$$630\frac{dT}{dt}+0.8T=\dot{q}_{in}+20 \tag{3.4}$$

with forcing function (energy input to the iron) $\dot{q}_{in}=150u(t)\,W$ and initial condition $T(0)=25$ °C. Obtain the analytical solution of the model.

The corresponding homogeneous equation is

$$630\frac{dT_H}{dt}+0.8T_H=0 \tag{3.42}$$

and the characteristic equation,

$$630r+0.8=0$$

with root $r=-0.00127$. Therefore, the corresponding homogeneous solution is

$$T_H=Ce^{-0.00127t} \tag{3.43}$$

Because of the form of the forcing function, the particular solution is

$$T_P=A_0$$

and $\frac{dT_P}{dt}=0$. Making the usual substitution of T_P, and $\frac{dT_P}{dt}$ gives

$$0.8A_0=170 \text{ and } A_0=212.5$$

The particular solution is then

$$T_P=212.5 \tag{3.44}$$

The complete solution is obtained by adding the corresponding homogeneous and particular solutions,

$$T=212.5+Ce^{-0.00127t}$$

and applying the initial condition $C=-187.5$. The analytical solution is then

$$T=212.5-187.5e^{-0.00127t} \tag{3.45}$$

Example 3.15

The voltage across the capacitor (v_C in Volts) in the electrical circuit shown in Figure 3.4 is described by the following model:

$$3.75\times10^{-6}\frac{d^2v_C}{dt^2}+3.0\times10^{-4}\frac{dv_C}{dt}+v_C=v_S \tag{3.5}$$

with forcing function (supplied voltage) $v_S=5u(t)\sin 600t\ V$, and initial conditions $\left.\frac{dv_C}{dt}\right|_{t=0}=0\,V/s$ and $v_C(0)=0\,V$. Obtain the analytical solution of the model.

The corresponding homogeneous equation is

$$3.75\times10^{-6}\frac{d^2v_{CH}}{dt^2}+3.0\times10^{-4}\frac{dv_{CH}}{dt}+v_{CH}=0 \tag{3.46}$$

and the characteristic equation,

$$3.75\times10^{-6}r^2+3.0\times10^{-4}r+1=0$$

with roots $r_1=-40-i515$ and $r_2=-40+i515$. Therefore, the corresponding homogeneous solution is

$$v_{CH}=e^{-40t}\left[C_1\cos515t+C_2\sin515t\right] \tag{3.47}$$

Because of the form of the forcing function, the particular solution is

$$v_{CP}=A_1\cos600t+A_2\sin600t$$

$$\frac{dv_{CP}}{dt}=-600A_1\sin600t+600A_2\cos600t$$

and

$$\frac{d^2v_{CP}}{dt^2}=-360000A_1\cos600t-360000A_2\sin600t$$

Substituting $v_{CP}, \frac{dv_{CP}}{dt}$, and $\frac{d^2v_{CP}}{dt^2}$ into the differential equation,

$$3.75\times10^{-6}\left(-360000A_1\cos600t-360000A_2\sin600t\right)+3.0\times10^{-4}$$
$$\left(-600A_1\sin600t+600A_2\cos600t\right)+\left(A_1\cos600t+A_2\sin600t\right)=5\sin600t$$

and from this equation,

$$A_1=-5.72 \text{ and } A_2=-11.23$$

The particular solution is then

$$v_{CP}=-5.72\cos600t-11.23\sin600t \tag{3.48}$$

The complete solution is the addition of the corresponding homogeneous and particular solutions,

$$v_C=e^{-40t}\left[C_1\cos515t+C_2\sin515t\right]-5.72\cos600t-11.23\sin600t$$

and applying the initial conditions $C_1=5.72$ and $C_2=13.53$. The analytical solution is then

$$v_C=e^{-40t}\left[5.72\cos515t+13.53\sin515t\right]-5.72\cos600t-11.23\sin600t \tag{3.49}$$

3.5.3 Multiple Forcing Functions

Consider the mechanical system shown in Figure 3.6; the cart is affected by two external forces, $f_1(t)$ and $f_2(t)$. Chapter 6 discusses these types of systems and shows that, for this particular system,

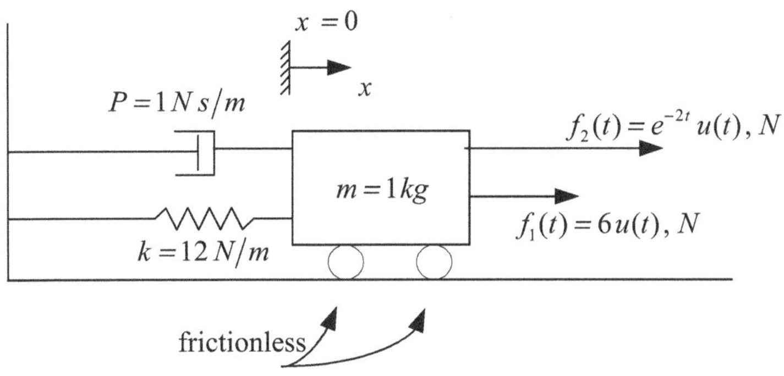

FIGURE 3.6 Cart with multiple forcing functions.

the model that describes how the displacement x of the cart is affected by the forcing functions is given by

$$x'' + x' + 12x = f_1(t) + f_2(t) \text{ with } x'(0) = 0\,\text{m/s} \text{ and } x(0) = 0\,\text{m} \tag{3.50}$$

This is the first time we face a differential equation with multiple forcing functions. However, the solution is not difficult because the equation is *linear,* and the principle of superposition discussed in Section 3.2 applies and states that the complete solution is

$$x = x_H + x_{P1} + x_{P2} \tag{3.51}$$

where x_{P1} is the particular solution for the first forcing function $f_1(t)$, and x_{P2} is the particular solution for the second forcing function $f_2(t)$. Essentially, the principle of superposition allows us to obtain the effect of each forcing function separately and then add the results to obtain the final solution; this is only applicable to linear differential equations.

The corresponding homogeneous solution is

$$x_H = e^{-0.5t}\left[C_1 \cos 3.43t + C_2 \sin 3.43t\right] \tag{3.52}$$

Using the undetermined coefficients method,

$$x_{P1} = A_0 \text{ and } x_{P2} = B_0 e^{-2t}$$

Following the same procedure previously shown several times,

$$x_{P1} = 0.5 \text{ and } x_{P2} = 0.071e^{-2t} \tag{3.53}$$

And using the initial conditions yields

$$x = 0.5 + 0.071e^{-2t} - e^{-0.5t}\left[0.571 \cos 3.43t + 0.042 \sin 3.43t\right] \tag{3.54}$$

Thus, the procedure to obtain the analytical solution of a linear differential equation when forced by k forcing functions is the same as before, except that now we apply the undetermined coefficients method k times, one for each of the forcing functions, to obtain the individual particular solutions; we then add the particular solutions to the homogeneous solution to obtain the complete or final solution.

3.6 VARIATION OF PARAMETERS

The undetermined coefficients method shown in Section 3.5 is a simple mathematical technique requiring only differential calculus and algebra. The forcing functions shown in Table 3.1 are most common in practical applications. The disadvantage of the method is that it can only be applied to those functions shown in the table. If a different forcing function is applied, that is, $\ln(t)$, $\csc(t)$, and so on, the method will not work because it has not been developed for these functions. For these instances, the variation of parameters method is available and can handle any type of forcing function; in this sense, variation of parameters is more general than undetermined coefficients.

3.6.1 Second-Order Systems

Let us apply the variation of parameters method to a second-order ordinary differential equation with constant coefficients. Consider

$$a_2\frac{d^2y}{dt^2}+a_1\frac{dy}{dt}+a_0y=f(t) \tag{3.55}$$

The initial steps are the same as those in the previous section; that is,

$$y=y_H+y_P \tag{3.28}$$

Solving for the corresponding homogeneous solution,

$$a_2\frac{d^2y_H}{dt^2}+a_1\frac{dy_H}{dt}+a_0y_H=0 \tag{3.56}$$

from which using the characteristic equation yields

$$y_H=C_1y_1+C_2y_2 \tag{3.57}$$

Up to this point, we have followed the same previous procedure. C_1 and C_2 are still constant terms and are obtained applying the initial conditions at the end.

The variation of parameters method assumes that the particular solution is the same as the corresponding homogeneous solution *but varies the parameters, making them a function of the independent variable*; that is,

$$y_P=c_1(t)y_1+c_2(t)y_2 \tag{3.58a}$$

or

$$y_p=c_1y_1+c_2y_2 \tag{3.58b}$$

For simplicity, we have dropped the notation (t) for the functions c_1 and c_2. Obviously, although not shown explicitly, y_P, y_1, and y_2 are also functions of the independent variable t. We now need to obtain the functions c_1 and c_2; to do so, we need two equations.

Taking the derivative with respect to time of Equation 3.58b gives

$$y_p'=c_1'y_1+c_1y_1'+c_2'y_2+c_2y_2' \tag{3.59}$$

The variation of parameters method states that whatever c_1 and c_2 end up being, the following is satisfied:

$$c_1' y_1 + c_2' y_2 = 0 \tag{3.60}$$

Thus, Equation 3.59 becomes

$$y_p' = c_1 y_1' + c_2 y_2' \tag{3.61}$$

Equation 3.60 is one of the two equations used to find terms c_1 *and* c_2. This equation is pretty much arbitrary, but, by choosing it, the method drops the derivatives of the c_i functions, c_1' and c_2', from the y_p' expression, thus avoiding higher derivatives of these coefficients and simplifying the solution.

Taking the derivative of Equation 3.61 with respect to time

$$y_p'' = c_1' y_1' + c_1 y_1'' + c_2' y_2' + c_2 y_2'' \tag{3.62}$$

Substituting Equations 3.58b, 3.61, and 3.62 into the differential equation yields

$$a_2\left(c_1' y_1' + c_1 y_1'' + c_2' y_2' + c_2 y_2''\right) + a_1\left(c_1 y_1' + c_2 y_2'\right) + a_0\left(c_1 y_1 + c_2 y_2\right) = f(t) \tag{3.63}$$

Rearranging this equation yields

$$c_1\left(a_2 y_1'' + a_1 y_1' + a_0 y_1\right) + c_2\left(a_2 y_2'' + a_1 y_2' + a_0 y_2\right) + a_2 c_1' y_1' + a_2 c_2' y_2' = f(t)$$

Note that the terms inside the parentheses are the corresponding homogeneous equations of the differential equation, and thus,

$$a_2 y_1'' + a_1 y_1' + a_0 y_1 = 0 \text{ and } a_2 y_2'' + a_1 y_2' + a_0 y_2 = 0$$

Therefore,

$$a_2 c_1' y_1' + a_2 c_2' y_2' = f(t) \tag{3.64}$$

Equation 3.64 is the second equation needed. So, we now have Equations 3.60 and 3.64 from where we can obtain c_1' and c_2', and integrating these yield c_1 and c_2. Let us do this and see what results.

From Equation 3.60,

$$c_2' = -\frac{y_1}{y_2} c_1'$$

Substituting into Equation 3.64 and solving for c_1',

$$c_1' = -\frac{y_2 f(t)}{a_2\left(y_1 y_2' - y_1' y_2\right)} \tag{3.65}$$

Following a similar procedure for c_2':

$$c_2' = \frac{y_1 f(t)}{a_2\left(y_1 y_2' - y_1' y_2\right)} \tag{3.66}$$

Note the negative sign in front of the c_1' expression.

Thus, once the corresponding homogeneous solution is obtained and the form of the particular solution is set, we can jump directly to the two equations, Equations 3.65 and 3.66, that provide the derivative of the coefficients of the particular solution, c_i', and integrating these provides the coefficients. As usual, the application of the initial conditions takes care of C_1 and C_2.

Let us look at three examples.

Example 3.16

Obtain the analytical solution of

$$0.25\frac{d^2y}{dt^2} + y = t + 2 \tag{3.67}$$

$$\text{with } \left.\frac{dy}{dt}\right|_{t=0} = y(0) = 0$$

The analytical solution of this differential equation can be easily obtained using undetermined coefficients; we do this purposely to show the reader the application of the variation of parameters method. The complete solution is

$$y = t + 2 - 2\cos 2t - \frac{1}{2}\sin 2t \tag{3.68}$$

The corresponding homogeneous equation is

$$0.25\frac{d^2y_H}{dt^2} + y_H = 0$$

and using the characteristic equation, we get

$$y_H = C_1\cos 2t + C_2\sin 2t \tag{3.69}$$

where $y_1 = \cos 2t$ and $y_2 = \sin 2t$, and because we know we are going to need the derivative of these,

$$y_1' = -2\sin 2t \text{ and } y_2' = 2\cos 2t$$

From Equation 3.65,

$$c_1' = -\frac{y_2 f(t)}{a_2\left(y_1 y_2' - y_1' y_2\right)}$$

$$c_1' = -\frac{(\sin 2t)(t+2)}{0.25\left[(-2\sin 2t)(\sin 2t) - (2\operatorname{con} 2t)(\cos 2t)\right]} = \frac{-2(\sin 2t)(t+2)}{\left(\sin^2 2t + \cos^2 2t\right)} = -2(\sin 2t)(t+2)$$

and after integration (using integration by parts),

$$c_1 = \cos 2t(t+2) - \frac{1}{2}\sin 2t$$

From Equation 3.66,

$$c_2' = \frac{y_1 f(t)}{a_2\left(y_1 y_2' - y_1' y_2\right)} = 2(\cos 2t)(t+2)$$

and after integration (using again integration by parts),

$$c_2 = \sin 2t(t+2) + \frac{1}{2}\cos 2t$$

Using Equation 3.58b (a bit of simple algebra, try it),

$$y_p = c_1 y_1 + c_2 y_2 = t + 2$$

The final solution is

$$y = y_H + y_p = C_1 \cos 2t + C_2 \sin 2t + t + 2$$

Using the initial conditions, we get, $C_1 = -2$ and $C_2 = -\frac{1}{2}$. Therefore,

$$y = t + 2 - 2\cos 2t - \frac{1}{2}\sin 2t \tag{3.70}$$

which is the same as Equation 3.68.

Example 3.17

Obtain the analytical solution of

$$\frac{d^2x}{dt^2} + 5\frac{dx}{dt} + 6x = e^{-2t}\ln(t) \tag{3.71}$$

$$\text{with } \left.\frac{dx}{dt}\right|_{t=0} = 0 \text{ and } x(0) = 3$$

The corresponding homogeneous equation is

$$\frac{d^2x_H}{dt^2} + 5\frac{dx_H}{dt} + 6x_H = 0$$

and using the characteristic equation yields

$$x_H = C_1 e^{-3t} + C_2 e^{-2t} \tag{3.72}$$

where $x_1 = e^{-3t}$ and $x_2 = e^{-2t}$, and its derivatives are $x_1' = -3e^{-3t}$ and $x_2' = -2e^{-3t}$.

The particular solution is

$$x_P = c_1 e^{-3t} + c_2 e^{-2t} \tag{3.73}$$

Using Equation 3.65 (changing the *y*'s for *x*'s),

$$c_1' = -\frac{x_2\, f(t)}{a_2\left(x_1 x_2' - x_1' y_2\right)} = -e^t \ln t \tag{3.74}$$

From Equation 3.66,

$$c_2' = \frac{x_1\, f(t)}{a_2\left(x_1 x_2' - x_1' y_2\right)} = -\ln t \tag{3.75}$$

From this last derivative,

$$c_2 = t\ (1 - \ln t) \tag{3.76}$$

From Equation 3.74,

$$c_1 = \int 0.2 e^t \ln t\, dt = e^t \ln t - \int \frac{e^t}{t} dt \tag{3.77}$$

The evaluation of the integral shown in Equation 3.77 is rather difficult. Often, what can be done is to represent the integrand term by a series and then integrate each term of the series. Example 3.19 shows how to use the Taylor series to do so. The result is given in Equation 3.78:

$$c_1 = e^t \ln(t) + \{\ln(t) + \frac{t}{1\times 1!} + \frac{t^2}{2\times 2!} + \frac{t^3}{3\times 3!} + \ldots\} = \ln(t)\left(1 - e^t\right) + \sum_i \frac{t^i}{i\times i!} \tag{3.78}$$

Substituting Equations 3.76 and 3.78 into Equation 3.73,

$$x_p = \left[0.2\ln(t)\left(1 - e^t\right) + 0.2\sum_i \frac{t^i}{i\times i!}\right] e^{-3t} + t\left(\ln(t) - 1\right)e^{-2t} \tag{3.79}$$

Adding the corresponding homogeneous and particular solutions,

$$x = C_1 e^{-3t} + C_2 e^{-2t} + \left[\ln(t)\left(1 - e^t\right) + \sum_i \frac{t^i}{i\times i!}\right] e^{-3t} + 0.2t\left(\ln(t) - 1\right)e^{-2t} \tag{3.80}$$

Finally, applying initial conditions provide C_1 and C_2.

Example 3.18

Obtain the analytical solution of

$$\frac{d^2x}{dt^2} + 4\frac{dx}{dt} + 13x = e^{-2t} cos^2 3t \tag{3.81}$$

$$\text{with } \left.\frac{dx}{dt}\right|_{t=0} = 0 \text{ and } x(0) = 0$$

The corresponding homogeneous equation is

$$\frac{d^2x_H}{dt^2} + 4\frac{dx_H}{dt} + 13x_H = 0$$

and using the characteristic equation the roots are $r_1 = -2 + i3$ and $r_2 = -2 - i3$. Therefore,

$$x_H = e^{-2t}\left[C_1 \cos(3t) + C_2 \sin(3t)\right] \tag{3.82}$$

where

$$x_1 = e^{-2t}\cos(3t) \text{ and } x_2 = e^{-2t}\sin(3t)$$

and their derivatives are

$$x_1' = -2e^{-2t}\cos(3t) - 3e^{-2t}\sin(3t) \text{ and } x_2' = -2e^{-2t}\sin(3t) + 3e^{-2t}\cos(3t)$$

The particular solution is

$$x_P = e^{-2t}\left[c_1 \cos(3t) + c_2 \sin(3t)\right] \tag{3.83}$$

From Equations 3.65 and 3.66, and following the necessary algebra, we get

$$c_1' = -\frac{1}{3}\sin(3t)\cos^2(3t) \text{ and } c_2' = \frac{1}{3}\cos^3(3t) \tag{3.84}$$

Integrating these two equations (and don't be ashamed of using a table of integrals!),

$$c_1 = \frac{1}{27}\cos^3(3t) \text{ and } c_2 = \frac{\sin(3t)}{9} - \frac{\sin^3(3t)}{27} \tag{3.85}$$

The particular solution is then

$$x_P = e^{-2t}\left[\frac{1}{27}\cos^4(3t) + \frac{\sin^2(3t)}{9} - \frac{\sin^4(3t)}{27}\right] \tag{3.86}$$

The complete solution is then

$$x = e^{-2t}\left[C_1 \cos(3t) + C_2 \sin(3t)\right] + e^{-2t}\left[\frac{1}{27}\cos^4(3t) + \frac{\sin^2(3t)}{9} - \frac{\sin^4(3t)}{27}\right]$$

Using the initial conditions, the constants are $C_1 = -\frac{1}{27}$ and $C_2 = 0$, and finally,

$$x = e^{-2t}\left[\frac{1}{27}\left(\cos^4(3t) - \cos(3t)\right) + \frac{\sin^2(3t)}{9} - \frac{\sin^4(3t)}{27}\right] \tag{3.87}$$

Let us present, before concluding this section, a notation that some textbooks use in obtaining the equations for c_1' and c_2'. We present once more the two equations from which the equations for c_1' and c_2' were developed:

$$c_1'y_1 + c_2'y_2 = 0 \tag{3.60}$$

and

$$a_2c_1^{'}y_1^{'} + a_2c_2^{'}y_2^{'} = f(t) \tag{3.64a}$$

or

$$c_1^{'}y_1^{'} + c_2^{'}y_2^{'} = \frac{f(t)}{a_2} \tag{3.64b}$$

These are two equations, Equations 3.60 and 3.64b, with two unknowns, and using the Cramer's rule, we can obtain $c_1^{'}$ and $c_2^{'}$(see the box for a review of Cramer's rule). Applying the rule,

$$c_1^{'}\frac{\begin{vmatrix} 0 & y_2 \\ f(t)/a^2 & y_2^{'} \end{vmatrix}}{\begin{vmatrix} y_1 & y_2 \\ y_1^{'} & y_2^{'} \end{vmatrix}} = \frac{-y_2{}^{f(t)}/a_2}{y_1y_2^{'} - y_2y_2^{'}}$$

The denominator is the determinant of a matrix with the first row formed by functions (those developed from the homogeneous solution), and the second row formed by the first derivatives of the functions (and so on if there was a need for it, as in a third- or higher-order differential equations). It happens that the determinant of this type of matrix is called the *Wronskian*, and it is represented by $W(y_1, y_2)$. Furthermore, the numerator is the determinant of the same matrix when the kth column is replaced by the column vector formed by the right hand-side of the equations, $\begin{bmatrix} 0 \\ f(t) \end{bmatrix}$ in this case. We refer to this determinant as the *Wronskian of the* k*th column* and represent it by $W_k(y_1, y_2)$. Therefore, we can write the formula for $c_1^{'}$ as

$$c_1^{'} = \frac{W_1(y_1, y_2)}{W(y_1, y_2)} = \frac{\begin{vmatrix} 0 & y_2 \\ f(t)/a^2 & y_2^{'} \end{vmatrix}}{\begin{vmatrix} y_1 & y_2 \\ y_1^{'} & y_2^{'} \end{vmatrix}} = \frac{-y_2\, f(t)/a_2}{y_1\, y_2^{'} - y_2\, y_1^{'}} \tag{3.88}$$

similarly, for $c_2^{'}$,

$$c_2^{'} = \frac{W_2(y_1, y_2)}{W(y_1, y_2)} = \frac{\begin{vmatrix} y_1 & 0 \\ y_1^{'} & f(t)/a^2 \end{vmatrix}}{\begin{vmatrix} y_1 & y_2 \\ y_1^{'} & y_2^{'} \end{vmatrix}} = \frac{y_1\, f(t)/a_2}{y_1\, y_2^{'} - y_2\, y_1^{'}} \tag{3.89}$$

Cramer's Rule

Cramer's rule is an easy way to solve for one variable of a set of linear algebraic equations without having to solve the entire set. To illustrate its use, consider the following set:

$$\begin{aligned} a_1x_1 + a_2x_2 + a_3x_3 &= b_1 \\ a_4x_1 + a_5x_2 + a_6x_3 &= b_2 \\ a_7x_1 + a_8x_2 + a_9x_3 &= b_3 \end{aligned} \quad ,$$

where the x's are the variables we need to obtain and the a's and b's are constants.

To solve for x_1, we use the following:

$$x_1 = \frac{\begin{vmatrix} b_1 & a_2 & a_3 \\ b_2 & a_5 & a_6 \\ b_3 & a_8 & a_9 \end{vmatrix}}{\begin{vmatrix} a_1 & a_2 & a_3 \\ a_4 & a_5 & a_6 \\ a_9 & a_8 & a_9 \end{vmatrix}}$$

Note that the denominator is the determinant of the coefficients of the variables (let us call it D), and the numerator is the same determinant replacing the coefficients of the x_1 column by the answer column values (let us call it D_1). Therefore,

$$x_1 = \frac{D_1}{D}$$

For x_2, we follow a similar procedure:

$$x_2 = \frac{\begin{vmatrix} a_1 & b_1 & a_3 \\ a_4 & b_2 & a_6 \\ a_9 & b_3 & a_9 \end{vmatrix}}{\begin{vmatrix} a_1 & a_2 & a_3 \\ a_4 & a_5 & a_6 \\ a_9 & a_8 & a_9 \end{vmatrix}} = \frac{D_2}{D}$$

Therefore, in general, we can use the following for solving for x_k:

$$x_k = \frac{D_k}{D}$$

3.6.2 Higher-Order Systems

The previous section presented the use of the variation of parameters method for obtaining the analytical solution of second-order differential equations. This section presents the application of the method to third-order differential equations. Once we do so, the method is extended to a nth-order equation.

Consider the following model:

$$a_3 \frac{d^3 y}{dt^3} + a_2 \frac{d^2 y}{dt^2} + a_1 \frac{dy}{dt} + a_0 y = f(t) \tag{3.90}$$

The corresponding homogeneous equation is

$$a_3 \frac{d^3 y_H}{dt^3} + a_2 \frac{d^2 y_H}{dt^2} + a_1 \frac{dy_H}{dt} + a_o y_H = 0 \tag{3.91}$$

from which using the characteristic equation yields

$$y_H = C_1 y_1 + C_2 y_2 + C_3 y_3 \tag{3.92}$$

There are three functions this time because this is a third-order system. Continuing with the assumption of the variation of parameters method that the particular solution is the same as the homogeneous solution *but varies the parameters making them a function of the independent variable*; that is,

$$y_p = c_1(t)\,y_1 + c_2(t)\,y_2 + c_3(t)\,y_1 \tag{3.93a}$$

or simply

$$y_p = c_1 y_1 + c_2 y_2 + c_3 y_3 \tag{3.93b}$$

This time there are three functions, c_1, c_2, and c_3, and therefore, we need three equations to solve for them.

Taking the differential of Equation 3.93b,

$$y_p' = c_1' y_1 + c_1 y_1' + c_2' y_2 + c_2 y_2' + c_3' y_3 + c_3 y_3' \tag{3.94}$$

Using again the statement of the variation of parameters method that whatever c_1, c_2, and c_3 end up being, the following is satisfied:

$$c_1' y_1 + c_2' y_2 + c_3' y_3 = 0 \tag{3.95}$$

Equation 3.95 is the first of the three equations needed to find terms c_1, c_2, and c_3. By choosing this equation, the method drops the derivatives of the c_i coefficients, c_1', c_2', and c_3', from the y_p' expression, and thus, avoiding higher derivatives of these coefficients and simplifying the solution. Equation 3.94 is then

$$y_p' = c_1 y_1' + c_2 y_2' + c_3 y_3' \tag{3.96}$$

Taking the derivative of y_p',

$$y_p'' = c_1' y_1' + c_1 y_1'' + c_2' y_2' + c_2 y_2'' + c_3' y_3' + c_3 y_3'' \tag{3.97}$$

Choosing a similar statement for the same reason as dropping the derivatives of the c_i functions:

$$c_1' y_1' + c_2' y_2' + c_3' y_3' = 0 \tag{3.98}$$

Equation 3.98 is the second equation needed to find the functions c_1, c_2, and c_3. The second derivative then becomes

$$y_p'' = c_1 y_1'' + c_2 y_2'' + c_3 y_3'' \tag{3.99}$$

Taking the differential of Equation 3.99,

$$y_p''' = c_1 y_1''' + c_1' y_1'' + c_2 y_2''' + c_2' y_2'' + c_3 y_3''' + c_3' y_3'' \tag{3.100}$$

Substituting Equations 3.96, 3.99, 3.100, and 3.93b into the differential equation:

$$\begin{aligned} &a_3\left(c_1 y_1''' + c_1' y_1'' + c_2 y_2''' + c_2' y_2'' + c_3 y_3''' + c_3' y_3''\right) + a_2\left(c_1 y_1'' + c_2 y_2'' + c_3 y_3''\right) \\ &\quad + a_1\left(c_1 y_1' + c_2 y_2' + c_3 y_3'\right) + a_0\left(c_1 y_1 + c_2 y_2 + c_3 y_1\right) = f(t) \end{aligned}$$

Rearranging,

$$c_3\left(a_3y_3''' + a_2y_3'' + a_1y_3' + a_0y_3\right) + c_2\left(a_3y_2''' + a_2y_2'' + a_1y_2' + a_0y_2\right)$$
$$+c_1\left(a_3y_1''' + a_2y_1'' + a_1y_1' + a_0y_1\right) + a_3\left(c_3'y_3'' + c_2'y_2'' + c_1'y_1''\right) = f(t)$$

Note that the first three terms in parenthesis are the corresponding homogeneous equations; thus,

$$a_3y_3''' + a_2y_3'' + a_1y_3' + a_0y_3 = 0; \qquad a_3y_2''' + a_2y_2'' + a_1y_2' + a_0y_2 = 0;$$
$$\text{and } a_3y_1''' + a_2y_1'' + a_1y_1' + a_0y_1 = 0$$

Therefore,

$$a_3\left(c_3'y_3'' + c_2'y_2'' + c_1'y_1''\right) = f(t)$$

$$\text{or } c_3'y_3'' + c_2'y_2'' + c_1'y_1'' = {f(t)}\big/{a_3} \tag{3.101}$$

This is the last equation we need to find the three functions c_1, c_2, and c_3.

We could proceed from this point on using substitutions and obtaining an equation for each of the three derivatives and then integrate. But certainly, obtaining the derivatives using Cramer's rule is easier; this is only an extension of what we previously did for the second-order differential equation. To help us with the presentation, let us write again the three equations:

$$c_1'y_1 + c_2'y_2 + c_3'y_3 = 0 \tag{3.95}$$
$$c_1'y_1' + c_2'y_2' + c_3'y_3' = 0 \tag{3.98}$$
$$\text{and } c_3'y''_3 + c_2'y''_2 + c_1'y''_1 = f(t)/a_3 \tag{3.101}$$

Because the matrix of the functions is formed by the functions and their derivatives the determinant is again the Wronskian. Using Cramer's rule, we obtain

$$c_1' = \frac{W_1(y_1,y_2,y_3)}{W(y_1,y_2,y_3)} = \frac{\begin{vmatrix} 0 & y_2 & y_3 \\ 0 & y_2' & y_3' \\ {f(t)}\big/{a_2} & y_2'' & y_3'' \end{vmatrix}}{\begin{vmatrix} y_1 & y_2 & y_3 \\ y_1' & y_2' & y_3' \\ y_1'' & y_2'' & y_3'' \end{vmatrix}} \qquad c_2' = \frac{W_2(y_1,y_2,y_3)}{W(y_1,y_2,y_3)} = \frac{\begin{vmatrix} y_1 & 0 & y_3 \\ y_1' & 0 & y_3' \\ y_1'' & {f(t)}\big/{a_2} & y_3'' \end{vmatrix}}{\begin{vmatrix} y_1 & y_2 & y_3 \\ y_1' & y_2' & y_3' \\ y_1'' & y_2'' & y_3'' \end{vmatrix}}$$

and

$$c_3' = \frac{W_3(y_1,y_2,y_3)}{W(y_1,y_2,y_3)} = \frac{\begin{vmatrix} y_1 & y_2 & 0 \\ y_1' & y_2' & 0 \\ y_1'' & y_2'' & {f(t)}\big/{a_2} \end{vmatrix}}{\begin{vmatrix} y_1 & y_2 & y_3 \\ y_1' & y_2' & y_3' \\ y_1'' & y_2'' & y_3'' \end{vmatrix}}$$

The similarity between the second- and third-order solutions allow us to write the following expression for the $c_k^{'}$ function of an nth order differential equation:

$$c_k^{'} = \frac{W_k\left(y_1, y_2, \cdots, y_n\right)}{W\left(y_1, y_2, \cdots, y_n\right)} \tag{3.102}$$

The variation of parameter procedure is not complex and can handle many different forcing functions. Without a doubt, the most difficult step is the integration steps of the $c_i^{'}$ function. As previously mentioned, (a) the use of a table of integrals will usually help, and (b) if the integral cannot be obtained, the integrand can be represented by a series and each term is then integrated. The next section discusses series.

3.7 HANDLING NONLINEARITIES AND VARIABLE COEFFICIENTS

The characteristic equation and undetermined coefficient methods presented in previous sections and the Laplace transform method presented in Chapter 4 are applicable to linear differential equations with constant coefficients. The variation of parameters presented in the previous section is applicable to linear differential equations; however, the coefficients do not have to be constants. Often, the equations describing physical systems are nonlinear and contain variable, nonstationary, coefficients. Thus, the methods presented thus far cannot provide a solution. There is a method for these cases that allows us to obtain an *approximate* analytical solution. We start the section with a presentation of the Taylor series followed by the solution method.

3.7.1 Taylor Series

Taylor series is probably the most common series used in engineering and physics applications:

$$f(x) = \sum\nolimits_{n=0}^{\infty} \frac{1}{n!} \left.\frac{d^n f(x)}{dx^n}\right|_{x=x_0} \left(x - x_0\right)^n \tag{3.103a}$$

or when the independent variable is time *t*,

$$f(t) = \sum\nolimits_{n=0}^{\infty} \frac{1}{n!} \left.\frac{d^n f(t)}{dt^n}\right|_{t=t_0} \left(t - t_0\right)^n \tag{3.103b}$$

or

$$f(t) = f(t_0) + \frac{1}{1!} f'(t_0)(t - t_0) + \frac{1}{2!} f''(t_0)(t - t_0)^2 + \frac{1}{3!} f'''(t_0)(t - t_0)^3 + \cdots$$

Equation 3.103 approximates $f(x/t)$"around $x/t = x_0/t_0$"; $f(x_0)$, or $f(t_0)$, and the derivative terms $\frac{1}{n!} \left.\frac{d^n f(x)}{dx^n}\right|_{x=x_0}$, or $\frac{1}{n!} \left.\frac{d^n f(t)}{dt^n}\right|_{t=t_0}$, are all constants.

Taylor series can also be used to approximate multivariable functions. For a two-variable function,

$$f(x,z)=\sum_{n=0}^{\infty}\frac{1}{n!}\left[\begin{array}{l}\left.\frac{\partial^n f(x,z)}{\partial x^n}\right|_{\substack{x=x_0\\z=z_0}}(x-x_0)^n+\left.\frac{\partial^n f(x,z)}{\partial z^n}\right|_{\substack{x=x_0\\z=z_0}}(z-z_0)^n\\ +\left.\frac{\partial^n f(x,z)}{\partial x^{n-1}\partial z^{n-1}}\right|_{\substack{x=x_0\\z=z_0}}(x-x_0)^{n-1}(z-z_0)^{n-1}+\cdots\end{array}\right] \tag{3.104}$$

When $x_0=0$, the Taylor series becomes $f(x)=\sum_{n=0}^{\infty}\frac{1}{n!}\left.\frac{d^n f(x)}{dx^n}\right|_{x=0}x^n$, and it is commonly known as the *Maclaurin series.*

Example 3.19

The integral $\int\frac{e^t}{t}dt$ is difficult to obtain; however, the integrand could be represented by a series, and then each term of the series could be integrated individually. Using Taylor series, obtain an approximation for the integral.

Instead of approximating the entire integrand, let us only approximate the numerator around $t_0=0$:

$$f(t)=f(t_0)+\frac{1}{1!}f'(t_0)t+\frac{1}{2!}f''(t_0)t^2+\frac{1}{3!}f'''(t_0)t^3+\cdots$$

$$e^t=e^t\Big|_{t_0=0}+\frac{1}{1!}e^t\Big|_{t_0=0}t+\frac{1}{2!}e^t\Big|_{t_0=0}t^2+\frac{1}{3!}e^t\Big|_{t_0=0}t^3+\cdots$$

Remember, any derivative of e^t is just e^t. Evaluating the exponentials,

$$e^t=1+\frac{1}{1!}t+\frac{1}{2!}t^2+\frac{1}{3!}t^3+\cdots$$

and

$$\frac{e^t}{t}=\frac{1}{t}+\frac{1}{1!}+\frac{1}{2!}t+\frac{1}{3!}t^2+\cdots$$

$$\int\frac{e^t}{t}dt=\int\frac{1}{t}dt+\frac{1}{1!}\int 1dt+\frac{1}{2!}\int tdt+\frac{1}{3!}\int t^2dt+\cdots$$

Finally,

$$\int\frac{e^t}{t}dt=\ln(t)+\frac{t}{1\times 1!}+\frac{t^2}{2\times 2!}+\frac{t^3}{3\times 3!}+\cdots$$

3.7.2 Linearization of Nonlinear Differential Equations

Taylor series expansion provides a way to linearize nonlinear terms and manipulate variable coefficients allowing obtaining an *approximate* analytical solution.

Before presenting the method to linearize, it is appropriate to discuss what is meant by linearization and why the term *approximate* is used. Consider Figure 3.7 showing the function $f(x)$ versus

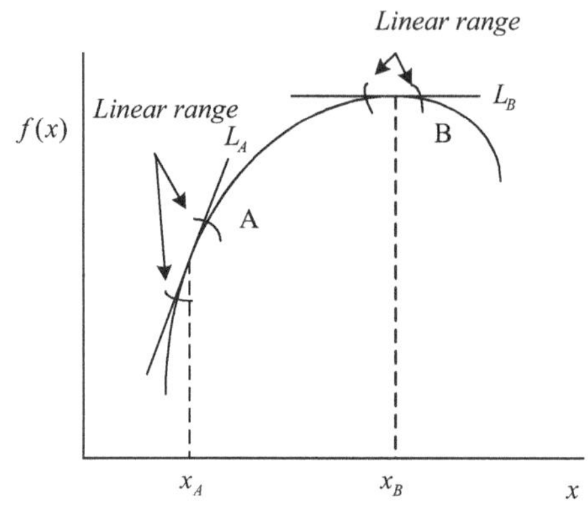

FIGURE 3.7 Linear approximation.

x; the figure shows a nonlinear relationship. The figure shows two tangent lines. These lines, L_A and L_B, approximate the function at two different values of x. The figure shows a range in which each tangent line provides a good approximation; outside this range, the approximation "breaks down." Obviously, the degree of nonlinearity affects the acceptable range. At point A, the function is "more linear" than at point B, and therefore, the acceptable range for L_A is wider than for L_B.

As previously presented, the Taylor series is

$$f(x) = \sum_{n=0}^{\infty} \frac{1}{n!} \frac{d^n f(x)}{dx^n}\bigg|_{x=x_0} (x - x_0)^n \tag{3.103a}$$

or

$$f(x) = f(x_0) + \frac{1}{1!} f'(x_0)(x - x_0) + \frac{1}{2!} f''(x_0)(x - x_0)^2 + \frac{1}{3!} f'''(x_0)(x - x_0)^3 + \cdots$$

Note that the first two terms on the right side are a constant and a linear term; from the third term on, the terms are nonlinear. Thus, neglecting the nonlinear terms,

$$f(x) \approx f(x_0) + \frac{1}{1!} f'(x_0)(x - x_0) \tag{3.105}$$

Equation 3.105 approximates the nonlinear function of x by a linear function of x; it is in the form of the well-known linear equation learned in algebra:

$$y = y_{\text{intercept}} + m(x - x_{\text{intercept}})$$

where m is the slope, $m = f'(x_0)$.

Before proceeding with the example, Equation 3.106 provides the linearization of a function of two variables (x, z):

$$f(x,z) \approx f(x_0, z_0) + \frac{\partial f(x,z)}{\partial x}\bigg|_{\substack{x=x_0 \\ z=z_0}} (x - x_0) + \frac{\partial f(x,z)}{\partial z}\bigg|_{\substack{x=x_0 \\ z=z_0}} (z - z_0) \tag{3.106}$$

The linearization is done around x_0, z_0. This linearization equation develops using only the first three terms of the series of Equation 3.104. The extension to three or more variables is easy (obvious?) using Equation 3.104 as a base. Remember that the term $f(x_0, z_0)$ and the partial derivatives are constants.

Figure 3.7 shows the nonlinear function *f*(*x*). Assuming that you may be operating at point *A* sometimes and at other times at point *B*, it would be realistic to have two approximating linearization and use the appropriate one to obtain the analytical solution.

Example 3.20 next shows the use of Equation 3.105 to linearize a nonlinear differential equation and obtaining an approximate analytical solution. Example 3.21 shows the use of the equation to handle nonconstant coefficients.

Example 3.20

Consider the differential equation describing the liquid level in the tank shown in Chapter 1,

$$146\frac{dh}{dt} + 166.7\sqrt{h} = w_1(t) + w_2(t) \tag{1.13b}$$

The second term on the left-hand side is a nonlinear function of *h*. Applying Equation 3.105 to that term,

$$f(h) \approx f(h_0) + f'(h_0)(h - h_0)$$

$$f(h) = \sqrt{h(t)} \approx \sqrt{h_0} + 0.5\left.\frac{1}{\sqrt{h}}\right|_{h=h_o}(h - h_o)$$

and selecting $h_0 = h(0) = 3.24\,\text{m}$ because that is where we know the value of *h* (the initial condition):

$$\sqrt{h(t)} \approx 1.8 + 0.278(h - 3.24) \tag{3.107}$$

Substituting this approximation to $\sqrt{h(t)}$ into Equation 1.13b yields

$$146\frac{dh}{dt} + 166.7[1.8 + 0.278(h - 3.24)] = w_1(t) + w_2(t)$$

or

$$146\frac{dh}{dt} + 46.34h = w_1(t) + w_2(t) - 150.14 \tag{3.108}$$

Equation 3.108 is a linear first-order homogeneous differential equation *with* constant coefficients and may be solved by any of the methods presented in Chapters 2, 3, and 4.

Figure 3.8 shows a comparison of the responses obtained by simulation of the actual model, Equation 1.13b, and of the approximating model, Equation 3.108. In this case flow w_2 was increased by 50 *kg/min* at times 5, 50, 80 and 120 min.

The figure shows that the responses are close at the very beginning when the conditions are the ones uscd to obtain the linearized model that is, when $w_2 = 100\,\text{kg/min}$ and $h = 3.24\,\text{m}$. As w_2 continue increasing and the operating conditions move away from the linearized conditions, the response of the linearized model degrades. There is also a difference in how fast the models reach the steady value (the dynamics provided by the models); however, this is not easily seen in the figure.

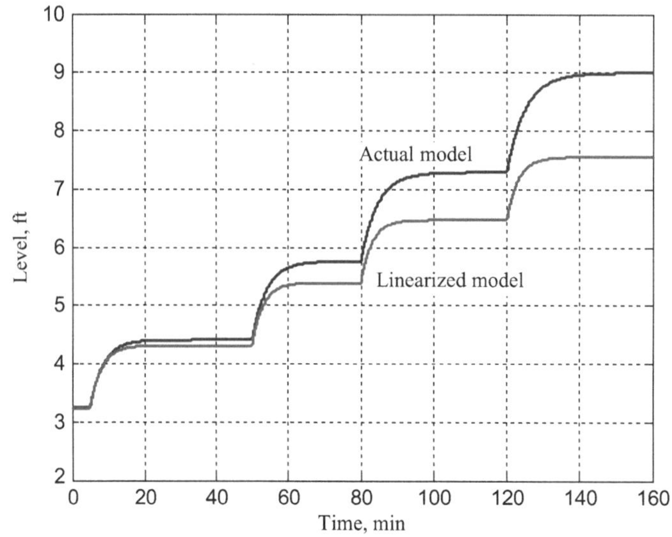

FIGURE 3.8 Comparison of actual and linearized models.

To complete this presentation, suppose that it is known that the operation will now be at $w_1 = 200\,\text{kg/min}$, $w_2 = 150\,\text{kg/min}$ and $h = 5.8\,\text{m}$, so it is desired to linearize the model around these conditions. Following the same linearization procedure, the new model is

$$146\frac{dh}{dt} + 34.67\,h = w_1(t) + w_2(t) - 200 \tag{3.109}$$

Figure 3.9 shows the new comparison. It is obvious that at the new linearizing conditions both models provide the same response; however, as the conditions move away, in both directions, the linearized model response degrades.

Example 3.21

Consider the tank shown in Figure 8.5. In this tank a concentrated solution of $NaOH$ and H_2O (stream 1) is diluted using pure water (stream 2); the concentrated solution contains 0.75 mass fraction of $NaOH$. Example 8.3 shows that the model that describes how the exit concentration of $NaOH$ varies when the concentration of stream 1 changes, and at the same time the flow of stream 1 changes is given by

$$341.1\frac{dx_3^{NaOH}}{dt} + (w_1 + w_2)x_3^{NaOH} = w_1 x_1^{NaOH} \tag{8.36}$$

Assume that $x_1^{NaOH} = 0.75 - 0.08\,u(t)$, $w_1 = 20 - 5\,u(t)\,\text{kg/min}$, and $w_2 = 7.273\,\text{kg/min}$. The initial condition is $x_3^{NaOH}(0) = 0.55$.

Obtaining the analytical solution is not easy because the coefficient of the second term on the left-hand side, $w_1 + w_2$, is a function of time. Thus, it is a nonstationary coefficient. Taylor series expansion provides a way to deal with this, and of course, the solution will be an approximation, but at least it provides an indication of the response.

Applying Equation 3.106 to the second term, and using $w_1 = x$, $x_3^{NaOH} = z$, $x_0 = 20$, and $z_0 = 0.55$,

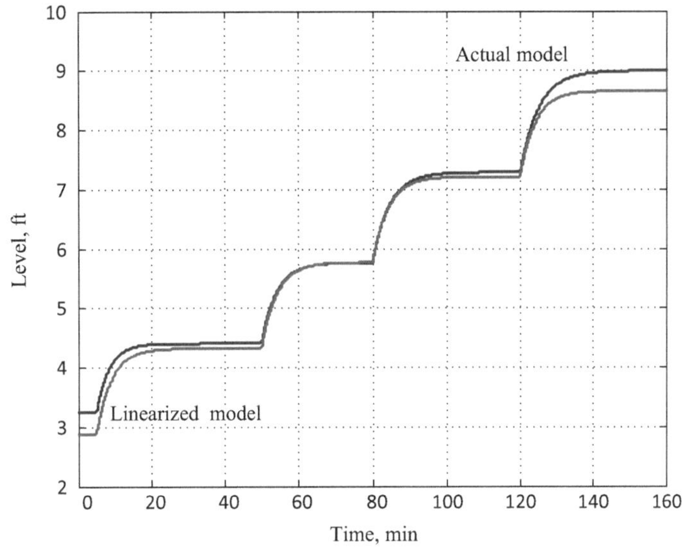

FIGURE 3.9 Comparison of actual and new linearized models.

$$f_o(w_1, x_3^{NaOH}) = (w_1 + w_2)x_3^{NaOH} \approx 15 + 0.55(w_1 - 20) + 27.273\left(x_3^{NaOH} - 0.55\right) \quad (3.110)$$

Similarly, applying Equation 3.106 to the forcing function term but this time with $w_1 = x$, $x_1^{NaOH} = z$, $x_0 = 20$, and $z_0 = 0.75$,

$$f\left(w_1, x_1^{NaOH}\right) = w_1 x_1^{NaOH} \approx 15 + 0.75(w_1 - 20) + 20\left(x_1^{NaOH} - 0.75\right) \quad (3.111)$$

Substituting Equations 3.110 and 3.111 into the differential equation and rearranging yields

$$341.1\frac{dx_3^{NaOH}}{dt} + 27.273x_3^{NaOH} = 0.2w_1 + 20x_1^{NaOH} - 4 \quad (3.112)$$

This is now a first-order linear nonhomogeneous differential equation *with* constant coefficients and may be solved by any of the methods presented in this chapter and Chapter 4.

Figure 3.10 shows the response of the actual and linearized models. Although small at this operating point, which is where we linearized the model, the figure shows the steady state and dynamic differences between responses.

Example 3.22

Chapter 9 develops the model of an adiabatic batch reactor; the model is given by the following two nonlinear differential equations:

$$\frac{dC^A}{dt} = -k_0 e^{-\frac{a}{T}} C^A \quad (3.113)$$

and

$$\frac{dT}{dt} = \frac{Lk_0}{\rho C} e^{-\frac{a}{T}} C^A \quad (3.114)$$

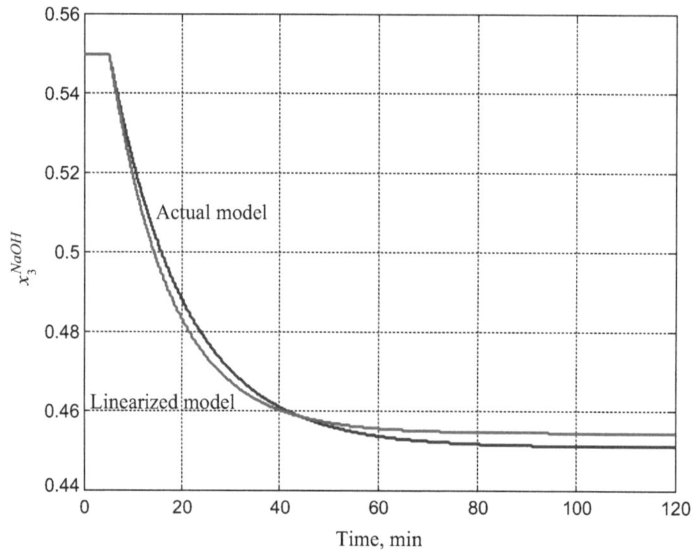

FIGURE 3.10 Comparison of actual and linearized models.

These are nonlinear equations, *and* the two dependent variables are multiplied together. The nonlinear term $f(T,C^A) = e^{-\frac{a}{T}} C^A$ is now linearized using Equation 3.106:

$$f(T,C^A) \approx f\left(T(0),C^A(0)\right) + \left.\frac{\partial f\left(T,C^A\right)}{\partial T}\right|_{\substack{T=T(0)\\C^A=C^A(0)}} [T-T(0)] + \left.\frac{\partial f\left(T,C^A\right)}{\partial C^A}\right|_{\substack{T=T(0)\\C^A=C^A(0)}} [C^A - C^A(0)]$$

where

$$f(T(0),C^A(0)) = e^{-\frac{a}{T(0)}} C^A(0)$$

$$\left.\frac{\partial f\left(T,C^A\right)}{\partial T}\right|_{\substack{T=T(0)\\C^A=C^A(0)}} = \frac{a}{T(0)^2} e^{-\frac{a}{T(0)}} C^A(0) \qquad \left.\frac{\partial f\left(T,C^A\right)}{\partial C^A}\right|_{\substack{T=T(0)\\C^A=C^A(0)}} = e^{-\frac{a}{T(0)}}$$

$$f(T,C^A) \approx e^{-\frac{a}{T(0)}} C^A(0) + \frac{a}{T(0)^2} e^{-\frac{a}{T(0)}} C^A(0)[T-T(0)] + e^{-\frac{a}{T(0)}} [C^A - C^A(0)]$$

Using the values given in Chapter 9 for the terms,

$$f(T,C^A) \approx 1.5\times10^{-7}T + 4.566\times10^{-9}C^A - 4.384\times10^{-5}$$

Including this linearized expression into the two original equations, and using again the numerical values,

$$\frac{dC^A}{dt} = 1058.74 - 3.62T - 0.11C^A \tag{3.115}$$

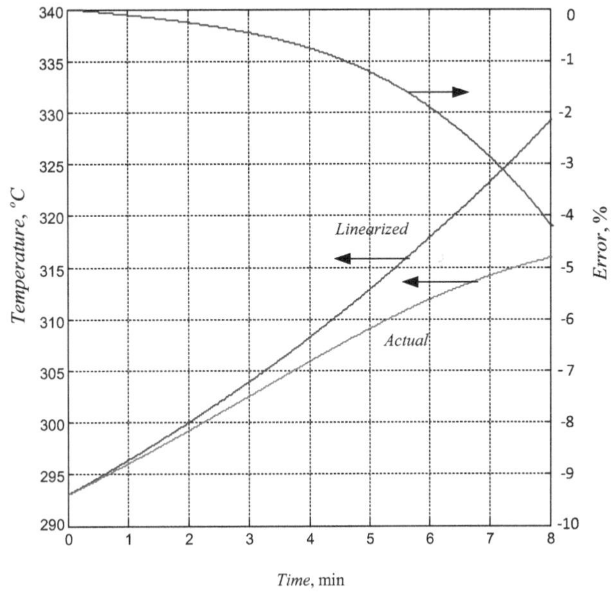

FIGURE 3.11 Comparison of actual and linearized models.

and

$$\frac{dT}{dt} = 0.19T + 0.006C^A - 55.46 \quad (3.116)$$

Figure 3.11 shows the solution of the temperature for the actual system, Equations 3.113 and 3.114, and for the linearized model, Equations 3.115 and 3.116; the figure also shows the percentage error at every instant of time. The linearization was done around $T = 293\,\text{K}$ and $C^A = 500\,\text{moles/m}^3$; therefore, as the system moves away from these values, the linearization "breaks down."

3.8 TRANSIENT AND FINAL RESPONSES

Up to now we have stressed obtaining the analytical solutions of differential equations. These solutions describe what happens to the dependent variable for the change in forcing function. That is, the analytical solution provides the "response" of the dependent variable. For example, the following differential equation and initial conditions describe the displacement of the cart, x, in Figure 3.1,

$$10\frac{d^2x}{dt^2} + 20\frac{dx}{dt} + 100x = f_A(t) \quad (3.2)$$

with $\left.\frac{dx}{dt}\right|_{t=0} = 0\,\text{m/s}$ and $x(0) = 0\,\text{m}$. For the forcing function $f_A(t) = 10u(t)\,\text{N}$, the solution is given by Equation 3.41:

$$x = 0.1 - e^{-t}\left[0.1\cos 3t + 0.033\sin 3t\right] \quad (3.41)$$

Thus, Equation 3.41 describes the displacement of the cart (the response of the cart) as a function of time when the forcing function is applied.

The following model describes the voltage across the capacitor in the electrical circuit shown in Figure 3.4:

$$3.75\times10^{-6}\frac{d^2 v_C}{dt^2}+3.0\times10^{-4}\frac{dv_C}{dt}+v_C=v_S \tag{3.5}$$

with initial conditions $\left.\frac{dv_c}{dt}\right|_{t=0}=0\,\mathrm{V}/\mathrm{s}$ and $v_c(0)=0\,\mathrm{V}$. The forcing function, v_S, the supply voltage, to the circuit is $v_S=5u(t)\sin 600t\ V$. Example 3.15 shows that the solution is

$$v_C=e^{-40t}[5.72\cos 515t+13.53\sin 515t]-5.72\cos 600t-11.23\sin 600t \tag{3.49}$$

Thus, Equation 3.49 describes the voltage across the capacitor (the response of the capacitor) as a function of time when the forcing function is applied.

Although equations such as Equations 3.41 and 3.49 are compact and not difficult to work with, a graph of the responses is most useful for engineering and scientific purposes. Figure 3.12 shows both responses. Let us analyze these responses in more detail.

To start, note that the roots of the corresponding characteristic equation in both cases are complex with negative real part. This is why both homogeneous solutions are composed of negative exponential and sine and cosine terms. The most significant part is the negative real part indicating a stable response.

Two important terms in the study and practice of dynamic analysis are *transient* and *final responses*. The *transient response* term refers to the response at the beginning values of the independent variable. *Final response* refers to the response after the effect of the transient response has disappeared. Figure 3.12 shows these two responses. Note that during the transient response changes are temporarily occurring to the dependent variable (displacement of cart, x, or voltage across the capacitor, v_C). After this transient period, the dependent variable stays either constant (as in Figure 3.12a) or it repeats itself (as in Figure 3.12b).

Let us dig more into these two responses. We repeat Equation 3.41 for convenience,

$$x=0.1-e^{-t}\left[0.1\cos 3t+0.033\sin 3t\right] \tag{3.41}$$

The first term in the equation is constant (this is the forced response), and the second term is composed of an exponential with a negative exponent and a sine and a cosine term (this is the natural response). Example 3.13 shows that the second term is the solution of the corresponding homogeneous equation, with roots $-1\pm 3\mathrm{i}$, which should be of no surprise; the first term is the particular solution which is of no surprise either because the forcing function is a constant. Because of the negative exponential term, the second term will eventually disappear. For example, when $t=5\mathrm{s}$ the exponential term is $e^{-5}=0.0067$, meaning that the contribution of the second term is negligible; as time continues increasing the term becomes more negligible until it disappears or "dies out." So, at just about $t=5\mathrm{s}$ the response has a value of 0.1 m. Figure 3.12a shows the entire response, confirming that indeed, at $t=5\mathrm{s}$ the response is almost a constant value of 0.1 m and that for $t<5\mathrm{s}$, it had increased from its initial value of 0 m.

The response of the voltage across the capacitor is given by Equation 3.49,

$$v_{CH}=e^{-40t}\left[5.72\cos 515t+13.53\sin 515t\right]-5.72\cos 600t-11.23\sin 600t \tag{3.49}$$

Analyzing this analytical solution (response equation) as we did with the previous one, we note that the first term, composed of a negative exponential and sine and cosine terms (roots of the corresponding homogeneous equation are $-40\pm 515\mathrm{i}$), will eventually decay to zero (its contribution to the response will disappear). The second and third terms, only composed of sine and cosine, will

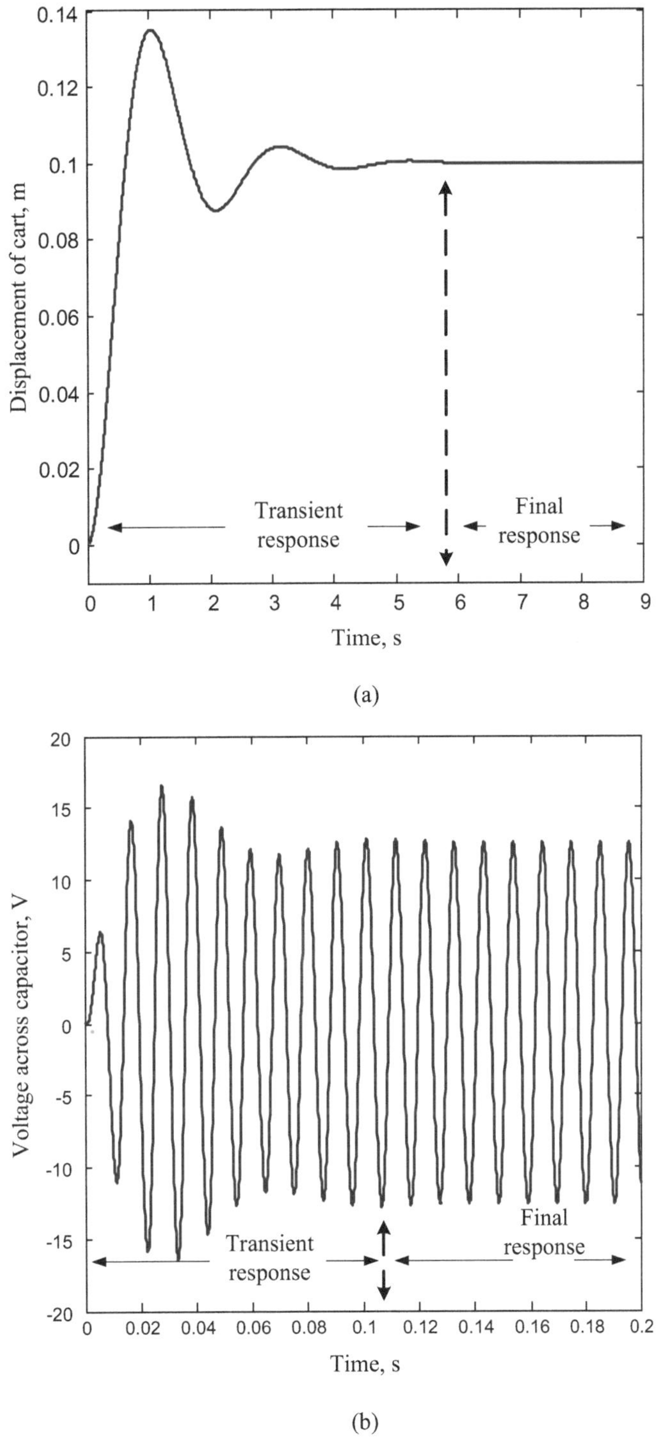

FIGURE 3.12 Graphs of Equations 3.41a and 3.49b.

not disappear but, rather, will continue repeating themselves. Once again, the terms that disappear, that is, *the transient response, depend on the corresponding homogeneous solution, and the final response depends on the particular solution*; this is the case for stable systems.

An important point in science and engineering is the size of the transient periods. Figure 3.12a indicates that for the mechanical system shown in Figure 3.1, the transients disappear at about $t = 5.7\,\text{s}$. Figure 3.12b shows that for the electrical system shown in Figure 3.4 the transients disappear at about $t = 0.102\,s$. The transients are much faster for electrical systems than for mechanical systems. The fluid and thermal systems that we will also study in this book have closer transients to the mechanical systems. Engineers and scientists often use the term *dynamics* to express transients. We say that systems having short transient periods have fast dynamics and those with long transient periods have slow dynamics. Thus, electrical systems have faster dynamics than mechanical, fluid, and thermal systems. We visit again this topic of dynamics in Chapter 5.

One final thought, what would you say about the transient and final responses if any root or roots of the corresponding homogeneous equation contain a positive real part?

3.9 SUMMARY

This chapter has presented classical methods to solve linear differential equations. The integrating factor method is used to solve first-order differential equations. The characteristic equation method is used to solve any order homogeneous or corresponding homogeneous differential equation that is linear and has constant coefficients, also referred to as time-invariant coefficients. The undetermined coefficient method is used to provide the particular solution of the nonhomogeneous differential equation. The variation of parameter method was presented and shown to be more general than the undetermined coefficient method. However, if the applied forcing function is one of those presented in Table 3.1, the undetermined coefficient method is easier to apply than the variation of parameters method. The Taylor series was also presented to linearize nonlinear differential equations and to handle nonstationary coefficients. In these cases, the method provides approximate solutions.

A very important discussion centered on the roots of the characteristic equation. The locations of these roots determine whether the response is oscillatory or monotonic and whether it is stable or unstable. A detailed discussion of the transient and final responses was also presented.

Problems

Problem 3.1

Solve each of the following by the integrating factor method.

a. $\dfrac{dy}{dt} + y = te^{-t} \quad y(0) = 1$

b. $\dfrac{dy}{dt} + \cos(t)y = \cos(t) \quad y(0) = 0$

c. $\dfrac{dy}{dt} + ty = e^{-t^2/2} \quad y(0) = 1$

d. $\dfrac{dy}{dt} + 3t^2 y = 3t^2 \quad y(0) = 0$

e. $\dfrac{dy}{dt} + 0.1y = e^{-0.1t} \quad y(0) = 0$

f. $\dfrac{dy}{dt} + 0.1y = 0.1t \quad y(0) = 0$

g. $\frac{dy}{dt}+3y=e^{-2t}\ y(0)=1$

h. $\frac{dy}{dt}+2y=t^2e^{-2t}\ y(0)=1$

i. $\frac{dy}{dt}+0.2y=2-2e^{-0.1t}\ y(0)=0$

j. $\frac{dy}{dt}+\left(2t+2\right)y=e^{-t^2}\ y(0)=0$

k. $\frac{dy}{dt}+2y=e^{-2t}\sin t\ y(0)=0$

l. $t\frac{dx}{dt}+2x=4t^2\ x(1)=0$

Problem 3.2

Solve each of the following initial value problems for $y(t)$.

a. $y''+y'-12y=0$	$y(0)=0$	$y'(0)=14$
b. $y''+6y'+10y=0$	$y(0)=2$	$y'(0)=0$
c. $y''+3y'+2y=0$	$y(0)=0$	$y'(0)=1$
d. $y''+4y=0$	$y(0)=0$	$y'(0)=2$
e. $y''+4y'+13y=0$	$y(0)=0$	$y'(0)=12$
f. $y''-6y'+9y=0$	$y(0)=-1$	$y'(0)=0$

Problem 3.3

Solve the following differential equations.

a. $4y''+4y'+17y=0$ with $y(0)=-1$ and $y'(0)=2$
b. $4y''+4y'=0$ with $y(0)=-1$ and $y'(0)=2$
c. $y''-4y'-5y=0$ with $y(1)=0$ and $y'(1)=2$
d. $y''+2y'+5y=0$ with $y(0)=2$ and $y'(0)=0$
e. $y''+4y'+4y=0$ with $y(0)=1$ and $y'(0)=4$

Problem 3.4

For each of the parts below, the complete solution.

a. $y''-9y=14-2x$ with $y(0)=1$ and $y'(0)=0$
b. $y''-9y=4-e^{-3x}$ with $y(0)=0$ and $y'(0)=0$
c. $y''+7y'+10y=2\sin 3t$ with $y(0)=-1$ and $y'(0)=1$
d. $y''+6y'+5y=2t$ with $y(0)=3$ and $y'(0)=-3$

Problem 3.5

For each of the following differential equations, (a) find the corresponding homogeneous solution, (b) find the particular solution and write the complete solution, and (c) use the initial conditions to provide the solution to the initial value problem.

a. $y''+3y'+2y=4te^{-3t}$	$y(0)=4$ and $y'(0)=-8$
b. $y''+2y'+y=6e^{-t}$	$y(0)=0$ and $y'(0)=1$

c. $y''-2y'+5y=16te^{-t}$ $y(0)=0$ and $y'(0)=0$
d. $y''+2y'+10y=10e^{-2t}$ $y(0)=1$ and $y'(0)=4$
e. $y''+6y'+9y=27+3e^{-2t}$ $y(0)=0$ and $y'(0)=0$
f. $y''+4y'+4y=3t+2e^{-t}$ $y(0)=0$ and $y'(0)=0$
g. $y''+9y=10\sin(2t)$ $y(0)=2$ and $y'(0)=-2$
h. $y''-y'-y=6e^{-t}$ $y(0)=0$ and $y'(0)=-1$
i. $y''+12y'+36y=72+4e^{-6t}$ $y(0)=0$ and $y'(0)=0$
j. $y''+2y'=4+2e^{-2t}$ $y(0)=0$ and $y'(0)=0$
k. $y''+5y'+6y=18+2e^{-2t}$ $y(0)=3$ and $y'(0)=0$

Problem 3.6

Suppose a differential equation has the roots shown in Figure P3.1. Offer the form of the natural response (corresponding homogeneous solution).

Problem 3.7

In Section 3.4, we mentioned that $y_1=C_1e^{r_1t}$, $y_2=C_2e^{r_2t}$, and $y=C_1e^{r_1t}+C_2e^{r_2t}$ are solutions. How could you prove that indeed this is the case?

Problem 3.8

Figure P3.2 shows a water treatment system consisting of two basins. Example 8.5 develops a model that describes the benzene concentration in each basin. You may assume that originally, neither basin contain benzene.

a. The model for the first basin is

$$39.27\frac{dC_1^{bz}(t)}{dt}+1278.1C_1^{bz}(t)=100C_0^{bz}$$

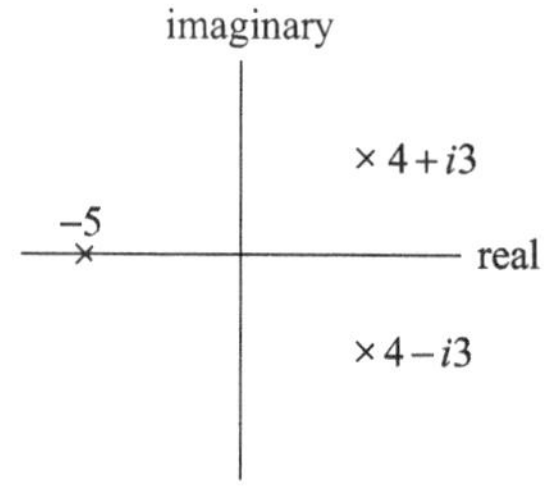

FIGURE P3.1 Roots for Problem 3.6.

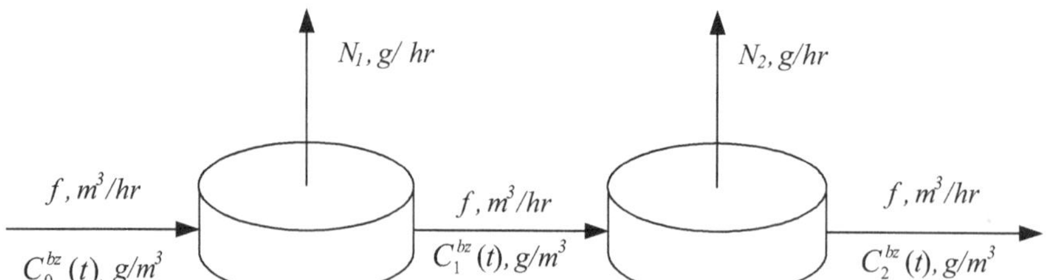

FIGURE P3.2 Water treatment basins for Problem 3.8.

where $C_1^{bz}(t)$ is the benzene concentration in the first basin, in $\mathrm{g/m^3}$, and C_0^{bz} is the entering concentration to the basin. At $t=0$ the concentration C_0^{bz} spikes to 5 g/m³ and remains there for 0.5 hours, at which point it drops back to zero; that is,

$$C_0^{bz} = \begin{cases} 5 & 0 \le t \le 0.5 \\ 0 & t \ge 0.5 \end{cases}$$

or simply,
$$C_0^{bz} = 5u(0.5-t)$$

Show that the analytical solutions providing $C_1^{bz}(t)$ are

$$C_1^{bz}(t) = 0.3912\left(1-e^{-32.546t}\right) \text{ for } 0 \le t \le 0.5$$
$$\text{and } C_1^{bz}(t) = 0.3912\, e^{-32.546\,(t-0.5)} \text{ for } t \ge 0.5$$

b. The model that describes the benzene concentration in the second basin is

$$39.27\frac{dC_2^{bz}(t)}{dt} + 1278.1C_2^{bz}(t) = 100C_1^{bz}$$

where $C_2^{bz}(t)$ is the benzene concentration in the second basin, in $\mathrm{g/m^3}$, and C_1^{bz} is the entering concentration coming from the first basin.

Using the solutions of part (a), show that the analytical solutions providing $C_2^{bz}(t)$ are

$$C_2^{bz}(t) = 0.0306\left(1-e^{-32.546t}\right) - 0.996\,t\,e^{-32.546t} \text{ for } 0 \le t \le 0.5$$

and $C_2^{bz}(t) = \left(0.996(t-0.5)+0.0306\right)e^{-32.546\,(t-0.5)}$ for $t \ge 0.5$.

Problem 3.9

Example 8.9 develops the following model for the glucose concentration in a human compartment:

$$\frac{dC^G}{dt} + 0.0224C^G = 2.016 + 5.513e^{-0.0453t}$$

with $C^G(0) = 90\,\mathrm{mg/dL}$. Obtain the analytical solution expressing $C^G(t)$.

Problem 3.10

The following model describes the temperature in a thermal system:

$$2\frac{d^2T}{dt^2} + 8\frac{dT}{dt} + 16T = T_i(t)$$

with $T_i(t) = 160 + 5t\,u(t)$, °C; $T(0) = 10$°C and $\left.\frac{dT}{dt}\right|_{t=0} = 0\ {}^{\circ}\mathrm{C}/_{\mathrm{min}}$.

a. Obtain the complete analytical solution.
b. Based on the roots of the corresponding homogeneous characteristic equation and on the form of the solution, do they agree? Yes, they agree, or no, they don't agree, and why?

Problem 3.11

Consider the spring–mass system in Figure P3.3. Chapter 6 shows that the model that describes the displacement x of the mass is given by the following differential equation:

$$m x'' + P x' + k x = 0 \text{ with } x'(0) = 0 \text{ and } x(0) = -0.3\text{m}$$

and $m = 3\text{kgm}$ and $k = 20\text{N/m}$.

a. Solve the initial value problem for the case of $P = 2\text{N}\cdot\text{s/m}$.
b. Repeat part (a) except use $P = 20\text{N}\cdot\text{s/m}$.

Problem 3.12

Equation 6.10 is the model of an undamped mechanical system. Prove that the solution is given by Equation 6.11.

Problem 3.13

Equation 6.21c is the model of a damped mechanical system. Prove that the solution is given by either Equation 6.22 or 6.23.

Problem 3.14

In Section 6.3 the model for the system shown in Figure P3.4 was developed using D'Alembert's principle, and it is given by

$$m\frac{d^2x}{dt^2} + P\frac{dx}{dt} + kx = f_A(t)$$

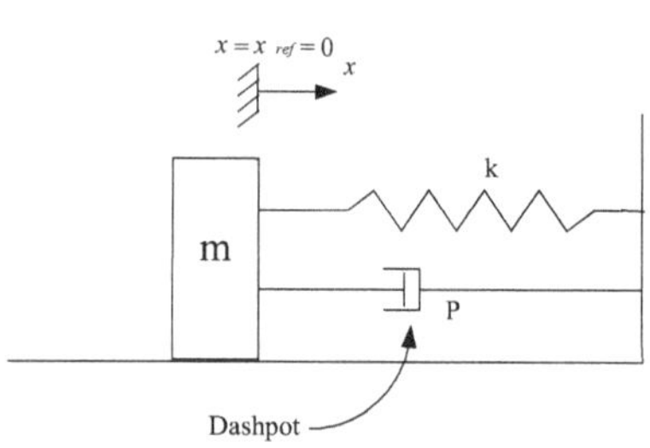

FIGURE P3.3 Spring–mass–dashpot system for Problem 3.11.

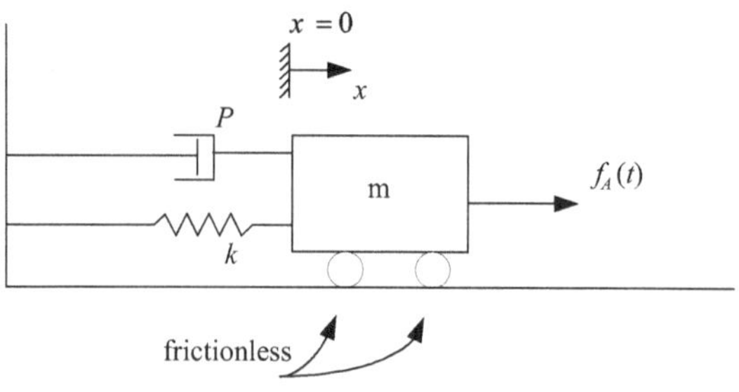

FIGURE P3.4 Spring-mass-dashpot system for Problem 3.14

Assuming $m = 10\,\text{kg}$, $P = 20\,\text{N}\cdot\text{s/m}$, $k = 100\,\text{N/m}$, that the force $f_A(t)$ increases from 0 to 10 N at time t = 0 s (or as Chapter 1 presented, $f_A(t) = 10u(t)\,\text{N}$), and that the initial conditions are $x(0) = \left.\frac{dx}{dt}\right|_{t=0} = 0$, prove that the analytical solution is

$$x = 0.1 - e^{-t}\left[0.1\cos 3t + 0.033\sin 3t\right]$$

Problem 3.15

Show that the solution of Equation 6.49 is Equation 6.50.

Problem 3.16

The model that describes the mechanical system of Figure P3.5 is

$$0.05\frac{d^2x}{dt^2} + 0.35\frac{dx}{dt} + x = 0.005\,f_A(t)$$

with $x(0) = \left.\frac{dx}{dt}\right|_{t=0} = 0$.

Solve the model to obtain the expression for the position x of the block as a function of time for $f_A(t) = 10u(t)$ N.

Problem 3.17

The model that describes the velocity of the block shown in Figure P3.6 is

$$2\frac{dv}{dt} + 4v = 16u(t)$$

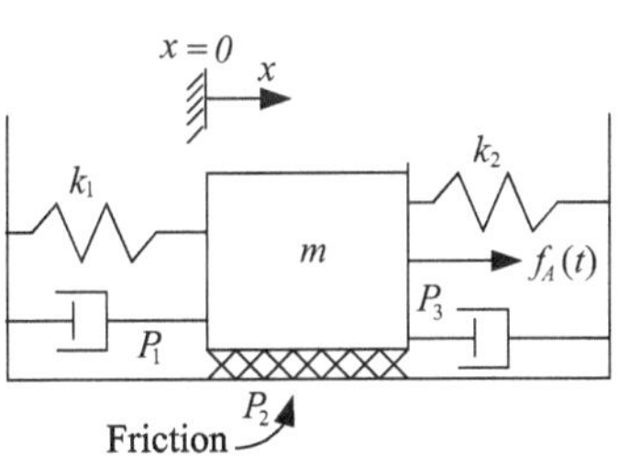

FIGURE P3.5 Mechanical system for Problem 3.16.

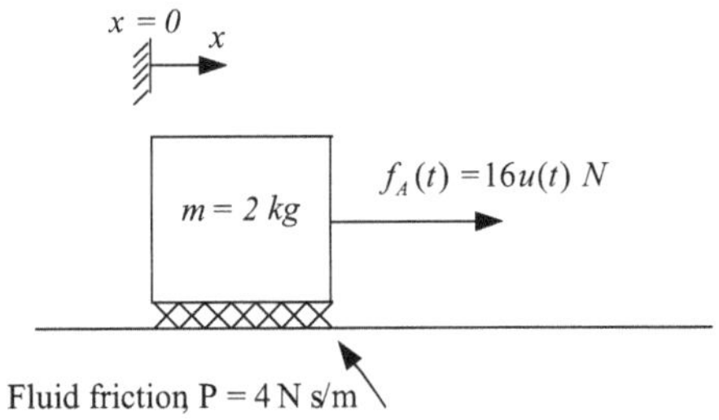

FIGURE P3.6 Mechanical system for Problem 3.17.

Assuming that the initial condition is $v(0) = 0\frac{\text{m}}{\text{s}}$, obtain the analytical equation that describes the velocity of the block.

Problem 3.18

The model that describes the position of the block shown in Figure P3.7 is

$$0.25\frac{d^2x}{dt^2} + \frac{dx}{dt} + x = 0$$

Assuming that initially the system is at a steady state and that the block is held at $x(0) = 0.1$ m and then let go, obtain the analytical equation that describes the position of the block.

Problem 3.19

The model that describes the position of the block shown in Figure P3.8 is

$$2\frac{d^2x}{dt^2} + 18x = 18\,y(t)$$

Assuming that the initial conditions are $x(0) = 0\,\text{m}$ and $\left.\frac{dx}{dt}\right|_{t=0} = 0\frac{\text{m}}{\text{s}}$, obtain the analytical equation that describes the position of the block for $y(t) = 0.1t\,u(t)\,\text{m}$.

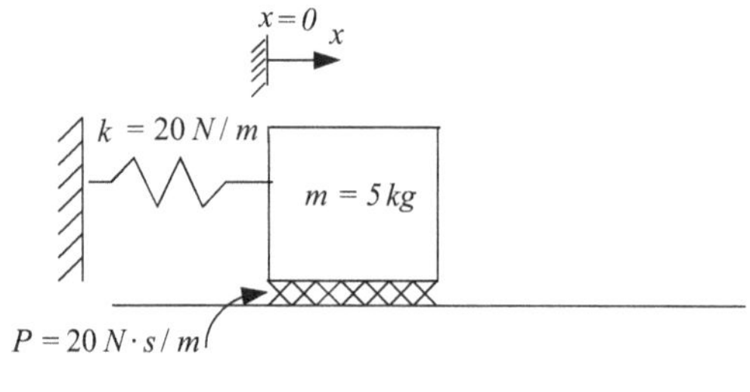

FIGURE P3.7 Mechanical system for Problem 3.18.

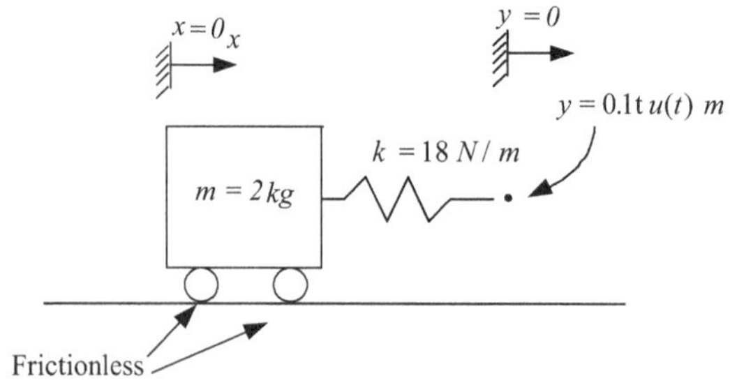

FIGURE P3.8 Mechanical system for Problem 3.19.

Problem 3.20

Example 10.1 considers the circuit shown in Figure P3.9. The example develops the following equation describing the voltage drop v_C across the capacitor:

$$RC\frac{dv_C}{dt}+v_C=v_S$$

Using $v_C(0)=10\,\text{V}$ $v_S=10u(t),\text{V}$ (where V stands for volts) show that the solution is given by

$$v_C=v_C(0)+\left[v_S-v_S(0)\right]\left(1-e^{-\frac{t}{RC}}\right)$$

or

$$v_C=10+15\left(1-e^{-\frac{t}{0.02}}\right)=25-15e^{-\frac{t}{0.02}}$$

Problem 3.21

The following equation describes the voltage across the capacitor shown in Figure P3.10. The voltage supply and switch combination can be described as $v_S(0)=10u(t),\text{V}$ (where V stands for volts). Using $v_C(0)=0\,\text{V}$ and $\left.\frac{dv_c}{dt}\right|_{t=0}=0\,\text{V/s}$, obtain the analytical solution.

$$3.75\times10^{-6}\frac{d^2v_C}{dt^2}+3\times10^{-4}\frac{dv_C}{dt}+v_C=v_S$$

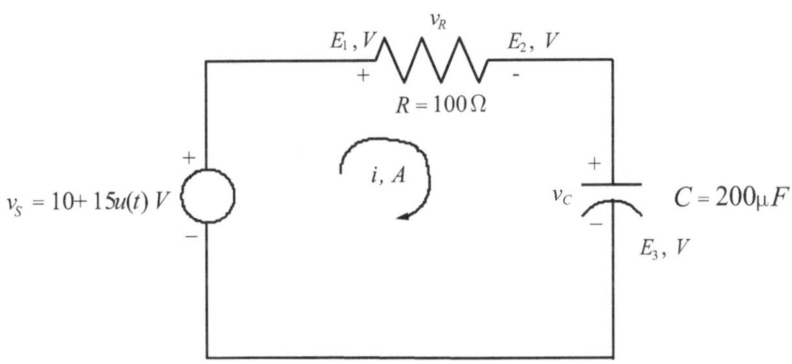

FIGURE P3.9 Electrical circuit for Problem 3.20

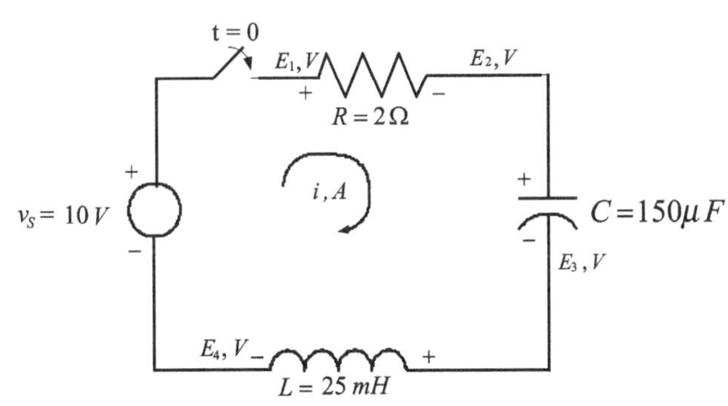

FIGURE P3.10 Electrical circuit for Problem 3.21.

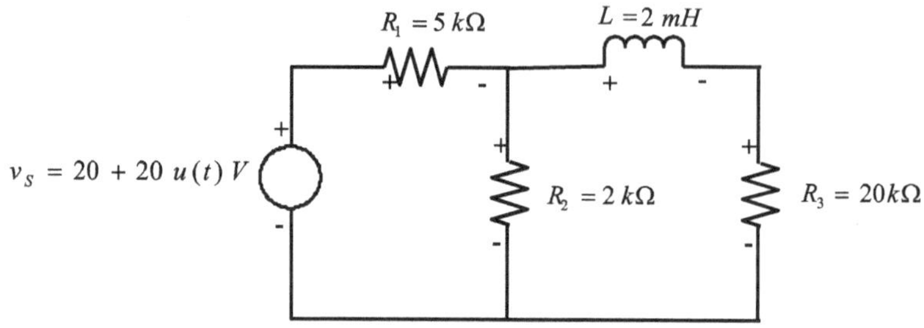

FIGURE P3.11 Electrical circuit for Problem 3.22

Problem 3.22

The model that describes the currents through resistances R_1 and R_3 in Figure P3.11 is given by the following two equations.

$$i_1 = 0.143 \times 10^{-3}\,v_S + 0.286\,i_2$$

and

$$2 \times 10^{-3}\frac{di_2}{dt} + 21428\,i_2 = 0.286\,v_S$$

The initial conditions are $i_2(0) = 2.66 \times 10^{-4}\ A$ and $i_1(0) = 2.93 \times 10^{-3}\ A$. For $v_S = 20 + 20u(t)$ V, solve the model to obtain expressions for $i_1(t)$ and $i_2(t)$.

Problem 3.23

The voltage E in in an electrical circuit is described by

$$0.3 \times 10^{-6}\frac{d^2E}{dt^2} + \frac{1}{1500}\frac{dE}{dt} + \frac{1}{1.5}E = 0$$

with $E(0) = 40$ V, and $\left.\frac{dE}{dt}\right|_{t=0} = 0\frac{\text{V}}{\text{s}}$. Obtain the analytical solution describing E.

Problem 3.24

The model that describes the current in a certain electrical circuit given by

$$0.025\frac{d^2i}{dt^2} + \frac{di}{dt} + i = 0$$

with $i(0) = 5\ A$ and $\left.\frac{di}{dt}\right|_{t=0} = 0\frac{\text{A}}{\text{s}}$. Obtain the analytical solution.

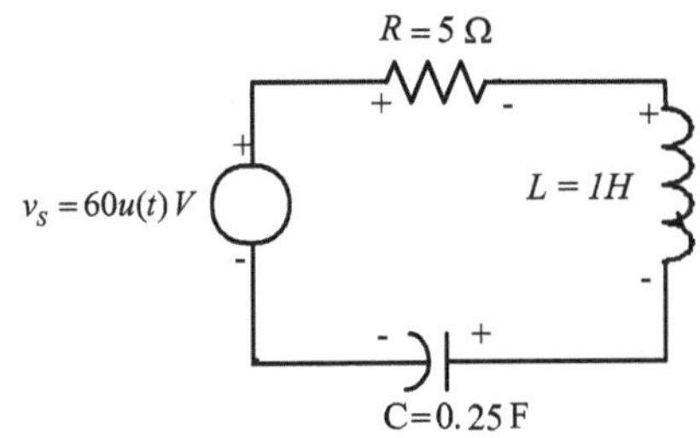

FIGURE P3.12 Electrical circuit for Problem 3.25.

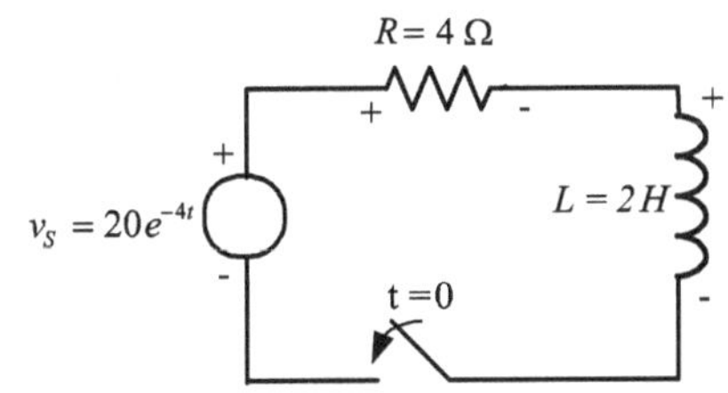

FIGURE P3.13 Electrical circuit for Problem 3.26.

Problem 3.25

The charge of the capacitor in Figure P3.12 is described by

$$0.25\frac{d^2q}{dt^2}+1.25\frac{dq}{dt}+q=0.25v_S$$

with $q(0)=0C$ and $\left.\frac{dq}{dt}\right|_{t=0}=0\frac{C}{s}$ and $v_S=60u(t)$ V. Obtain the analytical solution of the equation and the final charge of the capacitor.

Problem 3.26

The current in the circuit shown in Figure P3.13 is described by

$$0.5\frac{di}{dt}+i=0.25v_S$$

with $i(0)=0A$ and $v_S=20e^{-4t}u(t)$, V. Obtain the analytical solution.

Problem 3.27

The charge of the capacitor in the circuit shown in Figure P3.14 is described by

$$0.33\frac{d^2q}{dt^2}+1.33\frac{dq}{dt}+q=0$$

with $q(0)=5C$ and $\left.\frac{dq}{dt}\right|_{t=0}=0\frac{\text{C}}{\text{s}}$. Obtain the analytical solution.

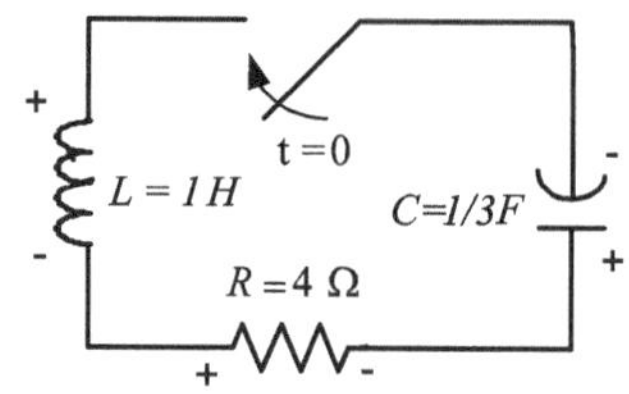

FIGURE P3.14 Electrical circuit for Problem 3.27.

Problem 3.28

Using the variation of parameters method, obtain the analytical solution of the following:

a. $y'' + 4y' + 4y = 3t \qquad y(0) = 0$ and $y'(0) = 0$

b. $y'' + 6y' + 9y = 3e^{-2t} \qquad y(0) = 0$ and $y'(0) = 0$

c. $y'' + 6y' + 9y = \dfrac{e^{-3t}}{t} \qquad y(1) = 3$ and $y'(1) = 0$

4 Laplace Transforms

The Laplace transform is a powerful tool for solving linear differential equations with constant coefficients and handling several of them simultaneously. This last property develops from the fact that in solving differential equations the Laplace transform method first converts them into algebraic equations, and the resulting equations are then manipulated algebraically before obtaining the result. Algebraic manipulations are much easier, and more often possible, than dealing directly with differential equations.

It is obvious at this stage that the reader may be thinking that the word *Laplace* denotes the name of the individual that developed the method, and this is indeed the case. But why the term *transform*? The reason for this term was hinted in the previous paragraph when we wrote "the Laplace transform method first converts them into algebraic equations," the word *converts* could be changed for *transforms*. The best way to explain the meaning of a "transformation" is by recalling the development and use of logarithms. Logarithms were independently developed by John Napier and Joost Burgi in the early 17th century to *simplify many mathematical calculations* in astronomy and navigation; Napier later collaborated with Henry Briggs, who developed most of the logarithm tables after Napier's death. For example, consider the multiplication of 43567 times 99876 with the result of 4351297692. This is an easy enough calculation with calculators, but back in the 17th century there were no calculators, computers, or even slide rules! (By the way, slide rules are based on the logarithm scale). With the use of logarithms, this calculation is much easily obtained by the following:

$$\mathrm{X} = \log(43567) + \log(99876)$$
$$\mathrm{X} = 4.639157\ldots + 4.999461\ldots = 9.63861$$
$$\text{Answer} = \text{antilog}(\mathrm{X}) = \text{antilog}(9.63861\ldots) = 4351297692$$

which is the correct value. So, the log of each number is obtained – *this is the same as "transforming" each number to a "different domain"* (in this case, the logarithm domain) – we then add the result of the transformation – *addition is a much simpler calculation than multiplication, or was in the 17th century*, – and finally we "transform back" (antilog or "inverse logarithm") to the original domain. Certainly, this calculation was possible, and "simple," because Briggs had previously developed the logarithm tables.

The Laplace transform method is like the procedure with logarithms in that it first "transforms" the differential equation(s) into a "different domain" (the Laplace domain), it then works algebraically with the resulting equation(s), and, finally, it "transforms back" or "inverse Laplace" the equation(s) to the original domain.

This chapter presents the Laplace transform method to solve differential equations and the important topic of transfer functions.

4.1 DEFINITION OF THE LAPLACE TRANSFORM

The Laplace transform of a function $f(x)$ where x is the independent variable is defined by the following formula:

$$F(s) = L[f(x)] = \int_0^\infty f(x)e^{-sx}dx \tag{4.1a}$$

DOI: 10.1201/9781003619161-4

where

$F(s)$ = the Laplace transform of $f(x)$ and

s = the Laplace transform variable. For the unit of the power of the exponential term to be dimensionless, the unit of s is the inverse of the unit of the independent variable x.

The limits of integration show that the Laplace transform contains information on the function $f(x)$ for only positive values of the independent variable x.

In the analysis of system dynamics, the independent variable is time, t, as in almost all the models developed in this book. The Laplace transform of a function of time, $f(t)$, is then

$$F(s) = L[f(t)] = \int_0^{\infty} f(t)e^{-st}dt \tag{4.1b}$$

and consequently, in this case the unit of s is time inverse (time^{-1}).

The following example uses the definition of the Laplace transform to develop the transforms of a few common forcing functions.

Example 4.1

Figure 4.1 shows four functions commonly applied as inputs to systems to study their responses. We now use the definition of the Laplace transform to derive their transforms.

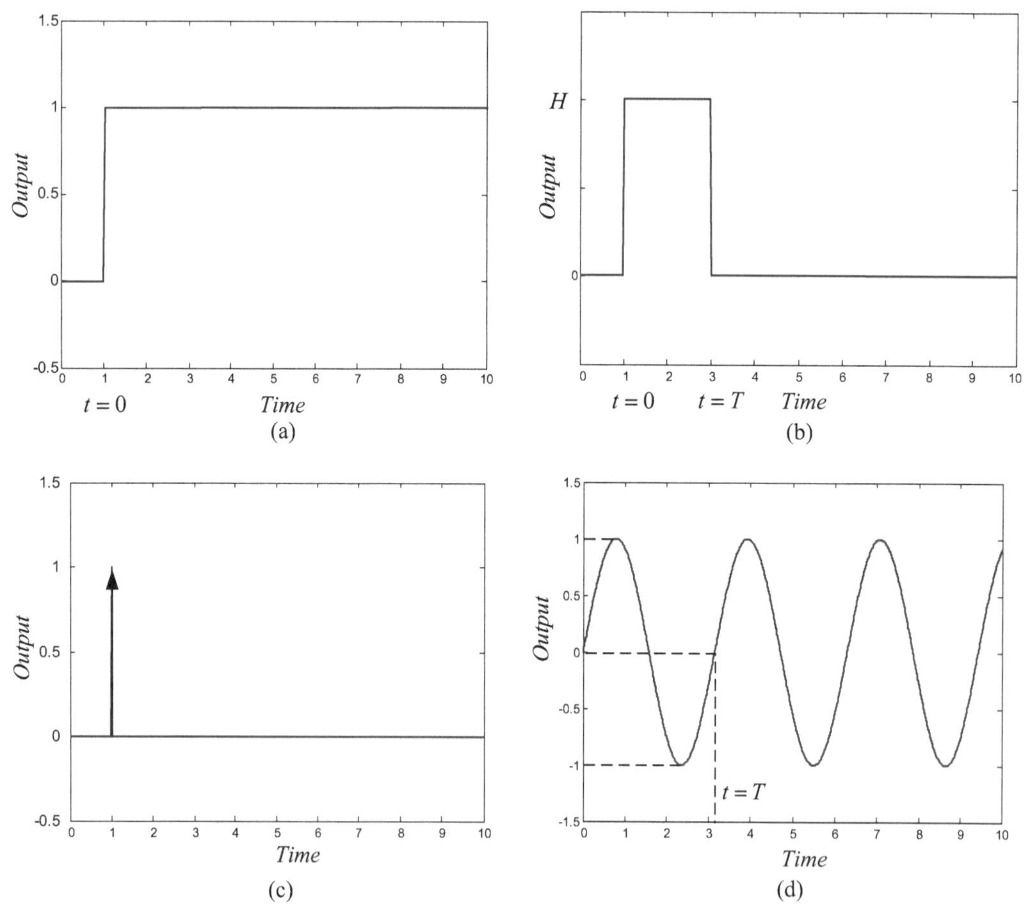

FIGURE 4.1 Common input signals. (a) Unit step function. (b) Pulse. (c) Unit impulse function. (d) Sine wave.

(a) Unit Constant

Let $f(t) = 1$

$$F(s) = \int_0^{\infty} f(t)e^{-st}dt = \int_0^{\infty} 1e^{-st}dt = -\frac{1}{s}e^{-st}\Big|_0^{\infty}$$

$$F(s) = \frac{1}{s}$$

(b) Unit Step Function

This is a sudden change of unit magnitude as sketched in Figure 4.1a. As presented in Chapter 1, we can use the term $u(t-a)$ to indicate when this step change occurs that is, we could write $f(t) = 1u(t) = u(t)$ to mathematically represent a unit step change at $t = 0$

Using the definition of the transform,

$$L[u(t)] = \int_0^{\infty} \underbrace{u(t)}_{\text{equals 1 for } t \geq 0} e^{-st}dt = \int_0^{\infty} 1e^{-st}dt = -\frac{1}{s}e^{-st}\Big|_0^{\infty} = -\frac{1}{s}(0-1) = \frac{1}{s}$$

$$L[u(t) = \frac{1}{s}$$

Note that the Laplace transform of a constant 1 and of a unit step function when it occurs at $t = 0$, $u(t)$ are exactly the same that is $\frac{1}{s}$. This is obvious because, as indicated in the previous equation, the integration starts at 0 and at that moment $u(t)$ becomes 1; that is, $u(t)$ is unity in the entire integration interval, and thus, it behaves as constant 1.

(c) A Pulse of Magnitude *H* and Duration *T*

The pulse sketched in Figure 4.1*b* is represented by

$$f(t) = \begin{cases} 0 & t < 0, t \geq T \\ H & 0 \leq t \leq T \end{cases}$$

Substituting into Equation 4.1b yields

$$L[f(t)] = \int_0^{\infty} f(t)e^{-st}dt = \int_0^{T} He^{-st}dt = -\frac{H}{s}e^{-st}\Big|_0^{T} = -\frac{H}{s}\left(e^{-sT}-1\right) = H\left[\frac{1-e^{-sT}}{s}\right]$$

$$L[f(t)] = H\left[\frac{1-e^{-sT}}{s}\right]$$

(d) A Unit Impulse Function

This unit impulse function, also known as the *Dirac delta function* and represented by $\delta(t)$, is sketched in Figure 4.1*c*. It is an ideal pulse with zero duration and unit area. All its area is concentrated at time zero. Because the function is always zero except at zero, and the term e^{-st} in Equation 4.1b is equal to unity at $t = 0$, the Laplace transform is

$$L[\delta(t)] = \int_0^{\infty} \delta(t) e^{-st} dt = 1$$

Note that the result of the integration, 1, is the area of the impulse.

(e) A Sine Wave of Unity Amplitude and Frequency ω

The sine wave is sketched in Figure 4.1*d*, and it is represented in exponential form by,

$$\sin \omega t = \frac{e^{i\omega t} - e^{-i\omega t}}{2i}$$

where $i = \sqrt{-1}$ is the symbol for imaginary numbers. Using Equation 4.1b yields

$$\begin{aligned}
L[\sin\omega t] &= \int_0^{\infty} \frac{e^{i\omega t} - e^{-i\omega t}}{2i} e^{-st} dt \\
&= \frac{1}{2i} \int_0^{\infty} \left[e^{-(s-i\omega)t} - e^{-(s+i\omega)t} \right] dt \\
&= \frac{1}{2i} \left[-\frac{e^{-(s-i\omega)t}}{s-i\omega} + \frac{e^{-(s+i\omega)t}}{s+i\omega} \right]_0^{\infty} \\
&= \frac{1}{2i} \left[-\frac{0-1}{s-i\omega} + \frac{0-1}{s+i\omega} \right] \\
L[\sin \omega t] &= \frac{1}{2i} \frac{2i\omega}{s^2 + \omega^2} \\
&= \frac{\omega}{S^2 + \omega^2}
\end{aligned}$$

Table 4.1 contains a short list of the Laplace transforms of some common functions; they are all derived as presented in the previous example.

4.2 PROPERTIES AND THEOREMS OF THE LAPLACE TRANSFORM

This section presents the properties and theorems of Laplace transforms in order of their usefulness in analyzing systems. The linearity property and the real differentiation and integration theorems are necessary for transforming differential equations into algebraic equations. The final value theorem is useful for predicting the final value of a function from its Laplace transform, and the real translation theorem is useful for dealing with functions delayed in time.

4.2.1 Linearity Property

Laplace transformation is a linear operation. This means that if a is a constant,

$$L[af(t)] = aL[f(t)] = aF(s) \tag{4.2}$$

The distributive property of addition also follows from the linearity property:

$$L[af(t) + bg(t)] = aL[f(t)] + bL[g(t)] = aF(s) + bG(s) \tag{4.3}$$

where a and b are constants. You can easily derive both formulas using the definition of the Laplace transform.

TABLE 4.1
Laplace Transforms of Common Functions

$f(t)$	$F(s) = \mathrm{L}[f(t)]$
$\delta(t)$	1
1 (constant) →	$\frac{1}{s}$
$u(t)$ →	
t	$\frac{1}{s^2}$
t^n	$\frac{n!}{s^{n+1}}$
e^{-at}	$\frac{1}{s+a}$
te^{-at}	$\frac{1}{(s+a)^2}$
$t^n e^{-at}$	$\frac{n!}{(s+a)^{n+1}}$
$\sin \omega t$	$\frac{\omega}{s^2+\omega^2}$
$\cos \omega t$	$\frac{s}{s^2+w^2}$
$e^{-at} \sin \omega t$	$\frac{\omega}{(s+a)^2+\omega^2}$
$e^{-at} \cos \omega t$	$\frac{s+a}{(s+a)^2+\omega^2}$

4.2.2 Real Differentiation Theorem

This theorem establishes a relationship between the Laplace transform of a function and that of its derivatives, and it is most important in transforming differential equations into algebraic equations. It states that

$$L\left[\frac{df(t)}{dt}\right] = sF(s) - f(0) \tag{4.4}$$

From the definition of the Laplace transform, Equation 4–1.1b,

$$L\left[\frac{df(t)}{dt}\right] = \int_0^\infty \frac{df(t)}{dt} e^{-st} dt$$

Integration by parts yields

$$u(t) = e^{-st} \qquad dv = \frac{df(t)}{dt} dt$$

$$du = -se^{-st} dt \qquad v = f(t)$$

$$\begin{aligned} L\left[\frac{df(t)}{dt}\right] &= \left[f(t)e^{-st}\right]_0^\infty - \int_0^\infty f(t)\left(-se^{-st} dt\right) \\ &= \left[0 - f(0)\right] + s\int_0^\infty f(t)e^{-st} dt \\ &= sF(s) - f(0) \end{aligned}$$

The extension to higher derivatives is

$$\begin{aligned}
L\left[\frac{d^2 f(t)}{dt^2}\right] &= L\left[\frac{d}{dt}\left(\frac{df(t)}{dt}\right)\right] \\
&= sL\left[\frac{df(t)}{dt}\right] - \left.\frac{df}{dt}\right|_{t-0} \\
&= s\left[sF(s) - f(0)\right] - \left.\frac{df}{dt}\right|_{t=0} \\
L\left[\frac{d^2 f(t)}{dt^2}\right] &= s^2 F(s) - sf(0) - \left.\frac{df}{dt}\right|_{t=0}
\end{aligned} \tag{4.5}$$

In general,

$$L\left[\frac{d^n f(t)}{dt^n}\right] = s^n F(s) - s^{n-1} f(0) - s^{n-2}\left.\frac{df}{dt}\right|_{t=0} - s^{n-3}\left.\frac{d^{n-2} f}{dt^{n-2}}\right|_{t=0} - \cdots - \left.\frac{d^{n-1} f}{dt^{n-1}}\right|_{t=0} \tag{4.6}$$

As we have previously mentioned in earlier chapters, often initially (initial condition) the variable is at steady state, *meaning that all time derivatives are zero*, and the variable itself is at some value that it is defined as zero. For this common case the preceding expression reduces to

$$L\left[\frac{d^n f(t)}{dt^n}\right] = s^n F(s) \tag{4.7}$$

More discussion about these zero initial conditions is in Sections 4.4 and 4.6.

4.2.3 Real Integration Theorem

This theorem establishes the relationship between the Laplace transform of a function and that of its integral. It states that

$$L\left[\int_0^t f(t)\,dt\right] = \frac{1}{s} F(s) \tag{4.8}$$

The proof of this theorem is carried out integrating the definition of the Laplace transform by parts, like the previous theorem. The Laplace transform of the nth integral of a function is the transform of the function divided by s^n.

4.2.4 Real Translation Theorem

This theorem deals with the translation of a function in the time axis, as shown in Figure 4.2. The translated function is the original function delayed in time. The theorem states that

$$L\left[f\left(t - t_0\right)\right] = e^{-st_0} F(s) \tag{4.9}$$

Because the Laplace transform does not contain information about the original function for negative time, the delayed function must be zero for all times less than the time delay (see Figure 4.2). This condition is satisfied if the initial condition(s) of the dependent variable(s) is (are) zero, or if not, the

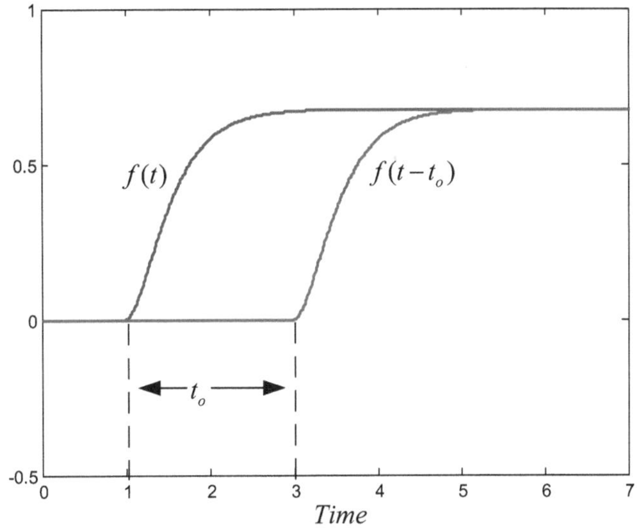

FIGURE 4.2 Function delayed in time is zero for all times less than the time delay t_0.

variable(s) is (are) expressed as deviation(s) from the initial steady-state condition(s) as explained in Section 4.6.

From the definition of the Laplace transform, Equation 4.1b,

$$L\left[f\left(t-t_0\right)\right]=\int_0^{\infty} f\left(t-t_0\right) e^{-st} dt$$

Let $\tau=t-t_0$ (or $t=t_0+\tau$) and substitute:

$$\begin{aligned}
L\left[f\left(t-t_0\right)\right] &= \int_{\tau=-t_0}^{\infty} f(\tau) e^{-s(t_0+\tau)} d\left(t_0+\tau\right) \\
&= \int_{\tau=0}^{\infty} f(\tau) e^{-st_0} e^{-s\tau} d\tau \\
&= e^{-st_0} \int_0^{\infty} f(\tau) e^{-s\tau} d\tau \\
&= e^{-st_0} F(s)
\end{aligned}$$

Note that in this proof we made use of the fact that $f(\tau)=0$ for $\tau<0$ ($t<t_0$).

4.2.5 Final Value Theorem

This theorem allows us to obtain the final value of a function from its transform. If the limit of $f(t)$ as t approaches ∞ exists, the final value can be found from its Laplace transform as follows:

$$\lim_{t \to \infty} f(t) = \lim_{s \to 0}\left[s F(s)\right] \tag{4.10}$$

4.2.6 Complex Differentiation Theorem

This theorem is useful for evaluating the transforms of functions involving multiplications of the independent variable t. It states that

$$L[tf(t)] = -\frac{d}{ds}F(s) \tag{4.11}$$

4.2.7 Complex Translation Theorem

This theorem is useful for evaluating transforms of functions involving exponential functions of the independent variable t. It states that

$$L\left[e^{at}f(t)\right] = F(s-a) \tag{4.12}$$

4.2.8 Initial Value Theorem

This theorem allows the calculation of the initial value of a function from its transform. It would provide another check of the validity of derived transforms were it not for the fact that often the initial conditions of the variables are zero. The theorem states that

$$\lim_{t\to 0} f(t) = \lim_{s\to\infty}\left[sF(s)\right] \tag{4.13}$$

The following examples illustrate the use of the properties and theorems of Laplace transforms we have just presented.

Example 4.2

This example shows the Laplace transform of a couple of types of forcing functions we may find in engineering applications. Obtain the Laplace transform of the following functions:

(a) $f(t) = e^{-6t}u(t)$
(b) $f(t) = 5 + 10tu(t)$

(a) This expression indicates that at any time before zero (0) the function has a value of zero (0) but, at $t = 0$ it becomes an exponential decay, from an initial value of one (1) to a final value of zero (0). Applying the Laplace definition,

equals 1 for $t \geq 0$

$$F(s) = \int_0^\infty e^{-6t}u(t)e^{-st}dt = \int_0^\infty e^{-6t}e^{-st}dt$$

Because $u(t)$ is always unity in the integration interval, the transform of $e^{-6t}u(t)$ is equal to the transform of e^{-6t}. Therefore, using Table 4.1

$$F(s) = \frac{1}{s+6}$$

(b) This expression indicates that at any time before zero (0) the function has a value of 5 but, at $t = 0$ it starts increasing as a ramp. The function can also be written as

$$f(t) = 5 + 10t\,u(t) = f_1 + f_2$$

Applying the Laplace definition,

$$F(s) = L[f_1] + L[f_2] = F_1(s) + F_2(s)$$

Using Table 4.1, $F_1(s) = \dfrac{5}{s}$

Using the definition of Laplace transform for the transform of f_2,

$$F_2(s) = \int_0^{\infty} 10t\underbrace{u(t)}_{\text{equals 1 for } t \geq 0} e^{-st}dt = \int_0^{\infty} 10t\, e^{-st}dt$$

Thus, like part (a), the Laplace transform of $10tu(t)$ is the same as the transform of $10t$. Using Table 4.1,

$$F_2(s) = \frac{10}{s^2}$$

Therefore,

$$F(s) = \frac{5}{s} + \frac{10}{s^2} = \frac{5s + 10}{s^2}$$

Sometimes $u(t)$ may not be the correct expression because the function may not enter the model at $t = 0$ but rather, at another time. In that case, we must use $u(t - a)$, where a is the correct time. Example 4.4 shows how to use the real translation theorem to handle this situation.

Example 4.3

Derive the Laplace transform of the following differential equation:

$$9\frac{d^2y(t)}{dt^2} + 6\frac{dy(t)}{dt} + y(t) = 2x(t)$$

with initial conditions $y''(0) = y'(0) = y(0) = 0$.

The distributive property allows us to take the Laplace transform of each term.

$$9L\left[\frac{d^2y(t)}{dt^2}\right] + 6L\left[\frac{dy(t)}{dt}\right] + L[y(t)] = 2L[x(t)]$$

Applying the real differentiation theorem,

$$9s^2y(s) - 9s\,y(0) - 9\left.\frac{dy}{dt}\right|_{t=0} + 6sy(s) - 6y(0) + y(s) = 2x(s)$$

and using the given initial conditions,

$$9s^2y(s) + 6sy(s) + y(s) = 2x(s) \tag{4.14a}$$

Finally, solving for $y(s)$,

$$y(s) = \frac{2}{9s^2 + 6s + 1}x(s) \tag{4.14b}$$

The preceding example shows how the Laplace transform converts the original differential equation into an algebraic equation in s, Equation 4.14a, which can then be algebraically rearranged to solve for the Laplace of the dependent variable $y(s)$, Equation 4.14b.

Example 4.4

Obtain the Laplace transform of the following function:

$$c(t)=u(t-3)\left[1-e^{-(t-3)/4}\right]$$

Note: The term $u(t-3)$ in this expression states that the function is zero for $t < 3$. We recall that $u(t-3)$ is a change from zero to one at $t = 3$, which means that the expression in brackets is multiplied by zero until $t = 3$, and multiplied by unity thereafter. Thus, the presence of the unit step function does not alter the rest of the function for $t > 3$.

Let us first assume there is no delay time; that is, $t-3=t$:

$$f(t)=u(t)\left[1-e^{-t/4}\right]$$

Using the entries from Table 4.1, with $a = ¼$,

$$f(s)=\frac{1}{s}-\frac{1}{s+\frac{1}{4}}=\frac{1}{s(4s+1)}$$

Next, consider the delay time by applying the real translation theorem, Equation 4.9,

$$c(s)=L\left[f(t-3)\right]=e^{-3s}f(s)=\frac{e^{-3s}}{s(4s+1)}$$

We can check the validity of this answer by using the final value theorem, Equation 4.10.

$$c(\infty)=\lim_{t\to\infty}c(t)=\lim_{t\to\infty}u(t-3)\left[1-e^{-(t-3)/4}\right]=1$$

$$c(\infty)=\lim_{s\to 0}s\,c(s)=\lim_{s\to 0}s\frac{e^{-3s}}{s(4s+1)}=1$$

Check!

4.3 SOLUTION OF DIFFERENTIAL EQUATIONS USING LAPLACE TRANSFORM

The procedure for solving a differential equation by Laplace transforms consists of the following three steps:

1. Transform the differential equation into an algebraic equation in the Laplace variable s.
2. Solve algebraically for the transform of the dependent variable.
3. Invert the resulting transform to obtain the response of the dependent variable.

Example 4.3 consisted of the first two steps however, we repeat them again.

Consider the following second-order differential equation:

$$a_2\frac{d^2y(t)}{dt^2}+a_1\frac{dy(t)}{dt}+a_0y(t)=bx(t) \tag{4.15}$$

The first step is to take the Laplace transform of the equation. The linearity property of Laplace transforms allows us to take the Laplace transform of each term separately:

$$a_2L\left[\frac{d^2y(t)}{dt^2}\right]+a_1L\left[\frac{dy(t)}{dt}\right]+a_0L[y(t)]=bL[x(t)] \tag{4.16}$$

The indicated Laplace transforms are obtained using the real differentiation theorem.

$$L\left[\frac{d^2y(t)}{dt^2}\right]=s^2y(s)-s\,y(0)-\left.\frac{dy}{dt}\right|_{t=0};\quad L\left[\frac{dy(t)}{dt}\right]=sy(s)-y(0)$$

$$L[y(t)]=y(s);\quad L[x(t)]=x(s)$$

Next substitute these terms into Equation 4.16 and algebraically rearrange it to obtain

$$\left(a_2s^2+a_1s+a_0\right)y(s)-\left(a_2s+a_1\right)y(0)-a_2\left.\frac{dy}{dt}\right|_{t=0}=by(s)$$

The second step is to manipulate this algebraic equation to solve for the transform of the dependent variable, $y(s)$.

$$y(s)=\frac{bx(s)+\left(a_2s+a_1\right)y(0)+a_2\left.\dfrac{dy}{dt}\right|_{t=0}}{a_2s^2+a_1s+a_0} \tag{4.17}$$

This equation shows the effect of the forcing function, $x(s)$, and of the initial conditions, on the dependent variable.

The third and final step is to invert the transform to obtain the time function $y(t)$. Inversion is the opposite operation to taking the Laplace transform. Before inversion, we must select a specific input function for $x(t)$. A common function, because of its simplicity, is the unit step function, $u(t)$, which was introduced in Example 4.1. From that example, or from Table 4.1, for $x(t) = u(t)$, $x(s) = 1/s$. Substitute into Equation 4.17 to obtain

$$y(t)=L^{-1}[y(s)]=L^{-1}\left[\frac{b\dfrac{1}{s}+\left(a_2s+a_1\right)y(0)+a_2\left.\dfrac{dy}{dt}\right|_{t=0}}{a_2s^2+a_1s+a_0}\right] \tag{4.18}$$

where the symbol L^{-1} stands for the inverse Laplace transform. The response to a step input is called the *step response* for short.

The inversion could easily be carried out if we could find the expression within the brackets in Table 4.1 or in a more extensive table of Laplace transforms. We will not be able to find complex expressions on such a table. The mathematical technique of partial fractions expansion, introduced in Chapter 2, is designed to expand the transform of the term in the brackets into a sum of simpler terms. We can then invert these simpler terms separately by matching entries in Table 4.1.

Assuming that $dy/dt\,|_{t=0} = 0$, and $y(0) = 0$, Equation 4.17 reduces to

$$y(s)=\left[\frac{b}{a_2s^2+a_1s+a_0}\right]x(s) \tag{4.19}$$

The form of Equation 4.19 allows us to break the transform of the dependent variable, $y(s)$, into the product of two terms, the term in brackets, known as the *transfer function*, and the transform of the input variable or forcing function, $x(s)$. The transfer function and its parameters characterize the system and determine how the dependent variable responds to the forcing function. Section 4.4 discusses transfer functions in detail.

4.3.1 Inversion by Partial Fractions Expansion

Section 2.8 presented the use of partial fractions expansion for factoring rational expressions into a sum of smaller terms and in so doing, simplifying further mathematical manipulations; these expansions are vital in obtaining the inversion of Laplace expressions. The section starts with a brief review of the method when the roots are real. We then present the inversion when the roots are complex, which were not presented in Chapter 2. The section presents several detailed examples. Take your time, study, and understand them.

4.3.1.1 Real Roots

Example 4.5

Obtain the analytical solution of

$$2\frac{d^2q}{dt^2}+8\frac{dq}{dt}+6\,q=5\,u(t) \text{ with } q(0)=0 \text{ and } \left.\frac{dq}{dt}\right|_{t=0}=0$$

Obtaining the Laplace transform,

$$2\left[s^2q(s)-sq(0)-\left.\frac{dq}{dt}\right|_{t=0}\right]+8[sq(s)-q(0)]+6q(s)=\frac{5}{s}$$

and using the initial conditions, $2s^2q(s)+8sq(s)+6q(s)=\dfrac{5}{s}$

Solving for the dependent variable, and using partial fractions expansion,

$$q(s)=\frac{5}{s(2s^2+8s+6)}=\frac{5}{2s(s^2+4s+3)}=\frac{2.5}{s(s^2+4s+3)}$$

$$q(s)=\frac{2.5}{s(s^2+4s+3)}=\frac{2.5}{s(s+3)(s+1)}=\frac{A_1}{s}+\frac{A_2}{s+3}+\frac{A_3}{s+1}$$

Applying Equation 2.29 for obtaining the A coefficients,

$$A_1=\lim_{s\to 0}[sq(s)]=\lim_{s\to 0}\left[\frac{s2.5}{s(s^2+4s+3)}\right]=\lim_{s\to 0}\left[\frac{2.5}{(s^2+4s+3)}\right]=0.833$$

$$A_2=\lim_{s\to -3}[(s+3)q(s)]=\lim_{s\to -3}\left[\frac{(s+3)2.5}{s(s+3)(s+1)}\right]=\lim_{s\to -3}\left[\frac{2.5}{s(s+1)}\right]=0.416$$

$$A_3=\lim_{s\to -1}[(s+1)q(s)]=\lim_{s\to -1}\left[\frac{(s+1)2.5}{s(s+3)(s+1)}\right]=\lim_{s\to -1}\left[\frac{2.5}{s(s+3)}\right]=-1.25$$

Therefore,

$$q(s) = \frac{2.5}{s(s^2+4s+3)} = \frac{2.5}{s(s+3)(s+1)} = \frac{A_1}{s} + \frac{A_2}{s+3} + \frac{A_3}{s+1}$$

Inverting,

$$q(t) = L^{-1}[q(s)] = L^{-1}\left[\frac{0.833}{s}\right] + L^{-1}\left[\frac{0.416}{s+3}\right] + L^{-1}\left[\frac{-1.25}{s+1}\right]$$

Now matching each term in Table 4.4,

$$q(t) = 0.833u(t) + 0.416e^{-3t} - 1.25e^{-t}$$

Example 4.6

Obtain the inverse transform of

$$q(s) = \frac{5}{s(s+3)^3}$$

Using partial fractions expansion for repeated roots,

$$q(s) = \frac{5}{s(s+3)^3} = \frac{A_1}{(s+3)^3} + \frac{A_2}{(s+3)^2} + \frac{A_3}{s+3} + \frac{A_4}{s}$$

Applying Equation 2.31 provides A_1, and Equation 2.29 gives A_4:

$$A_1 = \lim_{s\to -3}\left[(s+3)^3 q(s)\right] = \lim_{s\to -3}\left[\frac{(s+3)^3 5}{s(s+3)^3}\right] = -1.667$$

$$A_4 = \lim_{s\to 0}\left[sq(s)\right] = \lim_{s\to 0}\left[\frac{s5}{s(s+3)^3}\right] = 0.185$$

Equation 2.32 provides the other two coefficients:

$$A_2 = \lim_{s\to -3}\frac{1}{1!}\frac{d}{ds}\left[(s+3)^3 q(s)\right] = \lim_{s\to -3}\frac{d}{ds}[(s+3)^3 q(s)] = \lim_{s\to -3}\frac{d}{ds}\left[\frac{5}{s}\right] = \lim_{s\to -3}\left(-\frac{5}{s^2}\right) = -0.556$$

$$A_3 = \lim_{s\to -3}\frac{1}{2!}\frac{d^2}{ds^2} = \left[(s+3)^3 q(s)\right] = \lim_{s\to -3}\frac{1}{2}\frac{d^2}{ds^2}\left[\frac{5}{s}\right] = \lim_{s\to -3}\frac{1}{2}\left(\frac{10s}{s^4}\right) = -0.185$$

Therefore,

$$q(t) = L^{-1}\left[q(s)\right] = L^{-1}\left[\frac{-1.667}{(s+3)^3}\right] + L^{-1}\left[\frac{-0.556}{(s+3)^2}\right] + L^{-1}\left[\frac{-0.185}{s+3}\right] + L^{-1}\left[\frac{-0.185}{s}\right]$$

and using Table 4.1,

$$q(t) = 0.185u(t) - 0.185e^{-3t} - 0.556te^{-3t} - \frac{1.667}{2}t^2e^{-3t}$$

Make sure you understand all the steps in reaching the solution.

4.3.1.2 Complex Roots

As presented in Chapter 2, the roots may be complex that is, for a second-order equation the roots may be $r_1 = \alpha + i\beta$ and $r_2 = \alpha - i\beta$ where α is the real component and β the imaginary component. In this case there are two possible ways to apply partial fraction expansion. We next present both ways.

4.3.1.2.1 First-order Expansion

Although the following presentation was presented in Chapter 2, we present it once more to help the reader. In this case we proceed as usual:

$$y(s) = \frac{1}{(a_1s + a_0)(a_2s^2 + a_3s + a_4)} = \frac{1}{a_1a_2(s - r_1)(s - \alpha - i\beta)} = \frac{A_1}{s - r_1} + \frac{A_2}{s - \alpha - i\beta} + \frac{A_3}{s - \alpha + i\beta}$$

The A_1 coefficient is obtained as before:

$$A_1 = \lim_{s \to r_1}[(s - r_1)y(s)] = \frac{1}{a_1(a_2r_1^2 + a_3r_1 + a_4)}$$

The A_2 and A_3 coefficients are also obtained as before, except that a bit more algebra is required because of the complex numbers – be careful this is where mistakes occur! That is,

$$A_2 = \lim_{s \to \alpha + i\beta}[(s - \alpha - i\beta)y(s)] = \lim_{s \to \alpha + i\beta}\left[\frac{1}{a_1a_2(s - r_1)(s - \alpha + i\beta)}\right]$$

$$A_2 = \frac{1}{a_1a_2(\alpha + i\beta - r_1)(\alpha + i\beta - \alpha + i\beta)} = \frac{1}{a_1a_2(\alpha - r_1 + i\beta)(i2\beta)} = \frac{1}{a_1a_2(-2\beta^2 + i2\beta)(\alpha - r_1)}$$

The denominator is a complex expression; this is not useful. To change this, we multiply the numerator and the denominator of the entire expression by the complex conjugate of the denominator.

$$A_2 = \frac{(-2\beta^2 - i2\beta(\alpha - r_1))}{a_1a_2(-2\beta^2 + i2\beta(\alpha - r_1))(-2\beta^2 - i2\beta(\alpha - r_1))} = \frac{(-2\beta^2 - i2\beta(\alpha - r_1))}{a_1a_2(4\beta^4 + 4\beta^2(\alpha - r_1)^2]} = \sigma + i\gamma$$

where $\sigma = \dfrac{-2\beta^2}{a_1a_2[4\beta^4 + 4\beta^2(\alpha - r_1)^2]}$ and $\gamma = \dfrac{-2\beta(\alpha - r_1))}{a_1a_2[4\beta^4 + 4\beta^2(\alpha - r_1)^2]}$

Following the same procedure for A_3,

$$A_3 = \lim_{s \to \alpha - i\beta}[(s - \alpha + i\beta)Y(s)] = \sigma - i\gamma$$

In these cases of complex roots, the A coefficients will always be complex conjugates. Then,

$$y(s) = \frac{1}{(a_1s + a_0)(a_2s^2 + a_3s + a_4)} = \frac{A_1}{s - r_1} + \frac{\sigma + i\gamma}{s - \alpha - i\beta} + \frac{\sigma - i\gamma}{s - \alpha + i\beta}$$

Using the table of transforms, we invert this expression back into the time domain:

$$y(t) = A_1 e^{r_1 t} + \left[(\sigma + i\gamma)e^{(\alpha+i\beta)t} + (\sigma - i\gamma)e^{(\alpha-i\beta)t}\right]$$

$$y(t) = A_1 e^{r_1 t} + \left[(\sigma + i\gamma)e^{\alpha t}e^{i\beta t} + (\sigma - i\gamma)e^{\alpha t}e^{-i\beta t}\right]$$

And noting that the term $e^{\alpha t}$ is common to the last two terms,

$$y(t) = A_1 e^{r_1 t} + e^{\alpha t}\left[(\sigma + i\gamma)e^{i\beta t} + (\sigma - i\gamma)e^{-i\beta t}\right]$$

This is the response equation for when complex roots are present. However, it is rather difficult to obtain a good qualitative indication of this response because of the complex exponential powers. A better expression, avoiding complex numbers, is obtained using Euler's identity as we did in Chapter 3.

$$y = A_1 e^{r_1 t} + e^{\alpha t}\left[C_1 \cos\beta t + iC_2 \sin\beta t\right] \tag{4.20}$$

where

$$C_1 = A_2 + A_3 \text{ and } C_2 = A_2 - A_3$$

The following box shows the development of the equation.

Consider the following:

$$y(t) = e^{\alpha t}\left[A_1 e^{i\beta t} + A_2 e^{-i\beta t}\right] \tag{4.21}$$

using Euler's identity $e^{i\beta t} = \cos\beta t + i\sin\beta t$:

$$y(t) = e^{\alpha t}\left[A_1(\cos\beta t + i\sin\beta t) + A_2(\cos\beta t - i\sin\beta t)\right]$$

$$y(t) = e^{\alpha t}\left[(A_1 + A_2)\cos\beta t + i(A_1 - A_2)\sin\beta t\right]$$

$$y(t) = e^{\alpha t}\left[C_1 \cos\beta t + iC_2 \sin\beta t\right] \tag{4.22}$$

where

$$C_1 = A_1 + A_2 \text{ and } C_2 = A_1 - A_2 \tag{4.23}$$

Note that C_1 is the summation of the coefficient of the exponent with the positive power $(i\beta t)$ in Equation 4.21 plus the coefficient of the exponent with the negative power $(-i\beta t)$. C_2 is the difference of the coefficient of the exponent with the positive power $(i\beta t)$ minus the coefficient of the exponent with the negative power $(-i\beta t)$. The term C_2 will be an imaginary number but, once multiplied by i in Equation 4.22, results in a real number.

4.3.1.2.2 Second-Order Expansion

We have so far factored the denominator function into first-order terms. When the roots of a second-order term are complex, it may be more convenient to obtain the solution by keeping the term as a second order. For example,

$$y(s) = \frac{1}{(a_1 s + a_0)(a_2 s^2 + a_3 s + a_4)} = \frac{1}{a_1(s - r_1)(a_2 s^2 + a_3 s + a_4)} = \frac{A_1}{s - r_1} + \frac{A_2 s + A_3}{a_2 s^2 + a_3 s + a_4}$$

Note that now the denominator of the second term of the expansion is of second order; in this case, the numerator must be of first order as shown. Partial fraction allows that the denominator of an expansion term may be of any order, but its numerator must be at the most one order less.

The next step is obtaining the coefficients A_1, A_2, and A_3. This is easily done by setting the right-hand side under the common denominator, as we did in Section 2.8:

$$y(s) = \frac{1}{a_1(s-r_1)\left(a_2 s^2 + a_3 s + a_4\right)} = \frac{(A_1 a_2 + A_2)s^2 + (A_1 a_3 - A_2 r_1 + A_3)s + (A_1 a_4 - A_3 r_1)}{(s-r_1)\left(a_2 s^2 + a_3 s + a_4\right)}$$

Considering the numerators of both sides, equating equal coefficients we obtain

$$A_1 a_2 + A_2 = 0;\ A_1 a_3 - A_2 r_1 + A_3 = 0;\ A_1 a_4 - A_3 r_1 = 1$$

providing three equations with three unknowns.

Once the A coefficients are obtained, the following allows us to obtain the solution *without having to deal with complex numbers.*

$$y(t) = L^{-1}\left[\frac{A_1}{s} + \frac{A_2 s + A_3}{s^2 + c_1 s + c_0}\right] = L^{-1}\left[\frac{A_1}{s} + \frac{A_2 s + A_3}{s^2 + c_1 s + c_0}\right] = L^{-1}\left[\frac{A_1}{s}\right] + L^{-1}\left[\frac{d_2(s+d_0) + d_3}{(s+d_0)^2 + d_4}\right]$$

The last term in the inverse transform, $\frac{d_2(s+d_0)+d_3}{(s+d_0)^2+d_4}$, was obtained by applying the Completing the square method to the term $\frac{A_2 s + A_3}{s^2 + c_1 s + c_0}$ (see the next box for a review of the completing the square method).

COMPLETING THE SQUARE

As you may remember (oh . . . memories from high school . . . Algebra 1!), Completing the square is a method for obtaining the roots of second-order polynomials. Let us briefly review the method. Consider

$$x^2 + bx + c = 0$$

The following are the steps:

Step 1: $x^2 + bx = -c$

Step 2: $x^2 + bx + \left(\frac{b}{2}\right)^2 = -c + \left(\frac{b}{2}\right)^2$

Step 3: $\left(x + \frac{b}{2}\right)^2 = -c + \left(\frac{b}{2}\right)^2$

and

$$x^2 + bx + c = \left(x + \frac{b}{2}\right)^2 + \left(c - \left(\frac{b}{2}\right)^2\right)$$

Continuing,

$$y(t) = L^{-1}\left[\frac{A_1}{s} + \frac{d_2(s+d_0)+d_3}{(s+d_0)^2+d_4}\right] = L^{-1}\left[\frac{A_1}{s}\right] + L^{-1}\left[\frac{d_2(s+d_0)+d_3}{(s+d_0)^2+d_4}\right]$$

Using the complex translational theorem shown in Section 4.2, which states that $L\left[e^{-at}f(t)\right] = F(s+a)$, we can write the inverse as $L^{-1}\{F(s+a)\} = e^{-at}L^{-1}\left[f(t)\right]$, obtaining in this case

$$y(t) = L^{-1}\left[\frac{A_1}{s}\right] + e^{-d_0 t}L^{-1}\left[\frac{d_2 s + d_3}{s^2+d_4}\right]$$

Expanding now the second term,

$$y(t) = L^{-1}\left[\frac{A_1}{s}\right] + e^{-d_0 t}\left(L^{-1}\left[\frac{d_2 s}{s^2+d_4}\right] + L^{-1}\left[\frac{d_3}{s^2+d_4}\right]\right)$$

Using Table 4.1,

$$y(t) = A_1 u(t) + e^{-d_0 t}\left(d_2 \cos\sqrt{d_4 t} + \frac{d_3}{\sqrt{d_4}}\sin\sqrt{d_4 t}\right)$$

As mentioned, we obtain this solution without dealing with complex numbers; please see Example 4.7c.

We use the next example to show how to use this second expansion for reaching the analytical solution without dealing with complex numbers. The example is designed to numerically illustrate the partial fractions expansion procedure and the entire inversion process.

Example 4.7

Given the quadratic differential equation,

$$9\frac{d^2y(t)}{dt^2} + a_1\frac{dy(t)}{dt} + y(t) = 2\,x(t)$$

with zero initial conditions, $y(0) = 0$ and $\frac{dy}{dt}|_{t=0} = 0$, using $x(t) = 2\,u(t)$, obtain the analytical solution $y(t)$ for the three different values of a_1 of 10, 6, and 3.

Taking the Laplace transform of the differential equation, using the initial conditions, and solving for the Laplace of the dependent variable yields

$$Y(s) = \frac{2}{s\,(9s^2+as+1)}$$

a. Unrepeated Real Roots
   Letting $a_1 = 10$,

$$y(t) = L^{-1}[Y(s)] = L^{-1}\left[\frac{2}{s\left(9s^2+10s+1\right)}\right]$$

The roots, from the quadratic equation, are: $r_1 = -1/9$ and $r_2 = -1$. The denominator is factored as

$$Y(s) = \frac{2}{9\,s(s^2 + \frac{10}{9}s + \frac{1}{9})} = \frac{2}{9\,s\left(s + \frac{1}{9}\right)(s+1)} = \frac{A_1}{s} + \frac{A_2}{s + \frac{1}{9}} + \frac{A_3}{s+1}$$

Note that as previously explained, we first factored out 9 from the quadratic equation. The coefficients are calculated using Equation 4.21:

$$A_1 = \lim_{s\to 0} \frac{2}{9\left(s + \frac{1}{9}\right)(s+1)} = 2;\quad A_2 = \lim_{s\to -1/9} \frac{2}{9(s+1)s} = -2.25;\quad A_3 = \lim_{s\to -1} \frac{2}{9\left(s + \frac{1}{9}\right)s} = 0.25$$

Then,

$$Y(s) = \frac{2}{s} - \frac{2.25}{s + \frac{1}{9}} + \frac{0.25}{s+1}$$

$$y(t) = L^{-1}\left[Y(s)\right] = 2\,L^{-1}\left[\frac{1}{s}\right] - 2.25\,L^{-1}\left[\frac{1}{s + \frac{1}{9}}\right] + 0.25\,L^{-1}\left[\frac{1}{s+1}\right]$$

Each term in brackets can be found in Table 4.1; therefore, invert by matching entries to obtain the analytical solution.

$$y(t) = 2u(t) - 2.25e^{-t/9} + 0.25e^{-t}$$

b. Repeated Roots
Letting $a_1 = 6$, the roots are $r_1 = r_2 = -1/3$, and the expansion is

$$Y(s) = \frac{2}{9\,s\left(s + \frac{1}{3}\right)^2} = \frac{A_1}{s} + \frac{A_2}{\left(s + \frac{1}{3}\right)^2} + \frac{A_3}{s + \frac{1}{3}}$$

The A_1 coefficient is obtained using Equation 4.21, and A_2 using Equation 4.27,

$$A_1 = \lim_{s\to 0}\left[sY(s)\right] = \lim_{s\to 0}\left[\frac{2}{9\left(s + \frac{1}{3}\right)^2}\right] = 2;\quad A_2 = \lim_{s\to -1/3}\left[\left(s + \frac{1}{3}\right)^2 Y(s)\right] = \lim_{s\to -1/3} \frac{2}{9s} = -\frac{2}{3}$$

Finally, the A_3 coefficient is obtained using Equation 4.28,

$$A_3 = \lim_{s\to -1/3} \frac{1}{1!}\frac{d}{ds}\left[\left(s + \frac{1}{3}\right)^2 Y(s)\right] = \lim_{s\to -1/3} \frac{d}{ds}\left[\frac{2}{9s}\right] = \lim_{s\to -1/3} -\left(\frac{2}{9s^2}\right) = -2$$

Then

$$Y(s) = \frac{2}{s} - \frac{2/3}{\left(s+\frac{1}{3}\right)^2} - \frac{2}{s+\frac{1}{3}}$$

$$y(t) = L^{-1}\left[Y(s)\right] = 2\,L^{-1}\left[\frac{1}{s}\right] - \frac{2}{3}\,L^{-1}\left[\frac{1}{\left(s+\frac{1}{3}\right)^2}\right] - 2\,L^{-1}\left[\frac{1}{s+\frac{1}{3}}\right]$$

The analytical solution is obtained by matching entries in Table 4.1:

$$y(t) = 2u(t) - \frac{2}{3}te^{-t/3} - 2e^{-t/3} = 2u(t) - \left(\frac{2}{3}t + 2\right)e^{-t/3}$$

c. Pair of Complex Conjugate Roots

Letting $a_1 = 3$ the roots are $r_{1,2} = -0.167 \pm 0.289i$. As you may recall, there are two ways to handle the partial fraction expansion when the roots are complex conjugates: the first-order and second-order expansions. Let us obtain the solution using each of them.

First-order expansion

The expansion is

$$Y(s) = \frac{2}{9(s+0.167-i0.289)(s+0.167+i0.289)s}$$

$$Y(s) = \frac{A_1}{s} + \frac{A_2}{s+0.167-i0.289} + \frac{A_3}{s+0.167+i0.289}$$

Once more the coefficients are calculated by Equation 4.21:

$$A_1 = \lim_{s\to 0}\left[\frac{2}{s(9s^2+3s+1)}\right] = 2$$

$$A_2 = \lim_{s\to -0.167+i0.289}\left[\frac{2(s+0.167-i0.289)}{9(s+0.167-i0.289)(s+0.167+i0.289)s}\right]$$

$$A_2 = \lim_{s\to -0.167+i0.289}\left[\frac{2}{9(s+0.167+i0.289)s}\right] = -1 + i0.577$$

and

$$A_3 = \lim_{s\to -0.167-i0.289}\left[\frac{2(s+0.167+i0.289)}{9(s+0.167-i0.289)(s+0.167+i0.289)s}\right]$$

$$A_3 = \lim_{s\to -0.167-i0.289}\left[\frac{2}{9(s+0.167-i0.289)s}\right] = -1 - i0.577$$

Please make sure you fully understand how to obtain the A coefficients (go through the calculations!). Note that the fact that the numbers are complex does not affect this part of the procedure; it just takes a bit of manipulations with complex numbers (remember, although it is only algebra, it is here where most mistakes happen!). Also note that A_2 and A_3 are complex conjugates; this will always be the case.

Following the same procedure previously shown,

$$y(t) = L^{-1}[y(s)] = L^{-1}\left[\frac{2}{s}\right] + L^{-1}\left[\frac{-1+i0.577}{s+0.167-i0.289}\right] + L^{-1}\left[\frac{-1-i0.577}{s+0.167+i0.289}\right]$$

The inverse response is again obtained by matching entries in Table 4.1.

$$y(t) = 2u(t) + (-1+i0.577)e^{(-0.167t+i0.289t)} + (-1-i0.577)e^{(-0.167t-i0.289t)}$$
$$y(t) = 2u(t) + (-1+i0.577)e^{-0.167t}e^{i0.289t} + (-1-i0.577)e^{-0.167t}e^{-i0.289t}$$

or

$$y(t) = 2u(t) + e^{-0.167t}\left[(-1+i0.577)e^{i0.289t} + (-1-i0.577)e^{-i0.289t}\right]$$

Applying Equations 4.22 and 4.23 to this equation,

$$y(t) = 2u(t) + e^{-0.167t}\left[C_1 \cos 0.289t + C_2 \sin 0.289t\right]$$

and

$$C_1 = A_2 + A_3 = -1+i0.577-1-i0.577 = -2$$
$$C_1 = A_2 - A_3 = -1+i0.577+1+i0.577 = i1.154$$

or

$$y(t) = 2u(t) - e^{-0.167t}\left[2\cos 0.289t + 1.154 \sin 0.289t\right]$$

Second-order expansion

Let us now solve this part (c) again but this time using the second-order expansion.

$$y(t) = L^{-1}\left[\frac{2}{s(9s^2+3s+1)}\right] + L^{-1}\left[\frac{A_1}{s} + \frac{A_2 s + A_3}{9s^2+3s+1}\right]$$

To obtain the A coefficients,

$$\frac{2}{s(9s^2+3s+1)} = \frac{A_1}{s} + \frac{A_2 s + A_3}{9s^2+3s+1} = \frac{(9A_1+A_2)s^2 + (3A_1+A_3)s + A_1}{s(9s^2+3s+1)}$$
$$9A_1 + A_2 = 0; \quad 3A_1 + A_3 = 0; \quad A_1 = 2$$

Therefore,

$$A_1 = 2, \quad A_2 = -18, \quad A_3 = -6$$

$$y(t) = L^{-1}\left[\frac{A_1}{s} + \frac{A_2 s + A_3}{9s^2+3s+1}\right] = L^{-1}\left[\frac{2}{s}\right] - L^{-1}\left[\frac{18s+6}{9s^2+3s+1}\right] = L^{-1}\left[\frac{2}{s}\right] - \frac{1}{9}L^{-1}\left[\frac{18s+6}{s^2+1/3s+1/9}\right] \quad (4.36)$$

Completing the square of the denominator of the second term,

$$y(t) = L^{-1}\left[\frac{2}{s}\right] - \frac{1}{9}L^{-1}\left[\frac{18s+6}{(s+1/6)^2+0.0833}\right]$$

Note that the denominator of the second term contains $\left(s+\frac{1}{6}\right)$, but the numerator is only s. To make them the same, the numerator is changed as

$$y(t) = L^{-1}\left[\frac{2}{s}\right] - \frac{1}{9}L^{-1}\left[\frac{18(s+1/6)+3}{(s+1/6)^2+0.0833}\right]$$

Using the Complex Translational Theorem,

$$y(t) = L^{-1}\left[\frac{2}{s}\right] - \frac{1}{9}L^{-1}\left[\frac{18(s+1/6)+3}{(s+1/6)^2+0.0833}\right] = L^{-1}\left[\frac{2}{s}\right] - \frac{1}{9}e^{-0.167t}L^{-1}\left[\frac{18s+3}{s^2+0.0833}\right]$$

Expanding the second term and continuing obtaining the inverse using Table 4–1,

$$y(t) = L^{-1}\left[\frac{2}{s}\right] - \frac{1}{9}e^{-0.167t}\left(L^{-1}\left[\frac{18s}{s^2+0.0833}\right] + L^{-1}\left[\frac{3}{s^2+0.0833}\right]\right)$$

$$y(t) = 2u(t) - e^{-0.167t}\left(2\cos 0.289t + 1.154\sin 0.289t\right)$$

which is the same as the previous result. Once more, using this method we reach the solution without dealing with complex numbers.

Example 4.8

Obtain the solution of

$$y'' + 8y' + 15y = 30 + 60u(t) \text{ with } y(0) = 2 \text{ and } y'(0) = 0$$

Before solving the example, note that the forcing function could have been written $f(t) = 30 + 60u(t)$ as well.

Obtaining the Laplace of each term yields

$$\left(s^2 y(s) - s\,y(0) - y'(0)\right) + 8\left(s\,y(s) - y(0)\right) + 15\,y(s) = \frac{90}{s}$$

$$\left(s^2 y(s) - 2s\right) + 8\left(s\,y(s) - 2\right) + 15\,y(s) = \frac{90}{s}$$

$$(s^2+8s+15)\,y(s) = \frac{90}{s} + 2s + 16 = \frac{2s^2+16s+90}{s}$$

$$y(s) = \frac{2s^2+16s+90}{s(s^2+8s+15)}$$

Applying partial fractions and using Table 4.1,

$$y(s) = \frac{2s^2+16s+90}{s(s+3)(s+5)} = \frac{A_1}{s} + \frac{A_2}{s+3} + \frac{A_3}{s+5}$$

$$A_1 = \lim_{s \to 0}\left[\frac{2s^2+16s+90}{(s+3)(s+5)}\right] = 6; \; A_2 = \lim_{s \to -3}\left[\frac{2s^2+16s+90}{s(s+5)}\right] = -10$$

$$A_3 = \lim_{s \to -5}\left[\frac{2s^2+16s+90}{s(s+3)}\right] = 6$$

$$y = A_1 + A_2 e^{-3t} + A_3 e^{-5t} = 6 - 10e^{-3t} + 6e^{-5t}$$

Chapter 3 presented several examples showing the use of the classical solution methods; we now use some of these examples to further show the use of Laplace transforms. *(Suggestion: We have already seen several examples on how to use Laplace transforms to obtain the analytical solution. In the next seven examples, read the statement and then solve the example without looking at the answer. When you finish check, your answers with those in the book.)*

Example 4.9

Obtain the solution of

$$y'' + y' - 12y = e^{2t} \text{ with } y(0) = 3 \text{ and } y'(0) = 0$$

This is the same as Example 3.9.

Obtaining the Laplace of each term yields

$$\left(s^2 y(s) - s\,y(0) - y'(0)\right) + \left(s\,y(s) - y(0)\right) - 12\,y(s) = \frac{1}{s-2}$$

$$s^2 y(s) - 3s + s\,y(s) - 3 - 12\,y(s) = \frac{1}{s-2}$$

$$\left(s^2 + s - 12\right) y(s) = \frac{1}{s-2} + 3s + 3$$

$$y(s) = \frac{3s^2 - 3s - 5}{\left(s^2+s-12\right)\left(s-2\right)}$$

Applying partial fractions expansion,

$$y(s) = \frac{3s^2-3s-5}{\left(s^2+s-12\right)\left(s-2\right)} = \frac{3s^2-3s-5}{\left(s+4\right)\left(s-3\right)\left(s-2\right)} = \frac{A_1}{s+4} + \frac{A_2}{s-3} + \frac{A_3}{s-2}$$

$$A_1 = \lim_{s \to -4}\left[(s+4)Y(s)\right] = \lim_{s \to -4}\left[\frac{3s^2-3s-5}{(s-3)(s-2)}\right] = 1.309$$

$$A_2 = \lim_{s \to 3}\left[(s-3)Y(s)\right] = \lim_{s \to 3}\left[\frac{3s^2-3s-5}{(s+4)(s-2)}\right] = 1.857$$

$$A_3 = \lim_{s \to 2}\left[(s-2)Y(s)\right] = \lim_{s \to 2}\left[\frac{3s^2-3s-5}{(s+4)(s-3)}\right] = -\frac{1}{6}$$

$$y(s) = \frac{1.309}{s+4} + \frac{1.857}{s-3} - \frac{1/6}{s-2}$$

Finally, using Table 4.1,

$$y = 1.309\,e^{-4t} + 1.857\,e^{3t} - \frac{1}{6}e^{2t}$$

This is the same result as in Example 3.9. By the way, could you comment about the stability of this response?

Example 4.10

Obtain the solution of

$$y'' + y' - 12y = e^{3t} \text{ with } y(0) = 3 \text{ and } y'(0) = 0$$

This is the same as Example 3.10.

Obtaining the Laplace of each term yields

$$\left(s^2\,y(s) - s\,y(0) - y'(0)\right) + \left(s\,y(s) - y(0)\right) - 12\,y(s) = \frac{1}{s-3}$$

$$s^2\,y(s) - 3s + s\,y(s) - 3 - 12\,y(s) = \frac{1}{s-3}$$

$$\left(s^2 + s - 12\right)y(s) = \frac{1}{s-3} + 3s + 3$$

$$y(s) = \frac{3s^2 - 6s - 8}{\left(s^2 + s - 12\right)\left(s-3\right)}$$

Applying partial fractions expansion,

$$y(s) = \frac{3s^2 - 6s - 8}{\left(s^2 + s - 12\right)\left(s-3\right)} = \frac{3s^2 - 6s - 8}{\left(s+4\right)\left(s-3\right)\left(s-3\right)} = \frac{3s^2 - 6s - 8}{\left(s+4\right)\left(s-3\right)^2} = \frac{A_1}{(s-3)^2} + \frac{A_2}{s-3} + \frac{A_3}{s+4}$$

Using Equation 4.21,

$$A_3 = \lim_{s\to -4}\left[(s+4)\,y(s)\right] = \lim_{s\to -4}\left[\frac{3s^2 - 6s - 8}{(s-3)^2}\right] = 1.306$$

$$A_1 = \lim_{s\to 3}\left[(s-3)^2\,y(s)\right] = \lim_{s\to 3}\left[\frac{3s^2 - 6s - 8}{(s+4)}\right] = 0.143$$

Using Equation 4.28,

$$A_2 = \lim_{s\to 3}\frac{1}{(2-1)!}\frac{d}{ds}\left[(s-3)^2\,y(s)\right] = \lim_{s\to 3}\frac{d}{ds}\left[\frac{3s^2 - 6s - 8}{(s+4)}\right]$$

$$A_2 = \lim_{s\to 3}\left[\frac{(s+4)(6s-6) - (3s^2 - 6s - 8)(1)}{(s+4)^2}\right] = 1.694$$

$$y(s) = \frac{0.143}{(s-3)^2} + \frac{1.694}{s-3} + \frac{1.306}{s+4}$$

Finally, using Table 4.1,

$$y = 1.694\,e^{3t} + 1.306\,e^{-4t} + 0.143\,t\,e^{3t}$$

This is the same result as in Example 3.10.

Example 4.11

Obtain the solution of

$$x'' + 16x = 4\sin\omega t \text{ with } \omega \neq 4,\ x(0) = x_0, \text{ and } x'(0) = 0$$

This is the same as Example 3.11.

Obtaining the Laplace of each term yields

$$\left(s^2\,x(s) - s\,x(0) - x'(0)\right) + 16\,x(s) = \frac{4\omega}{s^2 + \omega^2}$$

$$\left(s^2\,x(s) - s\,x_0\right) + 16\,x(s) = \frac{4\omega}{s^2 + \omega^2}$$

$$x(s) = \frac{x_0\,s^3 + x_0\omega^2 s + 4\omega}{(s^2 + 16)(s^2 + \omega^2)}$$

Both denominator terms yield imaginary roots. Also, both are already in the form of a complete square. Thus,

$$x(s) = \frac{x_0\,s^3 + x_0\omega^2 s + 4\omega}{(s^2 + 16)(s^2 + \omega^2)} = \frac{A_1 s + A_2}{s^2 + 16} + \frac{A_3 s + A_4}{s^2 + \omega^2}$$

Setting the right-hand-side terms under common denominator and equating the coefficients of like terms in the resulting numerators, allow obtaining the A coefficients,

$$x_0\,s^3 + x_0\omega^2 s + 4\omega = (A_1 + A_3)s^3 + (A_2 + A_4)s^2 + (A_1\omega^2 + 16A_3)s + (A_2\omega^2 + 16A_4)$$

Matching equal terms yields

$$A_1 = x_0;\quad A_2 = \frac{4\omega}{\omega^2 - 16};\quad A_3 = 0;\quad A_4 = \frac{4\omega}{16 - \omega^2}$$

and

$$x(t) = L^{-1}\left[\frac{x_0 s + \dfrac{4\omega}{\omega^2 - 16}}{s^2 + 16}\right] + L^{-1}\left[\frac{\dfrac{4\omega}{16 - \omega^2}}{s^2 + \omega^2}\right] = L^{-1}\left[\frac{x_0 s}{s^2 + 16}\right] + L^{-1}\left[\frac{\dfrac{4\omega}{\omega^2 - 16}}{s^2 + 16}\right] + L^{-1}\left[\frac{\dfrac{4\omega}{16 - \omega^2}}{s^2 + \omega^2}\right]$$

Using Table 4.1,

$$x = x_0\cos 4t + \frac{\omega}{\omega^2 - 16}\sin 4t - \frac{4}{\omega^2 - 16}\sin\omega t$$

This is the same result as in Example 3.11 when $\omega \neq 4$.

Example 4.12

Obtain the solution of

$$y''+7y'+12y=f(t) \text{ with } y(0)=3 \text{ and } y'(0)=0$$

for the following two forcing functions:

a. $f(t)=6u(t)$
b. $f(t)=12t-72t^2$

This is the same as Example 3.12.

Obtaining the Laplace of each term yields

$$\left(s^2\,y(s)-s\,y(0)-y'(0)\right)+7\left(s\,y(s)-y(0)\right)+12\,y(s)=f(s)$$
$$\left(s^2\,y(s)-3s\right)+7\left(s\,y(s)-3\right)+12\,y(s)=f(s)$$
$$y(s)(s^2+7s+12)=f(s)+3s+21$$

a. $f(s)=\dfrac{6}{s}$

$$y(s)(s^2+7s+12)=\frac{6}{s}+3s+21=\frac{3s^2+21s+6}{s}$$
$$y(s)=\frac{3s^2+21s+6}{s(s^2+7s+12)}=\frac{3s^2+21s+6}{s(s+4)(s+3)}=\frac{A_1}{s}+\frac{A_2}{s+4}+\frac{A_3}{s+3}$$

Using Equation 4.21,

$$A_1=\lim_{s\to 0}\left[\frac{3s^2+21s+6}{(s+4)(s+3)}\right]=0.5;\; A_2=\lim_{s\to -4}\left[\frac{3s^2+21s+6}{s(s+3)}\right]=-7.5;$$
$$A_3=\lim_{s\to -3}\left[\frac{3s^2+21s+6}{s(s+4)}\right]=10$$

Finally,

$$y(s)=\frac{0.5}{s}-\frac{7.5}{s+4}+\frac{10}{s+3}$$
$$y=0.5-7.5e^{-4t}+10e^{-3t}$$

This is the same solution as in Example 3.12a.

b. $f(s)=\dfrac{12}{s^2}-\dfrac{144}{s^3}$

$$y(s)(s^2+7s+12)=\frac{12}{s^2}-\frac{144}{s^3}+3s+21=\frac{3s^4+21s^3+12s-144}{s^3}$$
$$y(s)=\frac{3s^4+21s^3+12s-144}{s^3(s+4)(s+3)}=\frac{A_1}{s^3}+\frac{A_2}{s^2}+\frac{A_3}{s}+\frac{A_4}{s+4}+\frac{A_5}{s+3}$$

$$A_1 = \lim_{s\to 0}\left[\frac{3s^4+21s^3+12s-144}{(s+4)(s+3)}\right] = -12;\ A_2 = \lim_{s\to 0}\frac{d}{ds}\left[\frac{3s^4+21s^3+12s-144}{(s+4)(s+3)}\right] = 8$$

$$A_3 = \lim_{s\to 0}\frac{1}{2}\frac{d^2}{ds^2}\left[\frac{3s^4+21s^3+12s-144}{(s+4)(s+3)}\right] = -3.67$$

$$A_4 = \lim_{s\to -4}\left[\frac{3s^4+21s^3+12s-144}{s(s+3)}\right] = -12;\ A_5 = \lim_{s\to -3}\left[\frac{3s^4+21s^3+12s-144}{s(s+4)}\right] = 18.67$$

Therefore,

$$y = A_3 + A_2\,t + A_1\,t^2 - A_4\,e^{-4t} + A_5\,e^{-3t}$$
$$y = -3.67 + 8t - 6t^2 - 12e^{-4t} + 18.67e^{-3t}$$

This is the same result as in Example 3.12b. What about the stability of this response? Stable or unstable, and why?

Example 4.13

Obtain the analytical solution of the model of the mechanical system shown in Figure 3.1. The model, initial conditions, and forcing function are

$$10\frac{d^2x}{dt^2} + 20\frac{dx}{dt} + 100x = f_A(t) \tag{3.2}$$

$$\left.\frac{dx}{dt}\right|_{t=0} = 0\ \mathrm{m/s} \text{ and } x(0) = 0\ \mathrm{m}, \text{ and } f_A(t) = 10u(t)\ \mathrm{N}$$

Obtaining the Laplace of each term yields

$$10s^2\,x(s) + 20s\,x(s) + 100x(s) = f_A(s)$$

In this case, $f_A(s) = \dfrac{10}{s}$:

$$x(s) = \frac{1}{10s^2+20s+100}f_A(s) = \frac{10}{s(10s^2+20s+100)} = \frac{1}{s(s^2+2s+10)}$$

$$x(s) = \frac{10}{s(10s^2+20s+100)} = \frac{1}{s(s^2+2s+10)}$$

The roots are $r_1 = -1+i3$ and $r_2 = -1-i3$. Let us find the solution using both methods presented. Using the first-order expansion,

$$x(s) = \frac{1}{s(s^2+2s+10)} = \frac{1}{s(s+1-i3)(s+1+i3)} = \frac{A_1}{s} + \frac{A_2}{s+1-i3} + \frac{A_3}{s+1+i3}$$

$$A_1 = \lim_{s\to 0}\left[\frac{1}{(s^2+2s+10)}\right] = 0.1$$

$$A_2 = \lim_{s \to -1+i3}\left[\frac{1}{s(s+1+i3)}\right] = \frac{1}{-18-i6} = -0.05 + i0.0167$$

$$A_3 = \lim_{s \to -1-i3}\left[\frac{1}{s(s+1-i3)}\right] = \frac{1}{-18+i6} = -0.05 - i0.0167$$

$$x(s) = \frac{0.1}{s} + \frac{-0.05+i0.0167}{s+1-i3} + \frac{-0.05-i0.0167}{s+1+i3}$$

$$x(t) = 0.1u(t) + e^{-t}\left[(-0.05+i0.0167)e^{i3t} + (-0.05-i0.0167)e^{-i3t}\right]$$

and using Equations 4.33, 4.34, and 4.35, we simplify the term in bracket:

$$x(t) = 0.1u(t) - e^{-t}[0.1\cos(3t) + 0.033\sin(3t)] \tag{4.37}$$

This is the same result as in Equation 3.41 when the model was solved by the classical technique.

Let us now obtain the solution using completing the square.

$$x(t) = L^{-1}\left[\frac{1}{s(s^2+2s+10)}\right] = L^{-1}\left[\frac{A_1}{s}\right] + L^{-1}\left[\frac{A_2 s + A_3}{(s^2+2s+10)}\right]$$

Obtaining the A coefficients,

$$\frac{1}{s(s^2+2s+10)} = \frac{A_1}{s} + \frac{A_2 s + A_3}{(s^2+2s+10)} = \frac{(A_1+A_2)s^2 + (2A_1+A_3)s + 10A_1}{s(s^2+2s+10)}$$

$$A_1 + A_2 = 0; \quad 2A_1 + A_3 = 0; \quad 10A_1 = 1$$

and

$$A_1 = 0.1, \quad A_2 = -0.1, \quad A_3 = -0.2$$

Then

$$x(t) = L^{-1}\left[\frac{0.1}{s}\right] - L^{-1}\left[\frac{0.1s+0.2}{(s^2+2s+10)}\right] = L^{-1}\left[\frac{0.1}{s}\right] - L^{-1}\left[\frac{0.1s+0.2}{(s+1)^2+9}\right]$$

The last term is the result of applying the completing the square method. Continue by rearranging the numerator of the second term, using the complex translational theorem, and using Table 4.1.

$$x(t) = L^{-1}\left[\frac{0.1}{s}\right] - L^{-1}\left[\frac{0.1s+0.2}{(s+1)^2+9}\right] = L^{-1}\left[\frac{0.1}{s}\right] - L^{-1}\left[\frac{0.1(s+1)+0.1}{(s+1)^2+9}\right] = L^{-1}\left[\frac{0.1}{s}\right] - e^{-t}L^{-1}\left[\frac{0.1s+0.1}{s^2+9}\right]$$

Note these terms

$$x(t) = L^{-1}\left[\frac{0.1}{s}\right] - e^{-t}\left(L^{-1}\left[\frac{0.1s}{s^2+9}\right] + L^{-1}\left[\frac{0.1}{s^2+9}\right]\right) \quad x(t) = 0.1u(t) - e^{-t}\left(0.1\cos 3t + 0.033\sin 3t\right)$$

This, of course, is the same result.

Example 4.14

Obtain the analytical solution of the model of the thermal system shown in Figure 3.3. The model, initial conditions and forcing function are

$$630\frac{dT}{dt}+0.8T=\dot{q}_{in}+20 \tag{3.4}$$

with forcing function (energy to the iron) $\dot{q}_{in}=150u(t)\,W$ and $T(0)=25$ °C.

Note that because the solution only makes sense for time equal or greater than zero ($t\geq 0$), we could simply write the right side of the model as $\dot{q}_{in}$ where $\dot{q}_{in}=20+150u(t)$; we select this last form.

Obtaining the Laplace of each term yields

$$630\,sT(s)-630T(0)+0.8T(s)=\dot{q}_{in}(s)$$

$$T(s)=\frac{630T(0)}{630s+0.8}+\frac{\dot{q}_{in}(s)}{630s+0.8}=\frac{15750}{630s+0.8}+\frac{\dot{q}_{in}(s)}{630s+0.8}$$

$$T(s)=\frac{19687.5}{787.5s+1}+\frac{1.25\dot{q}_{in}(s)}{787.5s+1}$$

and for $\dot{q}_{in}(s)=\dfrac{170}{s}$,

$$T(s)=\frac{19687.5s+212.5}{s(787.5s+1)}=\frac{A_1}{s}+\frac{A_2}{787.5s+1}$$

$$A_1=\lim_{s\to 0}[sT(s)]=\lim_{s\to 0}\left[\frac{19687.5s+212.5}{787.5s+1}\right]=212.5$$

$$A_2=\lim_{s\to -\frac{1}{787.5}}[(787.5s+1)T(s)]=\lim_{s\to -\frac{1}{787.5}}\left[\frac{19687.5s+212.5}{s}\right]=-147656.25$$

Inverting back to the time domain yields

$$T(t)=212.5-187.5e^{-t/787.5}=212.5-187.5e^{-0.00127t}$$

This is the same result as in Equation 3.45 when the model was solved by classical technique.

Example 4.15

The voltage across the capacitor (v_C in Volts) in a circuit is described by the following model:

$$0.375\frac{d^2v_C}{dt^2}+3.0\frac{dv_C}{dt}+v_C=v_S \tag{3.5}$$

with forcing function (supplied voltage) $v_S=5u(t)\sin 5t$ V and initial conditions $\left.\dfrac{dv_C}{dt}\right|_{t=0}=0\ \mathrm{V/s}$ and $v_C(0)=0$ V.

Obtaining the Laplace of each term yields

$$0.35s^2 v_C(s) + 3s\, v_C(s) + v_C(s) = v_s(s)$$

$$v_C(s) = \frac{1}{0.375s^2 + 3s + 1} v_s(s) = \frac{2.67}{s^2 + 8s + 2.67} v_s(s)$$

In this case, $v_S(s) = \frac{5(5)}{s^2 + 25} = \frac{25}{s^2 + 25}$; thus,

$$v_C(s) = \frac{2.67(25)}{(s^2 + 8s + 2.67)(s^2 + 25)} = \frac{66.75}{(s + 7.65)(s + 0.35)(s^2 + 25)}$$

The term $s^2 + 25$ gives complex roots, imaginary roots, so let us use the completing the square method.

$$v_C(s) = \frac{66.75}{(s + 7.65)(s + 0.35)(s^2 + 25)} = \frac{A_1}{s + 7.65} + \frac{A_2}{s + 0.35} + \frac{A_3 s + A_4}{s^2 + 25}$$

To obtain the A coefficients,

$$\frac{66.75}{(s + 7.65)(s + 0.35)(s^2 + 25)} = \frac{A_1}{s + 7.65} + \frac{A_2}{s + 0.35} + \frac{A_3 s + A_4}{s^2 + 25}$$

$$\frac{66.75}{(s + 7.65)(s + 0.35)(s^2 + 25)} = \frac{(A_1 + A_2 + A_3)s^3 + (0.35A_1 + 7.65A_2 + 8A_3 + A_4)s^2}{(s + 7.65)(s + 0.35)(s^2 + 25)}$$

$$\frac{+(25A_1 + 25A_2 + 2.67A_3 + 8A_4)s + (8.75A_1 + 191.28A_2 + 2.67A_4)}{(s + 7.65)(s + 0.35)(s^2 + 25)}$$

Equating equal coefficients in the numerator on both sides yields

$$A_1 + A_2 + A_3 = 0;\ 0.35A_1 + 7.65A_2 + 8A_3 + A_4 = 0$$

$$25A_1 + 25A_2 + 2.67A_3 + 8A_4 = 0;\ 8.75A_1 + 191.28A_2 + 2.67A_4 = 66.75$$

$$A_1 = -0.1094;\quad A_2 = 0.3639;\quad A_3 = -0.2544;\quad A_4 = -0.7101$$

Taking the inverse transform,

$$v_C(t) = L^{-1}\left[\frac{66.75}{(s + 7.65)(s + 0.35)(s^2 + 25)}\right] = L^{-1}\left[\frac{-0.1094}{s + 7.65}\right] + L^{-1}\left[\frac{0.3639}{s + 0.35}\right] - L^{-1}\left[\frac{0.2544s + 0.7101}{s^2 + 25}\right]$$

Note that there is no need to use the completing the square method on the last term because it is already in the necessary form to obtain the inverse. Continuing,

$$v_C(t) = -L^{-1}\left[\frac{0.1094}{s + 7.65}\right] + L^{-1}\left[\frac{0.3639}{s + 0.35}\right] - L^{-1}\left[\frac{0.2544s}{s^2 + 25}\right] - L^{-1}\left[\frac{0.7101}{s^2 + 25}\right]$$

Inverting back to the time domain using Table 4.1 gives

$$v_C(t) = 0.3639e^{-0.35t} - 0.1094e^{-7.65t} - 0.2544 \cos 5t - 0.1420 \sin 5t$$

4.3.2 Handling Time Delays

The technique of partial fractions expansion is restricted for use with Laplace transforms that can be expressed as the ratio of two polynomials. When the model contains time delays, an exponential function of s appears in the transform. Because the exponential is a transcendental function, we must appropriately modify the inversion procedure. We can handle exponential terms in the numerator of the transform, as we shall now see.

Consider the case in which there is a single exponential term that can be factored as follows:

$$Y(s) = Y_1(s)e^{-st_0} \tag{4.38}$$

The correct procedure is to expand in partial fractions the portion of the transform that does not contain the exponential term.

$$Y_1(s) = \frac{A_1}{s-r_1} + \frac{A_2}{s-r_2} + \cdots + \frac{A_n}{s-r_n} \tag{4.39}$$

Then invert this expression.

$$y_1(t) = A_1e^{r_1t} + A_2e^{r_2t} + \cdots + A_ne^{r_nt} \tag{4.40}$$

Making use of the real translation theorem, Equation 4.9,

$$\begin{aligned} y(t) &= L^{-1}\left[e^{-st_0}Y_1(s)\right] = u(t-t_0)\,y_1\left(t-t_0\right) \\ &= u(t-t_0)\left[A_1e^{r_1(t-t_0)} + A_2e^{r_2(t-t_0)} + \cdots + A_ne^{r_n(t-t_0)}\right] \end{aligned} \tag{4.41}$$

Note that the exponential term is excluded from the partial fractions expansion procedure, and then the real translational theorem is used to account for the delay time. Also, note that the unit step $u(t - t_0)$ must multiply the exponential term to show that $y(t) = 0$ for $t < t_0$.

Next, let us consider the case of multiple delays. When there is more than one delay term in the numerator of the transform, proper algebraic manipulation will convert the transform into a sum of terms, each having a single exponential function:

$$Y(s) = Y_1(s)e^{-st_{01}} + Y_2(s)e^{-st_{02}} + \cdots \tag{4.42}$$

Expand each of the sub-transforms – $Y_1(s)$, $Y_2(s)$, and so on – in partial fractions and invert them separately, leaving out the exponential terms. Finally, apply Equation 4.9 to each term to produce the result:

$$y(t) = u(t-t_{01})y_1\left(t-t_{01}\right) + u(t-t_{02})y_2\left(t-t_{02}\right) + \cdots \tag{4.43}$$

The following example illustrates this procedure.

Example 4.16

Given the differential equation

$$\frac{dc(t)}{dt} + 2c(t) = f(t)$$

with $c(0) = 0$, find the analytical solution for

a. a unit step change at $t = 1$: $f(t) = u(t - 1)$.
b. a staircase function of unit steps at every unit of time.

$$f(t) = u(t-1) + u(t-2) + u(t-3) + \cdots$$

The functions are sketched in Figure 4.3.

a. Transform the differential equation, solve for $C(s)$, and substitute $F(s) = (1/s)e^{-s}$.

$$C(s) = \frac{1}{s+2}F(s) = \frac{1}{s+2}\frac{1}{s}e^{-s}$$

Let $C(s) = C_1(s)e^{-s}$, $C_1(s) = \dfrac{1}{s+2}\dfrac{1}{s}$; then invert $C_1(s)$.

$$C_1(s) = \frac{1}{s+2}\frac{1}{s} = \frac{-0.5}{s+2} + \frac{0.5}{s}$$

Invert by matching entries in Table 4.1.

$$c_1(t) = -0.5e^{-2t} + 0.5u(t) = 0.5u(t)\left[1 - e^{-2t}\right]$$

Apply Equation 4.9.

$$c(t) = L^{-1}\left[C_1(s)e^{-s}\right] = c_1(t-1) = 0.5u(t-1)\left[1 - e^{-2(t-1))}\right]$$

Note that the unit step $u(t - 1)$ must multiply the exponential term to show that $c(t) = 0$ for $t < 1$.

b. For the staircase function,

$$\begin{aligned} C(s) &= \frac{1}{s+2}\left(\frac{e^{-s}}{s} + \frac{e^{-2s}}{s} + \frac{e^{-3s}}{s} + \cdots\right) \\ &= \frac{1}{(s+2)s}\left(e^{-s} + e^{-2s} + e^{-3s} + \cdots\right) \\ &= C_1(s)e^{-s} + C_1(s)e^{-2s} + C_1(s)e^{-3s} + \cdots \end{aligned}$$

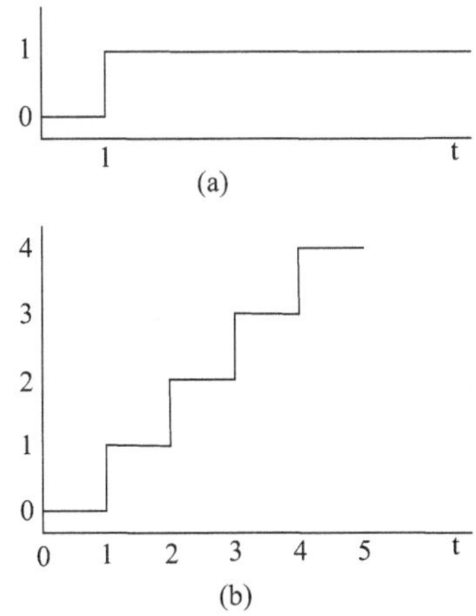

FIGURE 4.3 Input functions for Example 4.15. (a) Delayed unit step, $u(t - 1)$. (b) Staircase of unit steps.

We note that $C_1(s)$ is the same as for part (a), and therefore, $c_1(t)$ is the same. Applying Equation 4.9 to each term results in

$$\begin{aligned} c(t) &= c_1(t-1) + c_1(t-2) + c_1(t-3) + \cdots \\ &= 0.5u(t-1)\left[1-e^{-2(t-1)}\right] + 0.5u(t-2)\left[1-e^{-2(t-2)}\right] + 0.5u(t-3)\left[1-e^{-2(t-3)}\right] + \cdots \end{aligned}$$

The preceding example illustrates how to handle time delays in the input function. The same procedure can be applied when the time delay appears in the transfer function of the system, which we present next.

4.4 TRANSFER FUNCTIONS

The concept of transfer functions is *fundamental* in the study of system dynamics; at this time, we consider them in detail. The preceding section showed that we can express the Laplace of the system dependent variable as the product of two terms, a transfer function, which is characteristic of the system, and the transform of the forcing function. To refresh our memory, recall Equation 4.19:

$$Y(s) = \left[\frac{b}{a_2 s^2 + a_1 s + a_0}\right] X(s) \tag{4.19}$$

Dividing both sides by the Laplace transform of the forcing function, $X(s)$, gives

$$\frac{Y(s)}{X(s)} = \frac{b}{a_2 s^2 + a_1 s + a_0}$$

where the right side of the equation is the transfer function. Thus, we *define a transfer function as the ratio of the Laplace transform of the dependent variable (or output variable)* Y(s) *to the Laplace-transform of the forcing function (or input variable)* X(s). Transfer functions solely depend on the system itself.

In general, transfer functions are usually represented by

$$G(s) = \frac{Y(s)}{X(s)} = \frac{K\,(a_m s^m + a_{m-1} s^{m-1} + a_{m-2} s^{m-2} + \ldots + a_1 s + 1)\, e^{-t_o s}}{(b_n s^n + b_{n-1} s^{n-1} + b_{n-2} s^{n-2} + \ldots + b_1 s + 1)} \tag{4.44}$$

where
$G(s)$ = general representation of a transfer function,
$Y(s)$ = Laplace transform of the dependent or output variable,
$X(s)$ = Laplace transform of the forcing function or input variable,
K, a's, b's = constants, and
t_o = delay time.

Equation 4.44 shows the most general and best way to write a transfer function. When written in this way (the coefficient of s^0 is 1 in both numerator and denominator polynomials), K will have as units the units of $Y(t)$ over the units of $X(t)$. The other constants, a's and b's, will have as units $(time)^j$, where j is the power of the Laplace variable s multiplied by the particular constant; this will render a dimensionless term inside the parentheses because the unit of s is 1/time.

In transfer functions of real physical systems, the highest power of s in the numerator is never higher than that in the denominator; in other words, $n \geq m$.

To obtain the transfer function from a differential equation, *all initial conditions must be zero.* Otherwise, the nonzero initial conditions would contribute additional terms to the transform of the dependent variable, and the transfer function could not be obtained. As an example, consider the following differential equation with its initial conditions:

$$2\frac{d^2x}{dt^2}+5\frac{dx}{dt}+2x=f(t) \text{ with } x(0)=3 \text{ and } \left.\frac{dx}{dt}\right|_{t=0}=0$$

Following the procedure to obtain the Laplace transform yields

$$\begin{aligned}2s^2X(s)-2s\,x(0)-2x'(0)+5s\,X(s)-5x(0)+2X(s)&=F(s)\\ 2s^2X(s)+5s\,X(s)+2X(s)-6s-15&=F(s)\\ (2s^2+5s+2)X(s)&=F(s)+6s+15\end{aligned}$$

and

$$\frac{X(s)}{F(s)}=\frac{1}{(2s^2+5s+2)}+\frac{6s+15}{(2s^2+5s+2)F(s)}$$

Thus, this equation does not satisfy the definition of transfer function because of the last term on the right-hand side of the equation, which develops from the initial conditions.

Therefore, the transfer function relates to the transforms of the dependent variable and the forcing function when all initial conditions are zero. If initially the system is at steady state (or "at a rest condition"), the initial conditions of all the derivatives of the dependent variable with respect to time will be zero. But, what if the initial condition of the dependent variable itself is not zero? There is an easily applied method that drives this initial condition to zero, and it is presented in Section 4.6.

The importance of the roots of the *characteristic equation* to describe the qualitative behavior of the response was discussed in Section 3.4.2. Let us now look at the relation between the characteristic equation and the transfer function. Consider the following second-order differential equation:

$$y''+4y'+3y=f(t) \text{ with } y''(0)=y'(0)=y(0)=0$$

Following the procedure learned in Chapter 3, the subsequent is the resulting characteristic equation, and its roots, for the corresponding homogeneous equation:

$$r^2+4r+3=(r+1)(r+3)=0 \tag{4.45}$$

Now, following the procedure learned to obtain the transfer function yields

$$\frac{Y(s)}{F(s)}=\frac{1}{s^2+4s+3} \tag{4.46}$$

Taking the denominator of this transfer function and equating it to zero gives

$$s^2+4s+3=(s+1)(s+3)=0 \tag{4.47}$$

Equations 4.45 and 4.47 are exactly alike, and indeed, *the denominator of a transfer function set equal to zero is the characteristic equation.* As Section 3.4.3 showed, the locations of the roots of the characteristic equation indicate the qualitative response of any system. To briefly review, we copy for convenience what was said about the locations of the roots: "*For roots with negative real parts, the response is stable; for roots with positive real parts, the response is unstable.* Furthermore, *for real roots the response is monotonic, and for complex roots the response is oscillatory.*

Thus, *the qualitative behavior of the system does not depend on the type of forcing function,* f(t), *but rather, only on the characteristics of the system itself.*" Because the roots of Equation 4.47 are at −1 and −3, the response for a bounded input is stable and monotonic.

The reader is strongly encouraged reading Sections 3.4.2 and 3.4.3 again to review the qualitative response of systems.

4.5 ALGEBRAIC MANIPULATIONS USING LAPLACE TRANSFORMS

The very first paragraph of this chapter states "the Laplace transform method first converts them [differential equations] into algebraic equations, and the resulting equations are then manipulated algebraically before obtaining the final result." So far, we have only presented using the transform to solve a single equation that uses a simple algebra step. However, this important attribute of the transform is very powerful when dealing with multiple coupled differential equations. Let us use an example from Chapter 6 to demonstrate the use of Laplace transforms to obtain the solution of multiple coupled differential equations.

Consider the cart system shown in Figure 4.4. Example 6.1 shows that the displacement of the carts, x_1 and x_2 in meters (m), when the external force $f_A(t) = 5u(t)$ in Newton (N) is applied, is described by Equations 6.36 and 6.37,

$$3\frac{d^2x_1}{dt^2} + 5\frac{dx_1}{dt} + 120x_1 = 60x_2 + 5\frac{dx_2}{dt} \tag{6.36}$$

$$5\frac{d^2x_2}{dt^2} + 5\frac{dx_2}{dt} + 60x_2 = f_A(t) + 60x_1 + 5\frac{dx_1}{dt} \tag{6.37}$$

with initial conditions $\frac{dx_1}{d_t}\big|_{t=0} = \frac{dx_2}{dt}\big|_{t=0} = 0$ and $x_1(0) = x_2(0) = 0\text{ m}$. Obtain the transfer function relating the transform of the displacement x_1 to the transform of the force $f_A(t)$, and the one relating the transform of x_2 to the transform of the same force.

We proceed directly obtaining the Laplace transform of each equation. From Equation 6.36,

$$3s^2X_1(s) + 5sX_1(s) + 120X_1(s) = 60X_2(s) + 5sX_2(s)$$

and rearranging it algebraically,

$$X_1(s) = \frac{5s + 60}{3s^2 + 5s + 120}X_2(s) \tag{4.48}$$

From Equation 6.37,

$$5s^2X_2(s) + 5sX_2(s) + 60X_2(s) = F_A(s) + 60X_1(s) + 5sX_1(s)$$

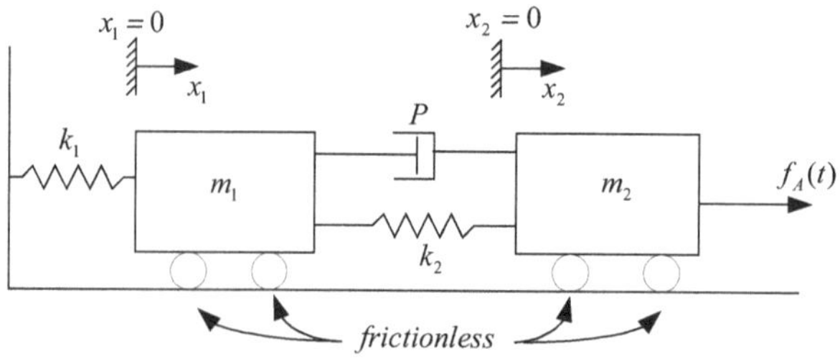

FIGURE 4.4 Frictionless carts with an external force.

and

$$X_2(s) = \frac{5s+60}{5s^2+5s+60} X_1(s) + \frac{1}{5s^2+5s+60} F_A(s) \tag{4.49}$$

At this point, we have two algebraic equations, Equations 4.48 and 4.49; therefore, we may proceed algebraically. Substituting Equation 4.49 into 4.48,

$$X_1(s) = \left(\frac{5s+60}{3s^2+5s+120}\right)\left[\frac{5s+60}{5s^2+5s+60} X_1(s) + \frac{1}{5s^2+5s+60} F_A(s)\right]$$

and solving for $X_1(s)$ yields

$$X_1(s) = \frac{5s+60}{15s^4+40s^3+780s^2+300s+3600} F_A(s) \tag{4.50a}$$

or

$$\frac{X_1(s)}{F_A(s)} = \frac{5s+60}{15s^4+40s^3+780s^2+300s+3600} \tag{4.50b}$$

Substituting Equation 4.48 into 4.49 and solving for $X_2(s)$ yields

$$X_2(s) = \frac{3s^2+5s+120}{15s^4+40s^3+780s^2+300s+3600} F_A(s) \tag{4.51a}$$

or

$$\frac{X_2(s)}{F_A(s)} = \frac{3s^2+5s+120}{15s^4+40s^3+780s^2+300s+3600} \tag{4.51b}$$

Equation 4.50b is the transfer function relating the transform of the dependent variable x_1 to the transform of the forcing function $f_A(t)$; Equation 4.51b is the transfer function relating the transform of the dependent variable x_2 to the transform of the forcing function $f_A(t)$. As learned in the previous section, the denominator of these transfer function equaled to zero – $15s^4+40s^3+780s^2+300s+3600=0$– is the characteristic equation of the system. As learned in Chapter 3, the roots of the characteristic equation (the denominator of the transfer function equated to zero) indicate the qualitative response of the system. The roots are

$$r_1 = -1.258 + i6.6987;\ r_2 = -1.258 - i6.6987;\ r_3 = -0.0752 + i2.2717;\ r_4 = -0.0752 - i2.2717$$

Thus, the response will be stable (due to the negative real part of all roots) and oscillatory (due to at least a complex conjugate root). Having the transfer functions and the forcing function, the analytical solution describing the responses of x_1 and x_2 can be obtained.

Note that the denominators of Equations 4.50b and 4.51b are the same. Essentially, this means that the characteristic equations for both carts are the same and that, therefore, the qualitative behavior or response is the same. This makes sense because both carts are connected by a spring and dashpot. It does not make sense that one cart be stable and the other cart unstable or that one oscillates with one frequency and the other one with a different frequency.

Following the method presented in this chapter, the analytical solutions (response of the carts) are given by (we strongly encouraged the reader to confirm these equations),

$$x_1(t) = 0.0833u(t) + e^{-1.258t}(0.0112\cos 6.6987t + 0.0002\sin 6.6987t) \\ - e^{-0.0752t}(0.0944\cos 2.2717t - 0.0024\sin 2.27177t) \quad (4.52)$$

$$x_2(t) = 0.1667u(t) - e^{-1.258t}(0.04\cos 6.6987t - 0.001\sin 6.6987t) \\ - e^{-0.0752t}(0.1626\cos 2.271t + 0.0104\sin 2.2717t) \quad (4.53)$$

Chapter 11 shows the comparison of the responses provided by Equations 4.52 and 4.53 to that of a simulation of Equations 6.36 and 6.37; the reader is encouraged to read the Example 11.7 in Section 11.7.

This section has shown the power of Laplace transform in handling several coupled differential equations at a time; this example only shows two coupled equations, but any number of them can be equally handled.

4.6 DEVIATION VARIABLES

Section 4.4 presented transfer functions, and that it is imperative to have all initial conditions equal to zero to obtain these functions. Very often, this is not the case. However, there is available an algebraic method to "force" the initial condition of the dependent variable to zero. Consider the following differential equation:

$$a_2 \frac{d^2y(t)}{dt^2} + a_1 \frac{dy(t)}{dt} + a_0 y(t) = bx(t) \quad (4.54)$$

The method assumes that initially the system described by the equation is at a steady state. This assumption makes all the initial values of the time derivatives equal to zero but not necessarily the initial value of the dependent variable itself, which in this case assume $y(0)$. Next, we write the model at the initial steady state:

$$a_2 \frac{d^2y(0)}{dt^2} - a_1 \frac{dy(0)}{dt} + a_0\, y(0) = b\,x(0) \quad (4.55)$$

Note that although the value of the derivatives is zero, we still included them in the equation. Subtracting Equation 4.55 from Equation 4.54 results in

$$a_2 \frac{d^2[y(t) - y(0)]}{dt^2} - a_1 \frac{d[y(t) - y(0)]}{dt} + a_0[y(t) - y(0)] = b[x(t) - x(0)] \quad (4.56)$$

Defining the following *deviation variables*,

$$Y(t) = y(t) - y(0) \quad (4.57a)$$

and

$$X(t) = x(t) - x(0) \quad (4.57b)$$

Substituting these variables into Equation 4.56,

$$a_2 \frac{d^2\,Y(t)}{dt^2} + a_1 \frac{d\,Y(t)}{dt} + a_0\, Y(t) = b\, X(t) \quad (4.58)$$

The initial condition of the dependent variable $Y(t)$ is zero $Y(0) = y(0) - y(0) = 0$; it is now possible to obtain the transfer function $\dfrac{Y(S)}{X(S)}$. Variables $Y(t)$ and $X(t)$ are called *deviation variables* because their numerical value indicates the deviation from their initial steady-state values. Equation 4.58 is essentially the same as Equation 4.54 except that it is written in deviation form. The analytical solution of Equation 4.58 yields the deviation variable $Y(t)$; to obtain the "actual" variable $y(t)$, we make use of Equation 4.57a.

Example 4.17

Consider the mechanical system shown in Figure 4.5. Section 6.3 develops the following model that describes the displacement x of the cart.

$$m\frac{d^2x}{dt^2} + P\frac{dx}{dt} + kx = f_A(t) \tag{4.59}$$

where $m = 10\,\text{kg}$, $P = 20\,\text{N}\cdot\text{s/m}$, $k = 100\,\text{N/m}$. Initially the cart is at steady state, and its displacement is 0.1 m ($x(0) = 0.1\,\text{m}$). Obtain the transfer function that relates the displacement to the applied force, and the response equation when the applied force changes from 10 N to 30 N, or $f_A(t) = 10 + 20u(t)\,\text{N}$.

The model is linear, and thus, obtaining its Laplace transform is possible. However, as the initial condition is not zero, the transform will not yield a transfer function. Having all the initial conditions equal to zero is imperative to obtain the transfer function. We then force the initial condition to zero using the concept of deviation variables.

The model applied at the initial steady state, $\left.\dfrac{d^2x(t)}{dt^2}\right|_{t=0} = 0\dfrac{\text{m}}{\text{s}^2}; \left.\dfrac{dx}{dt}\right|_{t=0} = 0\dfrac{\text{m}}{\text{s}}$ which is the same as writing $\dfrac{d^2x(0)}{dt^2} = 0\dfrac{\text{m}}{\text{s}^2}; \dfrac{dx(0)}{dt} = 0\dfrac{\text{m}}{\text{s}}$ is given by

$$kx(0) = f_A(0)$$

or

$$m\frac{d^2x(0)}{dt^2} + P\frac{dx(0)}{dt} + kx(0) = f_A(0) \tag{4.60}$$

Subtracting Equation 4.60 from Equation 4.59 gives

$$m\frac{d^2[x - x(0)]}{dt^2} + P\frac{d[x - x(0)]}{dt} + k[x - x(0)] = f_A(t) - f_A(0)$$

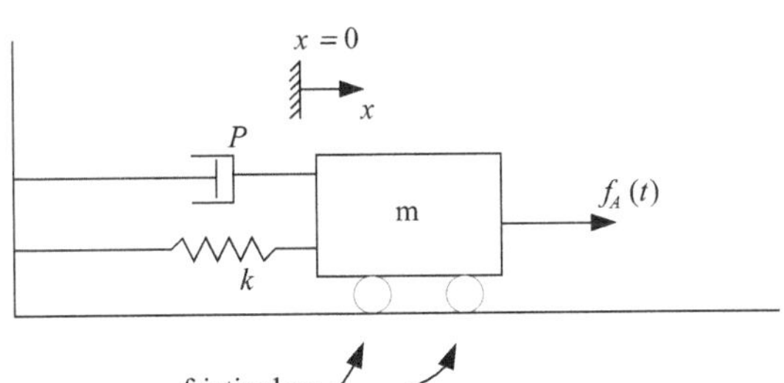

FIGURE 4.5 Mass-spring-dashpot system

$$m\frac{d^2X(t)}{dt^2}+P\frac{dX(t)}{dt}+k\,X(t)=F_A(t) \tag{4.61a}$$

or

$$10\frac{d^2X(t)}{dt^2}+20\frac{dX(t)}{dt}+100\,X(t)=F_A(t) \tag{4.61b}$$

where

$$X(t)=x(t)-x(0) \tag{4.62a}$$

and

$$F_A(t)=f_A(t)-f_A(0) \tag{4.62b}$$

are the deviation variables and, of course, $X(0)=0$ and $F_A(0)=0$.

Obtaining the Laplace transform of Equation 4.61b, and algebraically rearranging yields the desired transfer function,

$$\frac{X(s)}{F_A(s)}=\frac{1}{10s^2+20s+100} \tag{4.63a}$$

or

$$\frac{X(s)}{F_A(s)}=\frac{0.01}{0.1s^2+0.2s+1} \tag{4.63b}$$

Equations 4.63a and 4.63b (in this equation the coefficient of s^0 is 1) are two different forms of the desired transfer function; Equation 4.63b is most often used because, as explained in Section 4.4, it is easier to know the units of the constants 0.01 (units of $x(t)$ over units of $f_A(t)$), of 0.1 (time^2), and of 0.2 (time).

Now let us obtain the response equation when the forcing function $f_A(t)$ changes from 10 N to 30 N. We express this change as $f_A(t)=10+20u(t)$, and using the deviation variable, Equation 4.62b,

$$F_A(t)=f_A(t)-f_A(0)=10+20u(t)-10$$
$$F_A(t)=20u(t)$$

The Laplace transform of this forcing function is

$$F_A(s)=\frac{20}{s}$$

Solving for $X(s)$ in Equation 4.63b

$$X(s)=\frac{0.01}{0.1s^2+0.2s+1}F_A(s)=\frac{0.2}{s\left(0.1s^2+0.2s+1\right)}$$

Continuing with the method of partial fraction expansion to obtain the inverse transform gives

$$X(t) = 0.2 - e^{-t}\left[0.2 \cos 3t + 0.067 \sin 3t\right] \tag{4.64}$$

This expression provides the displacement of the deviation variable $X(t)$ from its initial condition, and not the actual displacement variable $x(t)$. We make use of Equation 4.62a to obtain the expression for the actual variable:

$$x(t) = X(t) + x(0) = 0.2 - e^{-t}\left[0.2 \cos 3t + 0.067 \sin 3t\right] + x(0)$$
$$x(t) = 0.3 - e^{-t}\left[0.2 \cos 3t + 0.067 \sin 3t\right] \tag{4.65}$$

The previous example showed the development of the deviation variables to force the initial conditions to be zero and, thus, be able to develop the transfer functions. The example has also shown how to proceed in obtaining the final expression for the response, in other words solving the model of the actual variable. Obviously, if the initial conditions of the actual variables are already zero, this step is not necessary, and the procedure can go directly into obtaining the Laplace transforms and transfer functions.

Example 4.18

Consider the electrical circuit shown in Figure 4.6. Example 10.2 develops the following model that describes the current i in the circuit.

$$L\frac{di}{dt} + Ri = v_S \tag{4.66}$$

or

$$2\frac{di}{dt} + 10i = v_S \tag{4.67}$$

The initial value of the voltage source v_S is 10 V, $v_S(0) = 10\,\text{V}$, and $i(0) = 1\,\text{A}$. At time $t = 0$, the voltage source jumps to 25 V thus, $v_S = 10 + 15u(t)$ V. Obtain the transfer function that relates the current in the loop to the voltage source v_S, and the response equation when the source changes.

Because the initial condition of the dependent variable (current) is not zero, we use deviation variables to force this to be zero. The model for the initial steady state, $\dfrac{di(0)}{dt} = 0$ A/s, is given by

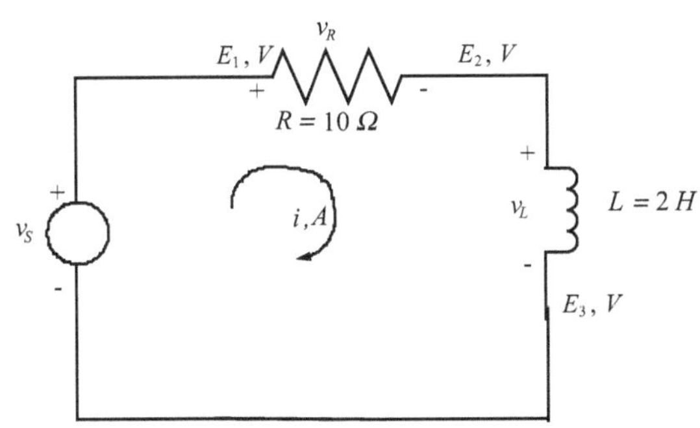

FIGURE 4.6 Electrical circuit.

$$2\frac{di(0)}{dt}+10\,i(0)=v_S(0) \tag{4.68}$$

Subtracting Equation 4.68 from Equation 4.67 gives

$$2\frac{d[i-i(0)]}{dt}+10[i-i(0)]=v_S-v_S(0)$$

$$2\frac{dI}{dt}+10I=V_S \tag{4.69}$$

where

$$I=i-i(0) \tag{4.70a}$$

and

$$V_S=v_S-v_S(0)=15\,u(t) \tag{4.70b}$$

Obtaining the Laplace transform of Equation 4.69 and algebraically rearranging yields the desired transfer function,

$$\frac{I(s)}{V_S(s)}=\frac{0.1}{0.2\,s+1} \tag{4.71}$$

Because the voltage source changes as a step change of 15 V of magnitude,

$$V_S(s)=\frac{15}{s}$$

and from Equation 4.71,

$$I(s)=\frac{0.1}{0.2\,s+1}V_S(s)=\frac{1.5}{s(0.2\,s+1)}$$

Continuing with the method of partial fraction expansion to obtain the inverse transform gives

$$I=1.5(1-e^{-\frac{t}{5}}) \tag{4.72a}$$

This expression provides the current in the loop as a deviation variable, I, from its initial condition. We make use of Equation 4.70a to obtain the expression for the actual variable:

$$i=i(0)+I=i(0)+1.5(1-e^{-\frac{t}{5}})=1+1.5(1-e^{-\frac{t}{5}}) \tag{4.72b}$$

4.7 SUMMARY

This chapter presented the techniques of Laplace transforms for obtaining the analytical solution of linear differential equations with constant coefficients. The concept of transfer function was presented and related to the characteristic equation of Chapter 3 to study the qualitative response of systems. It also showed the power of the transform in handling multiple coupled differential equations.

Problems

Problem 4.1

Using the definition of the Laplace transform, derive the transforms *F*(*s*) of the following functions.

a. $f(t) = t$
b. $f(t) = e^{-at}$, where a is a constant.
c. $f(t) = \cos\omega t$, where ω is a constant.
d. $f(t) = e^{-at}\cos\omega t$, where a and ω are constant.

Note: In parts (*c*) and (*d*), you will need the trigonometric identity

$$\cos x \equiv \frac{e^{ix} + e^{-ix}}{2}$$

Check your answers against the entries in Table 4.1.

Problem 4.2

Using a table of Laplace transforms and the properties of the transform, find the transforms *F*(*s*) of the following functions.

a. $f(t) = u(t) + 2t + 3t^2$
b. $f(t) = e^{-2t}[u(t) + 2t + 3t^2]$
c. $f(t) = u(t) + e^{-2t} - 2e^{-t}$
d. $f(t) = u(t) - e^{-t} + te^{-t}$
e. $f(t) = u(t-2)\left[1 - e^{-2(t-2)}\sin(t-2)\right]$

Problem 4.3

Check the validity of your results to Problem 4.2 by application of the initial and final value theorems. Do these theorems apply in all the cases?

Problem 4.4

In Example 4.1b, the Laplace transform of a pulse was obtained by application of the definition of the transform. Show that the same transform can be obtained by application of the real translation theorem. Notice that the pulse is the difference between two identical step changes of size *H* with the second one delayed by the duration of the pulse, *T*.

$$f(t) = Hu(t) - Hu(t-T)$$

Problem 4.5

For the general first-order differential equation, Equation 5.2, obtain the response to

a. an impulse, $X(t) = \delta(t)$
b. the pulse sketched in Figure 4.1b.

Sketch the graph of the response, $Y(t)$, for each case.

Problem 4.6

Invert the following transforms to obtain $y(t)$.

(a.) $Y(s) = \dfrac{5s-13}{s^2-5s+6}$

(b.) $Y(s) = \dfrac{2s+11}{s^2+8s+25}$

(c.) $Y(s) = \dfrac{2s+10}{s^2+8s+20}$

(d.) $Y(s) = \dfrac{4s+2}{s^2+4s+8}$

(e.) $Y(s) = \dfrac{s+10}{s^2+4s+8}$

(f.) $Y(s) = \dfrac{2s+10}{s^2+10s+41}$

(g.) $Y(s) = \dfrac{7s+15}{s^2+4s+3}$

(h.) $Y(s) = \dfrac{s}{(s+1)(s^2+4)}$

(i.) $Y(s) = \dfrac{6}{s(s+2)(s+3)}$

(j.) $Y(s) = \dfrac{9}{s^2(s+3)}$

(k.) $Y(s) = \dfrac{12}{s(s^2+7s+12)}$

(l.) $Y(s) = \dfrac{11s-6}{(s+4)(s^2+9)}$

Problem 4.7

Heat transfer from a fuel rod is represented by the differential equation

$$100\frac{dT}{dt} = 30e^{-2t} - 10(T-20)$$

with $T(0) = 20$.

(a). Take the Laplace transform of the equation and solve it for $T(s)$.
(b). Invert $T(s)$ to obtain the expression for $T(t)$.

Problem 4.8

Equation 3.2 is the model of the cart shown in Figure 3.1. Show that the solution is given by Equation 3.41.

Problem 4.9

Equation 6.10 is the model of an undamped mechanical system. Using Laplace transforms, show that the solution is given by Equation 6.11.

Problem 4.10

Equation 6.21c is the model of a damped mechanical system. Using Laplace transforms shows that the solution is given by either Equation 6.22, 6.23, or 6.24.

Problem 4.11

Equation 6.59 is the model of the system shown in Figure 6.30. Using $y(0) = 0.109\,\text{m}$, show that the solution of the equation is given by Equation 6.60.

Problem 4.12

The model that describes the displacement of a cart in meters, *m*, is

$$0.05\frac{d^2x}{dt^2} + 0.35\frac{dx}{dt} + x = 0.005 f_A(t)$$

with $x(0) = 0\,\text{m}$ and $\left.\frac{dx}{dt}\right|_{t=0} = 0\frac{\text{m}}{\text{s}}$ Using the Laplace transform method, obtain the transfer function $\frac{X(s)}{F_A(s)}$, and solve the model to obtain the expression for the position x of the block as a function of time for $f_A(t) = 10u(t)$.

Problem 4.13

The model that describes the position of the cart in Problem 6.5 is

$$10\frac{d^2x}{dt^2} + 10\frac{dx}{dt} + 40x = f_A(t)$$

Assuming that the cart starts from rest, and that the applied force is $f_A(t) = 20u(t)\,\text{N}$, using the Laplace transform method obtain the transfer function $\frac{X(s)}{F_A(s)}$ and the analytical equations that describe the displacement and velocity of the cart.

Problem 4.14

Example 8.7 develops the following model describing the concentrations of reactant S (C^S in mole/m^3) and product P (C^P in mole/m^3) in a reversible reaction:

$$\frac{dC^S}{dt} + 0.085C^S = 0.025C^P$$

and

$$\frac{dC^P}{dt} + 0.025C^P = 0.085C^S$$

with initial conditions $C^S(0) = 5\,\text{mole/m}^3$ and $C^P(0) = 0\,\text{mole/m}^3$. Obtain the analytical solution giving $C^S(t)$ and $C^P(t)$

Problem 4.15

The model that describes the positions of the blocks in meters, *m*, in Problem 6.7 is

$$10\frac{d^2x_1}{dt^2} + 40\frac{dx_1}{dt} + 20x_1 = f_A(t) + 10x_2 + 20\frac{dx_2}{dt}$$

and

$$10\frac{d^2x_2}{dt^2} + 40\frac{dx_2}{dt} + 20x_2 = 10x_1 + 20\frac{dx_1}{dt}$$

Assume that the initial conditions are $x_1(0) = x_2(0) = 0\,\text{m}$ and $\left.\frac{dx_1}{dt}\right|_{t=0} = \left.\frac{dx_2}{dt}\right|_{t=0} = 0\frac{\text{m}}{\text{s}}$ and that the applied force is $f_A(t) = 20u(t)\,\text{N}$. Using the Laplace transform method obtain the transfer function $\frac{X_2(s)}{F_A(s)}$. Based on this transfer function, do you expect the response of the system to be stable or unstable, monotonic or oscillatory, and why?

Problem 4.16

The model that describes the positions of the blocks in Problem 6.8 is

$$5\frac{d^2x_1}{dt^2} + 90\frac{dx_1}{dt} + 100x_1 = 20\frac{dx_2}{dt}$$

and

$$5\frac{d^2x_2}{dt^2} + 20\frac{dx_2}{dt} + 100x_2 = f_A(t) + 20\frac{dx_1}{dt}$$

Assume that the initial conditions are $x_1(0) = x_2(0) = 0\,\text{m}$ and $\left.\frac{dx_1}{dt}\right|_{t=0} = \left.\frac{dx_2}{dt}\right|_{t=0} = 0\frac{\text{m}}{\text{s}}$ and that the applied force is $f_A(t) = 10u(t)\,\text{N}$. Using the Laplace transform method obtain the transfer function $\frac{X_1(s)}{F_A(s)}$. Based on this transfer function, do you expect the response of the system to be stable or unstable, monotonic or oscillatory, and why?

Problem 4.17

The model that describes the position of a block is

$$5\frac{d^2x}{dt^2} + 7\frac{dx}{dt} = f_A(t)$$

Assuming that the initial conditions are $x(0) = 0\,\text{m}$ and $\left.\frac{dx}{dt}\right|_{t=0} = 0\frac{\text{m}}{\text{s}}$ and that the applied forces is $f_A(t) = 30u(t)$, using the Laplace transform method obtain the analytical equation that describes the position of the block.

Problem 4.18

The model that describes the position of the block shown in Problem 6.16 is

$$5\frac{d^2x}{dt^2}+5\frac{dx}{dt}+80x=f_A(t)+49$$

Assuming that the initial conditions are $x(0)=0.612\,\text{m}$; $\left.\frac{dx}{dt}\right|_{t=0}=0\frac{\text{m}}{\text{s}}$ and that the applied force is $f_A(t)=30u(t)\,\text{N}$, using the Laplace transform method obtain the analytical equation that describe the position of the block.

Problem 4.19

The model that describes the position of the block in Problem 6.13 is

$$2\frac{d^2x}{dt^2}+8\frac{dx}{dt}+40\,x=20.0\,u(t)$$

Assuming that the initial conditions are $x(0)=0\,\text{m}$ and $\left.\frac{dx}{dt}\right|_{t=0}=0\frac{\text{m}}{\text{s}}$, obtain the analytical equation that describes the position of the block.

Problem 4.20

The model that describes the velocity of the block in Problem 6.3 is

$$2\frac{dv}{dt}+4\,v=16u(t)$$

Assuming that the initial condition is $v(0)=0\,\text{m/s}$, obtain the analytical equation that describes the velocity of the block.

Problem 4.21

The model that describes the position of the block in Problem 6.1 is

$$0.25\frac{d^2x}{dt^2}+\frac{dx}{dt}+x=0$$

Assuming that initially the system is at steady state, meaning that $\left.\frac{dx}{dt}\right|_{t=0}=0\,\text{m/s}$ and that the block is held at $x(0)=0.1\text{m}$ and then let go, obtain the analytical equation that describes the position of the block.

Problem 4.22

The model that describes the position of the block in Problem 6.4 is

$$2\frac{d^2x}{dt^2}+18\,x=18\,y(t)$$

Assuming that the initial conditions are $x(0) = 0\,\text{m}$ and $\left.\frac{dx}{dt}\right|_{t=0} = 0\frac{\text{m}}{\text{s}}$, obtain the analytical equation that describes the position of the block for $y(t) = 0.1t\,\text{m}$

Problem 4.23

Chapter 8 develops the following model for the mass fraction of NaOH in a mixing tank:

$$12.5\frac{d\,x_3^{NaOH}}{dt} + x_3^{NaOH} = 0.73\,x_1^{NaOH}$$

with $x_3^{NaOH}(0) = 0.55$ and $x_1^{NaOH} = 0.75 - 0.08\,u(t)$ Using Laplace transform, show that the solution is

$$x_3^{NaOH} = 0.55 - 0.0625\left(1 - e^{-0.08t}\right)$$

Problem 4.24

Equation 10.13 is the model of the circuit shown in Example 10.1. Using the initial conditions given in that example, show that the solution of the model is given by Equation 10.14b.

Problem 4.25

Equation 10.27b is the model for the circuit of Example 10.3. Using the information given in the example, show that the solution of the model is given by Equation 10.28.

Problem 4.26

Equation 10.119 is the model that describes the current in the circuit shown in Figure 10.44 when a sinusoidal voltage source, $v_S = V_o \sin(\omega t)u(t)$; for $t < 0$, there is no current flow, and the capacitor is fully discharged. Show that Equation 10.120 is the solution for this model.

Problem 4.27

Response of an Integrating Process. The response of the liquid level in a tank is given by the first-order differential equation

$$A\frac{dh(t)}{dt} = f(t)$$

where $h(t)$, m, is the level in the tank, $f(t)$, m^3/s, is the flow of liquid into the tank, and A, m^2, is the constant area of the tank. You may assume that initially the tank is empty. Obtain the transfer function for the tank and the response of the level to a unit step in flow, $F(t) = u(t)$. Sketch the graph of the level response, $H(t)$. Why do you think we call this result the response of an *integrating process*?

Problem 4.28

The model that describes the currents through resistances R_1 and R_3 in Problem 10.22 are given by

$$i_1 = 0.143x10^{-3}v_S + 0.286i_2$$

and

$$2x10^{-3}\frac{d\,i_2}{dt} + 21428\,i_2 = 0.286\,v_S$$

The initial conditions are $i_1(0) = 2.93 \times 10^{-3}\ A$ and $i_2(0) = 2.67 \times 10^{-4}\ A$. For $v_S = 20 + 20u(t)\ V$, solve the model to obtain expressions for $i_1(t)$ and $i_2(t)$.

Problem 4.29

Consider the following differential equations, which model a sequential chemical reaction A => B => C:

$$\frac{dc_A}{dt} = -k_1 C_A \qquad \frac{dc_B}{dt} = k_1 C_A - k_2 C_B$$

where C_A and C_B are the concentrations of species A and B and k_1 and k_2 are known constants. Assume $C_A(0) = 1$ and $C_B(0) = 0$. Find expressions for $C_A(t)$ and $C_B(t)$ using the Laplace transform method. After that, find the solution for $C_B(t)$ for the special case of $k_2 = k_1$.

5 Response of First- and Second-Order Systems

Most of the models developed in this book are either first- or second-order linear differential equations with constant coefficients; this is a common case for physical/industrial models. The tools presented in the three previous chapters allow us to solve the models and study their responses. Although some analysis of the responses has been done in a few examples, first- and second-order systems are so important that it is worthwhile to extend the analysis more formally in this chapter. Actually, this discussion also allows us to study the similarities among different systems. In general, first- and second-order models can be written as

$$a\frac{dy}{dt}+by(t)=f(t) \tag{5.1}$$

$$a\frac{d^2y}{dt^2}+b\frac{dy}{dt}+cy(t)=f(t) \tag{5.2}$$

where a, b, and c are constant coefficients that depend on the system model, y is the dependent variable, t is time (the independent variable), and $f(t)$ is the forcing function, which may or may not be a function of time. The origin of the coefficients is discussed in Chapters 6–10. The forcing function is external to the system and can be thought of as any activity or event that forces the dependent variable y to change from one value to another. There are many important types of forcing functions experienced in physical and industrial settings: step, impulse, rectangular pulse, ramp, and periodic (such as a sinusoid). In this chapter, we will explore how both first- and second-order ordinary differential equations (ODEs) respond to the following types of forcing functions: Step function, impulse, rectangular pulse, and sinusoid. We will also explore the response of coupled first-order and coupled second-order linear ODEs.

5.1 FIRST-ORDER SYSTEMS

All linear first-order ODEs with constant coefficients can be written in the so-called standard form, which is independent of the physics underlying the model:

$$\tau\frac{dy}{dt}+y(t)=Kf(t) \tag{5.3}$$

Compared to Equation 5.1, we find

- $y(t)$ is the dependent variable.
- t is the dependent variable.
- $f(t)$ is the forcing function.
- $\tau = a/b$ is the time constant and always has units of time.
- $K = 1/b$ is the system gain and has units of the dependent variable over the forcing function.

The following are some examples of models described by first-order differential equations:

DOI: 10.1201/9781003619161-5

- Motion:

$$m\frac{dv_y}{dt} + \mu_f v_y = F_{ext} \tag{5.4}$$

- Heating/Cooling:

$$mc_p\frac{dT}{dt} = -hA\left(T - T_\infty\right) \tag{5.5}$$

- Mass Balance:

$$\dot{V}\frac{dc}{dt} = -kA(c - c_{in}) \tag{5.6}$$

- Series RC Circuit

$$RC\frac{dv_c}{dt} + v_c = v_{in} \tag{5.7}$$

To rewrite any of these equations in standard form, first move all terms with the dependent variable to the left-hand side of the equation and move the forcing function to the right-hand side of the equation. According to Equation 5.3, the dependent variable must not have a preceding coefficient. Hence, after rearranging, divide all terms by the coefficient immediately preceding the dependent variable.

Example. Rewrite the equation for Heating/Cooling in standard form

$$mc_p\frac{dT}{dt} = -hA\left(T - T_\infty\right)$$

To move the dependent variables to the left-hand side of the equation, we need to first identify the dependent variable. The dependent variable is found in the numerator of the derivative. In this case, it is temperature, or T. Moving all dependent variables to the left-hand side, the governing equation becomes

$$mc_p\frac{dT}{dt} + hAT = hAT_\infty$$

The forcing function, T_∞, which is not a function of the dependent variable, remains on the right-hand side. That said, T_∞ can be a function of the independent variable, which in this example is time t. Now we must divide both sides by hA to put the equation into standard form:

$$\frac{mc_p}{hA}\frac{dT}{dt} + T = T_\infty$$

If $m = 12\,\text{kg}$, $c_p = 3\frac{\text{J}}{\text{kg}\cdot\text{K}^2}$, $h = 5\frac{\text{J/s}}{\text{m}^2\cdot\text{K}}$, and $A = 5\text{m}^2$, the time constant (which must have units of time) is

$$\tau = \frac{mc_p}{hA} = \frac{12\text{ kg} \times 3\frac{\text{J}}{\text{kg}\cdot\text{K}}}{5\frac{\text{J/s}}{\text{m}^2\cdot\text{K}} \times 5\text{ m}^2} = 1.44\text{ s}$$

And the gain is

$$K = 1\,[\text{dimensionless}]$$

Note here that both the forcing function (T_∞) and the dependent variable (T) have the units Kelvin, and hence, the gain is dimensionless. This is not necessarily always the case. If we take the equation of motion, for instance, the gain is $K = 1/\mu_f$, and the gain would have to have units of velocity over force, such as m/(s·N) in the SI unit system.

Therefore, all first-order linear ODEs with constant coefficients can be put in the same standard form no matter the underlying physics. Hence, they have the same type of response given the same forcing function. We now look at the significance of the two parameters: the time constant τ and the system gain K.

What is the meaning of tau? The characteristic time can be thought of as a measurement of the system inertia. *Inertia* is defined as the tendency of a system to resist change. The larger the value of τ, the more sluggish a system is to a change in the forcing function, and the longer it takes for the system to reach a steady state. The time constant τ depends only on the system itself and not on the forcing function.

What is the meaning of the system gain? The system gain provides the relationship between the dependent variable and the forcing function after steady state has been achieved. If the forcing function is a constant $f(t) = A$, then the steady state value of the dependent variable y is always $y(\infty) = KA$. If the forcing function is a varying signal in time, then both the time constant τ and the gain K will determine the steady state behavior of y.

5.1.1 Response to a Step Function

One of the most important types of forcing functions is the step function. One frequently encounters forcing functions (or inputs) that are initially zero and then switch to a nonzero constant value at some arbitrary time. To treat such forcing functions mathematically, we define the unit step function using the Heaviside notation. The Heaviside unit step function takes the value of 0 when its argument is negative and 1 when its argument is zero or positive.

$$u(t) = \begin{cases} 1, & t \geq 0 \\ 0, & t < 0 \end{cases} \tag{5.8}$$

An example, for instance, is a series RC circuit that when turned on, the voltage is switched from 0 to 5V. Another example is when the force applied to an object is suddenly changed. Let's assume you have an object originally subjected to a force of $F_{ext,0}$ for a sufficiently long time followed by a step change in force ΔF_{ext} applied at time $t = 0$.

TABLE 5.1
Relevant Variables, Coefficients, and Expressions for the Time Constant τ and Gain K for the Models in Equations 5.4–5.7.

Model	Dependent variable	Independent variable	a	b	f(t)	τ	K
Motion	v_y	t	m	μ_f	F_{ext}	m/μ_f	$1/\mu_f$
Heating/ Cooling	T	t	mc_p	hA	T_∞	mc_p/hA	1
Mass balance	c	t	$\dot{V}$	kA	c_{in}	$\dot{V}/kA$	1
Series RC Circuit	v_c	t	RC	1	v_{in}	RC	1

$$\tau \frac{dv_y}{dt} + v_y = K\left(\Delta F_{ext} \cdot u(t) + F_{ext,0}\right) \tag{5.9}$$

Figure 5.1 shows two examples of a step change of the forcing function. In Figure 5.1a, $F_{ext,0} = 0$, and the step change is $\Delta F_{ext} = 10\,\mathrm{N}$. In the second example, Figure 5.1b, $F_{ext,0} = 5\,\mathrm{N}$, and the step change is $\Delta F_{ext} = 5\,\mathrm{N}$. In both cases, after the step change, $F_{ext} = 10\,\mathrm{N}$.

What is the response to a step change in the external force? The ODE can be solved by any number of methods, including separation of variables, variation of parameters, and a Laplace transform. They all lead to the same solution:

$$v_y(t) = K\Delta F_{ext}\left(1 - e^{-t/\tau}\right) + v_y(0) \tag{5.10}$$

where ΔF_{ext} is the step change in the external force and $v_y(0) = KF_{ext,0}$ is the initial condition. Here, the transient is the $e^{-t/\tau}$ term. If we use a gain value of $K = \dfrac{1}{\mu_f} = 0.5\dfrac{\mathrm{m}}{\mathrm{N \cdot s}}$, the response of v_y to the change in forcing function is given in Figure 5.2.

As $t \to \infty$, the transient term goes to zero, which defines the new steady state condition

$$v_y(\infty) = K\Delta F_{ext} + v_y(0) \tag{5.11}$$

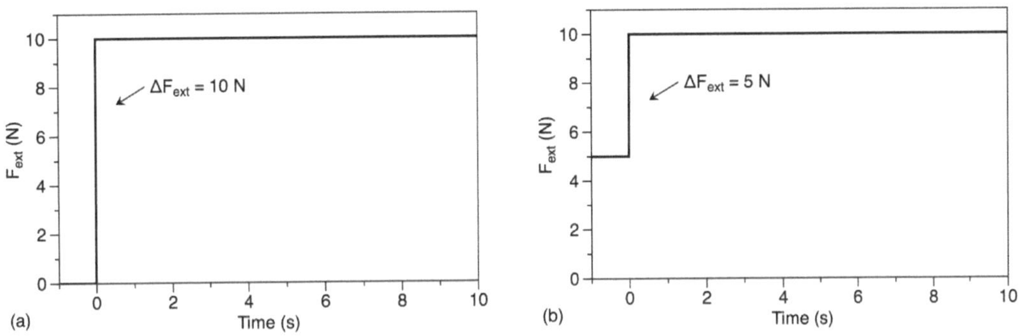

FIGURE 5.1 Step function $\Delta F_{ext} \cdot u(t) + F_{ext,0}$ for two cases. In (a), the external force is zero before the application of 10 N of force at $t = 0$. In (b), the external force is 5 N before the application of 5 N of force at $t = 0$.

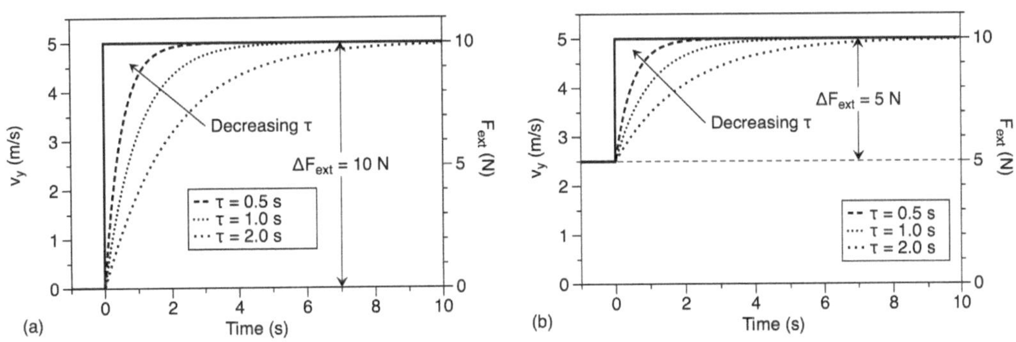

FIGURE 5.2 Response to a step change for a range of time constants. In both cases (a) and (b), the gain is $K = 0.5\,\mathrm{m/(N \cdot s)}$.

Normally, the initial condition $v_y(0)$ must also be provided. As we have waited a sufficiently long time before the step change ΔF_{ext}, the initial condition is the same as the steady state solution for $F_{ext,0}$. To perhaps more clearly elucidate, the solution to v_y can be expressed in terms of a complementary and a particular solution:

$$v_y(t) = v_{y,c} + v_{y,p} \tag{5.12}$$

The general solution to which is

$$v_y(t) = c_1 e^{-t/\tau} + KF_{ext} \tag{5.13}$$

Obviously, as enough time goes by, $e^{-t/\tau} \to 0$, and the steady state value of v_y depends only on the external force F_{ext}:

$$v_y(\infty) = KF_{ext} \tag{5.14}$$

Accordingly, the initial condition before the step change ΔF_{ext} is

$$v_y(0) = KF_{ext,0} \tag{5.15}$$

If $F_{ext,0} = 0$, then $v_y(0) = 0$. By comparison, if $F_{ext,0} = 5\text{N}$ and $K = 0.5\dfrac{\text{m}}{\text{N}\cdot\text{s}}$, then $v_y(0) = 2.5\text{m/s}$, as shown in Figure 5.2. To find the constant of integration in Equation 5.13, we substitute $t = 0$:

$$v_y(0) = c_1 + KF_{ext} \tag{5.16}$$

Hence, $c_1 = v_y(0) - KF_{ext}$. Therefore, the solution becomes

$$v_y(t) = \left[v_y(0) - KF_{ext}\right]e^{-t/\tau} + KF_{ext} \tag{5.17}$$

Equations 5.10 and 5.17 appear to be different, but they are equivalent. Equation 5.10 is written in terms of the step change ΔF_{ext}, whereas Equation 5.17 is written in terms of the final value of the step change F_{ext}.

Example. Prove that Equations 5.10 and 5.17 are equivalent by rewriting Equation 5.17 in terms of ΔF_{ext}.

Recognizing that $F_{ext} = \Delta F_{ext} + F_{ext,0}$, we substitute this expression into Equation 5.17:

$$v_y(t) = v_y(0)e^{-t/\tau} - K\Delta F_{ext}e^{-t/\tau} + K\Delta F_{ext} - KF_{ext,0}e^{-t/\tau} + KF_{ext,0}$$

To eliminate $F_{ext,0}$, we note that $v_y(0) = KF_{ext,0}$. Hence, $F_{ext,0} = v_y(0)/K$:

$$v_y(t) = v_y(0)e^{-t/\tau} - K\Delta F_{ext}e^{-t/\tau} + K\Delta F_{ext} - v_y(0)e^{-t/\tau} + v_y(0)$$

We are now able to recover Equation 5.10.

$$v_y(t) = K\Delta F_{ext}\left(1 - e^{-t/\tau}\right) + v_y(0)$$

What is the meaning of gain for a step change? As previously mentioned, the system gain provides the relationship between the dependent variable and the forcing function after the transient has died off.

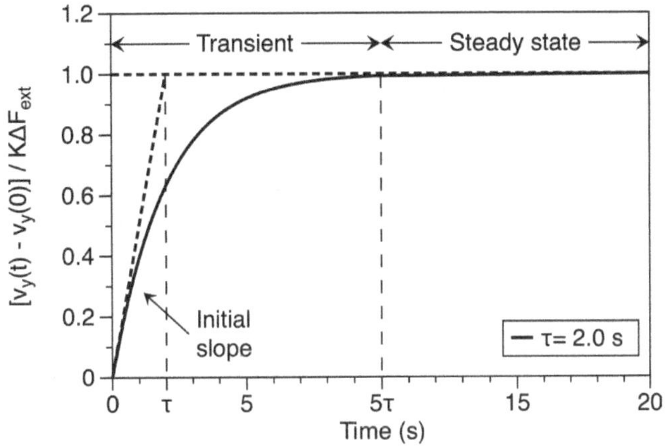

FIGURE 5.3 General response to a step change in the forcing function or input. When $t=\tau$, the response has reached 63.2% of its steady state value. When $t \geq 5\tau$, the transient has died off and the system is at steady state.

$$K = \frac{v_y(\infty) - v_y(0)}{\Delta F_{ext}} = \frac{\Delta v_y}{\Delta F_{ext}} \tag{5.18}$$

What is the meaning of the time constant for a step change? The time constant is related to how quickly the system responds to a step change in the forcing function. According to Equation 5.10, the complete response is the sum of a transient and a steady state component, the details of which are shown in Figure 5.3. The steady state response is the part of the solution that remains after the transient term has disappeared or died off. The rate at which the transient disappears is related to τ. Immediately after the step change (or $t=0$), the initial slope of the response is

$$\left.\frac{dv_y}{dt}\right|_{t=0} = \frac{K\Delta F_{ext}}{\tau} \tag{5.19}$$

which decays exponentially to zero as $t \to \infty$ as long as $\tau > 0$. When $t=\tau$ (or when the time equals one time constant), the response reaches 63.2% of its total change. When $t=2\tau$, the response reaches 86.5% of its total change. When $t=5\tau$, the response reaches 99.3% of its total change. In other words, the response is essentially complete after five time constants; it is commonly accepted in most areas of engineering to use 5τ as the time it takes to reach steady state. Table 5.2 shows the tabulated values of the transient term $e^{-t/\tau}$, and the approach to steady state as $\frac{v_y(t) - v_y(0)}{K\Delta F_{ext}}$ (in terms of %). It is important to note that the denominator $K\Delta F_{ext} = v_y(\infty) - v_y(0)$, which represents the total change of the output in response to the step change in the forcing function.

5.1.2 Response to an Impulse and a Rectangular Pulse

Two important forcing functions that are encountered in many engineering problems are the unit impulse and the unit rectangular pulse. We need to understand their properties and be able to find the response of a first-order system to both. Mathematically, the unit impulse is described by the Dirac delta function, which is used to model a pulse of infinitesimal width, the schematic representation of which is shown in Figure 5.4a. In reality, this representation is physically impossible; however, its idealization holds as long as the time constant τ of the ODE is much larger than the width

TABLE 5.2
Response to a Step Change for a First-Order ODE

t/τ	$e^{-t/\tau}$	$\frac{v_y(t)-v_y(0)}{K\Delta F_{ext}}$
0	1.000	0.00%
0.25	0.779	2.21%
0.50	0.607	3.94%
0.75	0.472	5.28%
1.0	0.368	63.2%
2.0	0.135	86.5%
3.0	0.050	95.5%
4.0	0.018	98.2%
5.0	0.007	99.3%
6.0	0.003	99.8%

Note that this is the type of step change response for any first-order ODE with constant coefficients. If modeling the voltage across the capacitor of a series RC circuit, for instance, the dependent variable becomes v_c and the change in the forcing function becomes Δv_{in}, but everything else remains the same.

of the pulse. In doing so, one not only simplifies the equations but is also able to predict the response without a detailed model of the pulse itself. While not a true function in the traditional sense, the Dirac delta function is zero everywhere except for an infinitesimal range at $t = 0$,

$$\delta(t) = 0 \quad t \neq 0 \tag{5.20}$$

and the integral of $\delta(t)$ over this infinitesimal range is unity:

$$\int_{-\infty}^{\infty} \delta(t)\,dt = 1 \tag{5.21}$$

Hence, according to the previous equation, the dimensions of $\delta(t)$ must be 1/time. The unit impulse is also the derivative of a step function, so that

$$\delta(t) = \frac{du(t)}{dt} \tag{5.22}$$

An example of a rectangular pulse is shown in Figure 5.4b. Here, in terms of the Heaviside step function, the mathematical representation of the forcing function $f(t)$ for a pulse of amplitude A and width a is

$$f(t) = A\left[u(t) - u(t-a)\right] \tag{5.23}$$

A physical model involving an impulse, for example, is a billiard ball that is struck by a cue stick. If P is the momentum transferred to the billiard ball as it is struck, the external force acting on the billiard ball will be

$$F_{ext} = P\delta(t) \tag{5.24}$$

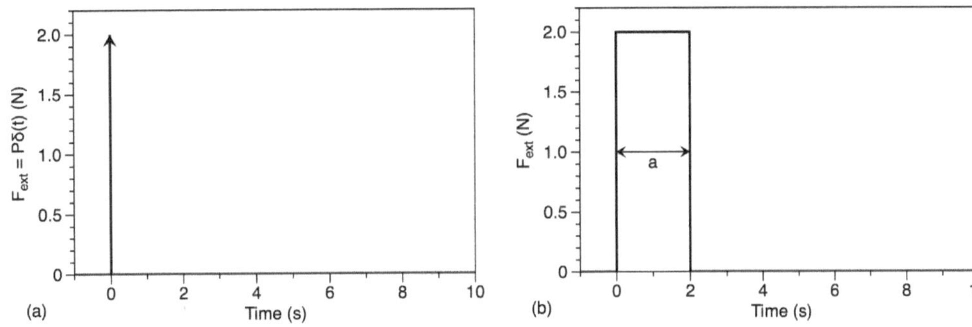

FIGURE 5.4 Impulse (a) and rectangular pulse (b) of width $a = 2s$.

The equation of motion becomes

$$\tau \frac{dv_y}{dt} + v_y = KP\delta(t) \tag{5.25}$$

Assuming the velocity of the billiard ball is zero before being hit by the cue stick, $v_y(0) = 0$, the ODE can readily be solved by a Laplace transform. The Laplace transform is

$$\tau\left[sV_y(s)\right] + V_y(s) = KP \tag{5.26}$$

Solving for v_y in Laplace space,

$$V_y(s) = \frac{KP}{\tau s + 1} = \frac{KP/\tau}{s + 1/\tau} \tag{5.27}$$

Converting back to the time domain gives the solution

$$v_y(t) = K\frac{P}{\tau}e^{-t/\tau} \cdot u(t) \tag{5.28}$$

In the case of a finite pulse, the ODE is written as

$$\tau \frac{dv_y}{dt} + v_y = KF_{ext}\left[u(t) - u(t-a)\right] \tag{5.29}$$

Similar to the case of the impulse, the ODE can be solved by a Laplace transform. The solution in terms of the Heaviside function is

$$v_y(t) = KF_{ext}\left[u(t) - u(t-a)\right] - KF_{ext}e^{-t/\tau} \cdot u(t) + KF_{ext}e^{-(t-a)/\tau} \cdot u(t-a) \tag{5.30}$$

This equation can be broken into two distinct solutions depending on whether the time t is less than the pulse width a or if it is equal to the pulse width or greater:

$$v_y(t) = \begin{cases} KF_{ext}\left(1 - e^{-t/\tau}\right) & 0 \le t \le a \\ KF_{ext}\left(e^{a/\tau} - 1\right)e^{-t/\tau} & t \ge a \end{cases} \tag{5.31}$$

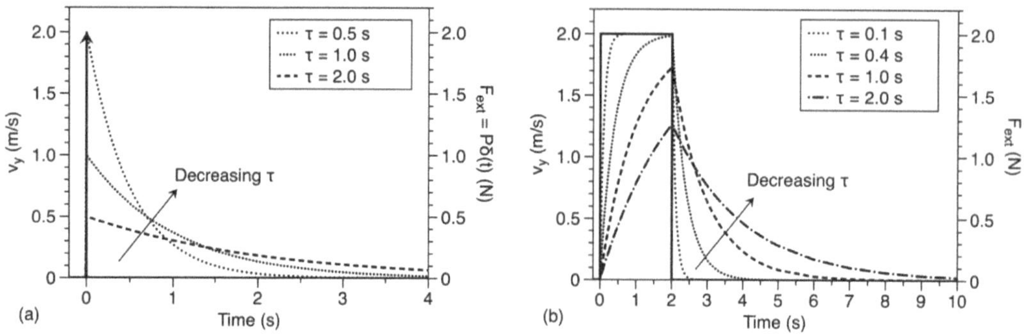

FIGURE 5.5 Response to an impulse (a) and a rectangular pulse (b) for a range of time constants. In both cases (a) and (b), the gain is $K = 1.0\,\mathrm{m}/(\mathrm{N}\cdot\mathrm{s})$.

Similar to the response to a step function, the time constant τ is a measure of how quickly the system responds either to an impulse or a rectangular pulse. As previously mentioned, the characteristic time can be thought of as a measurement of the system inertia or the resistance to change. Figure 5.5 shows both types of responses. In the case of the impulse, Figure 5.5a, the maximum velocity is reached instantaneously. In other words, the cue stick imparts maximum velocity to the billiard ball after contact. After contact, the velocity decays exponentially back to zero, as there is no longer a force acting on the billiard ball. We note two observations in accordance with Equation 5.28. First, the larger the time constant, the lower the maximum velocity. Second, the larger the time constant, the longer it takes for the billiard ball to stop moving (or for the billiard ball to return to zero velocity).

In contrast to an impulse, a rectangular pulse consists of two distinct step changes, where the forcing function switches from a low value of zero to a high value and then back down to zero after a finite time of a seconds. As expected, the response for the first a seconds is the same as for a step change. When $t \geq a$, the response decays exponentially to zero. It was already established in the previous section that a system reaches a steady state after 5τ. If the width or duration of the pulse a is much larger than the time constant, then the response of the system reaches its steady state value before the pulse returns to its low value. If, however, the duration of the pulse a is less than 5τ, the response will not reach its steady state value before the pulse returns to its low value. Figure 5.5b shows the response to a pulse of duration of 2 seconds for a system with a time constant ranging from 0.1 s to 2.0 s. The amplitude of the pulse is 2 N and the gain K is 1 m/N s. In the first case, where $\tau = 0.1\,\mathrm{s}$, steady state is reached after 0.5 s, and hence the velocity reaches its steady state value ($v_y = 2$ m/s) before termination of the pulse. The maximum time constant for a system to achieve a steady state for this pulse is $\tau = 0.4\,\mathrm{s}$, as $5 \times 0.4\,\mathrm{s} = 2\,\mathrm{s}$. If the time constant is larger than $\tau = 0.4\,\mathrm{s}$, the response will not reach a steady state before the end of the pulse.

In the limit when $\tau \gg a$, it was previously stated that a rectangular pulse can be approximated as an impulse. To understand why, it is instructive to rewrite the pulse response for $\tau \geq a$ with a Taylors series expansion of $e^{a/\tau}$. In other words,

$$e^{a/\tau} = 1 + \frac{a}{\tau} + \frac{1}{2}\left(\frac{a}{\tau}\right)^2 + \cdots \tag{5.32}$$

Substituting into Equation 5.31 for when $t \geq a$

$$v_y(t) = KF_{ext}\left[\frac{a}{\tau} + \frac{1}{2}\left(\frac{a}{\tau}\right)^2 + \cdots\right]e^{-t/\tau} \tag{5.33}$$

If the time constant is much larger than the duration of the pulse, or $\tau \gg a$, we can neglect all terms inside the brackets expect for the first term. The response then reduces to

$$v_y(t) \approx \frac{KF_{ext}a}{\tau} e^{-t/\tau} \tag{5.34}$$

Comparing to the response of the impulse (Equation 5.28), we find $P = F_{ext}a$. It is important to note that $F_{ext}a$ is the area underneath the input pulse. We could also consider input pulses that may not be rectangular but have a different shape. We would find that as long as $\tau \gg a$, the response depends only on the area underneath the pulse but not on the shape of the pulse.

5.1.3 Response to a Sinusoidal Signal

In the previous sections, we were concerned with the response of systems where the input to the forcing function occurred only once. Many systems, however, are subject to repetitive inputs, such as a sinusoid or a train of pulses with a specified period T. Here, we will show that both the time constant τ and the period T dictate the nature of the response. We consider the voltage across the capacitor v_c in a series RC circuit with a sinusoidal input that oscillates with a frequency ω:

$$\tau \frac{dv_c}{dt} + v_c = v_{in} \cos(\omega t) \tag{5.35}$$

The relationship between the period and frequency of the cosine wave is

$$\omega = \frac{2\pi}{T} \tag{5.36}$$

Like we have done before, we split the general solution into a complementary and particular solution:

$$v_c = v_{c,c} + v_{c,p} \tag{5.37}$$

where the complementary solution is the solution when the forcing function is equal to zero (the homogenous ODE)

$$v_{c,c} = c_1 e^{-t/\tau} \tag{5.38}$$

To find the particular solution, we use the method of undetermined coefficients, and we guess a solution of the form

$$v_{c,p} = A \cos(\omega t + \phi) \tag{5.39}$$

Note that this is equivalent to $v_{c,p} = A\cos(\omega t) + B\sin(\omega t)$; however, it is a bit easier to solve. Substituting the guess solution, Equation 5.39, into the governing equation, Equation 5.35:

$$-\tau\omega A \sin(\omega t + \phi) + A \cos(\omega t + \phi) = v_{in} \cos(\omega t) \tag{5.40}$$

Noting the following trigonometric relationships

$$\sin(\omega t + \phi) = \cos(\omega t)\sin\phi + \sin(\omega t)\cos\phi$$

$$\cos(\omega t+\phi)=\cos(\omega t)\cos\phi-\sin(\omega t)\sin\phi$$

Equation 5.40 reduces to

$$\left[v_{in}-A(\cos\phi-\omega\tau\sin\phi)\right]\cos(\omega t)+\left[\sin\phi+\omega\tau\cos\phi\right]\sin(\omega t)=0 \tag{5.41}$$

Since $\cos(\omega t)$ and $\sin(\omega t)$ are linearly independent, they must vanish to zero identically. In other words,

$$\left[v_{in}-A(\cos\phi-\tau\omega\sin\phi)\right]\cos(\omega t)=0$$
$$\left[\sin\phi+\tau\omega\cos\phi\right]\sin(\omega t)=0$$

For the $\sin(\omega t)$ term to vanish, the following must be true:

$$\sin\phi=-\tau\omega\cos\phi \tag{5.42}$$

Using the Pythagorean identity, we can determine both $\sin\phi$ and $\cos\phi$. Substituting the previous expression into the Pythagorean identity,

$$\sin^2\phi+\cos^2\phi=1\Rightarrow\left[(\tau\omega)^2+1\right]\cos^2\phi=1$$

we arrive at

$$\cos\phi=\frac{1}{\sqrt{1+(\tau\omega)^2}}$$

Substituting back into Equation 5.42,

$$\sin\phi=\frac{-\tau\omega}{\sqrt{1+(\tau\omega)^2}}$$

If we now look at the $\cos(\omega t)$ term in Equation 5.41, we see that

$$A=\frac{v_{in}}{\cos\phi-\tau\omega\sin\phi}=\frac{\sqrt{1+(\tau\omega)^2}}{1+(\tau\omega)^2}v_{in}=\frac{v_{in}}{\sqrt{1+(\tau\omega)^2}}$$

So the particular solution is

$$v_{c,p}=\frac{v_{in}}{\sqrt{1+(\tau\omega)^2}}\cos(\omega t+\phi) \tag{5.43}$$

where, recognizing that $\sin\phi/\cos\phi=\tan\phi$, the phase angle is

$$\phi=\tan^{-1}(-\tau\omega) \tag{5.44}$$

Hence, the general solution is

$$v_c(t) = \left[c_1 e^{-t/\tau} + \frac{v_{in}}{\sqrt{1+(\tau\omega)^2}} \cos(\omega t + \phi) \right] \tag{5.45}$$

If the initial condition is $v_c(0) = 0$, then $c_1 = -\frac{v_{in}}{\sqrt{1+(\tau\omega)^2}} \cos(\phi)$, and the final solution becomes

$$v_c(t) = \frac{v_{in}}{\sqrt{1+(\tau\omega)^2}} \left[\cos(\omega t + \phi) - \cos(\phi) e^{-\frac{t}{\tau}} \right] \tag{5.46}$$

Therefore, similar to a step change, the transient term dies off exponentially. After 5τ, only the steady state response remains, which is

$$v_c(t \geq 5\tau) = \frac{v_{in}}{\sqrt{1+(\tau\omega)^2}} \cos(\omega t + \phi) \tag{5.47}$$

Here, we note that $\tau\omega$ ranges from 0 and ∞; hence, ϕ ranges from 0° to −90° and represents a phase lag. A phase lag means that the response lags the input. It is obvious that both the amplitude of v_c and phase lag ϕ depend uniquely on $\tau\omega$. According to Equation 5.44 as $\tau\omega \to 0$, the phase lag is minimal, and the amplitude is the same as in the input voltage. On the other hand, as $\tau\omega \to \infty$, the phase lag approaches −90° and the amplitude goes to zero. For instance, Figure 5.6 shows the response of the system to an input value of $v_{in} = 1\text{V}$ and a frequency of $\omega = 1.57\,\text{rad/s}$ for two values of τ: 0.1 s and 2.0 s. In the first case, Figure 5.6a, $\tau\omega = 0.157$, and in the second case, Figure 5.6b, $\tau\omega = 3.14$. In both plots, the solid line is the forcing function, or the input voltage, and the dashed line is the output voltage response. Only in the first case does the output keep up with the input. To understand these observations, we return to the response behavior if the forcing function is a step function. Here, it takes the system a time of 5τ to reach steady state. Unlike a step function change, however, a sinusoid is constantly changing value. We can use its period T as a measure of how quickly the sinusoid is changing. If the system can respond much faster than the period ($\tau \ll T$), we expect the response to keep up with input. By comparison, as the time constant approaches the period, the response cannot keep up with the input.

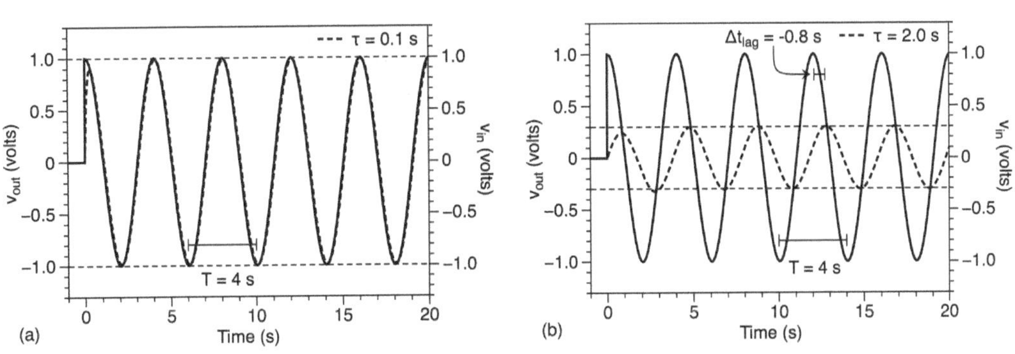

FIGURE 5.6 Rcsponse (dashed line) to an input cosine wave (solid line) of frequency $\omega = 1.57\,\text{rad/s}$, which corresponds to a period of $T = 4\,\text{s}$, for two different time constants (a) $\tau = 0.1\,\text{s}$ and (b) $\tau = 2.0\,\text{s}$ In both cases, the gain is $K = 1.0$.

For example, in Figure 5.6 the period of the input sinusoid with a frequency $\omega = 1.57\,\text{rad/s}$ is $T = 4\,\text{s}$. In Figure 5.6a, the time constant is $\tau = 0.1\,\text{s}$, and after a short transient time of 0.5 s, the response overlaps the input. In Figure 5.6b, where the time constant is $\tau = 2.0\,\text{s}$, the amplitude of the response is attenuated by 70% and lags the input by 0.8 s. From Equation 5.44, the phase lag ϕ is $\tan^{-1}(-3.14) = -1.26\,\text{rad}$, or equivalently -72.3°. To translate the phase lag into a lag time Δt_{lag} between the input and response, we use the following relationship:

$$\Delta t_{lag} = \frac{T\phi(rad)}{2\pi} \tag{5.48}$$

which, for this problem, is

$$\Delta t_{lag} = \frac{4\,\text{s} \times (-1.26\,\text{rad})}{2\pi} = -0.8\,\text{s} \tag{5.49}$$

We note that in the limit that $\tau\omega \gg 1$, the steady state response can be approximated as

$$v_c(t \geq 5\tau) \approx \frac{V_{in}}{\tau\omega}\cos(\omega t + \phi) \tag{5.50}$$

where $\phi \approx \frac{\pi}{2} - \frac{1}{\tau\omega}$. It is now clearly evident that if $\tau\omega$ is sufficiently large, the inertia inherent to the system prevents it from responding on the time scale (or period) of the signal, and therefore the system is much less responsive than when $\tau\omega \ll 1$.

5.1.4 Transfer Functions for First-Order ODEs

Section 4.4 presented the topic of transfer functions. It is straightforward to show that if the initial condition of the dependent variable is zero, $y(0) = 0$, the transfer function relating the dependent variable to the forcing function is given by

$$\frac{Y(s)}{X(s)} = \frac{K}{\tau s + 1} \tag{5.51}$$

where $X(s)$ is the Laplace transform of the forcing function. If the initial condition is not zero, then we follow the method presented in Section 4.6 to develop deviation variables forcing the initial condition to zero. In this case, the resulting transfer function is the same as Equation 5.51, except that the dependent variable and forcing function are now expressed in terms of the deviation variables, which was shown in Example 4.17. The poles of a transfer function are the values of s for which the denominator goes to zero. The values of the poles determine whether the system response is stable. If all the poles are negative, the transients will die off leading to a stable response. If any of the poles are positive, the transient will grow in time leading to an unstable response. The pole of a first-order ODE is always $s = -1/\tau$. Hence, as long as $\tau > 0$, the response will be stable.

5.2 SECOND-ORDER SYSTEMS

Just as with first-order systems, second-order systems can be written in the so-called standard form:

$$\tau^2\frac{d^2y}{dt^2} + 2\tau\zeta\frac{dy}{dt} + y(t) = Kf(t) \tag{5.52}$$

where, compared to Equation 5.2,

- $y(t)$ is the dependent variable.
- t is the dependent variable
- $f(t)$ is the forcing function.
- $\tau = \sqrt{a/c}$ is a characteristic time, and it always has units of time.
- $\zeta = \dfrac{b}{2\sqrt{ac}}$ is the damping ratio, and it is always dimensionless.
- $K = 1/c$ is the system gain and has units of the dependent variable over the forcing function.

The following are some examples of models described by first-order differential equations:

- Mass–spring–damper:

$$m\frac{d^2y}{dt^2} + \mu_f \frac{dy}{dt} + ky(t) = F_{ext} \tag{5.53}$$

- Series RLC circuit:

$$L\frac{d^2v_c}{dt^2} + R\frac{dv_c}{dt} + \frac{v_c(t)}{C} = \frac{v_{in}}{C} \tag{5.54}$$

Similar to first-order linear ODEs, to rewrite any of the preceding equations in standard form, first move all terms with the dependent variable to the left-hand side of the equation and move the forcing function to the right-hand side of the equation. According to Equation 5.52, the dependent variable must not have a preceding coefficient. Hence, after rearranging, divide all terms by the coefficient immediately preceding the dependent variable.

Also, similar to linear first-order ODEs, the solution can be broken into the sum of the transient (complimentary solution) and the steady state (particular solution) responses:

$$y(t) = y_c + y_p$$

The primary difference between first-order and second-order ODES, however, is that the transient response of a first-order ODE is always $c_1e^{-t/\tau}$. A second-order ODE, by comparison, can undergo three types of transient responses: overdamped, critically damped, and underdamped. The associated characteristic equation is

$$\tau^2 r^2 + 2\zeta\tau r + 1 = 0 \tag{5.55}$$

TABLE 5.3
List of Relevant Variables, Coefficients, and Expressions for the Time Constant and Gain for the Models in Equations 5.53–5.54

Model	Dependent Variable	Independent Variable	a	b	c	f(t)	τ	ζ	K
Mass-spring-damper	y	t	m	μ_f	k	F_{ext}	$\sqrt{\frac{m}{k}}$	$\frac{\mu_f}{2\sqrt{mk}}$	$\frac{1}{k}$
Series RLC	v_c	t	L	R	$\frac{1}{C}$	v_c	$\sqrt{LC}$	$\frac{R}{2}\sqrt{\frac{C}{L}}$	1

TABLE 5.4
Complementary Solutions to Second-order ODEs

	Determinant $\Delta = \zeta^2 - 1$	Damping ratio ζ	Complementary solution
Overdamped	$\Delta > 0$	$\zeta > 1$	$c_1 e^{r_1 t} + c_2 e^{r_2 t}$
Critically Damped	$\Delta = 0$	$\zeta = 1$	$c_1 e^{rt} + c_2 t e^{rt}$
Underdamped	$-1 \le \Delta < 0$	$0 \le \zeta < 1$	$e^{\alpha t}\left[c_1 \sin(\omega t) + c_2 \cos(\omega t)\right]$

TABLE 5.5
Parameters for the Complementary Solutions to Second-Order ODEs in Terms of the Undamped Frequency ω_0 and the Damping Ratio ζ

	Parameters
Overdamped	$r_1, r_2 = -\omega_0\left(\zeta \pm \sqrt{\zeta^2 - 1}\right)$
Critically Damped	$r = -\omega_0$
Underdamped	$\alpha = -\omega_0\zeta,\ \omega = \omega_0\sqrt{1 - \zeta^2}$

which has two roots. The type of transient response is determined by the determinant, as listed in Table 5.4 The associated parameters to the complementary solutions are given in Table 5.5 in terms of the undamped frequency $\omega_0 = 1/\tau$.

A system that is underdamped has a natural frequency $\omega = \omega_0\sqrt{1-\zeta^2}$, which is the frequency that the system wants to vibrate or oscillate at when first disturbed. A special case is $\zeta = 0$, such as an undamped harmonic oscillator or an idealized spring with no energy loss, wherein the system exhibits constant-amplitude oscillations with a constant frequency if perturbed. The phenomenon of resonance occurs when a forced (or external) vibration is close to the system's natural frequency. Critically damped and overdamped systems do not display a natural frequency and hence cannot undergo resonance.

5.2.1 Response to a Step Function

Here we consider

$$\tau^2 \frac{d^2 y}{dt^2} + 2\tau\zeta \frac{dy}{dt} + y(t) = K\left(\Delta F_{ext} \cdot u(t) + F_{ext,0}\right) \tag{5.56}$$

Like the first-order case, a step change in the forcing function of a damped second-order ODE leads to the same particular solution (See Equation 5.14), namely,

$$y(\infty) = KF_{ext} \tag{5.57}$$

where $F_{\text{ext}} = \Delta F_{ext} + F_{ext,0}$. This is because if a forcing function is a constant value, then its steady state solution must also be a constant value as all derivatives vanish when $t \to \infty$.

Example. Confirm that the particular solution is $y_p = KF_{ext}$ for the response of a second-order ODE to a step function $F_{ext} = \Delta F_{ext} \cdot u(t) + F_{ext,0}$

As the forcing function is a constant value $F_{ext} = \Delta F_{ext} + F_{ext,0}$ when $t \geq 0$, we use as the guess function for the particular solution $y_{p,guess} = a_0$ where a_0 is the undetermined coefficient. To find a_0, we determine its derivatives and substitute into Equation 5.56.

$$y'_{p,guess} = 0$$

$$y''_{p,guess} = 0$$

Hence, we are left with

$$a_0 = KF_{ext}$$

Inserting a_0 into the guess function for y_p, we get Equation 5.57.

To complete the solution to Equation 5.56, we need the complementary solution that depends on the damping ratio ζ according to Table 5.4. The general solution of the response of a second-order ODE to a step function is given in Table 5.6.

Figure 5.7 shows the response to a step change in the forcing function from a constant value of zero to 10 N. The initial conditions are $y(0) = 0$ and $y'(0) = 0$. Here, the characteristic time is

TABLE 5.6
General Solution to a Step Change in the Forcing Function to a Second-Order ODE Where the Forcing Function is $F_{ext} = \Delta F_{ext} \cdot u(t) + F_{ext,0}$

	General Solution
Overdamped	$y(t) = c_1 e^{-\omega_0 t\left(\zeta + \sqrt{\zeta^2 - 1}\right)} + c_2 e^{-\omega_0 t\left(\zeta - \sqrt{\zeta^2 - 1}\right)} + KF_{ext}$
Critically Damped	$y(t) = (c_1 + tc_2)e^{-\omega_0 t} + KF_{ext}$
Underdamped	$y(t) = e^{-\omega_0 \zeta t}\left[c_1 \sin\left(\omega_0 t\sqrt{1-\zeta^2}\right) + c_2 \cos\left(\omega_0 t\sqrt{1-\zeta^2}\right)\right] + KF_{ext}$
Undamped	$y(t) = c_1 \sin(\omega_0 t) + c_2 \cos(\omega_0 t) + KF_{ext}$

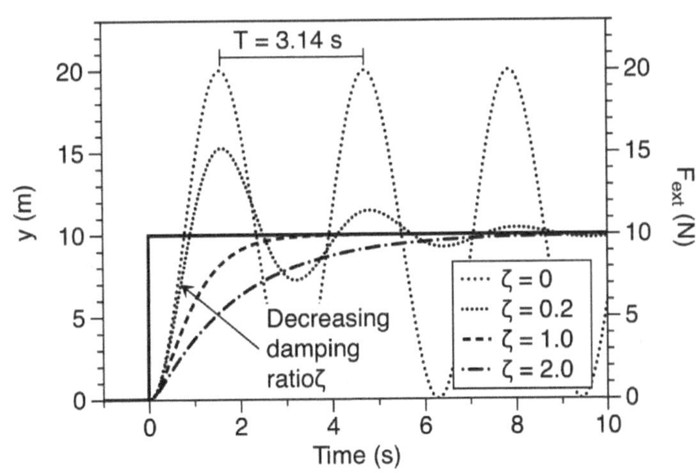

FIGURE 5.7 Response (dashed lines) to a step change (solid line) in the forcing function from 0 to 10 N for a range of damping ratios. The characteristic time is $\tau = 0.5\,\text{s}$ and the gain is $K = 1\,\text{m}/(\text{N} \cdot \text{s})$.

$\tau = 0.5\text{s}$, and the gain is $K = 1\text{m}/(\text{N}\cdot\text{s})$. Depending on the value of ζ, four types of stable responses are observed: undamped, underdamped, critically damped, and over damped.

5.2.1.1 Undamped Response

When $\zeta = 0$, the response is a perpetual sinusoidal wave that does not change or vary in its amplitude. This is a special case of an underdamped response. The frequency of this sinusoid is

$$\omega_0 = \frac{1}{\tau} \tag{5.58}$$

which is sometimes referred to as the undamped natural frequency. Note that the frequency has units of radians/time. The associated period of the oscillation is the time that it takes to complete an entire cycle and is

$$T = 2\pi\tau \tag{5.59}$$

As the characteristic time is $\tau = 0.5\text{s}$, the associated period is $T = 2\pi \times 0.5\text{s} = 3.14\text{s}$, which is shown in Figure 5.7. Note that the average of the undamped response equals the product of KF_{ext}. Here $KF_{ext} = 10\text{m/s}$, and the response oscillates between 0 and 20m/s.

5.2.1.2 Underdamped Response

When $0 < \zeta < 1$, the response also has a natural frequency which oscillates about KF_{ext}; however, the oscillations decay until KF_{ext} is reached. The natural frequency is smaller than in the undamped case; hence, the period of the oscillation is also longer:

$$\omega = \omega_0\sqrt{1-\zeta^2} \tag{5.60}$$

$$T = \frac{2\pi\tau}{\sqrt{1-\zeta^2}} \tag{5.61}$$

In the case of the undamped case, there are also four other parameters of interest: rise time, overshoot, decay ratio, and the settling time, as shown in Figure 5.8. Example values are given in Table 5.7.

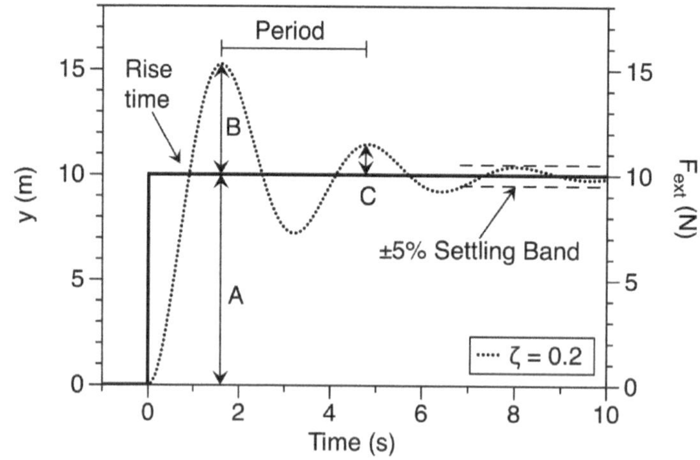

FIGURE 5.8 Response (dashed line) of an underdamped second-order ODE to a step function input (solid line) showing the rise time, period, and settling band. The ratio B/A is the overshoot, and the ratio C/B is the decay ratio.

TABLE 5.7
Second-Order Underdamped Step Response

Damping Ratio ζ	Rise time	% overshoot (B/A)	Decay ratio (C/B)
0.0	$(\pi/2)\tau$	100.0 %	1
0.215	1.8τ	50.0 %	1/4
0.344	2.0τ	29.3 %	1/10
0.707	3.3τ	4.3 %	1/500
1.0	N/A	0 %	$1/\infty$

1. **Rise time:** The rise time t_r is the time that it takes for the response to first reach the final steady state value. It is a function of τ and ζ:

$$t_r = \frac{\tau}{\sqrt{1-\zeta^2}}\left(\pi - \cos^{-1}\zeta\right)$$

2. **Overshoot:** The overshoot, sometimes referred to as the maximum overshoot, is the ratio B/A. It also can be related to τ and ζ:

$$\text{Overshoot} = \frac{B}{A} = \exp\left(\frac{-\pi\zeta}{\sqrt{\left(1-\zeta^2\right)}}\right)$$

3. **Decay ratio:** The decay ratio is the ratio of two successive peaks, so, for instance, C/B. The decay ratio is the square of the overshoot, or

$$\text{Decay Ratio} = \frac{C}{B} = \exp\left(\frac{-2\pi\zeta}{\sqrt{\left(1-\zeta^2\right)}}\right)$$

4. **Settling time:** The settling time is defined as the time required for the response to come within the prescribed band of the final steady state value and to remain in this band. Typical band limits are $\pm$ 5%, $\pm$3%, and $\pm$1%. The settling time for a band limit of $\pm$5% is approximately $3\tau/\zeta$, and the settling time for a band limit of $\pm$1% is approximately $5\tau/\zeta$.

Note that when $\zeta = 0$, the percent overshoot is 100%, and the decay ratio is 1 (meaning no decay), which is consistent with the response in Figure 5.7. As ζ approaches 1, both the overshoot and the decay ratio approach 0.

5.2.1.3 Critically Damped Response

When $\zeta = 1$, the period of oscillation diverges (it goes to infinity). In this state, the response reaches its steady state value as quickly as possible without oscillating back and forth about this value.

5.2.1.4 Overdamped Response

When $\zeta > 1$, the response is considered sluggish. Here, the solution has two distinct and real roots, r_1 and r_2, the smaller of which is known as the dominant root, as it dominates the transient response at sufficiently large values of ζ. For instance, when the damping ratio is $\zeta = 1.74$, the dominant root is

10 times larger than the nondominant root. Here, the transient associated with the nondominant root dies off quickly enough so that it does not affect the transient associated with the dominant root, and the time to reach the steady state can be approximated as $\dfrac{5\tau}{\zeta-\sqrt{\zeta^2-1}}$.

5.2.1.5 Unstable Responses

In the event that the damping ratio is negative ($\zeta < 0$), the response becomes unstable as the transient grows in time rather than dying out. There are two types of unstable systems, those that oscillate and those that do not. The division between the two is determined by the damping ratio:

- **Unstable oscillating:** $-1 < \zeta < 0$. Like the underdamped case, the roots are complex conjugates *but* with a positive real part ($\alpha > 0$). Thus, the transient will oscillate with a frequency $\omega = \omega_0\sqrt{1-\zeta^2}$; however, the amplitude will grow in time, producing an unstable response.
- **Unstable non-oscillating:** $\zeta \leq -1$. Here, the roots are distinct, real, and positive. Similar to the overdamped case, there is no oscillation, but the response grows in time as the arguments to the exponential terms are positive rather than negative.

5.2.2 Response to an Impulse and a Rectangular Pulse

Here, we consider

$$\tau^2\frac{d^2y}{dt^2}+2\tau\zeta\frac{dy}{dt}+y(t)=KP\delta(t) \tag{5.62}$$

Similar to the first-order ODE, we use a Laplace transform to solve. If the initial conditions are $y(0)=0$ and $y'(0)=0$, the Laplace transform reduces to

$$Y(s)=\frac{KP/\tau^2}{s^2+\dfrac{2\zeta}{\tau}s+\dfrac{1}{\tau^2}}=\frac{KP\omega_0^2}{s^2+2\zeta\omega_0 s+\omega_0^2} \tag{5.63}$$

The value of ζ determines the type of response. The solutions are given in Table 5.8.

Figure 5.9 shows the response of a second-order ODE to an impulse. Here, the gain $K = 0.5\,\text{m}/(\text{N}\cdot\text{s})$, the characteristic time $\tau = 0.5\,\text{s}$, and responses are given for the following cases: $\zeta = 0$ (undamped),

TABLE 5.8
Solution of a Second-Order ODE to an Impulse Defined by P with Initial Conditions $y(0) = 0$ and $y'(0) = 0$

	Solution
Overdamped	$y(t)=\dfrac{KP\omega_0}{2\sqrt{\zeta^2-1}}\left[e^{-\omega_0 t\left(\zeta-\sqrt{\zeta^2-1}\right)}-e^{-\omega_0 t\left(\zeta+\sqrt{\zeta^2-1}\right)}\right]$
Critically Damped	$y(t)=KP\omega_0^2te^{-\omega_0 t}$
Underdamped	$y(t)=\dfrac{KP\omega_0}{\sqrt{1-\zeta^2}}e^{-\omega_0\zeta t}\sin\left(\omega_0 t\sqrt{1-\zeta^2}\right)$
Undamped	$y(t)=KP\omega_0\sin(\omega_0 t)$

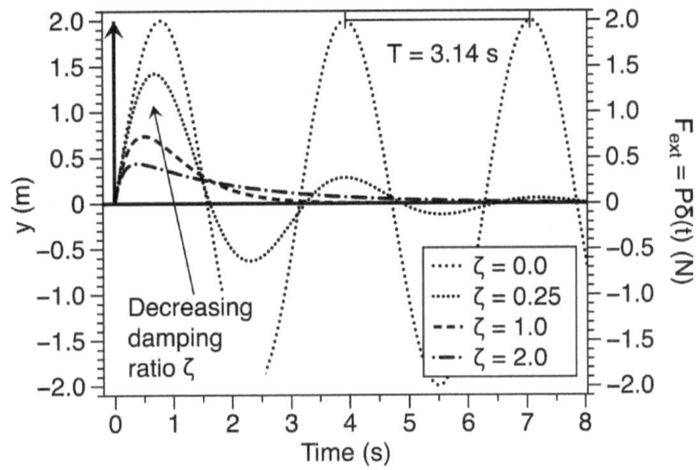

FIGURE 5.9 Response (dashed lines) to an impulse (solid line) for a range of damping ratios. The gain is $K = 0.5\,\mathrm{m}/(\mathrm{N\cdot s})$, and the characteristic time is $\tau = 0.5\,\mathrm{s}$.

$\zeta = 0.25$ (underdamped), $\zeta = 1.0$ (critically damped), and $\zeta = 2.0$ (overdamped). The momentum transferred by the impulse results in an upward movement of the mass followed by a return to the steady state value of 0 (as the impulse is zero everywhere except when $t = 0$). The maximum value of the displacement can be found by setting $dy/dt = 0$. For the undamped system, the maximum displacement is

$$y_{max} = \frac{KP}{\tau} \text{ undamped} \tag{5.66}$$

Just as we saw with the step function, the response oscillates in perpetuity with a frequency of $\omega = \omega_0$. As the damping ratio is increased but remains less than $\zeta = 1$, the response becomes underdamped, where now the oscillations eventually decay to zero. Two behaviors are noted: The maximum value of y is reduced compared to the undamped response and the oscillations have a frequency of $\omega = \omega_0\sqrt{1-\zeta^2}$. In the critically damped case ($\zeta = 1$), the response reaches $y = 0$ as quickly as possible without oscillating about this value. The maximum displacement occurs when $t = \tau$ and the maximum in y is

$$y_{max} = 0.368\frac{KP}{\tau} \text{ critically damped} \tag{5.67}$$

When the damping ratio exceeds $\zeta = 1$, the response becomes sluggish. Here, y does not rise as much, and it takes longer than the critically damped case for y to return to zero.

If the forcing function is a pulse with a duration of time a, the governing equation is

$$\tau^2\frac{d^2y}{dt^2} + 2\tau\zeta\frac{dy}{dt} + y(t) = KF_{ext}\left[u(t) - u(t-a)\right] \tag{5.68}$$

We can use the results for a step function to model the response (Table 5.6) provided that we divide the response into two parts: When the pulse is on ($0 \le t \le a$) and when the pulse is off ($t > a$). For instance, if the system is underdamped, we would use

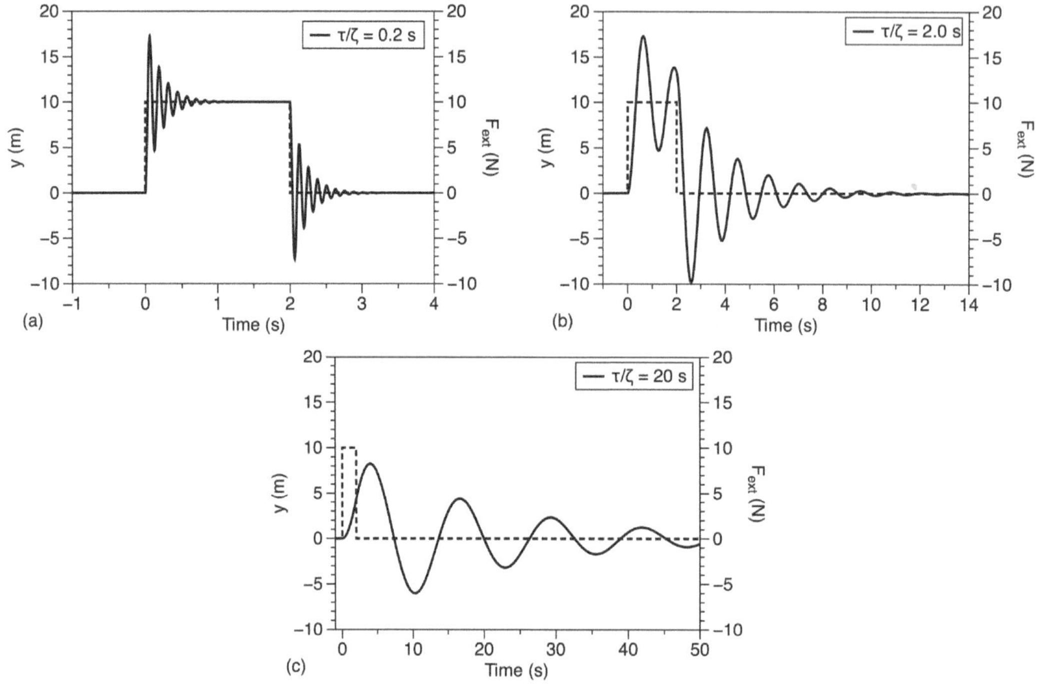

FIGURE 5.10 Response (solid line) of an underdamped second-order ODE ($\zeta = 0.1$) to a rectangular pulse (dashed line) of width $a = 2\,\text{s}$ for a range of damping ratios. The gain is $K = 1.0\,\text{m}/(\text{N}\cdot\text{s})$.

$$y_{on}(t) = e^{-\omega_0\zeta t}\left[c_1 \sin\left(\omega_0 t\sqrt{1-\zeta^2}\right) + c_2 \cos\left(\omega_0 t\sqrt{1-\zeta^2}\right)\right] + KF_{ext} \qquad 0 \le t \le a$$
$$y_{off}(t) = e^{-\omega_0\zeta(t-a)}\left[c_1 \sin\left(\omega_0 (t-a)\sqrt{1-\zeta^2}\right) + c_2 \cos\left(\omega_0 (t-a)\sqrt{1-\zeta^2}\right)\right] \qquad t \ge a$$

where the constants of integration for y_{off} are determined by the value and slope of y_{on} at the end of the pulse; that is, $y_{off}(a) = y_{on}(a)$ and $y'_{off}(a) = y'_{on}(a)$.

We learned that in the case of a first-order ODE, when $\tau \gg a$, the response could be described by an impulse. An analogous question is: Under what conditions can a pulse input be described as an impulse for a second-order ODE? We consider an underdamped system ($\zeta = 0.1$) with an initial condition of $y(0) = 0$ and $y'(0) = 0$. Using the results from the response to a step change, the settling time for an underdamped second-order ODE is $5\tau/\zeta$ with a band limit of $\pm 1\%$, which we can use to understand whether the response reaches steady state before the trailing edge (or the end) of the pulse. If $a > 5\tau/\zeta$, the response will settle to KF_{ext}. After termination of the pulse, the response settles back to zero with oscillations that mirror the leading edge of the pulse, as shown in Figure 5.10a. Here, the width of the pulse is $a = 2\,\text{s}$ and $\tau/\zeta = 0.2\,\text{s}$. Hence, $5\tau/\zeta = 5 \times 0.2\,\text{s} = 1\,\text{s}$, and the response reaches steady state before the end of the pulse. In Figure 5.10b, when $a = \tau/\zeta$, the response is asymmetric with respect to the leading and trailing edges of the pulse. In 5.10c, $a \ll \tau/\zeta$, and the response can be approximated by an impulse.

5.2.3 Response to a Sinusoidal Signal

We now consider

$$\tau^2 \frac{d^2y}{dt^2} + 2\tau\zeta\frac{dy}{dt} + y(t) = KF_{ext}\cos(\omega t) \tag{5.69}$$

Analogous to the first-order ODE, we will guess a particular solution of the following form and use the exact same procedure as detailed in Section 5.1.3:

$$y_p = A\cos(\omega t + \phi)$$

Letting $\beta = \tau\omega$ gives us

$$\left\{KF_{ext} - A\left[\left(\beta^2 - 1\right)\cos\phi + 2\beta\zeta\sin\phi\right]\right\}\cos(\omega t)$$
$$+\left\{A\left[\left(\beta^2 - 1\right)\sin\phi - 2\beta\zeta\cos\phi\right]\right\}\sin(\omega t) = 0$$

Because $\cos(\omega t)$ and $\sin(\omega t)$ are linearly independent, they must vanish to zero identically. Solving for $\sin(\omega t)$,

$$\tan\phi = \frac{2\beta\zeta}{\beta^2 - 1}$$

From this, we arrive at

$$\sin\phi = \frac{2\beta\zeta}{\sqrt{\left(\beta^2 - 1\right)^2 + 4\beta^2\zeta^2}}$$

$$\cos\phi = \frac{\beta^2 - 1}{\sqrt{\left(\beta^2 - 1\right)^2 + 4\beta^2\zeta^2}}$$

Finally,

$$A = \frac{KF_{ext}}{\left(\beta^2 - 1\right)\cos\phi + 2\beta\zeta\sin\phi} = \frac{KF_{ext}}{\sqrt{\left(\beta^2 - 1\right)^2 + 4\beta^2\zeta^2}}$$

Hence, the particular solution is

$$y_p(t) = \frac{KF_{ext}}{\sqrt{\left(\beta^2 - 1\right)^2 + 4\beta^2\zeta^2}}\cos(\omega t + \phi) \tag{5.70}$$

where

$$\tan^{-1}\phi = \frac{2\beta\zeta}{\beta^2 - 1} \tag{5.71}$$

Here we recall that $\beta = \tau\omega = \omega/\omega_0$ where ω_0 is the natural frequency of an undamped system. Therefore, β is a ratio of the driving frequency to the undamped natural frequency. The range of β is from 0 to ∞, and $\tan^{-1}\phi$ changes sign from a positive to negative value when $\beta = 1$. The phase value is usually taken to be between 0 ($\beta = 0$) and $-180°$ ($\beta \to \infty$), so it represents a phase lag for both positive and negative values of the arctan argument.

Unlike the first-order linear ODE, which can only be attenuated, the amplitude of the response can undergo an enhancement in amplitude. This is readily realized by setting $\beta = 1$. The amplitude becomes

$$A = \frac{KF_{ext}}{2\zeta} \tag{5.72}$$

when $\zeta < 0.5$, the amplitude supersedes KF_{ext}, which occurs when the frequency of the driving force effectively couples with the energy storage modes of the system. The frequency at which the amplitude is a maximum is called the resonance frequency. To find the resonance frequency, we take the derivative of Equation 5.70 with respect to β and set the derivative to zero.

$$\frac{dy_p}{dt} = 0$$

which occurs when

$$\beta = \frac{\omega}{\omega_0} = \sqrt{1-2\zeta^2} \tag{5.73}$$

Defining this frequency as the resonant frequency ω_r, the resonant frequency in terms of ω_0 is

$$\omega_r = \omega_0\sqrt{1-2\zeta^2} \tag{5.74}$$

Recall that the natural frequency of an underdamped system is $\omega_{und} = \omega_0\sqrt{1-\zeta^2}$; hence, the resonant frequency always lies between the natural frequency of the undamped system and the natural frequency of the underdamped system.

$$\omega_0 \le \omega_r < \omega_{und} \tag{5.75}$$

According to Equation 5.74, the maximum damping ratio at which resonance is observed is

$$\zeta = \frac{1}{\sqrt{2}} \tag{5.76}$$

For systems with a very small damping ratio and a driving frequency near the resonant frequency, the steady state oscillations become excessive and can cause damage in mechanical systems.

Bode plots provide a simple way to visualize both how a system amplifies or attenuates signals of different frequencies and how it alters the phase. For instance, the Bode plot representation of Equation 5.70 is shown in Figure 5.11. Bode plots are often used in control systems and signal processing and consist of two separate plots:

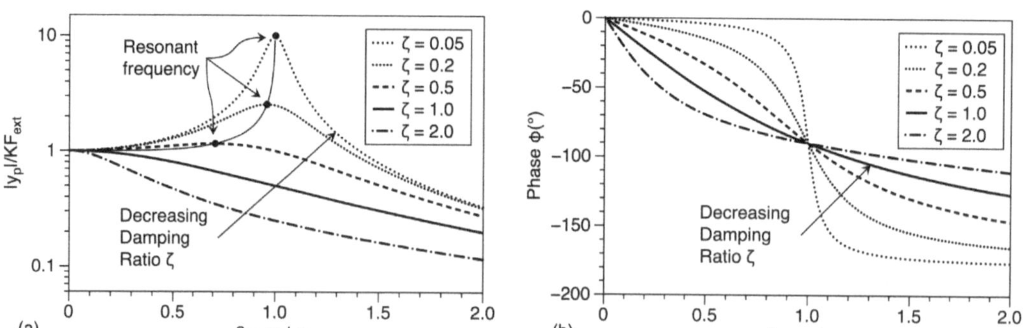

FIGURE 5.11 Bode plot representation of Equation 5.70 for a range of damping ratios. (a) The magnitude plot. (b) The phase plot. The resonant frequencies in (a) represent Equation 5.74.

- **Magnitude plot:** This shows how the magnitude (or amplitude) of the system's output changes with respect to frequency.
- **Phase plot:** This shows how the phase lag ϕ between the input and output signals varies with frequency.

To find the general solution, we need the complementary solution, which as we saw in the case of step function or impulse depends on the value of the damping ratio. If the system is underdamped, for example, the general solution is

$$y(t)=e^{-\omega_0\zeta t}\left[c_1\sin\left(\omega_0 t\sqrt{1-\zeta^2}\right)+c_2\cos\left(\omega_0 t\sqrt{1-\zeta^2}\right)\right]$$
$$+\frac{KF_{ext}}{\sqrt{\left[\left(\omega/\omega_0\right)^2-1\right]^2+4\left(\omega/\omega_0\right)^2\zeta^2}}\cos(\omega t+\phi)$$

5.2.4 Transfer Function

The transfer function for the standard form of a linear second-order differential equation is

$$\frac{Y(s)}{X(s)}=\frac{K}{\tau^2 s+2\zeta\tau s+1} \tag{5.77}$$

If the initial conditions are not zero, using deviation variables delivers the same transfer function. Just as in the case of a first-order ODE, the poles (or the values of s in which the denominator goes to zero) determine the type of transient response.

5.3 COUPLED FIRST-ORDER ODES

Coupled first-order ODEs are crucial in many fields of science and engineering, as they describe systems where multiple variables interact and evolve together over time. For instance, many physical processes involve interactions between different variables, such as in mechanics, where position and velocity are coupled through Newton's laws. In real-world systems, variables do not act in isolation. In an ecological model, for example, the population of one species affects and is affected by the population of another, leading to complex dynamics. Coupled first-order ODEs are fundamental in control theory, where the goal is to design systems that maintain desired behaviors by controlling inputs.

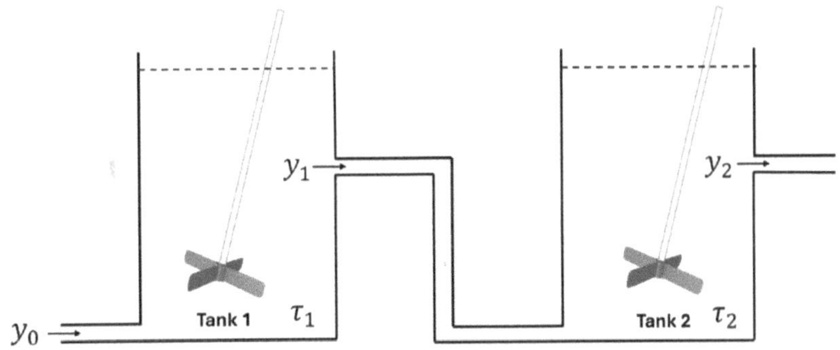

FIGURE 5.12 Two tanks in series as representative coupled first-order system.

Suppose we have two well-mixed tanks connected in a series where the mass balance on both tanks is described by a first-order ODE:

$$\tau_1 \frac{dy_1}{dt} + y_1 = K_1 y_0 \quad \text{Tank 1} \tag{5.78}$$

$$\tau_2 \frac{dy_2}{dt} + y_2 = K_2 y_1 \quad \text{Tank 2} \tag{5.79}$$

The mole fraction y_0 is the entering mole fraction (forcing function) for Tank 1, y_1 is the exiting mole fraction for Tank 1, and y_2 is the exiting mole fraction for Tank 2. Note that the dependent variable for Tank 1, y_1, is the forcing function for the mole balance on Tank 2. To express y_2 in terms of y_0 (while eliminating y_1), we take the derivative of the mass balance on Tank 2 with respect to t:

$$\tau_2 \frac{d^2 y_2}{dt^2} + \frac{dy_2}{dt} = K_2 \frac{dy_1}{dt} \tag{5.80}$$

We rearrange the mass balance on Tank 1,

$$\frac{dy_1}{dt} = \frac{K_1 y_0 - y_1}{\tau_1}$$

and substitute in Equation 5.80,

$$\tau_2 \frac{d^2 y_2}{dt^2} + \frac{dy_2}{dt} = \frac{K_2}{\tau_1}(K_1 y_0 - y_1)$$

Now we use the mass balance on Tank 2 to eliminate y_1:

$$\tau_2 \frac{d^2 y_2}{dt^2} + \frac{dy_2}{dt} = \frac{K_2}{\tau_1}(K_1 y_0) - \frac{K_2}{\tau_1}\left(\frac{\tau_2}{K_2}\frac{dy_2}{dt} + \frac{y_2}{K_2}\right)$$

Moving all dependent terms to the left-hand side of the equation,

$$\tau_2 \frac{d^2 y_2}{dt^2} + \frac{dy_2}{dt}\left(1 + \frac{\tau_2}{\tau_1}\right) + \frac{y_2}{\tau_1} = \frac{K_2}{\tau_1}(K_1 y_0)$$

and multiplying by τ_1, we find that the mass balance is second-order in terms of y_2

$$\tau_2 \tau_1 \frac{d^2 y_2}{dt^2} + (\tau_2 + \tau_1)\frac{dy_2}{dt} + y_2 = K_2 K_1 y_0 \tag{5.81}$$

Rewriting Equation 5.81 in standard form,

$$\tau_c^2 \frac{d^2 y_2}{dt^2} + 2\tau_c \zeta_c \frac{dy_2}{dt} + y_2 = K_c y_0 \tag{5.82}$$

where the subscript c refers to the characteristic time, damping ratio, and gain of the coupled system:

$$\tau_c = \sqrt{\tau_2 \tau_1} \tag{5.83}$$

$$\zeta_c = \frac{\tau_2 + \tau_1}{2\sqrt{\tau_2 \tau_1}} \tag{5.84}$$

$$K_c = K_1 K_2 \tag{5.85}$$

The characteristic time of the coupled system is the geometric mean of the individual characteristic times of the two tanks, the damping ratio is the ratio of the arithmetic mean to the geometric mean of the individual characteristic times, and the gain is the product of the gains of the two tanks. To understand the response of the system, we rewrite τ_2 in terms of τ_1

$$\tau_2 = \eta \tau_1 \tag{5.86}$$

where $0 \leq \eta < \infty$. Substituting into Equation 5.84, we find

$$\zeta_c = \frac{\tau_2 (1+\eta)}{2\sqrt{\tau_2^2 \eta}} = \frac{(1+\eta)}{2\sqrt{\eta}} \tag{5.87}$$

Figure 5.13 shows a plot of ζ versus η, where it is obvious that the damping ratio is always 1 or greater. When $\tau_1 = \tau_2$, the coupled system is critically damped, and if $\tau_1 \neq \tau_2$, the coupled system is overdamped. Interestingly, while the coupled first-order ODE leads to a second-order ODE, it will never be underdamped as long as the characteristic time τ of coupled first-order ODE is a positive, real value.

5.4 COUPLED UNDAMPED SECOND-ORDER ODES

Oscillation is the repetitive or periodic variation about a point of equilibrium. Familiar examples include a swinging pendulum, machine vibration, structure dynamics, and alternating current. Oscillations can also be used to model physically complex interactions such as those between atoms. Oscillations are found not only in mechanical systems but also in dynamic systems across virtually all fields of study. Examples include the beating of the human heart, stock market cycles, predator–prey population cycles, tone production in stringed instruments, and radio waves. Uncontrolled oscillations can lead to catastrophic consequences for buildings and bridges, excess wear in machines, and loss of control in feedback loops.

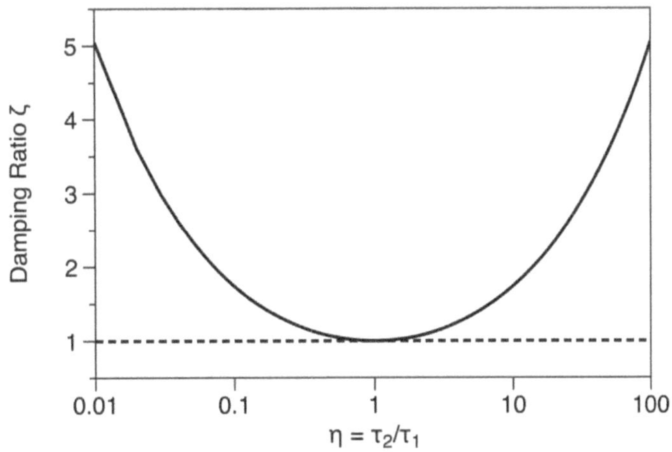

FIGURE 5.13 Damping ratio of coupled first-order ODEs in series as a function of the ratio τ_2/τ_1.

The simplest model is the mass-spring-damper, which is a lumped-element model that assumes that a physical or dynamical system can be represented by a single point mass. For a single mass, the equation of motion for this model is developed by summing the forces on the mass including all external forces. Taking x as the deviation (or variation) about the point of equilibrium, the force balance is

$$m\ddot{x} = -kx - \mu_f \dot{x} + \sum F_{ext} \tag{5.88}$$

Rewriting Equation 5.88 so that only the dependent terms are on the left-hand side,

$$m\ddot{x} + \mu_f \dot{x} + kx = \sum F_{ext}$$

In the absence of friction, $\mu_f = 0$, the force balance becomes

$$m\ddot{x} + kx = \sum F_{ext}$$

Rewriting this equation in standard form,

$$\frac{m}{k}\ddot{x} + x = \frac{1}{k}\sum F_{ext}$$

Noting that $\tau = \sqrt{m/k}$, $\zeta = 0$, and $K = 1/k$, the ODE can be written in its most generalized form:

$$\tau^2 \ddot{x} + x = K\sum F_{ext} \tag{5.89}$$

To find the natural frequency of the system, we solve for the case where $F_{ext} = 0$, which is the homogeneous solution or, equivalently, the complementary solution:

$$\tau^2 \ddot{x} + x = 0 \tag{5.90}$$

As the damping ratio is $\zeta = 0$, the general solution is (see Tables 5.4–5.5)

$$x(t) = c_1 \cos(\omega_0 t) + c_2 \sin(\omega_0 t) \tag{5.91}$$

where ω_0 is the natural frequency of the system and is $\omega_0 = 1/\tau = \sqrt{k/m}$. Hence, in the absence of external forces, if a mass–spring–damper is initially displaced by a small distance and/or given some initial velocity, it will perpetually oscillate at a single frequency $\omega_0 = \sqrt{k/m}$.

Example. If we subject a mass–spring system to an initial displacement of 1 cm with zero velocity, how will it oscillate in the absence of external forces?

The initial condition is $x(0) = 1\text{cm}$ and $\dot{x}(0) = 0$. Substituting the initial conditions into Equation 5.91 and its derivative, the constants of integration can be found:

$$x(0) = 1 = c_1 \cos(\omega_0 \times 0) + c_2 \sin(\omega_0 \times 0) \rightarrow c_1 = 1$$
$$\dot{x}(0) = 0 = -c_1\omega_0 \sin(\omega_0 \times 0) + c_2\omega_0 \cos(\omega_0 \times 0) \rightarrow c_2 = 0$$

Hence, the response after displacing the mass is

$$x(t) = 1\,\text{cm} \times \cos\left(\sqrt{\frac{k}{m}}t\right)$$

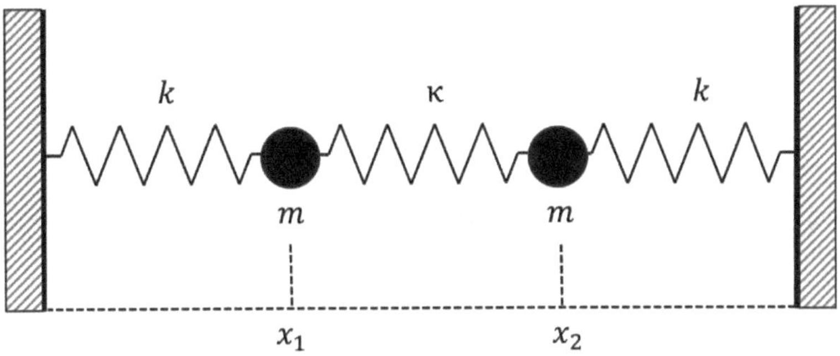

FIGURE 5.14 Mass–spring system with two coupled masses. At equilibrium, the displacement variables x_1 and x_2 are zero.

Most real systems display more than 1 degree of freedom and can be built up by placing mass–spring–dampers in series and in parallel. We will see that two oscillators coupled in a series allow for two natural frequencies, which are known as the normal modes. An illustration of the system is given in Figure 5.14, where two masses are coupled to fixed wall through springs. Let's say the masses are equivalent, and the springs between each wall and mass have a spring constant k. The spring attaching the two masses together, however, has a variable spring constant κ. Let x_1 be the displacement of the first mass from equilibrium and x_2 be the displacement of the second mass from equilibrium.

The sum of the forces on each mass due to motion can be written in the same form as Equation 5.88, where, now, the relative displacement from equilibrium between the two masses is $x_2 - x_1$.

$$m\ddot{x}_1 = -kx_1 + \kappa(x_2 - x_1) - \mu_f\dot{x}_1 + \sum F_{ext}$$
$$m\ddot{x}_2 = \kappa(x_1 - x_2) - kx_2 - \mu_f\dot{x}_2 + \sum F_{ext}$$

As we did with a single mass–spring–damper, we set the friction $\mu_f = 0$. To find the natural frequencies, we solve for the homogeneous case by setting $F_{ext} = 0$. Rewriting the previous equations,

$$m\ddot{x}_1 + \mu_f\dot{x}_1 + (k+\kappa)x_1 - \kappa x_2 = 0 \tag{5.92}$$
$$m\ddot{x}_2 + \mu_f\dot{x}_2 + (k-\kappa)x_2 - \kappa x_1 = 0 \tag{5.93}$$

where now it can be clearly seen that κ is measure of coupling between the two dependent variables x_1 and x_2. If $\kappa = 0$, there is no coupling and Equations 5.92 and 5.93 reduce to two independent equations. In the case that $\kappa \neq 0$, we can solve these two equations by adding and subtracting them:

$$m(\ddot{x}_1 + \ddot{x}_2) + \mu_f(\dot{x}_1 + \dot{x}_2) + k(x_1 + x_2) = 0$$
$$m(\ddot{x}_1 - \ddot{x}_2) + \mu_f(\dot{x}_1 - \dot{x}_2) + (k + 2\kappa)(x_1 - x_2) = 0,$$

where now the two equations are completely decoupled and can be solved independently. In the case that $\mu_f = 0$, the equations of motion reduce to

$$m(\ddot{x}_1 + \ddot{x}_2) + k(x_1 + x_2) = 0$$
$$m(\ddot{x}_1 - \ddot{x}_2) + (k + 2\kappa)(x_1 - x_2) = 0$$

Writing in standard form

$$\tau_1^2(\ddot{x}_1 + \ddot{x}_2) + (x_1 + x_2) = 0$$
$$\tau_2^2(\ddot{x}_1 - \ddot{x}_2) + (x_1 - x_2) = 0$$

where

$$\tau_1 = \sqrt{m/k}$$
$$\tau_2 = \sqrt{m/(k+2k)}$$

There are clearly now two distinct time constants as long as $\kappa \neq 0$. Both equations can be solved in the same vein as the single mass–spring system:

$$x_1 + x_2 = c_1 \cos(\omega_1 t) + c_2 \sin(\omega_1 t)$$
$$x_1 - x_2 = c_3 \cos(\omega_2 t) + c_4 \sin(\omega_2 t)$$

where $\omega_1 = \sqrt{\frac{k}{m}}$ and $\omega_2 = \sqrt{\frac{k+2\kappa}{m}}$. As the coupling coefficient is always greater than zero $(\kappa > 0)$, $\omega_2 > \omega_1$.

To recast the solution in terms of the individual displacements x_1 and x_2, we just simply add and subtract the previous solutions:

$$2x_1 = c_1 \cos(\omega_1 t) + c_2 \sin(\omega_1 t) + c_3 \cos(\omega_2 t) + c_4 \sin(\omega_2 t)$$
$$2x_2 = c_1 \cos(\omega_1 t) + c_2 \sin(\omega_1 t) - c_3 \cos(\omega_2 t) - c_4 \sin(\omega_2 t)$$

Recognizing that c_1, c_2, and so on are constants of integration, we can divide each side by a factor of 2 and then recast each constant $c_1 = c_1/2$, $c_2 = c_2/2$, and so on:

$$x_1(t) = c_1 \cos(\omega_1 t) + c_2 \sin(\omega_1 t) + c_3 \cos(\omega_2 t) + c_4 \sin(\omega_2 t) \tag{5.94}$$
$$x_2(t) = c_1 \cos(\omega_1 t) + c_2 \sin(\omega_1 t) - c_3 \cos(\omega_2 t) - c_4 \sin(\omega_2 t) \tag{5.95}$$

Therefore, in general, with any arbitrary displacement of both masses, both masses with oscillate about their equilibrium position with two distinct frequencies, a slow mode ω_1, and a fast mode ω_2. If the coupling constant is the same as the spring constant, $\kappa = k$, the frequency of the fast mode is three times the frequency of the slow mode.

Example. Find the system response of the coupled mass-spring system in the absence of an external force for the following three cases:

- Case 1: if the masses are displaced equally in the same direction
- Case 2: if the masses are displaced equally but in opposite directions
- Case 3: if only one of the masses is displaced

We will first solve for the arbitrary displacements of the two masses considering $\kappa = k$. Let's say the first mass is displaced by a distance A and the second mass is displaced by fA, where is f is the fraction that the second mass is displaced with respect to the first mass. The fraction f can be negative (meaning that the second mass is displaced in the opposite direction as the first mass), zero (meaning that the second mass is not displaced), and positive (meaning the second mass is displaced in the same direction as the first mass). If the initial velocity of both masses is zero, the initial conditions become

$$x_1(0) = A,\ \dot{x}_1(0) = 0 \quad \text{Mass 1}$$
$$x_2(0) = fA,\ \dot{x}_2(0) = 0 \quad \text{Mass 2}$$

which allows us to determine all four constants of integration in Equations 5.94 and 5.95:

$$x_1(0) = c_1 + c_3 = A$$
$$x_2(0) = c_1 - c_3 = fA$$

The first derivatives are written as

$$\dot{x}_1(t) = -c_1\omega_1\sin(\omega_1 t) + c_2\omega_1\cos(\omega_1 t) - c_3\omega_2\sin(\omega_2 t) + c_4\omega_2\cos(\omega_2 t)$$
$$\dot{x}_2(t) = -c_1\omega_1\sin(\omega_1 t) + c_2\omega_1\cos(\omega_1 t) + c_3\omega_2\sin(\omega_2 t) - c_4\omega_2\cos(\omega_2 t)$$

which after substituting $t = 0$, become

$$\dot{x}_1(0) = c_2\omega_1 + c_4\omega_2 = 0$$
$$\dot{x}_2(0) = c_2\omega_1 - c_4\omega_2 = 0$$

If we sum the last two equations,

$$2c_2\omega_1 = 0.$$

Hence, $c_2 = 0$. If $c_2 = 0$, then $c_4 = 0$. The only remaining constants are c_1 and c_3. They are, respectively,

$$c_1 = (A + fA)/2$$
$$c_3 = (A - fA)/2$$

Hence, the solution is

$$x_1(t) = \frac{A}{2}\left[(1+f)\cos(\omega_1 t) + (1-f)\cos(\omega_2 t)\right] \tag{5.96}$$
$$x_2(t) = \frac{A}{2}\left[(1+f)\cos(\omega_1 t) - (1-f)\cos(\omega_2 t)\right] \tag{5.97}$$

- **Case 1**: $f = 1$, which is equivalent to displacing both masses the same distance in the same direction. The coefficient $(1-f)$ in Equations 5.96 and 5.97 is zero and $x_1(t) = x_2(t)$:

  $$x_1(t) = x_2(t) = A\cos(\omega_1 t)$$

  This is the symmetric oscillation mode. Both masses will move in unison and in the same direction. They also oscillate at the frequency of the slow mode $\omega_1 = \sqrt{\frac{k}{m}}$.
- **Case 2:** $f = -1$, which is equivalent to displacing both masses the same distance but in opposite directions. The coefficient $(1+f)$ in Equations 5.95 and 5.96 is zero and $x_1(t) = -x_2(t)$:

  $$x_1(t) = -x_2(t) = A\cos(\omega_2 t)$$

  This is known as the antisymmetric mode. Both masses will move in unison but in opposite directions. They also oscillate at the frequency of the fast mode $\omega_1 = \sqrt{\frac{3k}{m}}$.
- **Case 3:** $f = 0$, which is equivalent to displacing the first mass but not the second. The response will be a mixture of both the slow and fast modes, and unlike Cases 1 and 2, the masses do not move in unison with respect to one another.

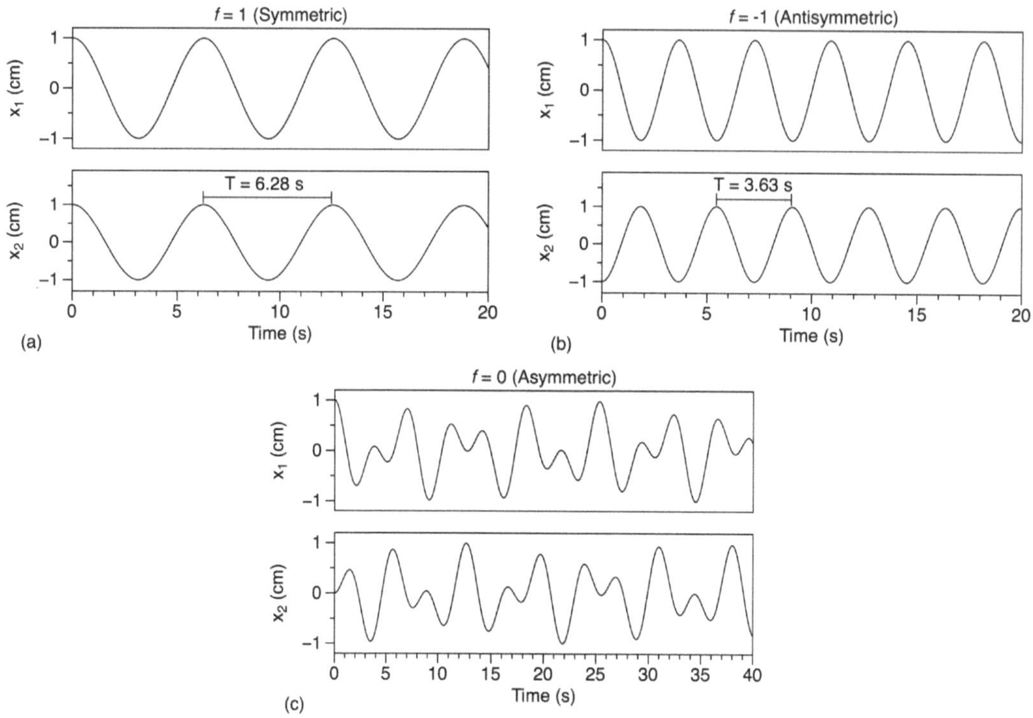

FIGURE 5.15 Three types of responses of coupled masses to the three cases of initial displacement: $f=+1$, $f=-1$, and $f=0$. In all cases, the amplitude A is 1 cm.

$$x_1(t)=\frac{A}{2}\left[\cos(\omega_1 t)+\cos(\omega_2 t)\right]$$

$$x_2(t)=\frac{A}{2}\left[\cos(\omega_1 t)-\cos(\omega_2 t)\right]$$

The three cases are shown graphically in Figure 5.15. The parameters used are $A=1\text{ cm}$, $m=0.5\text{ kg}$, and $k=0.5\text{ N/m}$. The period of the symmetric and antisymmetric cases is given by Equation 5.36, or

$$T=\frac{2\pi}{\omega}$$

The frequency of the slow mode is $\omega_1=\sqrt{\frac{k}{m}}$, and the frequency of the fast mode is $\omega_2=\sqrt{\frac{3k}{m}}$. To find the associated periods of the symmetric and antisymmetric responses,

$$T_{f=1}=2\pi\sqrt{\frac{m}{k}}=2\pi\sqrt{\frac{0.5\text{ kg}}{0.5\text{ kg/s}^2}}=6.28\text{ s} \qquad \text{slow mode}$$

$$T_{f=-1}=2\pi\sqrt{\frac{m}{3k}}=2\pi\sqrt{\frac{0.5\text{ kg}}{3\times 0.5\text{ kg/s}^2}}=3.63\text{ s} \qquad \text{fast mode}$$

5.4.1 Weakly Coupled Oscillators and Beats

In the limit that $\kappa\to 0$, the fast frequency ω_2 approaches the slow frequency ω_1 such that beats emerge in the output. Beats are important as they can generate frequencies well below the normal mode frequencies. A beat is an interference pattern between two periodic waves with slightly different frequencies. In sound generation, for instance, this is perceived by the ear as a variation in volume.

To understand the rise of such a phenomenon in the coupled mass-spring systems, let only the first mass be displaced from equilibrium such that $f = 0$. The solution to the equations of motion is

$$x_1(t) = \frac{A}{2}\left[\cos(\omega_1 t) + \cos(\omega_2 t)\right]$$
$$x_2(t) = \frac{A}{2}\left[\cos(\omega_1 t) - \cos(\omega_2 t)\right]$$

Using the trigonometric relations,

$$\cos(\omega_1 t) + \cos(\omega_2 t) = 2\cos\left(\frac{\omega_1 + \omega_2}{2} t\right)\cos\left(\frac{\omega_1 - \omega_2}{2} t\right)$$
$$\cos(\omega_1 t) - \cos(\omega_2 t) = 2\sin\left(\frac{\omega_1 + \omega_2}{2} t\right)\sin\left(\frac{\omega_1 - \omega_2}{2} t\right)$$

The displacements in x_1 and x_2 become

$$x_1(t) = \frac{A}{2}\cos(\omega_1 t) + \frac{A}{2}\cos(\omega_2 t) = \mathrm{A}\cos\left(\frac{\omega_1 + \omega_2}{2} t\right)\cos\left(\frac{\omega_1 - \omega_2}{2} t\right)$$
$$x_2(t) = \frac{A}{2}\cos(\omega_1 t) - \frac{A}{2}\cos(\omega_2 t) = \mathrm{A}\sin\left(\frac{\omega_1 + \omega_2}{2} t\right)\sin\left(\frac{\omega_1 - \omega_2}{2} t\right)$$

If we define the average frequency as

$$\omega_{avg} = \frac{\omega_1 + \omega_2}{2}$$

and the beat frequency as

$$\omega_{beat} = \frac{\omega_1 - \omega_2}{2}$$

the displacements x_1 and x_2 can be expressed as

$$x_1(t) = A\cos(\omega_{avg} t)\cos(\omega_{beat} t)$$
$$x_2(t) = A\sin(\omega_{avg} t)\sin(\omega_{beat} t)$$

Both of these expressions can be characterized as a carrier wave with frequency ω_{avg} whose amplitude A is modulated by an envelope with a frequency of ω_{beat}. If $\omega_{beat} \ll \omega_{avg}$, the period of the envelope is much larger than the period of the carrier wave, as shown in Figure 5.16.

Example. Calculate the duration of a beat given that $\omega_1 = 1.0\,\mathrm{rad/s}$ and $\omega_2 = 1.1\,\mathrm{rad/s}$.

The frequency of the beat envelope is

$$\omega_{beat} = \frac{\omega_1 - \omega_2}{2} = \frac{1.0 - 1.1}{2}\,\mathrm{rad/s} = -0.05\ \mathrm{rad/s}$$

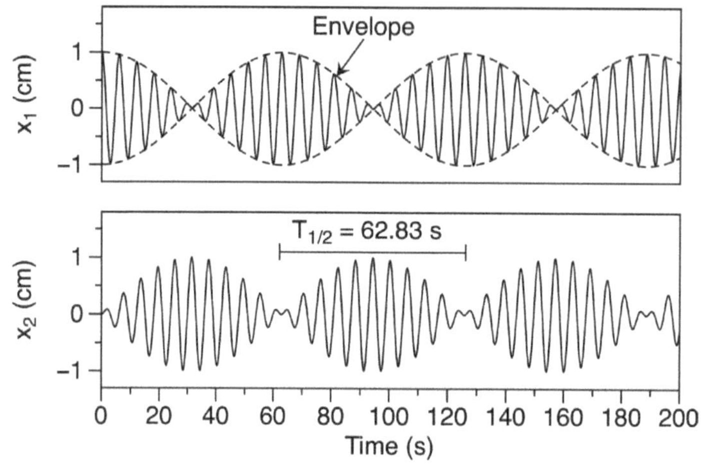

FIGURE 5.16 Response of weakly coupled oscillators with $f = 0$.

and the period of a beat envelope is (ignoring the negative sign)

$$T_{beat} = \frac{2\pi}{\omega_{beat}} = \frac{2\pi}{0.05\ \text{rad/s}} = 125.66\ \text{s}$$

There are two beats in the envelope; hence, the duration of a beat is one-half the period, or

$$T_{1/2} = \frac{T_{beat}}{2} = \frac{125.66\ \text{s}}{2} = 62.83\ \text{s}$$

5.5 EXAMPLES

This section presents four example problems concerning concepts presented in this chapter.

Example 5.1

Consider the mass-spring-damper system shown in Figure 5.17. The governing ODE (see section 6.3) is

$$m\frac{d^2x}{dt^2} + P\frac{dx}{dt} + kx(t) = f(t) \tag{5.98}$$

where the mass of the block is $m = 10\,\text{kg}$ and the damping factor of the dashpot is $P = 20\,\text{N}\cdot\text{s/m}$. It is desired to size the spring so that when the forcing function undergoes a step change from 0 N to 20 N, or $f_1(t) = 20\cdot u(t)$, the response will be critically damped. From Table 5.4, we know that the response is critically damped when the damping ratio is $\zeta = 1$. Hence, what is the value of k in order for the damping ratio to be 1?

We note that the governing ODE is given in coefficient form. To rewrite it in the form given in Equation 5.52,

$$\tau^2\frac{d^2x}{dt^2} + 2\tau\zeta\frac{dx}{dt} + x(t) = Kf(t)$$

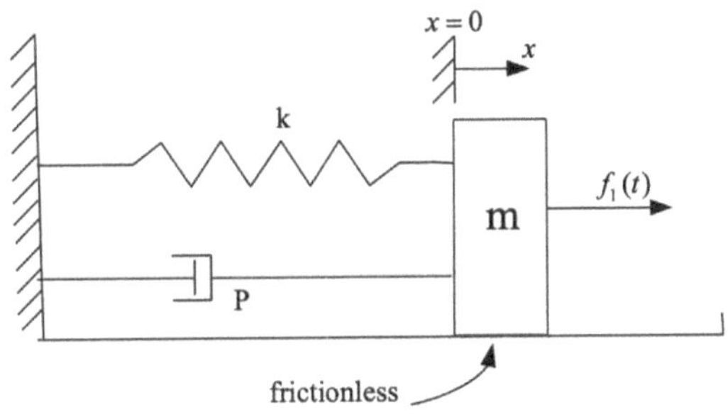

FIGURE 5.17 Mass-spring-damper system.

we divide through by k

$$\frac{m}{k}\frac{d^2x}{dt^2}+\frac{P}{k}\frac{dx}{dt}+x(t)=\frac{1}{k}f(t)$$

Hence,

$$\tau^2=\frac{m}{k}$$
$$2\tau\zeta=\frac{P}{k}$$
$$K=\frac{1}{k}$$

To find the damping ratio ζ,

$$\zeta=\frac{P}{2\tau k}=\frac{P}{2k}\sqrt{\frac{k}{m}}=\frac{P}{2\sqrt{km}}$$

which is the same as in Table 5.3, except here $P=\mu_f$. Solving for k,

$$k=\frac{P^2}{4\zeta^2 m}=\frac{(20\ \text{N}\cdot\text{s/m})^2}{4\times 1^2\times 10\ \text{kg}}=10\frac{\text{N}}{\text{m}}$$

If $k>10\frac{\text{N}}{\text{m}}$, the system becomes underdamped and if $k<10\frac{\text{N}}{\text{m}}$, the system becomes overdamped. Does this make sense? Well, from the definition of the damping ratio, as the spring constant increases, the damping ratio is reduced for a given value of P. There are two things to note. First, as the spring constant increases, the gain decreases, and the steady state displacement will also decrease. Second, as the system becomes stiffer, it cannot as effectively dissipate energy through damping. To understand this, the applied force will always partition between the three terms on the left-hand side of Equation 5.98, which are collectively the acceleration force, the damping force, and the elastic force. In other words,

$$F_{acceleration}=m\frac{d^2x}{dt^2}$$

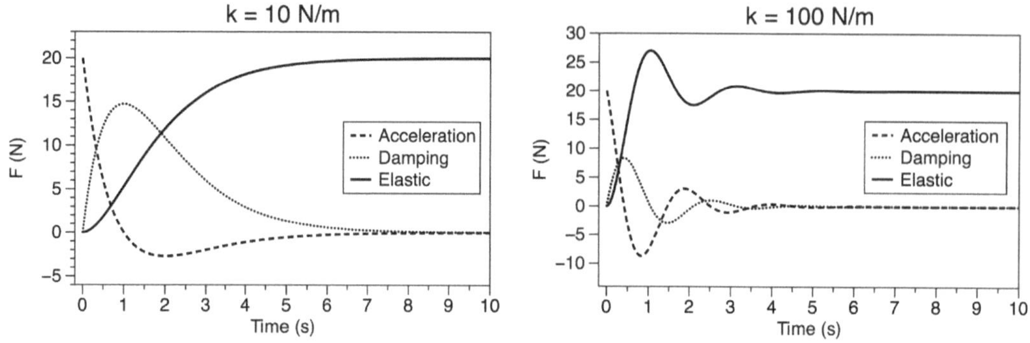

FIGURE 5.18 Individual forces associated with the left-hand side of Equation 5.98 for the mass–spring–damper system. All three forces must sum to the externally applied force, which is 20 N for both examples.

$$F_{damping} = \mathrm{P}\frac{dx}{dt}$$
$$F_{elastic} = kx$$

We know that from Equation 5.57 that when the mass comes to a rest, its steady state displacement will be $x(\infty) = KF_{ext}$ and the elastic force equals the external force $F_{elastic} = F_{ext}$. As the mass has ceased all motion (i.e. $d^2x/dt^2 = 0$ and $dx/dt = 0$) both $F_{acceleration} = 0$ and $F_{damping} = 0$.

In Figure 5.18, we plot the individual components of the force ($F_{acceleration}$, $F_{damping}$, and $F_{elastic}$) for the critically damped case ($k = 10\,\text{N/m}$). At $t = 0$, the mass is accelerated with a force of 20 N. During its approach to steady state, there is sufficient time for damping to adequately slow down the motion so that both $F_{acceleration}$ and $F_{damping}$ approach 0 as $F_{elastic}$ approaches F_{ext}. If the spring constant increases, the system stiffness increases, and damping can no longer sufficiently slow down the motion before $F_{elastic}$ reaches F_{ext}. If $k = 100\,\text{N/m}$, for example, when $F_{elastic}$ first reaches 20 N, $F_{acceleration}$ and $F_{damping}$ have equal (but nonzero) values of opposite signs. It takes several oscillations for the damping to diminish $F_{acceleration}$ and $F_{damping}$ to zero.

It should be emphasized that this finding has nothing to do with the nature of the forcing function; it only has to do with the system itself. Perhaps more importantly, it gives critical insight into how the response of system is related to its properties. If, for example, the spring constant is increased by a factor of 2, the damping coefficient will have to be increased by a factor of $\sqrt{2}$ to maintain the same type of response.

Example 5.2

Chapter 10 shows that the voltage across the capacitor (v_C in volts) in the circuit shown in Figure 5.19 is described by the following model

$$LC\frac{d^2v_c}{dt^2} + RC\frac{dv_c}{dt} + v_c = v_s \tag{5.99}$$

with a forcing function (supplied voltage) $v_s = 20\,\text{V}\cdot u(t)$, and the initial conditions are $v_c'(0) = 0$ and $v_c(0) = 0$. The resistor is $R = 2\Omega$, and the inductor is $L = 0.25\,\text{H}$. It is desired to size the capacitor to obtain a voltage response with a one-quarter decay ratio.

From Section 5.2.1.2, the decay ratio is

$$\text{Decay Ratio} = \frac{C}{B} = \exp\left(\frac{-2\pi\zeta}{\sqrt{(1-\zeta^2)}}\right)$$

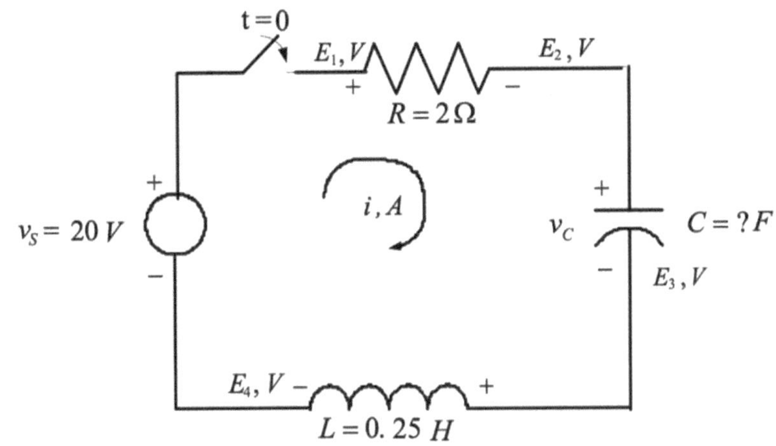

FIGURE 5.19 Electrical circuit.

Setting

$$\exp\left(\frac{-2\pi\zeta}{\sqrt{(1-\zeta^2)}}\right) = \frac{1}{4}$$

and solving for ζ, we find that $\zeta = 0.215$. Comparing Equation 5.99 to the standard form,

$$\tau^2 \frac{d^2y}{dt^2} + 2\tau\zeta\frac{dy}{dt} + y(t) = Kf(t)$$

$$\tau^2 = LC$$

$$2\tau\zeta = RC$$

$$K = 1$$

To find the damping ratio (as shown in Table 5.3),

$$\zeta = \frac{RC}{2\tau} = \frac{RC}{2}\frac{1}{\sqrt{LC}} = \frac{R}{2}\sqrt{\frac{C}{L}}$$

Hence,

$$C = \frac{4\zeta^2 L}{R^2} = \frac{4 \times 0.215^2 \times 0.25\text{H}}{(2\Omega)^2} = 0.012\text{F}$$

Example 5.3

Referring to Figure 5.20, suppose it is necessary to obtain the model that describes how the input variable, $I(t)$, affects the output variable, $O(t)$; the term *gpm* stands for gallons per minute and *ppm* for parts per million. You do not know much about the system, and therefore, you cannot develop the model starting from a physical law. What you can do, however, is manipulate the input, and you decide to change it in a step form and watch what happens to the output. The graph in Figure 5.20 shows the response to the step change. Can you obtain the model from this response? If so, provide the model with the value of all the terms.

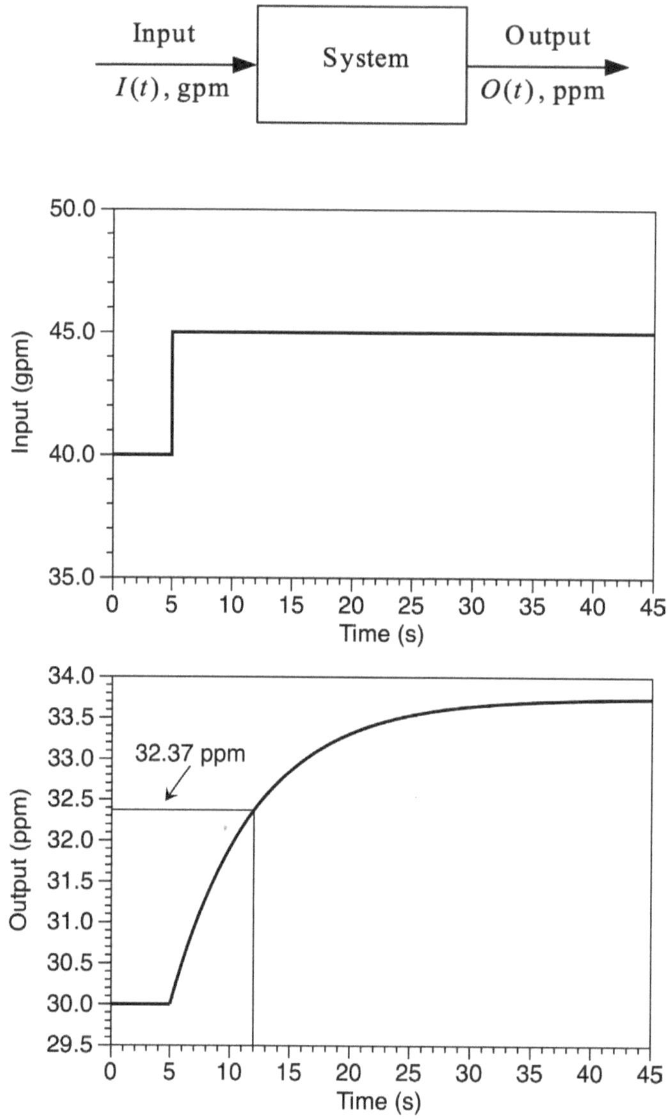

FIGURE 5.20 Information for Example 5.3.

The response of the system to the step change in input shows the steepest slope happening at the very beginning. Therefore, we can say that the mathematical model relating the output variable to the input variable is given by a first-order differential equation, or

$$\tau \frac{dO(t)}{dt} + O(t) = KI(t)$$

As we presented, the system gain provides the change in output per unit change in input or, using Equation 5.18 in this example,

$$K = \frac{\Delta O}{\Delta I} = \frac{(33.75 - 30)\,\text{ppm}}{(45 - 40)\,\text{gpm}} = 0.75 \text{ ppm/gpm}$$

To find τ, we use the results in Table 5.2. When $t=\tau$, the response has reached 63.2% of its final steady state value,

$$\frac{O(\tau)-O(0)}{O(\infty)-O(0)}=0.632$$

So, the value of O at $t=\tau$ is

$$O(\tau)=0.632(33.75\text{ ppm}-30\text{ ppm})+30\text{ ppm}=32.37\text{ ppm}$$

From Figure 5.20, we see that this value is reached in 12 minutes. However, note that the time at which the step change in input occurred is at the 5-minute indication. Therefore, the time constant τ is 7 minutes, and the model is

$$2\,\text{min}\frac{dO(t)}{dt}+\text{O}(t)=0.75\frac{\text{ppm}}{\text{gpm}}\text{I}(t)$$

The figure also shows that the initial condition is $O(t=0)=30\,\text{ppm}$. As τ is 7 minutes, we also know that the new steady state value is reached in $5\times 7\,\text{minutes}=35\,\text{minutes}$.

Example 5.4

A certain "black box" contains an RLC circuit and a power supply. It is necessary to find the size of the resistor (R), capacitor (C), and inductor (L); however, the box cannot be opened. An old file indicates that the resistor is $R=200\Omega$, but there is no indication for the capacitor and the inductor. Fortunately, there is a way to change the power supplied to the circuit and measure the voltage drop across the capacitor. The voltage drop across the capacitor in an RLC circuit is described by

$$LC\frac{d^2v_c}{dt^2}+RC\frac{dv_c}{dt}+v_c=v_s$$

where v_s is the power supplied to the circuit. A test was performed in the circuit by first making sure the capacitor was fully discharged, with the result shown in Figure 5.21. Assuming that indeed the resistance is $R=200\Omega$, provide an estimate for the values of C and L,

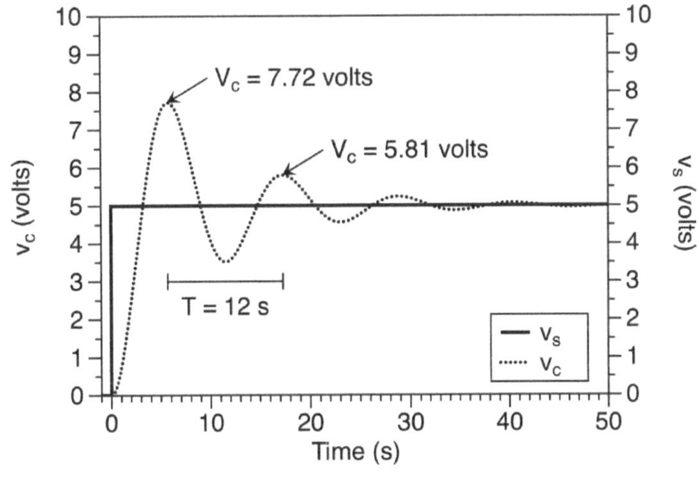

FIGURE 5.21 Information for Example 5.4.

The response of the voltage across the capacitor to the step change in supply voltage suggests a second-order underdamped model, or

$$\tau^2 \frac{d^2y}{dt^2} + 2\tau\zeta \frac{dy}{dt} + y(t) = Kf(t)$$

with $\zeta < 1$. Using the definition of decay ratio (Section 5.2.1.2) and measuring the peaks of the first and second oscillation of the response, we find

$$\text{Decay ratio} = \frac{7.72 - 5}{5.81 - 5} = 0.298$$

from which we can determine the damping ratio ζ:

$$\exp\left(\frac{-2\pi\zeta}{\sqrt{(1-\zeta^2)}}\right) = 0.298$$

which is $\zeta = 0.189$. Now, using the definition of the period, we can determine τ.

$$T = \frac{2\pi\tau}{\sqrt{1-\zeta^2}}$$

Rearranging,

$$\tau = \frac{T\sqrt{1-\zeta^2}}{2\pi} = \frac{12\,\text{s} \times \sqrt{1-.189^2}}{2\pi} = 1.88\,\text{s}$$

Now, finally $\tau^2 = LC$ and $2\tau\zeta = RC$. Recalling that $R = 200\Omega$, we can use these two expressions to find C and L.

$$C = \frac{2\tau\zeta}{R} = 0.00356\,\text{F} = 3.56\,\text{mF}$$

$$L = \frac{\tau^2}{C} = 987\,\text{H}$$

5.6 SOME CONCLUDING REMARKS (BEFORE THE SUMMARY)

Before closing this chapter, we would like to stress a couple of points.

Systems that are described by linear first-order ODEs have the following characteristic equation: $\tau r + 1 = 0$ and hence only allow 1 root, $r = -t/\tau$. This means that the transient (or the complementary solution) always decays as $e^{-t/\tau}$ toward a steady state. Hence, a first-order system cannot oscillate about its steady state value, and the amount of time for the transient to die off is generally accepted as 5τ. Linear second-order ODES have the characteristic equation $\tau^2 r^2 + 2\zeta\tau r + 1 = 0$ and allow two roots. Here, both τ and ζ characterize the transient, which may be underdamped, critically damped, or overdamped. In the case of an underdamped system, the transient will oscillate about its steady state value with a period of $T = \frac{2\pi\tau}{\sqrt{1-\zeta^2}}$ and decay as $e^{-t\zeta/\tau}$. Whether or not the transient oscillates depends entirely on the value of ζ. For values of $\zeta \geq 1$, the transient will not oscillate.

So a natural question that arises is how one can differentiate the response of a first-order ODE and the response of a critically or overdamped second-order ODE. First-order ODEs change instantaneously with a change in the forcing function, whereas the change to a second-order ODE is delayed. This is most obvious for a step change and impulse forcing functions. In the case of a step change (Figures 5.2 and 5.7), the slope of the response is steepest for the first-order ODE with the application of the step change ($t = 0$), whereas the steepest slope occurs at $t > 0$ for the second-order ODE. In the case of an impulse (Figures 5.5 and 5.9), the response of the first-order ODE jumps immediately. In contrast, the response of the second-order ODE does not jump immediately, but it continues to rise for a time.

5.7 SUMMARY

This chapter has presented in detail the response of systems described by linear first- and second-order ODEs with constant coefficients. Sections 5.1 and 5.2 showed how to convert first and second-order ODEs into a standard form so that the underlying physical model can be generalized to the same parameters. A first-order ODE is fully characterized by two parameters: The time constant τ and the system gain K. A second-order ODE is fully characterized by three parameters: the characteristic time τ, the damping ratio ζ, and the system gain K. The response consists of two parts: the transient or complementary solution and steady state or the particular solution. Whereas a first-order ODE only has one type of transient, the transient of a second-order ODE can be underdamped, critically damped, or overdamped. In the case of an underdamped response, the response fluctuates around its steady state value characterized by an overshoot and a decay ratio. After the transient has died off, you are left with the steady state response that depends on the forcing function. If the forcing function is a sinusoid, the steady state response will also be a sinusoid of the same frequency, although it is generally shifted by a phase lag. In the case of a first-order ODE, the amplitude of the response never exceeds the amplitude of the forcing function multiplied by its gain. In the case of a second-order ODE, however, the amplitude of its steady state response can become excessively large due to resonance. This only occurs in underdamped systems. The Bode plot was introduced to fully characterize the steady state response of the second-order ODE.

It is valuable to stress once more that all systems described by first-order differential equations with constant coefficients respond the same way to forcing functions. Sometimes the exact details or the shape of the forcing function is unknown, and an impulse can be substituted if the time constant is much larger than the duration of the pulse. Section 5.2 followed a similar pattern this time applied to second-order ODEs. Section 5.3 introduced coupled first-order ODEs, which lead to a second-order ODE. Section 5.4 introduced coupled second-order ODEs to show the rich range of responses that can emerge from relatively simple underlying physics. For instance, two masses connected through frictionless springs have two characteristic frequencies known as the slow mode and the fast mode.

Finally, Section 5.5 presented a few examples as to how a system can be engineered to obtain a desired response and how the response of an ODE to a forcing function can be used to identify the key parameters of the underlying system.

Problems

Problem 5.1

As presented in this chapter, the standard form of the linear first-order differential equation with constant coefficients is

$$\tau \frac{dy}{dt} + y(t) = Kf(t)$$

where for real physical systems the time constant τ is real and positive. The standard form of the linear second-order differential equation with constant coefficients is

$$\tau^2 \frac{d^2 y}{dt^2} + 2\tau\zeta \frac{dy}{dt} + y(t) = Kf(t)$$

where for real physical systems the characteristic time τ is real and positive. Answer the following qualitative questions to test your understanding:

a. Does a first-order system always reach steady state behavior after a time equal to 5τ independent of the forcing function?
b. Does a second-order system always reach steady state behavior after a time equal to 5τ independent of the forcing function?
c. For a step change from A to B in the forcing function, where A and B are constants, what is the steady state response of both a first and second-order ODE? Express the answer in terms of τ, ζ, and K, as necessary.
d. An impulse tests the transient response of an ODE, sometimes known as the natural response. Does the natural response of a first-order ODE ever oscillate? Is the same true of a second-order ODE?
e. For a sinusoidal forcing function, will a first-order ODE display resonance? Why or why not? Explain the conditions for it to have resonance.
f. For a sinusoidal forcing function, will a second-order ODE display resonance? Why or why not? Explain the conditions for it to have resonance.

Problem 5.2

A dye tracer can be used to find the residence time of a well-stirred tank with one inlet and one outlet. The mass balance of the dye tracer is a first-order ODE:

$$\frac{dm_{dye}}{dt} = \dot{m}_{in} - \dot{m}_{out}$$

Where m_{dye} is the total mass of the dye in the tank, $\dot{m}_{in}$ is the mass flow rate of dye entering the tank and $\dot{m}_{out}$ is the mass flow rate leaving the tank. We note that the relationship between the mass concentration c_{dye} in the tank and m_{dye} and $\dot{m}_{out}$ is

$$m_{dye} = Vc_{dye}$$
$$\dot{m}_{out} = \dot{V}c_{dye}$$

where V is the volume of the fluid in the tank and $\dot{V}$ is the volumetric flowrate at both the inlet and outlet.

Hence, the mass balance can be expressed as

$$V\frac{dc_{dye}}{dt} = \dot{m}_{in} - \dot{V}c_{dye}$$

a. Rewrite the mass balance in standard form. If $V = 10\,\text{m}^3$ and $\dot{V} = 0.02\,\text{m/s}$, what is the value of τ and K? Use correct units.
b. You inject the dye at a constant rate of 0.01 kg/min. What will be the steady state concentration based on the gain K that you calculated in (a) and how long will it take to reach the steady state value based on the time constant τ that you calculated in part (a)?
c. Do any of the answers in part (b) depend on the initial conditions? Explain why or why not.

d. Let's say you inject 0.01 kg of dye in 1 second. This is equivalent to a pulse input, which can be approximated as an impulse when the duration of injection is much less than the time constant τ, or

$$V\frac{dc_{dye}}{dt}+\dot{V}c_{dye}=m_{injected}\delta(t)$$

The initial condition should be the amount of dye injected divided by the volume of the tank, or $c_{dye}(0)=0.001\,\text{kg/m}^3$. Based on Equation 5.28, the response will be

$$c_{dye}(\text{t})=\frac{Km_{injected}}{\tau}e^{-(t/\tau)}$$

Show that $\frac{Km_{injected}}{\tau}$ is equal to 0.001 kg/m^3.

e. If 0.01 kg of dye is injected over 10 minutes, can you approximate the response as an impulse? Explain why or why not.

Problem 5.3

An antibiotic is injected quickly into a person with a blood volume of 5 L. The equation describing the clearance of the antibiotic is

$$\frac{dm_a}{dt}+km_a=0$$

where k is the rate of clearance and m_a is the mass of the antibiotic in the blood. Experimental data m_a versus time is shown in Figure P5.1.

a. Rewrite the ODE in standard form. What is τ in terms the clearance rate k?
b. Based on Figure P5.1, what is the initial condition for m_a?
c. Based on Figure P5.1, what is the value of the clearance rate k?
d. Based on your answer in part (c), how long will it take for the antibiotic to be cleared from the bloodstream?

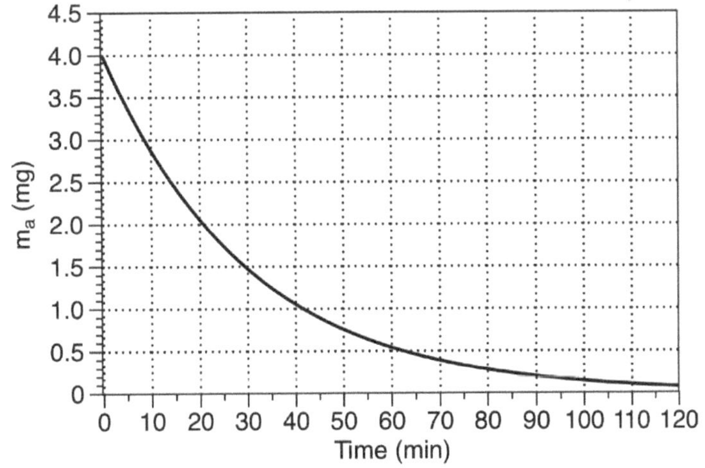

FIGURE P5.1 Experimental data of m_a versus time.

Problem 5.4

A cup of water is placed in a room in which the temperature of the water that can be described with the following ODE:

$$mc_p \frac{dT_{water}}{dt} = -UA\left(T_{water} - T_{room}\right)$$

where T_{room} is the room temperature. Let's say τ is 12 minutes. Answer the following questions:

a. If the water is originally at 50°F, and the room temperature is suddenly changed to 80°F, what will the water temperature be after 12 minutes?
b. How long will it take for the water to approximately reach steady state after the external temperature is changed to 10°F?
c. The room temperature is originally at 70°F, and then instantaneously changed to 100°F. After 1 minute, the temperature is instantaneously returned to 70°F. Sketch the forcing function and the expected behavior of T_{water}.
d. Repeat part (c) but now assume τ is 1 minute.

Problem 5.5

A pharmaceutical company has a new liquid product that must be heated in a highly specified way for product quality. The liquid is well mixed and is heated by an electric resistance with an input energy of $\dot{q}_{in}$. The model that describes the liquid temperature is

$$V\rho c_p \frac{dT}{dt} = \dot{q}_{in} - UA\left(T - T_A\right)$$

The volume of liquid in the tank is $V = 0.3\,\text{m}^3$, its density is $\rho = 1200\dfrac{\text{kg}}{\text{m}^3}$, and the heat capacity is $c_p = 450\dfrac{\text{J}}{\text{kg}\cdot{}^\circ\text{C}}$. The tank is not well insulated so there are losses to the ambient. It is known that the overall heat transfer coefficient is

$$U = 10\frac{\text{J}}{\text{s}\cdot\text{m}^2\cdot{}^\circ\text{C}}$$

and the heat transfer area is $A = 80\,\text{m}^2$. The ambient temperature is $T_A = 25^\circ\text{C}$. The units of $\dot{q}_{in}$ is J/s. The initial temperature of the liquid is 25°C and it is desired to heat the liquid to 75°C.

a. What is the value of τ for this process?
b. What is the value that $\dot{q}_{in}$ must be for the fluid to reach 75°C?
c. If $\dot{q}_{in}$ is changed instantly to the value calculated in part (b), how long will it take the fluid to reach 75°C?
d. Let's say that $\dot{q}_{in}$ changes linearly over the course of 40 minutes. Sketch the forcing function and the temperature of the fluid based only on the knowledge of τ and K.

Problem 5.6

The voltage across an *RC* series circuit is

$$RC\frac{dv_c}{dt} + v_c = v_s$$

where $v_s = 5\text{V}\cos(\omega t)\cdot u(t)$.
If $RC = 1\text{min}$, answer the following questions:

a. What is the period of the response, v_c, if the frequency of v_s is $\omega = 0.001\text{Hz}$?
b. Does the transient response depend on the frequency of the forcing function? Why or why not?
c. If $\omega = 10.0\text{Hz}$, will v_c ever reach (approximately) 5V? Why or why not?
d. If $\omega = 1\times10^{-10}$ Hz, will v_c ever reach (approximately) 5V? Why or why not?

Problem 5.7

The equation describing the current i through a series RL circuit is

$$L\frac{di}{dt} + Ri = v_s$$

where $L = 20\text{mH}$ and $R = 20\Omega$. Answer the following questions:

a. For this circuit, what is the value of τ and K?
b. If $i(0) = 0$, how long will it take for any transients to die out after the supply voltage is turned on? Does this time depend on the nature of the forcing function (i.e., whether it is a step change or a sinusoid?)
c. If $v_s = 15\text{A}\sin(\omega t)\cdot u(t)$, what is the amplitude of steady state value of i and how many seconds does it lag v_s? Here, $\omega = 1\,\text{rad/s}$.

Problem 5.8

The velocity v_y of an object in response to an applied external force F_{ext} is shown in Figure 5P.2 for two separate systems. The governing equation is

$$m\frac{dv_y}{dt} + \mu_f v_y = F_{ext}(t)$$

where m is the mass of the object and μ_f is the friction. We can easily measure the mass, but we do not know the friction.

a. Based on the response as depicted in Figure 5P.2a, what is the value of the friction μ_f? Here, $m = 0.1\,\text{kg}$.
b. Based on the response as depicted in Figure 5P.2b, what is the value of the friction μ_f? Here, $m = 5.0$ kg.

Problem 5.9

Suppose that you are planning to use an electric iron to heat a temperature-sensitive material. The governing ODE is

$$mc_p\frac{dT_{iron}}{dt} + hAT_{iron} = \dot{q}_{\text{in}} + hAT_\infty$$

The mass of the iron is $m = 1.75\,\text{kg}$, the heat capacity $c_p = 450\dfrac{\text{J}}{\text{kg}\cdot{}^\circ\text{C}}$, the exposed area $A = 0.05\,\text{m}^2$, the heat transfer coefficient $h = 20\dfrac{\text{J}}{\text{s}\cdot\text{m}^2\cdot{}^\circ\text{C}}$, and the ambient air temperature is $T_\infty = 25^\circ\text{C}$.

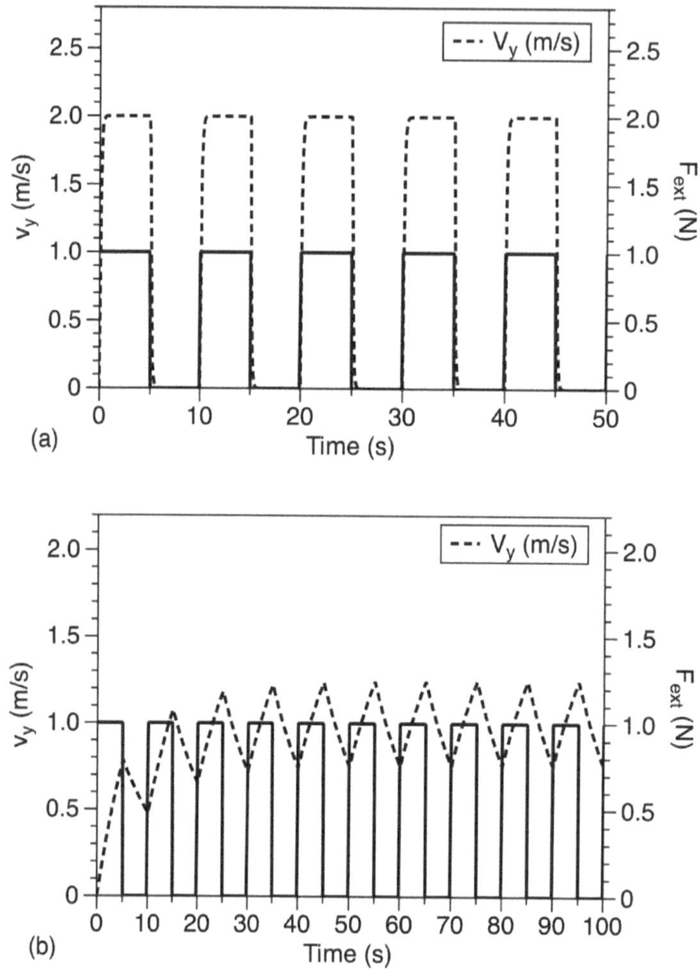

FIGURE P5.2 Experimental data of v_y versus time with the application of a force F_{ext}. The velocity v_y is represented by a dashed line, and the external force F_{ext} is represented by a solid line.

a. What is the value of τ for this problem? Use correct units.
b. As long as the temperature of the iron is at steady state, for any change in either $\dot{q}_{in}$ or T_∞, how long will it take the temperature of the iron T_{iron} to reestablish steady state?
c. In this problem, there are two forcing functions. What are they, and what are their associated gains? Use correct units.
d. Without obtaining the analytical solution, what must be the value of $\dot{q}_{in}$ in order for the iron to reach $T_{iron} = 100^\circ\text{C}$?
e. Let's say the ambient temperature drops to $T_\infty = 20^\circ\text{C}$, how much more input energy $\Delta\dot{q}_{in}$ is needed to maintain the iron at $T_{iron} = 100^\circ\text{C}$?

Problem 5.10

The following second-order ODE describes the displacement of a block y (in meters) connected to a spring.

$$m\frac{d^2y}{dt^2} + P\frac{dy}{dt} + ky = F_{ext} \cdot u(t)$$

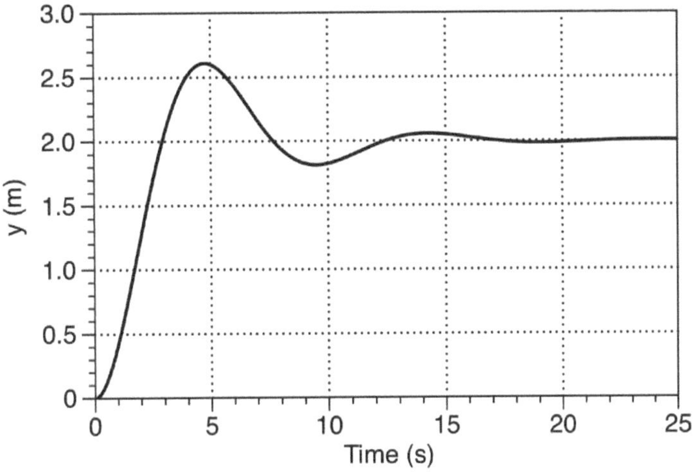

FIGURE P5.3 Displacement y of a block connected to a spring as a function of time.

The only parameter that is known in the problem is that the mass of the block is $m = 10\,\text{kg}$. An external force of $F_{ext} = 10\,\text{N}$ is applied to the block and the displacement is recorded, as shown in Figure 5P.3.

a. Express τ, ζ, and K in terms of m, P, and k.
b. Based on the response, what is the numerical value of the gain K?
c. Based on the response, what is the decay ratio?
d. Based on your answers in parts (b) and (c), what must be the values of P and k?
e. For the block to reach its steady state displacement as quickly as possible without any oscillations after applying F_{ext}, what needs to be the value of P?

Problem 5.11

The following second-order ODE describes a system that must be controlled to maintain a certain desired value of C for changes in R

$$a\frac{d^2C}{dt^2} + b\frac{dC}{dt} + (1+2G)C = 2GR(t)$$

The term G is a controller term that must be adjusted to obtain the desired performance. The terms a and b are constants.

a. Express τ, ζ, and K in terms of a, b, and G.
b. If $a = 25$ and $b = 10$, what must be the value of G so that the response is underdamped with a one-quarter (1/4) decay ratio?
c. Using this value of G, what is the value of C at the steady state if $R = 5$?
d. Let's say that R suddenly changes to 3. How long will it take for C to reach the $\pm 1\%$ settling band?

Problem 5.12

The current i through an RLC series circuit is recorded, as shown in Figure 5P.4. to a step change in the supply voltage v_s from 0 to 5 V. For this circuit, it is known that the inductance is $\text{L} = 1.25\,\text{H}$ and

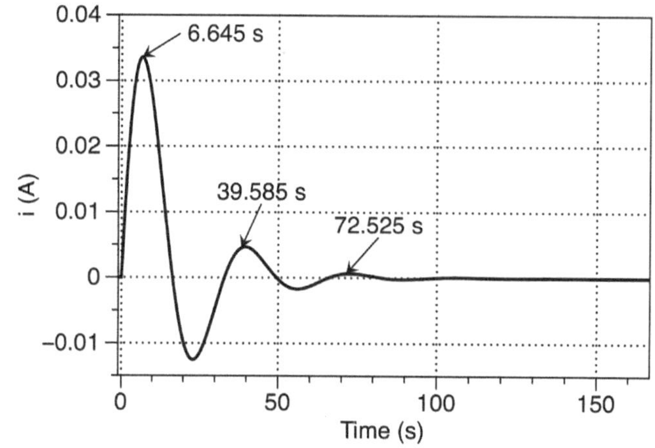

FIGURE P5.4 The current i through an RLC series circuit in response to a step change in the supply voltage v_s from 0 to 5 V.

that the capacitance is C = 20F. The following second-order ODE describes the current through the circuit

$$L\frac{d^2i}{dt^2}+R\frac{di}{dt}+\frac{i}{C}=\frac{dv_s}{dt}$$

a. Express τ, ζ, and K in terms of R, L, and C.
b. The forcing function is $v_s = 5\text{V}\cdot u(t)$. What is the derivative of this step function?
c. If the resistance R of the circuit is zero, what would be frequency of the current i in response to the step change? Would these oscillations ever decay?
d. Based on the figure, determine the actual frequency of the response.
e. Based on your answer in part (d), what is the resistance of the circuit?

Problem 5.13

The voltage across the resistor of an RLC circuit is described by the following second-order ODE:

$$LR\frac{d^2v_r}{dt^2}+RC\frac{dv_r}{dt}+v_r=RC\frac{dv_s}{dt}$$

where $v_s = A\cos(\omega t)\cdot u(t)$.

a. What is the derivative of dv_s/dt?
b. Show that the governing equation can be written as

$$\tau^2\frac{d^2v_r}{dt^2}+2\tau\zeta\frac{dv_r}{dt}+v_r=-2\tau\zeta\omega\sin(\omega t)$$

What are τ and ζ in terms of R, L, and C?

c. Using a guess function of $v_{r,guess} = \mathrm{A}\cos(\omega t + \phi)$ show that the amplitude and phase lag of the steady state response are

$$A = \frac{2\beta\zeta}{\sqrt{\left(\beta^2 - 1\right)^2 + 4\zeta^2\beta^2}}$$

$$\phi = \tan^{-1}\left(\frac{1 - \beta^2}{2\zeta\beta}\right)$$

where $\beta = \tau\omega$.

6 Mechanical Systems
Translational

This chapter focuses on mechanical systems formed by springs, masses, and dashpots (also sometimes referred to as pistons or dampers). Many systems are formed by these three elements. The chapter begins by looking at the physical law used, as well as element equations and other experimental facts necessary to develop the models.

In this and the following four chapters, many models will be developed. The solution of these models must be obtained to be of use to engineers/scientists; this is why we presented the analytical solution of differential equations early in the book. Having an equation describing the response of a system to a forcing function is useful; however, quite often a graph of the response may be even more useful. In several examples in this chapter, the analytical solution is used to graph the response. As discussed in Chapter 1, another way of obtaining the graph of the response is by solving the model using the computer – this is what we call simulation. Chapter 11 discusses simulation and shows the comparison with the graph obtained from the analytical solution. The reader is strongly encouraged to read and study Chapter 11.

6.1 MECHANICAL LAW, SYSTEMS COMPONENTS, AND FORCES

6.1.1 Mechanical Law

The physical law for the systems in this chapter is the same as that used in Chapter 2, namely, Newton's second law,

$$\vec{F} = m\vec{a}$$

where, as explained in Chapter 2, the arrows above F and a indicate that these quantities are vectors, meaning that their specifications must contain magnitude and direction and obey the parallelogram rule. Because the systems in this chapter will only move in the vertical direction, the y-direction, or the horizontal direction, the x-direction, we can write the law in its component form as

$$\sum F_y = m a_y \text{ for the vertical, or } y\text{, direction} \tag{6.1}$$

or

$$\sum F_x = m a_x \text{ for the horizontal, or } x\text{, direction} \tag{6.2}$$

6.1.2 System Components

1. *Springs.* Springs are shown in Figure 6.1.

If a force is applied to the spring (as simple as a mass hanging from the spring), the spring itself generates a force *opposing* the applied force (actually, this is Newton's third law that states: "If object A exerts a force F on object B, then object B exerts a force F on object A, equal in magnitude but opposite in direction to F, or $F_A = -F_B$"). The force generated by the spring is proportional to the displacement from some reference position:

DOI: 10.1201/9781003619161-6

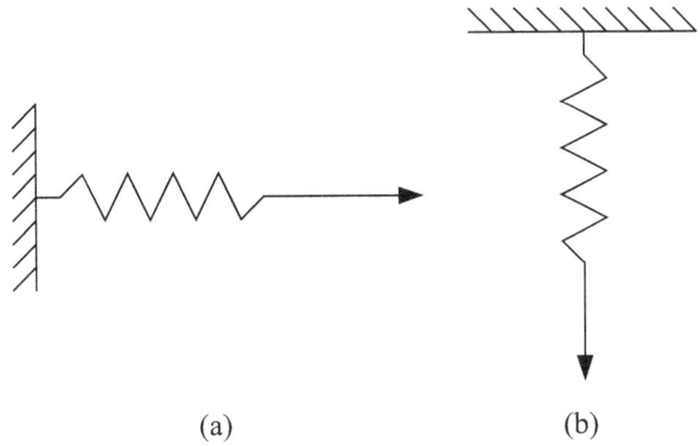

FIGURE 6.1 Springs.

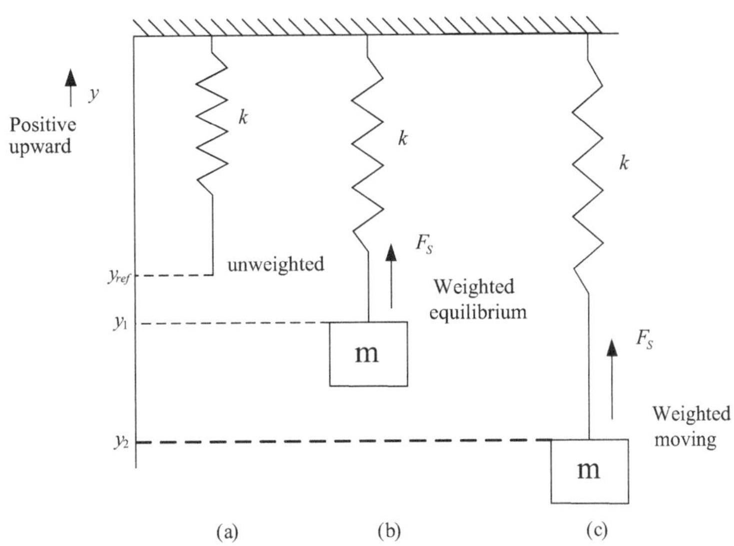

FIGURE 6.2 Forces generated by springs

$$F_{Spring} \sim \Delta\, x$$

where F_{Spring} is the force generated by the spring, the symbol $\sim$ denotes is "proportional to," and Δx is the displacement from the reference position (amount of deformation of the spring). The *reference position* is usually the natural position, unstretched length, of the spring, when it is neither compressed nor stretched, and therefore, no forces are generated; x_{ref} or y_{ref} signify this position.

To further explain this force, consider Figure 6.2; note that the "up" direction is defined as positive. At (a), the spring is at y_{ref} (y instead of x because it is in the vertical direction). At (b), a mass is connected to the spring, and the spring is somewhat extended. At this position, the force generated by the spring is upward, and therefore positive, opposing the weight force (force generated by the mass). This force is expressed by *Hooke's law*:

$$F_{Spring} = -k\, \Delta\, y = -k(y_1 - y_{ref}) \tag{6.3}$$

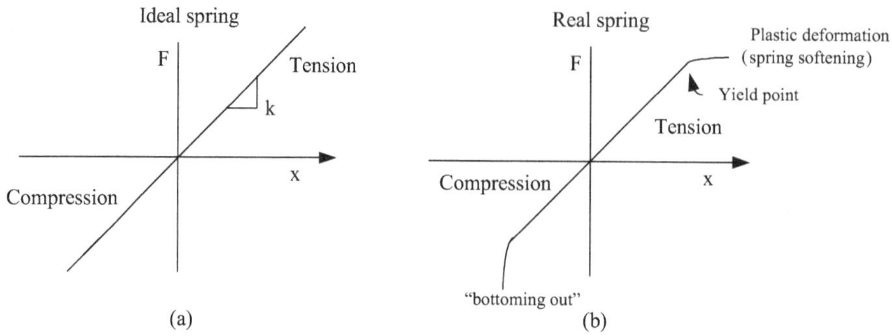

FIGURE 6.3 Ideal and real springs.

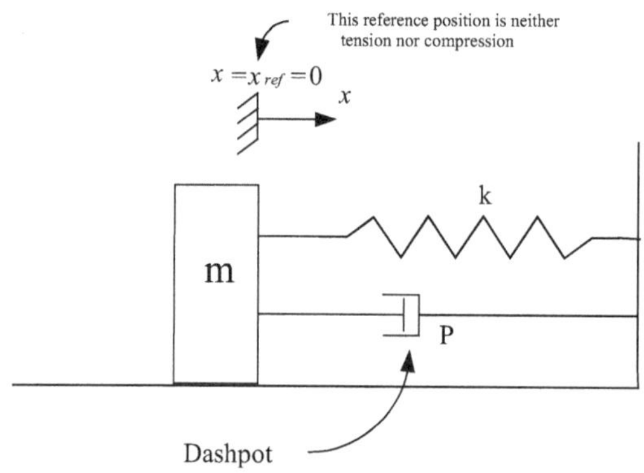

FIGURE 6.4 Spring–mass system showing dashpot.

where k is the proportionality constant, also called the *spring constant*. The negative sign is used because the term $(y_1 - y_{ref})$ is negative, but the force generated is positive, and negative times a negative gives a positive result. Likewise, the force generated by the spring at (c) is

$$F_{Spring} = -k(y_2 - y_{ref})$$

So far nothing has been said about k (can you think of its units?). A spring with a constant value of k is referred to as an *ideal spring*, and Figure 6.3a shows the relation between the displacement and the force. For *real springs*, the linear relation "breaks down" at high displacements from the reference position, as shown in Figure 6.3b.

2. *Dashpots (Pistons or Dampers).* Figure 6.4 shows a dashpot.

The force generated by a dashpot is always opposite to the direction of motion, and it is proportional to the velocity,

$$F_{Dashpot} \sim v$$

We write this force as

$$F_{Dashpot} = F_f = -Pv = -P\frac{dx}{dt} \tag{6.4}$$

Opposite to direction of motion
Damping coefficient
velocity

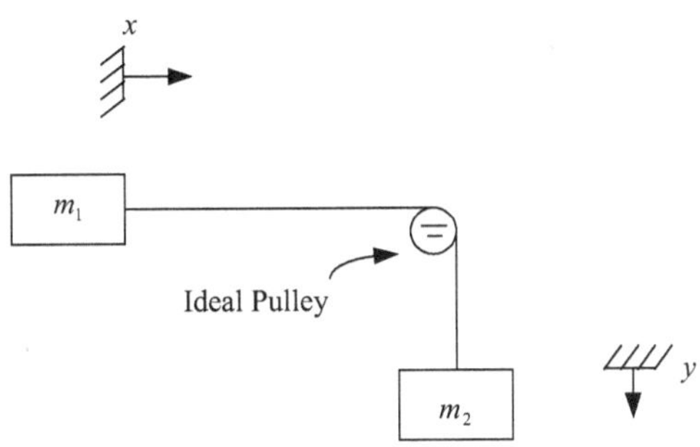

FIGURE 6.5 Mechanical system showing a pulley.

where P is the proportionality constant and referred to as the *damping coefficient*. Can you think of its units?

3. *Ideal Pulley*. Pulleys are commonly used to change the direction of motion of translational mechanical systems; Figure 6.5 shows a pulley. In this book, we will assume "ideal pulleys," meaning that they do not have friction with the connecting cable and that they are massless. Both restrictions can be relaxed for more realistic models.

6.1.3 Forces

Next, we list the forces we will encounter.

1. *Gravity force*. For systems that move vertically the force of gravity must be considered. As discussed in Chapter 2, this force is given by mg, where m is the mass of the object, and g is the local acceleration due to gravity. In developing the models, there is a need to select a convention for directions. We may, for example, choose the "up-direction" to be positive, and in that case, because the force of gravity is always pulling "downward," we express the force as

$$F_y = F_g = -\text{mg} \tag{2.3}$$

where the minus sign indicates that this force is acting down, or in the direction opposite to that chosen as positive. If the "down-direction" had been chosen as positive, then the sign in Equation 2.3 would have been positive.

2. *Damping forces*. There are several damping forces (sometimes referred to as *drag forces*) in any mechanical system; the most common ones are *friction* and *air resistance*.

   The frictional forces can be quite complex. In general, they can be divided into *dry* and *fluid* friction. To briefly explain the dry friction forces, consider Figure 6.6a showing a spring–mass system attached to a wall and sliding over a surface.

   The dry friction force is proportional to the normal force between the contact surfaces. Normal force N is the perpendicular force exerted by the supporting surface on the object it supports. Thus,

$$F_{dry\ friction} \sim N$$

or

$$F_{dry\ friction} = \mu_K N \tag{6.5}$$

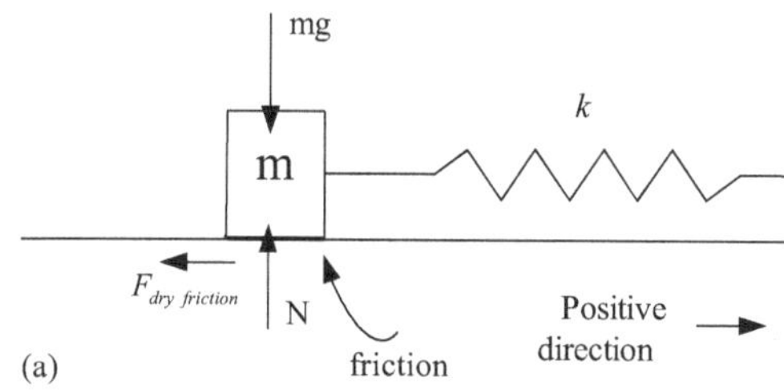

FIGURE 6.6A Dry friction force between two surfaces.

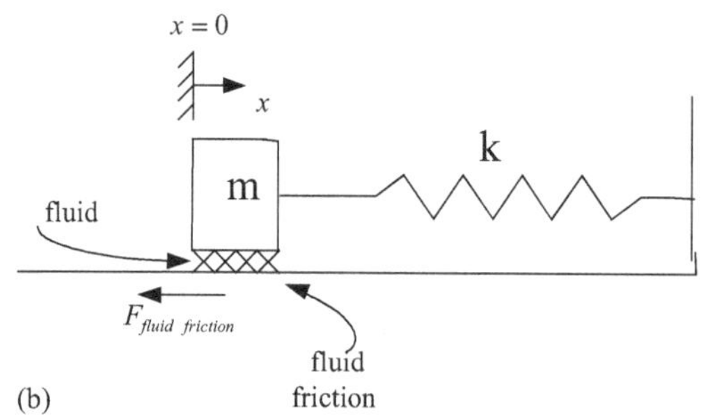

FIGURE 6.6B Fluid friction force between two surfaces.

where μ_K is the proportionality constant and called the *coefficient of kinetic friction.* $F_{dry\ friction}$ is usually independent of the speed and contact area, acts always opposite to the direction of motion, and mostly is a constant value. Equation 6.5 is sometimes referred to as *Coulomb's friction law.*

To briefly explain the fluid friction model, consider Figure 6.6b. In this case, a fluid (like a lubricant) separates the two surfaces. This fluid friction force is proportional to the velocity v:

$$F_{fluid\ friction} \sim v$$

Because $F_{fluid\ friction}$ and $F_{Dashpot}$ are both proportional to velocity, we will also use Equation 6.4 to describe the fluid friction force. $F_{fluid\ friction}$ is independent of the normal force and acts opposite to the direction of motion. Equation 6.4 is sometimes referred to as *viscous friction model.* What are the units of the damping coefficient P?

Note that the negative sign in Equation 6.4 is needed because the force due to the fluid friction (drag force) is opposite to the direction of motion. Because both the force and the velocity are vector quantities, and they are in opposite directions, the negative sign in the equation guarantees that they will have opposite signs. Figure 6.7 shows the directions and the resulting sign of the equation.

Other models are sometimes used if necessary to further approximate real systems. We will assume in this book that all the forces due to friction are of the fluid friction type (viscous forces) and that, therefore, Equation 6.4 applies.

The air resistance drag force is the same as we previously discussed in Chapter 2 and given by Equation 2.10, which happens to be the same as Equation 6.4.

6.2 TYPES OF SYSTEMS

This section looks at two different types of systems, at their differences, models, and solutions. The systems can be classified as follows:

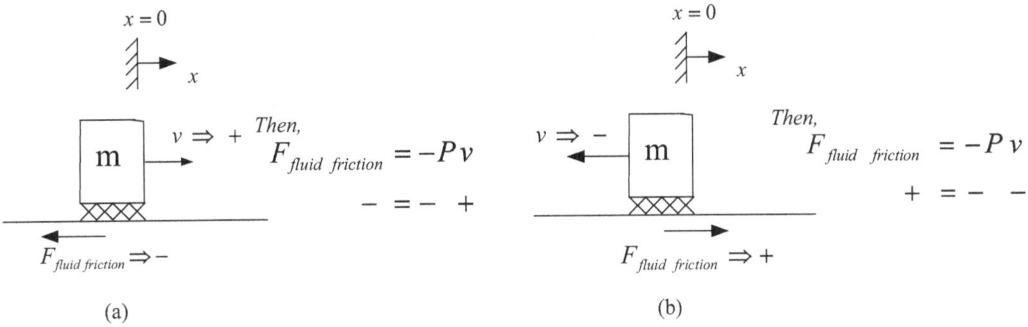

FIGURE 6.7 Demonstration of the negative sign in Equation 6.4.

1. An *undamped system* is one that does not come to a stop and just oscillating back and forth without stopping. In these systems, the energy put into setting them in motion is not dissipated.
2. A *damped system* is one that eventually comes to a rest at an equilibrium condition. In these systems, the energy put into setting them in motion is dissipated. Systems may be damped due mainly to friction (drag) forces.

6.2.1 Undamped System

Chapter 5 and Example 3.11 discussed the response of undamped systems. These discussions showed that a second-order differential equation where the coefficient of the first derivative is zero (there is no first derivative term in the equation) produces this type of response.

To further understand these systems, and any necessary condition (assumption) for their existence, let us develop the model and the solution that gives the displacement y as a function of time for the spring–mass system shown in Figure 6.8; the numerical values of the mass m of the object and of the spring constant k are both known. In this example we neglect air resistance.

Applying Newton's second law to the mass, as was done in Chapter 2,

$$\sum F_y = m\,a_y = m\frac{dv_y}{dt} = m\frac{d^2y}{dt^2} \tag{6.6}$$

1 equation 2 unknowns $\left[\sum F_y, y\right]$.

The forces are

$$\sum F_y = F_g + F_{Spring} = -m\,g + F_{Spring} \tag{6.7}$$

2 equations, 3 unknowns $\left[F_{Spring}\right]$.

The force due to the spring is given by

$$F_{Spring} = -k(y - y_{ref}) \tag{6.8}$$

3 equations, 3 unknowns.

At this moment, we are not considering y_{ref} as an unknown but still need its value and will show you shortly how to obtain it.

We now have the model for the system subject to the assumptions taken. The model is composed of three equations, but simple algebra allows us to reduce it to a single equation. Substituting Equations 6.7 and 6.8 into Equation 6.6 gives

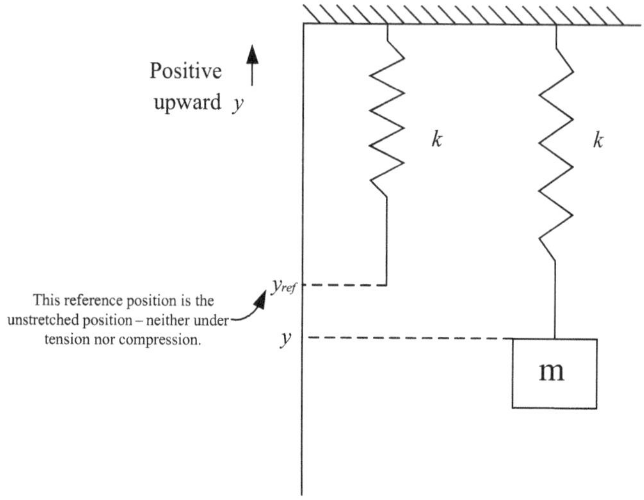

FIGURE 6.8 Mass–spring system.

$$m\frac{d^2y}{dt^2} = -mg - k(y - y_{ref}) \tag{6.9a}$$

or

$$m\frac{d^2y}{dt^2} + ky = -mg + ky_{ref} \tag{6.9b}$$

To obtain the value of y_{ref}, consider the system at equilibrium, which means that $\frac{d^2y}{dt^2} = 0$ and now define $y = 0$ at this equilibrium point; then from Equation 6.9b,

$$-mg + ky_{ref} = 0 \Rightarrow y_{ref} = \frac{mg}{k}$$

and substituting this expression for y_{ref} in Equation 6.9b,

$$m\frac{d^2y}{dt^2} + ky = 0 \tag{6.10}$$

which as shown in Example 3.11, is the *classic equation of an undamped system* (no first-order derivative term).

Following the method presented in Chapter 3 for the solution of second-order homogeneous linear differential equation with constant coefficients, the solution is

$$y = C_1 \cos\left(\sqrt{\frac{k}{m}}t\right) + C_2 \sin\left(\sqrt{\frac{k}{m}}t\right) \tag{6.11}$$

where the values of C_1 and C_2 are obtained using the initial conditions $y(0)$ and $\left.\frac{dy}{dt}\right|_{t=0}$.

For example, suppose that originally the mass is pulled down 0.1 m and at some time, call it $t = 0$, it is released. At that moment, $y(0) = 0.1$ m and $\left.\frac{dy}{dt}\right|_{t=0} = 0$ m/s. Using this information, the constants C_1 and C_2 are calculated.

The sine and cosine terms in Equation 6.11 clearly indicate that the response is continuously oscillatory. As you may remember, Example 3.11 mentioned that the frequency at which the mass will oscillate, $\sqrt{k/m}$ is called the *natural, or resonant frequency*. That is, the natural frequency is the frequency at which the system tends to oscillate in the absence of any damping and any external force. As it is written this frequency is in *radians per second*. Often frequencies are given in *cycles per second* or *hertz*, the relation between them is *hertz* $= 2\pi$ *(radians/s)*.

Because of the importance of these undamped systems in engineering design, recalling some of the previous comments on natural, or resonant, frequency and the phenomenon of resonance is worthwhile. To help us in doing so, recall the differential equation of Example 3.11,

$$m\,y'' + k\,y = F \sin \omega t \tag{6.12}$$

Following the methods learned in previous chapters,

$$\begin{aligned} y &= y_H + y_P \\ y_H &= C_1 \cos\sqrt{k/m}\,t + C_2 \sin\sqrt{k/m}t \end{aligned} \tag{6.13}$$

This corresponding homogeneous, or natural, solution indicates a continuous oscillatory response. The natural frequency is $\sqrt{k/m}$ radians/second, or $\omega_n = \sqrt{k/m}$ radians/s.

When the frequency of the forcing function is not equal to the natural frequency, $\omega \neq \omega_n$, the particular and complete solutions are

$$y_P = \frac{F}{k - \omega^2 m} \sin \omega t \tag{6.14}$$

and

$$y = C_1 \cos\sqrt{k/m}t + C_2 \sin\sqrt{k/m}t + \frac{F}{k - \omega^2 m} \sin \omega t \tag{6.15}$$

If, however, $\omega = \omega_n = \sqrt{k/m}$ radians/s, the particular and complete solutions are

$$y_P = -\frac{F}{2m\sqrt{k/m}}\, t \cos\sqrt{k/m}t \tag{6.16}$$

and

$$y = C_1 \cos\sqrt{k/m}t + C_2 \sin\sqrt{k/m}t - \frac{F}{2m\sqrt{k/m}}\, t \cos\sqrt{k/m}t \tag{6.17}$$

Note that when $\omega = \omega_n$, y becomes unbounded as t increases due to the last term. That is, the response oscillates with increasing amplitude; this type of response is referred to as *resonance*.

To show you graphically these results, assume $m = 5$ kg; $k = 80$ N/m, F = 4 N, $\omega = 2$ radians/s; $y(0) = 0$ m; $\left. dy/dt \right|_{t=0} = 0$ m/s; under these conditions Equation 6.15 becomes

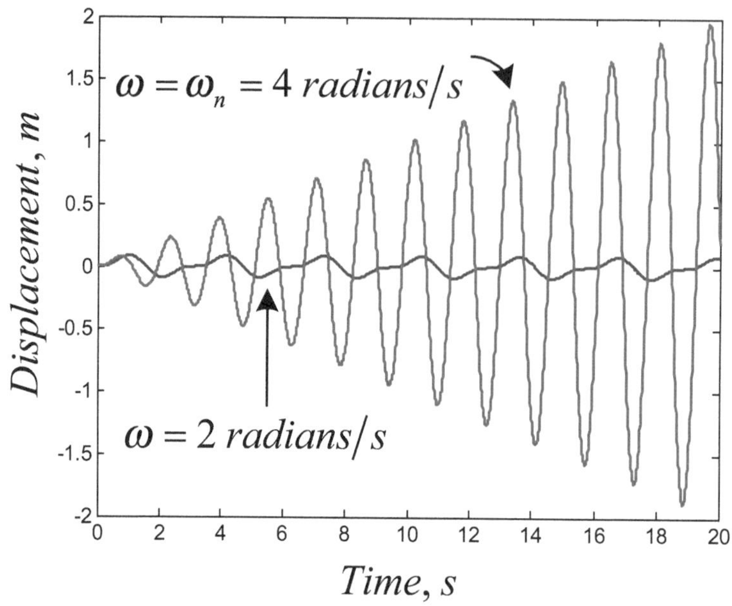

FIGURE 6.9 Response for undamped system – with and without resonance.

$$y = -0.0335\sin 4t + 0.067\sin 2t \tag{6.18}$$

and using $\omega = \sqrt{k/m} = 4\,\text{radians/s}$, Equation 6.17 becomes

$$y = 0.025\sin 4t - 0.1t\cos 4t \tag{6.19}$$

Figure 6.9 shows the responses.

Much damage may happen in mechanical systems when resonance occurs. Thus, designers of vibrating systems or structures must consider resonance when designing. Truly resonance may happen when all the following conditions exist:

a. No damping exists – in the cases we have shown, when $P = 0$.
b. A periodic forcing function acts on the system.
c. The frequency of the forcing function matches the natural frequency of the system; that is, $\omega_{forcing\ function} = \omega_{natural\ frequency\ of\ system}$.

6.2.2 Damped System

The assumption taken in developing the previous model, Equation 6.10, is that there is no air resistance. If we had considered this resistance the resulting model would have been (try it),

$$m\frac{d^2y}{dt^2} + P\frac{dy}{dt} + ky = 0 \tag{6.20}$$

The term $P\frac{dy}{dt}$ represents air resistance. In this case, the solution would have a negative exponential, and the displacement would reach a final value resulting in a damped (not undamped) response. The

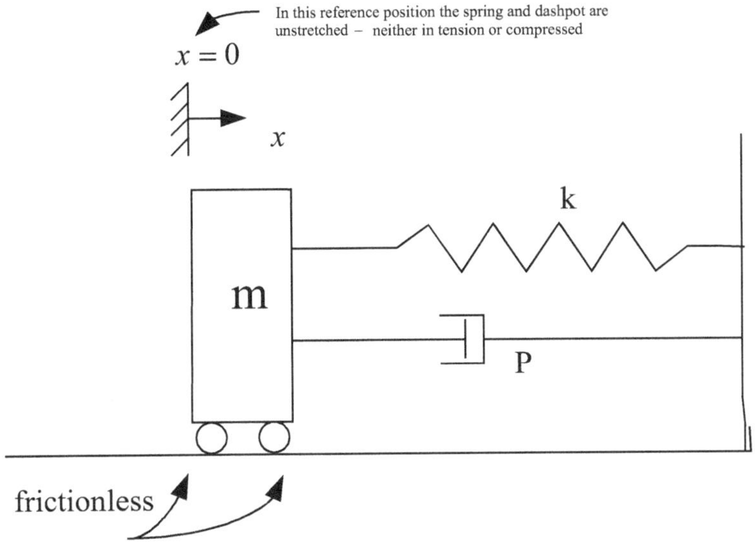

FIGURE 6.10 Spring–mass system.

energy that was put in to put it in motion would have been dissipated by the air resistance. Thus, it is under the assumption of no air resistance that this mechanical system is undamped.

As another example, let us develop the model and the solution that gives the displacement as a function of time for the mass–spring–dashpot system of Figure 6.10. The application of Newton's second law gives

$$ma_x = m\frac{d^2x}{dt^2} = \sum F_x \tag{6.21a}$$

1 equation, 2unknowns $\left[x, \sum F_x\right]$.

$$\sum F_x = F_{Spring} + F_{Dashpot} = -kx - P\frac{dx}{dt} \tag{6.21b}$$

2 equations, 2 unknowns.

Substituting Equation 6.21b into Equation 6.21a and rearranging yields

$$m\frac{d^2x}{dt^2} + P\frac{dx}{dt} + kx = 0 \tag{6.21c}$$

This equation is the model of the system. The figure indicates that the reference position ($x_{ref} = 0$) for the spring elongation is where the spring is neither under tension nor compressed. In the case of horizontal motion, gravity does not affect the motion (weight is not a factor).

Using the method presented in Chapter 3 for the solution of second order homogeneous linear differential equation with constant coefficients, depending on the numerical value of $P^2 - 4\,\text{mk}$ the solution is either

$$x = e^{\alpha t}[C_1 \cos \beta t + C_2 \sin \beta t] \tag{6.22}$$

where $\alpha = \dfrac{-P}{2\,m}$ and $\beta = \dfrac{\sqrt{|P^2 - 4\,\text{mk}|}}{2\,m}$, or

$$x = C_1 e^{\gamma_1 t} + C_2 e^{\gamma_2 t} \tag{6.23}$$

where $\gamma_1 = \dfrac{-P - \sqrt{P^2 - 4\,\text{mk}}}{2\,m}$ and $\gamma_2 = \dfrac{-P + \sqrt{P^2 - 4\,\text{mk}}}{2\,m}$,

or

$$x = C_1 e^{\sigma t} + C_2 t e^{\sigma t} \tag{6.24}$$

where $\sigma = \dfrac{-P}{2\,\text{m}}$ (repeated root). As presented in Chapter 3, Equation 6.22 is the appropriate solution when the value of $P^2 - 4\,\text{mk}$ is negative, Equation 6.23 is the solution when the value is positive, and Equation 6.24 is the solution when the value is zero. The constants C_1 and C_2 are obtained using the initial conditions.

The previous section discussed the system shown in Figure 6.8, and the resulting model is given by Equation 6.10, which we call the classic equation of undamped systems. As we are learning, the existence of purely undamped systems is an idealized case; some amount of damping is always present (such as air resistance), although it may be small. Then, the differential equation we would use, instead of Equation 6.10, looks like

$$m\,y'' + P\,y' + k\,y = F \sin \omega t \tag{6.25}$$

where P is the damping present. Assuming the damping P is small and consequently $P^2 - 4\text{mk} < 0$, the complete solution is

$$y = e^{\frac{-P}{2m}t}\left[C_1 \cos\left(\frac{\sqrt{|P^2 - 4mk|}}{2m}t\right) + C_2 \sin\left(\frac{\sqrt{|P^2 - 4mk|}}{2m}t\right)\right] + A\cos(\omega t) + B\sin(\omega t) \tag{6.26a}$$

Using the trigonometric identity

$$C\sin(\omega t) + D\cos(\omega t) = E\sin(\omega t + \theta) \quad \text{where} \quad E = \sqrt{C^2 + D^2} \text{ and } \theta = \tan^{-1}(C/D)$$

yields

$$y = e^{-\frac{P}{2m}t}\left[C_1 \cos\left(\frac{\sqrt{|P^2 - 4mk|}}{2m}t\right) + C_2 \sin\left(\frac{\sqrt{|P^2 - 4mk|}}{2m}t\right)\right] + \frac{F}{\sqrt{(P\omega)^2 + (k - \omega^2 m)^2}} \sin(\omega t + \theta) \tag{6.26b}$$

where $\theta = \tan^{-1}\left((k - \omega^2 m)/P\omega\right)$. As we have previously mentioned, θ is called the phase lag, and in this case, it is a function of ω.

As discussed in Section 3.8, the first term on the right-hand side is the transient response that will decay to zero after a long time; the second term is the final response (also referred to a "steady state" response). This final response produces a sine wave response with an amplitude and phase lag that are both functions of the forcing function frequency ω. A *pseudo resonance*, also sometimes referred to as a *practical resonance*, occurs at the frequency that produces the largest possible final amplitude. There is a problem at the end of the chapter on this topic.

6.3 D'ALEMBERT'S PRINCIPLE AND FREE-BODY DIAGRAMS

Often, engineers and scientists use D'Alembert's principle and free-body diagrams (FBDs) to model mechanical systems. As we have learned, Newton's second law is

$$\sum F_i = ma$$

Subtracting the right-hand term from both sides results in *D'Alembert's principle*,

$$\sum F_i - ma = 0 \tag{6.27a}$$

or

$$\sum F_j = 0 \tag{6.27b}$$

Thus, essentially, D'Alembert's principle states that the *sum of all forces acting on a body is zero*; this is referred to as *dynamic equilibrium*. The term $-ma = -m\frac{dv}{dt} = -m\frac{d^2x}{dt^2}$ is considered another force, and referred to as the *inertial force*, $F_{Inertial}$. This inertial force is truly not an *externally* applied force, and that is why it is sometimes referred to as pseudo force.

The FBD is a drawing showing all external forces acting on a body. The inertial force is usually shown in another diagram referred to as "mass acceleration diagram," or MAD. However, for simplicity, in this book, we will show the inertial force also in the FBD, and in this way, the drawing is the application of the D'Alembert's principle to the body.

Consider the system shown in Figure 6.11. To draw the FBD, we consider all the forces acting on the body and show them as arrows pointing in the direction in which they act; Figure 6.12 shows the diagram. Note that all the forces acting on the body, including the inertial force $-m\frac{d^2x}{dt^2}$, are opposing the applied force and the direction of movement (resisting the motion).

Applying D'Alembert's principle, Equation 6.27b,

$$\sum F_j = F_{Spring} + F_{Dashpot} + f_A(t) + F_{Inertial} = 0$$

or

$$\sum F_j = F_{Spring} + F_{Dashpot} + f_A(t) - m\frac{d^2x}{dt^2} = 0 \tag{6.28}$$

1 equation, 3 unknowns $\left[F_{Spring}, F_{Dashpot}, x\right]$.

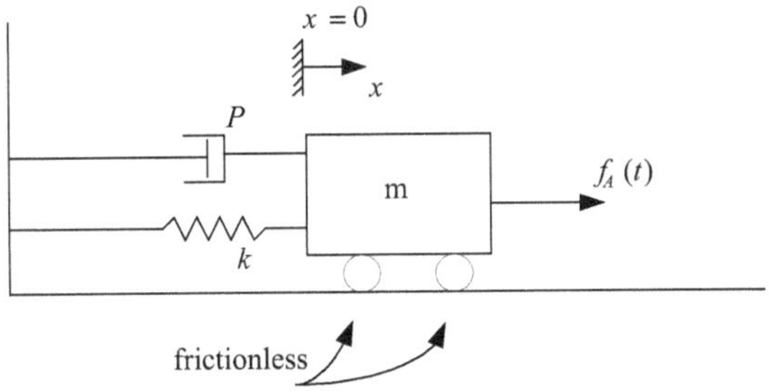

FIGURE 6.11 Mass–spring–dashpot system.

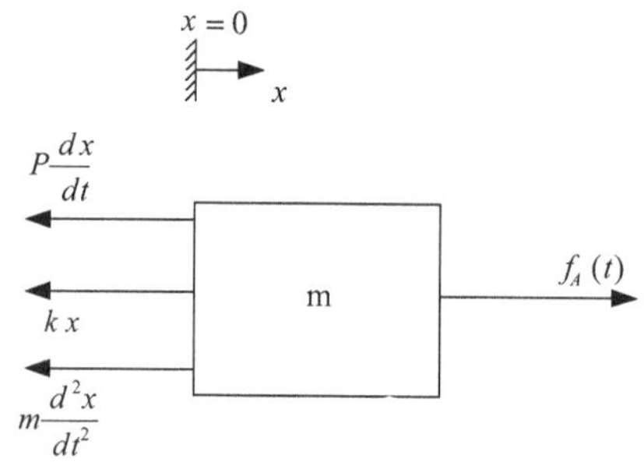

FIGURE 6.12 Free body diagram of system shown in Figure 6.11.

The applied force $f_A(t)$ is not considered an unknown because it is up to the modeler to decide on the type of forcing function; thus, it is known.

The forces generated by both the spring and the dashpot are

$$F_{Spring} = -kx \tag{6.29}$$

2 equations, 3 unknowns,

and

$$F_{Dashpot} = -P\frac{dx}{dt} \tag{6.30}$$

3 equations, 3 unknowns.

These last three equations constitute the model. Substituting the forces equations into Equation 6.28 gives

$$-kx - P\frac{dx}{dt} + f_A(t) - m\frac{d^2x}{dt^2} = 0 \tag{6.31a}$$

or

$$m\frac{d^2x}{dt^2} + P\frac{dx}{dt} + kx = f_A(t) \tag{6.31b}$$

Note that Equation 6.31a is shown in the FBD.

Assuming $m = 10$ kg, $P = 20\,\text{N}\cdot\text{s/m}$, $k = 100\,\text{N/m}$, that the force $f_A(t)$ increases from *0* to *10* N at time $t = 0$ s (or as Chapter 1 presented, $f_A(t) = 10u(t)\,\text{N}$), and that the initial conditions are $x(0) = 0$ m and $\left.\frac{dx}{dt}\right|_{t=0} = 0\,\text{m/s}$, Example 3.13 shows that the analytical solution is

$$x = 0.1 - e^{-t}\left[0.1\cos 3t + 0.033\sin 3t\right] \tag{3.41}$$

Using this solution, we graph the displacement of the cart (response of the system) when the force is applied as shown in Figure 6.13. The figure shows that the response is oscillatory, as indicated by the cosine and sine terms, and that as time becomes large the response settles at a final value as indicated by the exponential term that becomes zero as $t \to$ large; the final value is $x(\infty) = x_{final} = 0.1$ m.

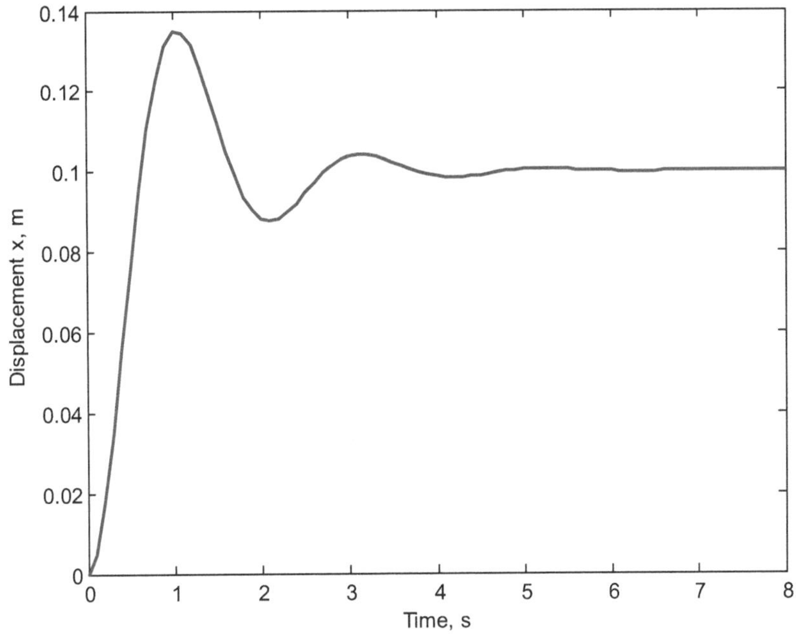

FIGURE 6.13 Response of cart shown in Figure 6.11.

This example has presented the application of D'Alembert's principle and FBD to a rather simple system. For more complex systems, such as those containing more than one object, the application calls for drawing the FBD for each object. The next section presents some of these systems.

Section 6.2 presented the modeling of an undamped and a damped system. In that section, we did not use D'Alembert's principle and FBD but, rather, just started from the diagram itself. That method and the FBD method are analogous and will obviously reach the same model. The FBD method may be best for those of us that are "visual," because it clearly shows all the forces involved. The method to use is just a matter of personal preference.

6.4 EXAMPLES

This section looks at the modeling of some more complex systems. Note that "more complex" does not necessarily mean "more difficult." There are a couple of things we need go over before attacking the examples in this section.

Hooke's law, Equation 6.3, indicates that the force generated by a spring is proportional to its displacement from a reference position (often the position where the spring is unstretched). Equation 6.29 shows that the force generated by the spring in Figure 6.11 is $F_{Spring} = -kx$, where x is the displacement from its unstretched position ($x = 0$ at the unstretched position). The other side of the spring is attached to the wall, which obviously does not move. But what if the spring is attached to two moving bodies? In this case, the displacement of each body affects the force generated by the spring itself *and* on the other body.

Figure 6.14 shows a system consisting of two blocks and connected to each other by a spring and a dashpot; block 1 is also connected to a wall by a spring.

The force generated on block 1 by the spring connected to the wall is given by $F_{1,k1} = -k_1\, x_1$, where $F_{1,k1}$ indicates the force generated on block 1 by spring 1 (denoted by k_1). The expression indicates that the force depends only on the displacement of block 1 because the wall does not move. However, the force generated by spring 2 on block 1 depends on the displacement of block 1 *and*

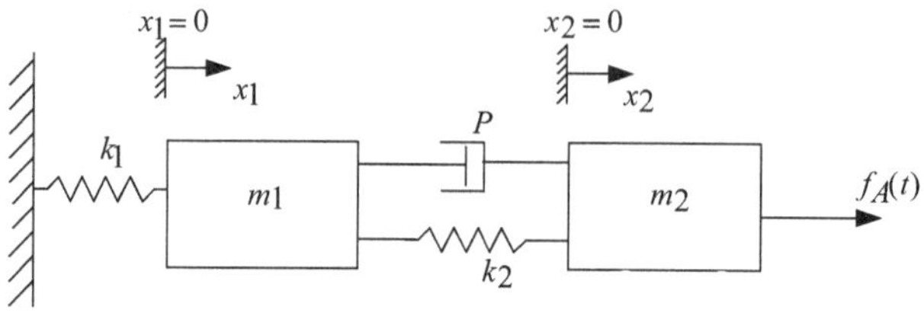

FIGURE 6.14 System consisting of two blocks.

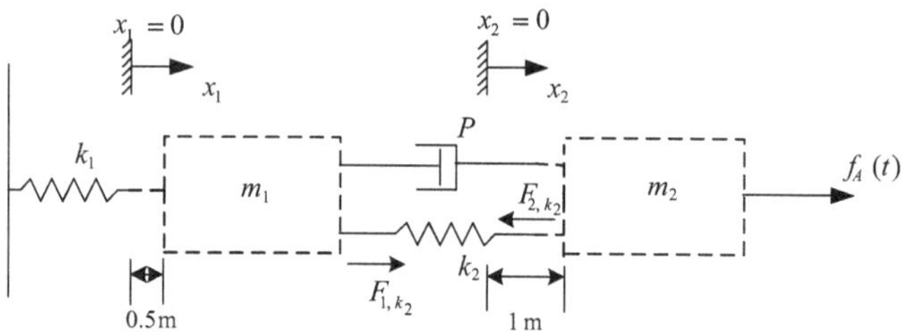

FIGURE 6.15 Displacement of blocks after applying force

on the displacement of block 2. In this case, the force generated by spring 2 on block 1 is given by $F_{1,k2} = -k_2(x_1 - x_2)$, that is, the force depends on the difference in *relative displacement* of the two blocks, $x_1 - x_2$. Think about this. Suppose that at the very beginning, before $f_A(t)$ is applied, the positions of blocks 1 and 2 are such that spring 2 finds itself in an unstretched position, meaning that the spring does not generate any force on either block. Now suppose that force $f_A(t)$ is applied and the result is that both blocks move 0.5 m to the right, the same displacement. The result is that spring 2 remains unstretched and continues generating no force on either block; the expression indicates that $F_{1,k2} = 0$. Now suppose that after applying $f_A(t)$ the result is that block 1 moves 0.5 m to the right, and block 2 moves 1 m to the right; this is shown in Figure 6.15. In this case, spring 2 generates a force in both blocks. The force generated on block 1 is to the right (the positive direction), and the force generated on block 2 is to the left (the negative direction). The sign in the expression of the force agrees with the physics:

$$F_{1,k2} = -k_2(x_1 - x_2) = -k_2(0.5 - 1) = 0.5\,\mathrm{k}_2$$

The force generated by spring 2 on block 2 also depends on the relative displacements, but this time is given by $F_{2,k2} = -k_2(x_2 - x_1)$, and $F_{2,k2}$ indicates the force generated on block 2 by spring 2 (denoted by k_2). The sign in the expression of the force again agrees with the physics,

$$F_{2,k2} = -k_2(x_2 - x_1) = -k_2(1 - 0.5) = -0.5k_2$$

where the negative sign indicates the opposite direction to that chosen as positive. Note that for block 1 the relative displacement is written as $x_1 - x_2$, and for block 2, it is written as $x_2 - x_1$. The point is that the first element in the difference is always the *position of interest*. Therefore, *the relative displacement is the difference in the position of the block of interest minus the position of the block attached.*

The forces generated by a dashpot when attached to moving bodies, or the friction between two moving bodies, is described using *relative velocities*. The force that the dashpot in

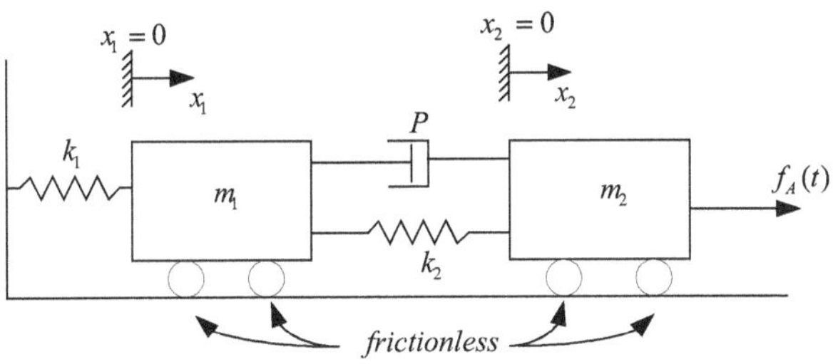

FIGURE 6.16 Frictionless carts with an external force.

Figure 6.15 generates on block 1 is given by $F_{1,P} = -P\left(\frac{dx_1}{dt} - \frac{dx_2}{dt}\right)$, and the force it generates on block 2 is given by $F_{2,P} = -P\left(\frac{dx_2}{dt} - \frac{dx_1}{dt}\right)$. Once again, the first element of the difference is related to the *position of interest*. The reader is encouraged to think through these equations of $F_{1,P}$ and $F_{2,P}$ to make sure they make physical sense.

Example 6.1

Consider the system shown in Figure 6.16 where $m_1 = 3\,\text{kg}$, $m_2 = 5\,\text{kg}$, $k_1 = k_2 = 60\,\text{N/m}$, and $P = 5\,\text{N}\cdot\text{s/m}$. The system is initially at a steady state, $\left.\frac{dx_1}{dt}\right|_{t=0} = \left.\frac{dx_2}{dt}\right|_{t=0} = 0\,\text{m/s}$, and the springs and dashpot are unstretched in their normal length, meaning $x_1(0) = x_2(0) = 0\,\text{m}$. Develop the models that describe the position of each cart; the applied force is $f_A = 5u(t)\,\text{N}$.

This is the first time we encountered two subsystems. To develop the models, take each subsystem one at a time. Consider cart 1 and applying Newton's second law,

$$\sum F_1 = m_1 a_1$$

or because we desire the position of each cart,

$$m_1 \frac{d^2 x_1}{dt^2} = \sum F_1 \tag{6.32}$$

1 equation, 2 unknowns $\left[x_1, \sum F_1\right]$.

The forces acting on cart 1 are

$$\sum F_1 = F_{1,k_1} + F_{1,k_2} + F_{1,P} = -k_1 x_1 - k_2(x_1 - x_2) - P\left(\frac{dx_1}{dt} - \frac{dx_2}{dt}\right) \tag{6.33}$$

2 equations, 3 unknowns $[x_2]$.

These are all the equations for cart 1, but we still have one unknown, x_2, which is the position of cart 2. Thus, we now consider cart 2.

$$\sum F_2 = m_2 a_2$$

or

$$m_2\frac{d^2x_2}{dt^2}=\sum F_2 \tag{6.34}$$

3 equations, 4 unknowns $\left[\sum F_2\right]$.

The forces acting on cart 2 are

$$\sum F_2 = f_A(t)+F_{2,k_2}+F_{2,P}=f_A(t)-k_2(x_2-x_1)-P\left(\frac{dx_2}{dt}-\frac{dx_1}{dt}\right) \tag{6.35}$$

4 equations, 4 unknowns.

Equations 6.32 through 6.35 constitute the model for the system. The model can be solved by either using Laplace transform presented in Chapter 4 or by simulation as shown in Chapter 11. To obtain the analytical solution, it would help to reduce the number of equations by combining Equations 6.32 and 6.33 into a single equation, as well as Equations 6.34 and 6.35. This algebra gives

$$m_1\frac{d^2x_1}{dt^2}=-k_1x_1-k_2(x_1-x_2)-P\left(\frac{dx_1}{dt}-\frac{dx_2}{dt}\right)$$

$$m_1\frac{d^2x_1}{dt^2}+P\frac{dx_1}{dt}+(k_1+k_2)x_1=k_2x_2+P\frac{dx_2}{dt}$$

or

$$3\frac{d^2x_1}{dt^2}+5\frac{dx_1}{dt}+120\,x_1=60\,x_2+5\frac{dx_2}{dt} \tag{6.36}$$

and

$$m_1\frac{d^2x_2}{dt^2}=f_A(t)-k_2(x_2-x_1)-P\left(\frac{dx_2}{dt}-\frac{dx_1}{dt}\right)$$

$$m_2\frac{d^2x_2}{dt^2}+P\frac{dx_2}{dt}+k_2x_2=f_A(t)+k_2x_1+P\frac{dx_1}{dt}$$

or

$$5\frac{d^2x_2}{dt^2}+5\frac{dx_2}{dt}+60x_2=f_A(t)+60x_1+5\frac{dx_1}{dt} \tag{6.37}$$

Before continuing looking at the responses, let us use what we learned about the roots of the characteristic equations in Chapter 3. From Equation 6.36, a nonhomogeneous equation, the characteristic equation of the corresponding homogeneous equation is $3r^2+5r+120=0$ with roots $-0.833\pm i6.27$ indicating that because of the imaginary roots the response will be oscillatory and because of the negative real root the response will be stable for an stable forcing function. From Equation 6.37, another nonhomogeneous equation, the characteristic equation of the corresponding homogeneous equation is $5r^2+5r+60=0$ with roots $-0.5\pm i3.43$, indicating that because of the imaginary roots, the response will be oscillatory and because of the negative real root, the response will be stable for an stable forcing function.

Note that these models, Equations 6.36 and 6.37, are "coupled"; that is, each response affects the other response (displacement x_1 affects displacement x_2 and vice versa). Chapter 4, Section 4.5, shows how to solve coupled equations, and actually, it uses these two equations to do so. The solutions are

$$\begin{aligned}x_1(t)=&\,0.0833u(t)+e^{-1.258t}(0.0112\cos 6.6987t+0.0002\sin 6.6987t)\\&-e^{-0.0754t}(0.0944\cos 2.2717t-0.0024\sin 2.2717t)\end{aligned} \tag{4.52}$$

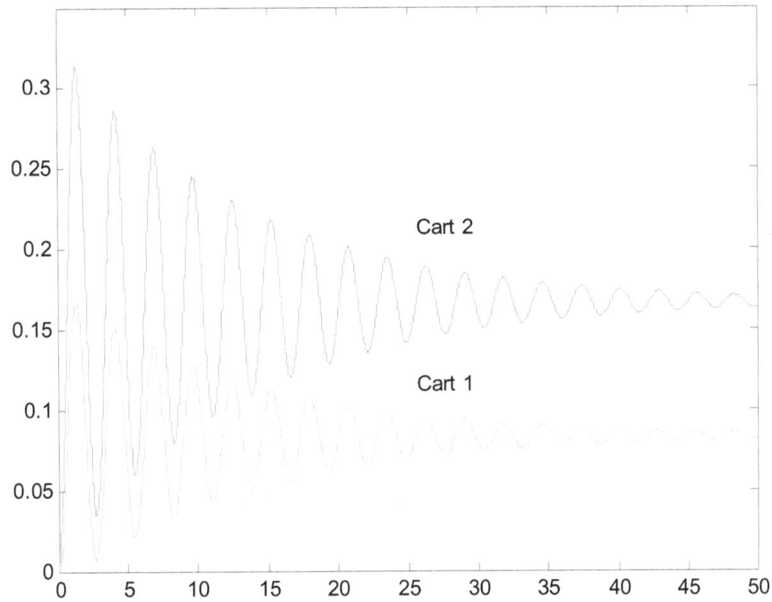

FIGURE 6.17 Response of carts.

and

$$x_2(t) = 0.1667u(t) - e^{-1.258t}(0.004\cos 6.6987t - 0.001\sin 6.6987t)$$
$$- e^{-0.0754t}(0.1626\cos 2.2717t + 0.0104\sin 2.2717t) \quad (4.53)$$

Figure 6.17 shows the responses of both carts obtained by graphing Equations 4.52 and 4.53. The qualitative responses are as expected, oscillatory and stable, as indicated by the roots.

Before completing this example, let us work it out again but now using the D'Alembert's principle and the FBD approach. Figure 6.18a shows the FBD of cart 1, and applying D'Alembert's principle results in

$$\sum F_1 = 0$$

and

$$\sum F_1 = F_{1,k_1} + F_{1,k_2} + F_{1,P} - F_{1_{inertial}} = -k_1x_1 - k_2(x_1 - x_2) - P\left(\frac{dx_1}{dt} - \frac{dx_2}{dt}\right) - m_1\frac{d^2x_1}{dt^2} = 0 \quad (6.38a)$$

or

$$m_1\frac{d^2x_1}{dt^2} + P\frac{dx_1}{dt} + (k_1 + k_2)x_1 = k_2x_2 + P\frac{dx_2}{dt}$$
$$3\frac{d^2x_1}{dt^2} + 5\frac{dx_1}{dt} + 120x_1 = 60x_2 + 5\frac{dx_2}{dt} \quad (6.38b)$$

1 equation, 2 unknowns $[x_1, x_2]$,

which, of course, is the same as Equation 6.36.

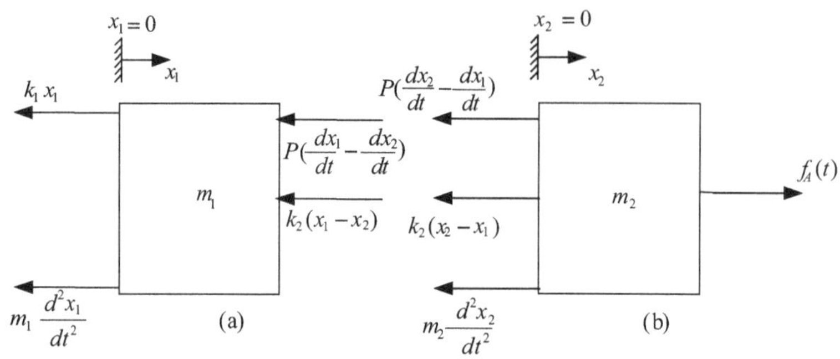

FIGURE 6.18 Free-body diagram for carts in Figure 6.16.

Figure 6.18b shows the FBD of cart 2, and applying D'Alembert's principle results in

$$\sum F_2 = 0$$

and

$$\sum F_2 = f_A(t) + F_{2,k_2} + F_{2,P} - F_{2_{inertial}} = f_A(t) - k_2(x_2 - x_1) - P\left(\frac{dx_2}{dt} - \frac{dx_1}{dt}\right) - m_2\frac{d^2x_2}{dt^2} = 0 \qquad (6.39a)$$

or

$$m_2\frac{d^2x_2}{dt^2} + P\frac{dx_2}{dt} + k_2x_2 = f_A(t) + k_2x_1 + P\frac{dx_1}{dt}$$

$$5\frac{d^2x_2}{dt^2} + 5\frac{dx_2}{dt} + 60x_2 = f_A(t) + 60x_1 + 5\frac{dx_1}{dt} \qquad (6.39b)$$

2 equations, 2 unknowns,

which is the same as Equation 6.37.

Analyzing Figure 6.18a, it is interesting to realize that no force is shown pulling to the right, or positive, side, except for the applied force $f_A(t)$. Furthermore, thinking about it, this does not seem to make sense because we know that due to $f_A(t)$ both carts will move to the right. The reason for this drawing of the FBD is because we have previously chosen to express all the forces generated by the elements joining two moving bodies by a negative sign and a difference in displacements/velocities with the first element of the difference that of the point of interest. We could have also drawn the FBD as shown in Figure 6.19.

Note that in going from Figure 6.18a to 6.19a, we have only used simple math, changing the order of the differences is like multiplying by −1, but then showing the forces to the right side is like multiplying by another −1, thus no change. Some readers may like this way of drawing the FBD because it explicitly shows the effect of the elements to each cart. *If you are familiar with simulation, simulate the mechanical system using Equations 6.36 and 6.37 and compare the results with Figure 6.17.*

Example 6.2

Consider the system shown in Figure 6.20. Different from the system in Figure 6.16, the second cart now has friction with the floor, $P_f = 3\ \text{N}\cdot\text{s/m}$.

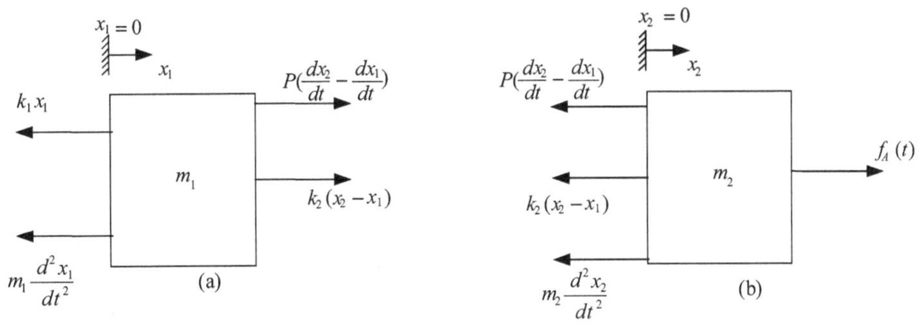

FIGURE 6.19 Free-body diagram for carts in Figure 6.16.

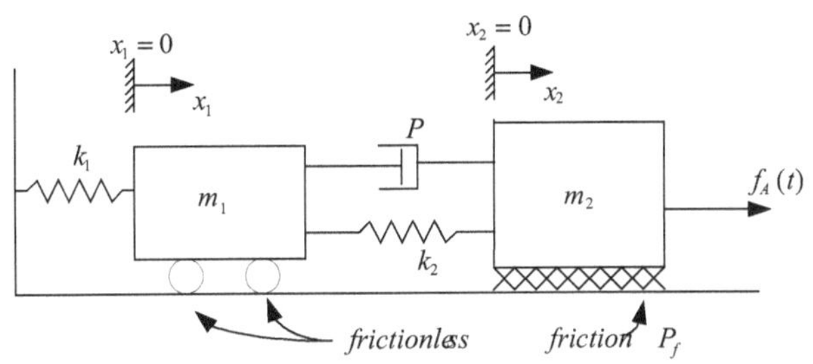

FIGURE 6.20 Cart with friction and an external force.

The model for the first cart is still given by Equation 6.32 in Example 6.1; nothing has changed in the system as far as cart 1 is concerned. For cart 2, however, the summation of the forces must contain the new friction force. Figures 6.21 and 6.22 show two ways to show the FBDs; Figures 6.21a and 6.22a show the same FBD for cart 1 as the one shown in Figures 6.18a and 6.19a, respectively. Figures 6.21b and 6.22b show the new friction force added for cart 2.

Applying D'Alembert's principle to cart 1 leads to the same model as before, which we repeat for convenience,

$$\sum F_1 = 0$$

and

$$\sum F_1 = F_{1,k_1} + F_{1,k_2} + F_{1,P} - F_{1_{inertial}} = -k_1 x_1 - k_2(x_1 - x_2) - P\left(\frac{dx_1}{dt} - \frac{dx_2}{dt}\right) - m_1 \frac{d^2x_1}{dt^2} = 0 \tag{6.38a}$$

or

$$m_1 \frac{d^2x_1}{dt^2} + P\frac{dx_1}{dt} + (k_1 + k_2)x_1 = k_2 x_2 + P\frac{dx_2}{dt}$$

$$3\frac{d^2x_1}{dt^2} + 5\frac{dx_1}{dt} + 120x_1 = 60x_2 + 5\frac{dx_2}{dt} \tag{6.38b}$$

1 equation, 2 unknowns $[x_1, x_2]$.

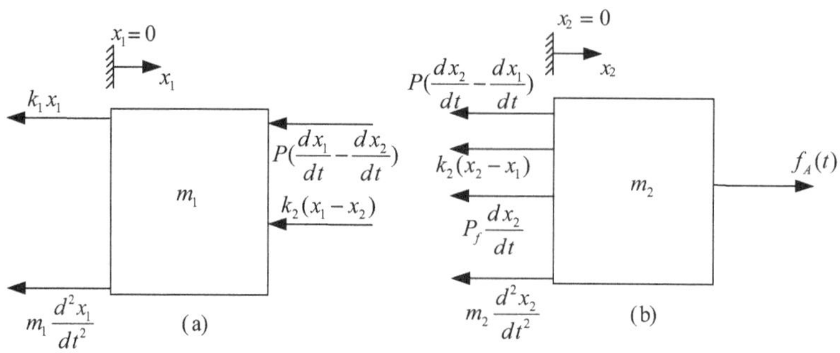

FIGURE 6.21 Free body diagrams for carts of Figure 6.20.

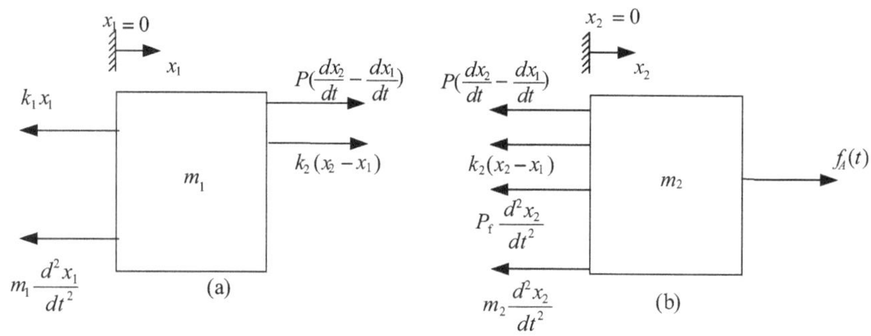

FIGURE 6.22 Free body diagrams for carts of Figure 6.20.

Applying D'Alembert's principle to cart 2

$$\sum F_2 = 0$$

$$\sum F_2 = f_A(t) + F_{2,k_2} + F_{2,P} + F_f - F_{2_{inertial}} = 0$$

$$\sum F_2 = f_A(t) - k_2(x_2 - x_1) - P\left(\frac{dx_2}{dt} - \frac{dx_1}{dt}\right) - P_f \frac{dx_2}{dt} - m_2 \frac{d^2x_2}{dt^2} = 0 \tag{6.40a}$$

or

$$m_2 \frac{d^2x_2}{dt^2} + (P + P_f)\frac{dx_2}{dt} + k_2 x_2 = f_A(t) + k_2 x_1 + P\frac{dx_1}{dt} = 0$$

$$m_2 \frac{d^2x_2}{dt^2} + (P + P_f)\frac{dx_2}{dt} + k_2 x_2 = f_A(t) + k_2 x_1 + P\frac{dx_1}{dt} = 0$$

$$5\frac{d^2x_2}{dt^2} + 8\frac{dx_2}{dt} + 60x_2 = f_A(t) + 60x_1 + 5\frac{dx_1}{dt} \tag{6.40b}$$

2 equations, 2 unknowns.

Equation 6.40b is the new model for cart 2.

Example 6.3

Consider the system shown in Figure 6.23, where $m_1 = 15\,\text{kg}, m_2 = 7\,\text{kg},\ k_1 = k_2 = 6\,\text{N/m}$, $P_1 = 5\,\text{N}\cdot\text{s/m}$, and $P_2 = P_3 = 2\,\text{N}\cdot\text{s/m}$. Originally, the system is at steady state, and the

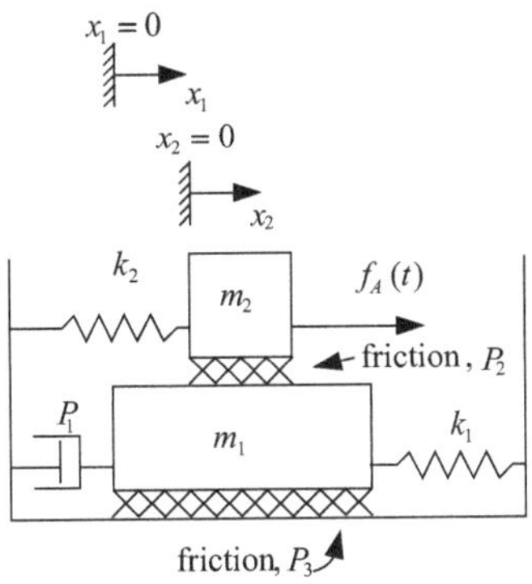

FIGURE 6.23 Two blocks with friction

springs and dashpot are unstretched in their normal length; thus, $\left.\frac{dx_1}{dt}\right|_{t=0} = \left.\frac{dx_2}{dt}\right|_{t=0} = 0\,\mathrm{m/s}$, and $x_1(0) = x_2(0) = 0\,\mathrm{m}$. Develop the model that provides the velocity and position of each block.

We start modeling applying Newton's second law to block 1.

$$\sum F_1 = m_1 \frac{d^2 x_1}{dt^2} \tag{6.41}$$

1 equation, 2 unknowns $\left[\sum F_1, x_1\right]$.

The forces acting on block 1 are

$$\sum F_1 = F_{1,k_1} + F_{1,P_1} + F_{f,P_2} + F_{f,P_3}$$

or

$$\sum F_1 = -k_1\, x_1 - P_1 \frac{dx_1}{dt} - P_2 \left(\frac{dx_1}{dt} - \frac{dx_2}{dt}\right) - P_3 \frac{dx_1}{dt} \tag{6.42}$$

2 equations, 3 unknowns $[x_2]$,

and using the definition of velocity

$$v_1 = \frac{dx_1}{dt} \tag{6.43}$$

3 equations, 4 unknowns $[v_1]$.

We still have 1 degree of freedom, and realizing that one of the unknowns is the displacement of block 2, x_2, we focus our attention on this block and apply Newton's second law,

$$\sum F_2 = m_2 \frac{d^2 x_2}{dt^2} \tag{6.44}$$

4 equations, 5 unknowns $\left[\sum F_2\right]$.

The forces acting on block 2 are

$$\sum F_2 = f_A(t) + F_{2,k_2} + F_{f,P_2}$$

or

$$\sum F_2 = f_A(t) - k_2 x_2 - P_2\left(\frac{dx_2}{dt} - \frac{dx_1}{dt}\right) \tag{6.45}$$

5 equations, 5 unknowns,

and using the definition of velocity

$$v_2 = \frac{dx_2}{dt} \tag{6.46}$$

6 equations, 6 unknowns $[v_2]$.

Equations 6.41 through 6.46 constitute the model of the system. Chapter 4 has provided the analytical tools to solve these equations, and Chapter 11 provides the simulation tools.

6.5 SPRINGS AND DAMPERS IN SERIES OR PARALLEL

Up to now, we have considered only springs and dampers (dashpots) connected to a mass or wall. This section considers these devices connected to each other in series or in parallel.

6.5.1 Springs in Parallel

Consider the system shown in Figure 6.24a. Two springs are connected to a fixed wall and to a moving mass; there is also an external force connected to the mass. The question is, Is there an "equivalent spring" that can substitute the two springs, as shown in Figure 6.24b, and provide the same performance?

When no force f is applied, both springs are at their normal lengths. It is easy to realize that anytime the body moves, the two springs will move by the same amount that is, at any time $x_1 = x_2 = x$. At that time, the forces generated by the springs are

$$f_1 = -k_1 x_1 = -k_1 x \text{ and } f_2 = -k_2 x_2 = -k_2 x$$

For the equivalent system in Figure 6.24b, moving the same distance will generate the same force,

$$f = -k_{Eq} x$$

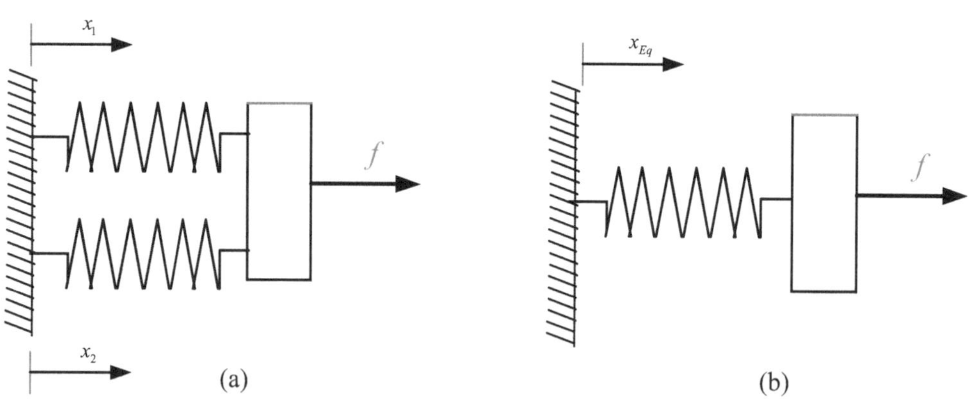

FIGURE 6.24 Springs in parallel.

where k_{Eq} is the coefficient of the one spring that would provide the same effect as the two other springs; this is what we refer to as the "equivalent spring," Back to Figure 6.24a, it is also easy to realize $f = f_1 + f_2$. Then, substituting the previous three equations into this last equation gives,

$$-k_{Eq}x = -k_1x - k_2x$$

or

$$k_{Eq} = k_1 + k_2$$

So, the equivalent spring constant will be equal to the sum of the individual spring constants. Following the same development, if there were three springs in parallel, the equivalent spring constant would be

$$k_{Eq} = k_1 + k_2 + k_3$$

In general, for n springs in parallel, the equivalent spring constant is given by

$$k_{Eq} = \sum_{i=1}^{n} k_i \tag{6.47}$$

6.5.2 Springs in Series

Consider the system shown in Figure 6.25a. The question of interest is similar to the previous subsection that is, is there an "equivalent spring" that can substitute the two springs, as shown in Figure 6.25b, and provide the same performance?

When no force f is applied, both springs are at their normal lengths with no extension or compression. When a positive force is applied the system moves to a new location to the right (positive); both springs are elongated. At that time, the force generated by spring 2 (spring connected to the mass) is the one holding the mass; thus, it is the same as the force applied to the mass, or $f_2 = -k_2x_2$. Because both springs are connected, spring 1 is holding spring 2. This means that the force generated by spring 1 must be the same as that generated by spring 2 or $f_1 = -k_1x_1$. The forces generated by each spring and the one pulling on the mass are the same $f = f_1 = f_2$. Although the forces are the same, the change in length of each spring is different. The total change in length of the system is

$$x_T = x_1 + x_2$$

The equivalent spring, the one shown in Figure 6.25b, is the one that would generate the same force while moving the same total length, or

$$f = -k_{Eq}x_T$$

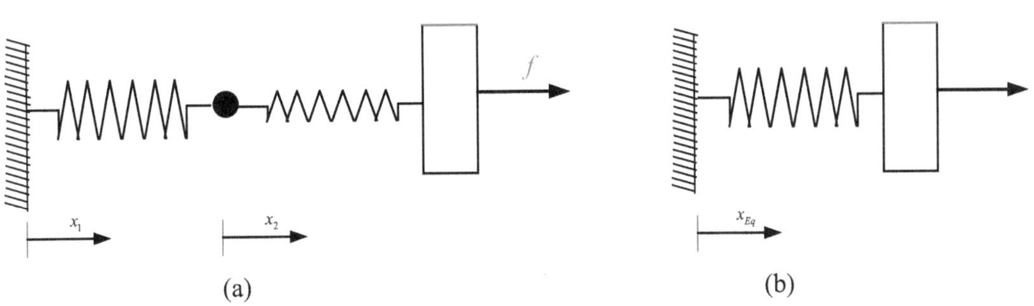

FIGURE 6.25 Springs in series.

Using the equations for the individual forces into the equation for the total change in length,

$$-\frac{f}{k_{Eq}} = -\frac{f}{K_1} - \frac{f}{k_2} \text{ or } \frac{1}{k_{Eq}} = \frac{1}{K_1} + \frac{1}{k_2}$$

or

$$\mathrm{k}_{Eq} = \frac{1}{\frac{1}{k_1} + \frac{1}{k_2}} = \frac{k_1 k_2}{k_1 + k_2}$$

In general, for n springs in series, the equivalent spring constant is given by

$$k_{Eq} = \frac{\prod_{i=1}^{n} k_i}{\sum_{i=1}^{n} k_i} \tag{6.48}$$

6.5.3 Dampers in Parallel

Figure 6.26a shows two dampers in parallel, and Figure 6.26b shows the equivalent damper. The question of interest is the same as in the previous subsections, what is the "equivalent damper" that can substitute the two dampers and provide the same performance?

Following a similar development as previously shown, the coefficient for the equivalent damper is,

$$P_{Eq} = P_1 + P_2$$

In general, for n dampers in parallel, the equivalent damper constant is given by

$$P_{Eq} = \sum_{i=1}^{n} P_i \tag{6.49}$$

The development of this equation is the subject of a problem at the end of the chapter.

6.5.4 Dampers in Series

Figure 6.27a shows two dampers in series, and Figure 6.26b shows the equivalent damper. The question of interest is the same as in the previous subsections, what is the "equivalent damper" that can substitute the two dampers and provide the same performance?

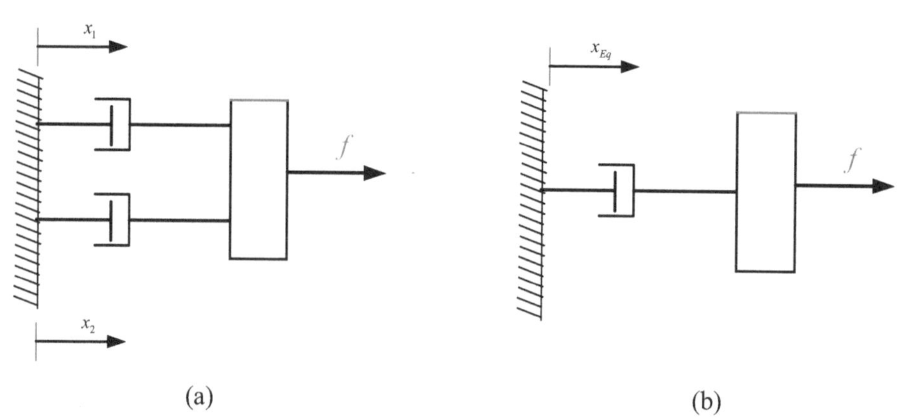

FIGURE 6.26 Dampers in parallel.

Following a similar development as previously shown, the coefficient for the equivalent damper is

$$P_{Eq} = \frac{1}{\frac{1}{P_1} + \frac{1}{P_2}} = \frac{P_1 P_2}{P_1 + P_2}$$

In general, for n springs in series, the equivalent spring constant is given by

$$P_{Eq} = \frac{\prod_{i=1}^{n} P_i}{\sum_{i=1}^{n} P_i} \quad (6.50)$$

The development of this equation is the subject of a problem at the end of the chapter.

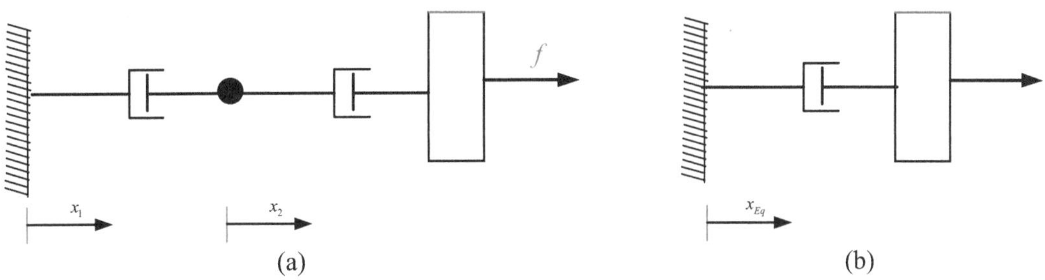

FIGURE 6.27 Dampers in series.

6.5.5 Springs and Dampers in Series

Figure 6.28 shows a damper connected to a spring and then the spring to a mass. Let us look at the analysis of this system to obtain the model describing the mass response to a change in the external force.

Applying Newton's second law to the mass,

$$ma = m\frac{dx}{dt} = \sum F = f + f_k = f - k(x - x_2) \quad (6.51)$$

1 equation 2 unknowns $[x, x_2]$.

We still need one more equation, however, there is no object (mass) that can be used to continue with the model development. We use a simple trick. Assume that at any place between the damper and spring exists a "massless object," as shown in dashed lines in Figure 6.28.

Apply Newton's second law to this new object.

$$m_{mo}\, a_{mo} = m_{mo}\frac{dx_2}{dt} = \sum F = f_P + f_k = -P\frac{dx_1}{dt} - k(x - x_2)$$

But, because the new object is massless, $m_{mo} = 0$, the equation becomes

$$0 = -P\frac{dx_1}{dt} - k(x - x_2) \quad (6.52)$$

2 equations, 3 unknowns $[x_1]$.

We have finished writing all the equations involving the object, spring, and damper but still need another equation. Looking over Figure 6.27, note the change in location of the object, x, is equal to the change in x_1 *plus* x_2, or

$$x_1 + x_2 = x \tag{6.53}$$

3 equations, 3 unknowns.

$$DOF = 0!!$$

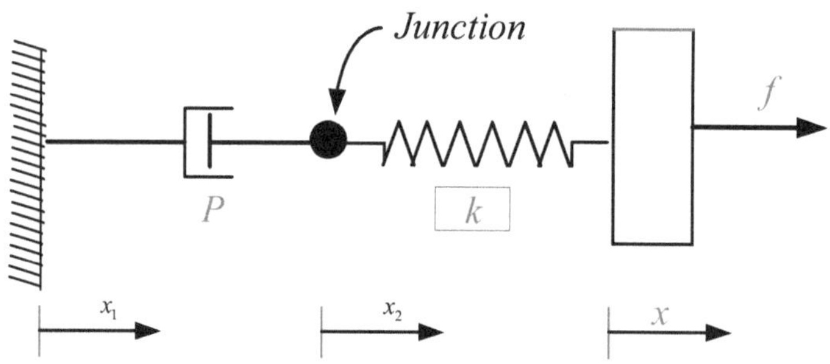

FIGURE 6.28 Mechanical system.

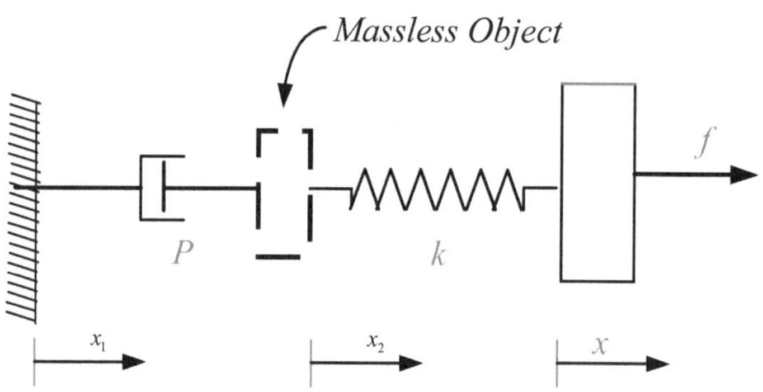

FIGURE 6.29 Mechanical system.

6.6 VERTICAL SYSTEMS

The mechanical systems presented thus far move horizontally, and in these cases, gravity does not have any effect. Gravity, however, affects vertical systems and has an impact on what we referred to as $y = 0$. In systems with horizontal components, where gravity does not have any effect, we consider the reference position for displacement the location, where the springs and dashpots are not stretched either under compression or tension; in this case, $x_{ref} = x = 0$. However, in vertical systems, the initial displacement, $y(0)$, before an external force is applied, is not equal to the position where the springs and dashpots are not stretched. As soon as a body is attached to these elements, they become stretched, and a new steady-state position is reached; Figure 6.2 refers exactly to this issue. Therefore, it is worthwhile to look at one example.

Example 6.4

Consider the spring shown in Figure 6.30a. We may refer to the location at the end of the spring, when it is unstretched, as y_{ref}. Now consider that a body of $m = 1\,\text{kg}$ is attached to the spring, as shown in Figure 6.30b. At that moment, the spring stretches because of the pull (force) of gravity and reaches a new final location. Actually, as we have already mentioned once before, attaching a body to a spring, or dashpot, is exactly the same as attaching any other external force. Thus, we can use what we have learned so far to obtain the response of the new system. Note that in this example the down direction is positive (+).

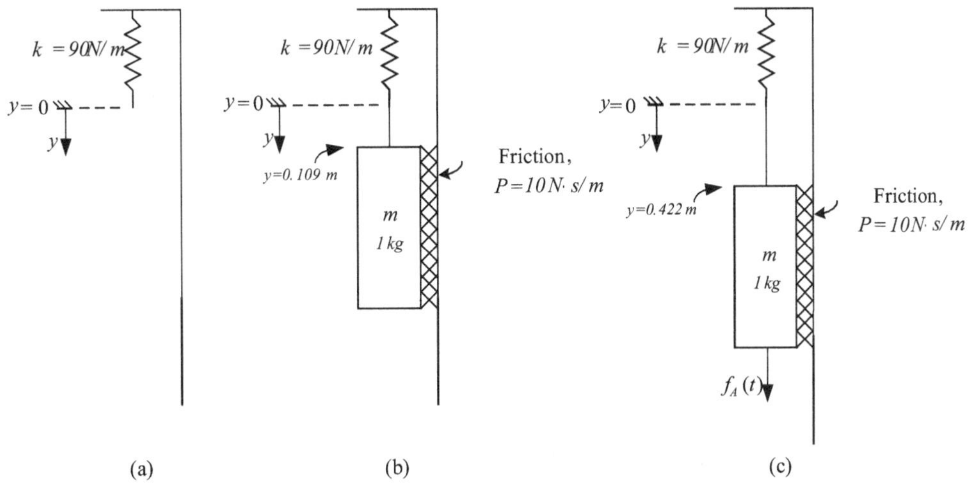

FIGURE 6.30 Vertical mechanical system.

The model is obtained by applying Newton's second law again,

$$\sum F_y = m\, a_y$$

or

$$m\frac{d^2y}{dt^2} = \sum F_y \tag{6.54}$$

1 equation, 2 unknowns $\left[y, \sum F_y\right]$

and

$$\sum F_y = F_k + F_P + F_g$$

or

$$\sum F_y = -k(y - y_{ref}) - P\frac{dy}{dt} + mg$$

where $f_g = $ mg is the force of gravity. Letting $y_{ref} = y(0) = 0$,

$$\sum F_y = -ky - P\frac{dy}{dt} + mg \tag{6.55}$$

2 equations, 2 unknowns.

Substituting Equation 6.55 into 6.54, including the values, and rearranging gives,

$$\frac{d^2y}{dt^2} + 10\frac{dy}{dt} + 90\, y = g \tag{6.56}$$

The initial conditions are $\left.\frac{dy}{dt}\right|_{t=o} = 0\ \text{m/s}$ and $y(0) = 0\ \text{m}$. The analytical solution of this model is

$$y = 0.109 - e^{-5t}[0.109\cos 8.062t + 0.060\sin 8.062t] \tag{6.57}$$

The final value of the displacement is 0.109 m.

Suppose now that once the final displacement is reached, it is desired to study the effect of applying an external force to the block, $f_A(t) = 30u(t)$ N as shown in Figure 6.30c. For this case, the initial value of the displacement is equal to the final value of the previous case when the block was connected to the spring, $y(0) = 0.109\ \text{m}$, and $\left.\frac{dy}{dt}\right|_{t=0} = 0\ \text{m/s}$. Equation 6.54 is again the first equation of the model. Equation 6.55 is the one that changes to

$$\sum F_y = F_k + F_P + f_A(t) = -k(y - 0.109) - P\frac{dy}{dt} + f_A(t) \tag{6.58}$$

2 equations, 2 unknowns.

Note that there is no need to account for the force of gravity because this was used to calculate the initial condition of the displacement. Note also that the force generated by the spring, due to only the applied force is given by $F_k = -k(y - 0.109)$ because $(y - 0.109)$ is the displacement due to the force.

Equations 6.54 and 6.58 constitute the model. If only one equation is desired to obtain the analytical solution, substituting Equation 6.58 into 6.54, substituting the values of the terms, and rearranging gives

$$\frac{d^2y}{dt^2} + 10\frac{dy}{dt} + 90y = 90(0.109) + f_A(t) \tag{6.59}$$

and the analytical solution is

$$y = 0.442 - e^{-5t}[0.333\cos 8.062t + 0.207\sin 8.062t] \tag{6.60}$$

Figure 6.31 shows the response of the system when the block is first attached to the spring (first part of the figure), and when the force is applied to the block (second part of the figure). The figure shows that the first part of the response is oscillatory, and that the final displacement is 0.109 m. Obviously, this same information can be obtained from Equation 6.57; also, the roots of the characteristic equation, $-5 \pm i8.062$, indicate a stable oscillatory response. Because the roots are the same for the second part (as presented in Chapters 3 and 4, the roots only depend on the system, which is the same in both cases, and not on the forcing function) the response is also oscillatory. The final value of the displacement is 0.442 m. Note that this is the final displacement of the spring due to the mass of the block *plus* the external force $f_A(t)$. The displacement due to only the mass of the block was 0.109 m, and that due to only $f_A(t)$ is 0.333 m.

In the previous example, we considered the initial condition for the displacement due to the external applied force to be the final displacement due to the mass of the object, or $y = 0.109\ \text{m}$. Note that this 0.109 m is the displacement *relative to* the position where the spring was not stretched. Obviously, the final value of 0.442 m is also *relative to* the same position. All of this makes perfect common sense.

When developing the model for the effect of the external applied force, we could have defined the 0.109 m position as $y = 0\ \text{m}$. In this case, the force exerted by the spring would have been given by $F_k = -k\,y$, resulting in the model

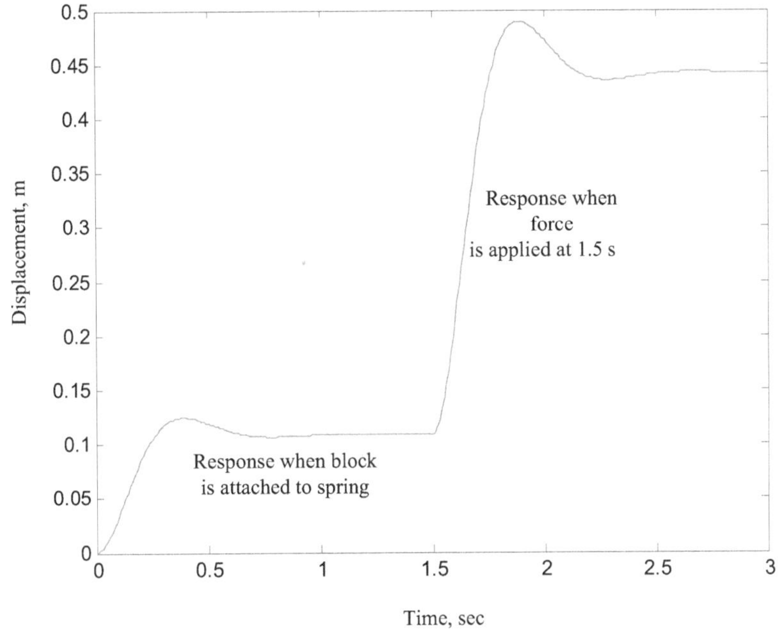

FIGURE 6.31 Displacement of system of Example 6.4

$$\frac{d^2y}{dt^2}+10\frac{dy}{dt}+90y=f_A(t)$$

and the resulting analytical solution is

$$y=0.333-e^{-5t}\left[0.333\cos 8.062t+0.207\sin 8.062t\right]$$

The displacement y given by this solution is *relative to* the 0.109 m position. The final displacement relative to the 0.109m position is 0.333 m, or 0.442m (0.109m + 0.333m) relative to the unstretched position of the spring.

6.7 SUMMARY

This chapter has focused on translational mechanical systems. Newton's second law was shown to be the fundamental principle to use for developing the models. The experimental relations for springs, dashpots, and frictions were used to develop the models. D'Alembert's principle and FBDs were presented showing how they apply in the modeling of systems. Chapter 7 continues with modeling rotational mechanical systems.

Problems

Problem 6.1

Consider the spring-mass system shown in Figure P6.1. The mass is released from rest at $x=0.1$ m and $t=0$.

a. Develop a model that describes the position of the block and shows the initial conditions.
b. Without obtaining the complete solution, graph the qualitative response of the position from its initial to the final value. Explain why the response is as you propose.

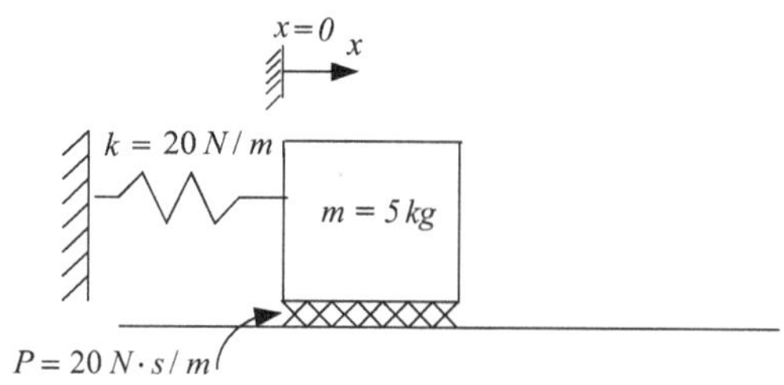

FIGURE P6.1 Mechanical system for Problem 6.1.

Problem 6.2

a. A block is suspended vertically on a spring as shown in Figure P6.2. The mass of the block is 3 kg and at equilibrium, the spring extends 0.02 m from its un-weighted length. Find the spring constant k.
b. Now the same block and spring are placed on a flat surface as shown. The reference position is where the spring is neither compressed nor stretched and there is no friction between the block and surface. At time zero, the block is pulled 0.04 m to the right and released from rest. Write the differential equation and initial conditions that model the motion of the block and obtain the analytical solution. What does the solution indicate about the response?

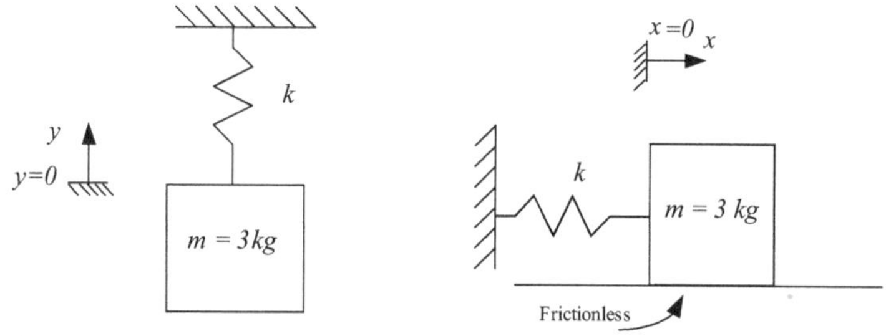

FIGURE P6.2 Mechanical system for Problem 6.2.

Problem 6.3

Consider the block shown in Figure P6.3. There is fluid friction (with damping coefficient $P = 4$ Ns/m) between the block and the surface. The 2-kg block is initially at rest at $x = 0$. The applied force is $f_A(t) = 16u(t)$ N. Write the differential equation and initial conditions that models the velocity of the block. Without solving the model, what velocity will the block ultimately reach?

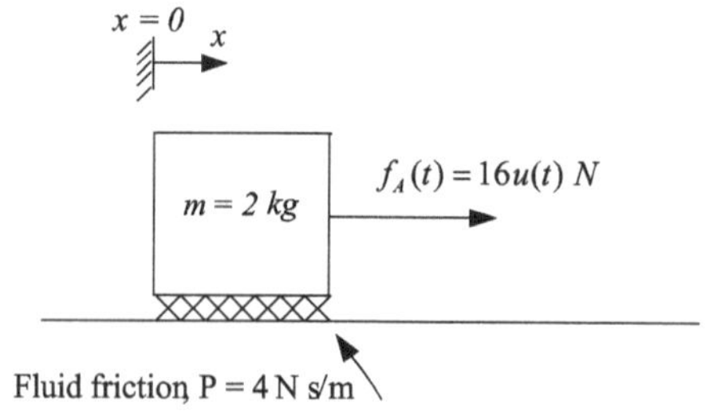

FIGURE P6.3 Mechanical system for Problem 6.3.

Problem 6.4

Consider the cart shown Figure P6.4. There is no friction or damping in the system. The position of the cart is represented by x and the position of the end of the spring by y. The spring is under neither compression nor tension when both x and y are zero.

The initial position and velocity of the cart are zero. Starting at time zero, the *right* end of the spring is given a motion $y = 0.1t\,u(t)\,\text{m}$. The cart has a mass of 2 kg, and the spring constant is 18 N/m. Write a differential equation and initial conditions that models the motion of the cart and obtain its analytical solution.

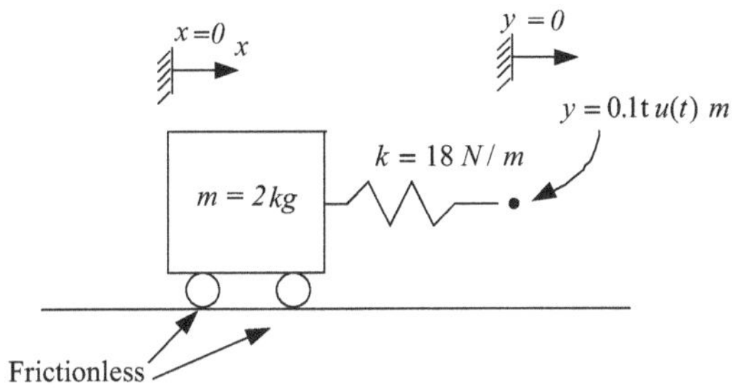

FIGURE P6.4 Mechanical system for Problem 6.4.

Problem 6.5

Figure P6.5 shows a single mass attached to a dashpot and a spring; the bearings are frictionless.

Let $m = 10$ kg, $P = 10$ N s / m, and $k = 40$ N / m

a. Develop a model that describes how the position of the block, x, varies as a function of time for $f_A(t) = 20u(t)$ N.
b. Obtain the analytical solution of the model describing the response of the system.
c. Graph the response.

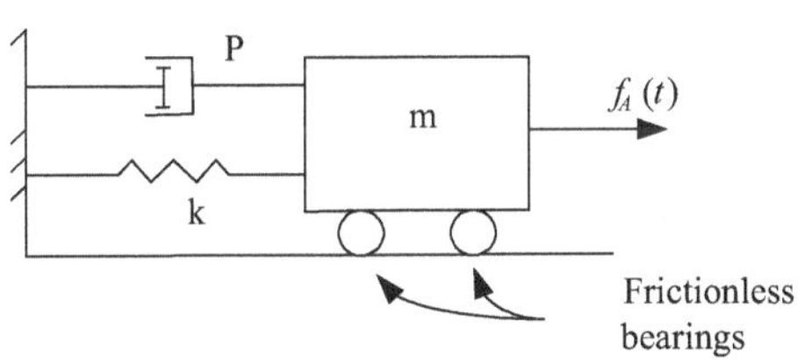

FIGURE P6.5 Mechanical system for Problem 6.5.

Problem 6.6

Consider the mechanical system shown in Figure P6.6.

With $m = 10$ kg, $P_1 = P_3 = 20$ N·s / m, $P_2 = 30$ N·s / m, $k_1 = k_2 = 100$ N / m.

The force $f_A(t)$ increases from *0* to *10* N at time $t = 0$ s, or $f_A(t) = 10u(t)$ N.

a. Develop a model that describes how the position of the block, x, varies as a function of time.
b. Obtain the analytical solution of the model describing the response of the system and graph the response.

c. *If you are familiar with simulation, you can use it for this next part. It is desired to find the value of P_3 that would result in the fastest approach to the final value, without any oscillation (critically damped response). What is the value and graph the response.*

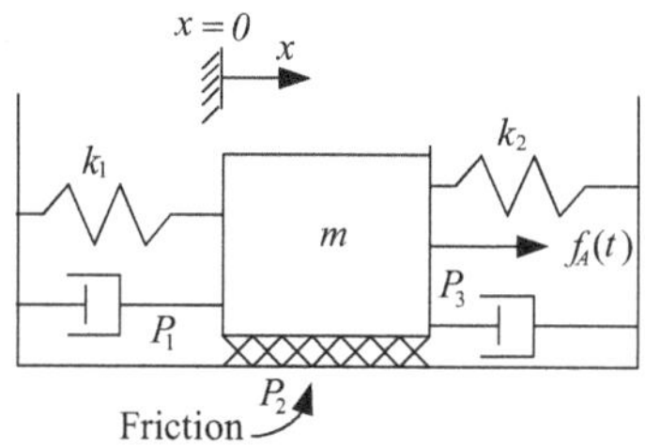

FIGURE P6.6 Mechanical system for Problem 6.6.

Problem 6.7

For the mechanical system shown in Figure P6.7: $m_1 = m_2 = 10$ kg, $P_1 = P_2 = P_3 = 20\ \text{N}\cdot\text{s/m}$, $k_1 = k_2 = k_3 = 10\ \text{N/m}$, and $f_A(t) = 20u(t)$ N.

a. Develop the model to provide the displacements and the velocities of both blocks.
b. Obtain the analytical solutions of the model describing the responses of the system.
c. Graph the responses.

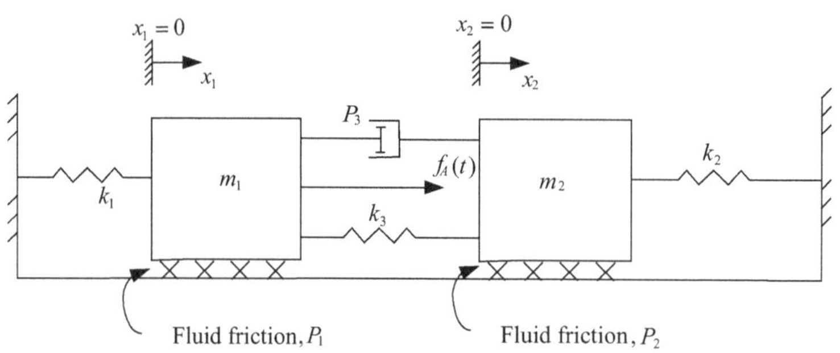

FIGURE P6.7 Mechanical system for Problem 6.7.

Problem 6.8

For the two-mass system shown in Figure P6.8, the information is the following:

$$m_1 = m_2 = 5\,\text{kg},\ P_1 = P_2 = 20\ \text{N}\cdot\text{s/m},\ P_3 = 50\ \text{N}\cdot\text{s/m},\ k_1 = k_2 = 100\ \text{N/m, and}$$
$$f_A(t) = 10u(t-1)\ \text{N}$$

a. Develop a model that describes the position and velocity of both blocks.
b. *If you are familiar with simulation, simulate the response of the blocks when applying the forcing function. In this case, the solutions will indicate that both blocks will oscillate. These oscillations may present a problem if the top block (block 2) oscillates more (longer) than block 1, meaning it may get off from block 1. If this were the case, block 1 would have to be built "long enough" to provide a base to block 2 all the time. Decide and explain what lengths you would size each block.*

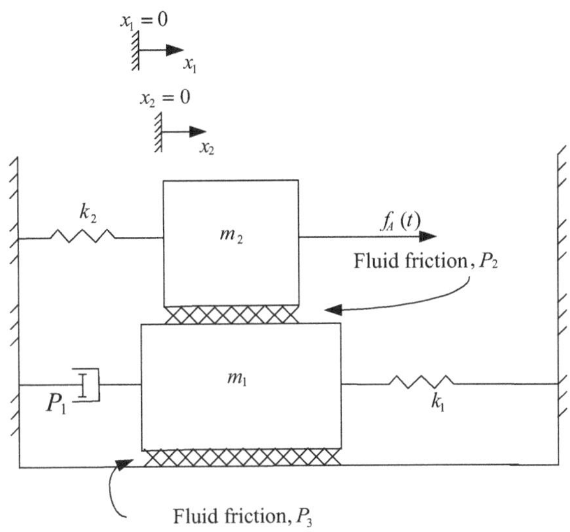

FIGURE P6.8 Mechanical system for Problem 6.8.

Problem 6.9

Consider the system shown in Figure P6.9. There is fluid friction (with damping coefficient $P_1 = 20$ N s/m) between block 1 and the surface. There is also a fluid friction (with damping coefficient $P_2 = 15$ N s/m) between blocks 1 and 2. Spring constants are $k_1 = 30$ N/m, $k_2 = 40$ N/m, $k_3 = 150$ N/m, $m_1 = m_2 = 10$ kg, and $f_A(t) = 40u(t)$ N. Reference positions for each block are their initial positions where the springs are neither compressed nor stretched.

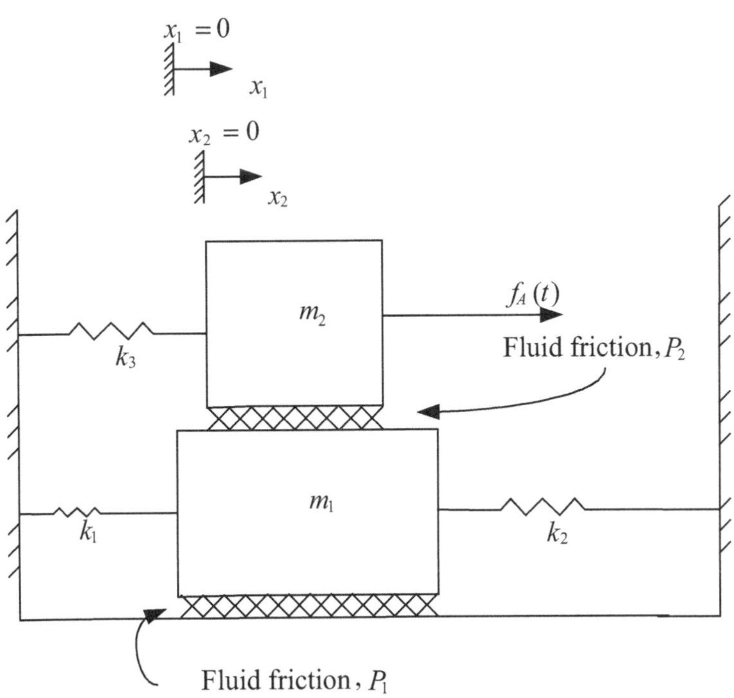

FIGURE P6.9 Mechanical system for Problem 6.9.

a. Develop a model that describes the position and velocity of both blocks.
b. What are the final positions of the displacements?

Problem 6.10

Consider the mechanical system shown in Figure P6.10.

The numerical value of the mass, damping terms, and spring constant are

$$m = 10\text{ kg};\ P_1 = 10\text{ N}\cdot\text{s/m};\ P_2 = 10\text{ N}\cdot\text{s/m};\ k_1 = 100\text{ N/m};\ k_2 = 130\text{ N/m}$$

Force $f_1(t)$ is 0 for the first 10 seconds, and then it becomes 20 N. Force $f_2(t)$ is 0 for the first 40 seconds, and then it becomes 25 N.

Develop the model that describes the position and velocity of the block and the junction A versus time. (**Hint:** You may treat junction A as a body with no mass.)

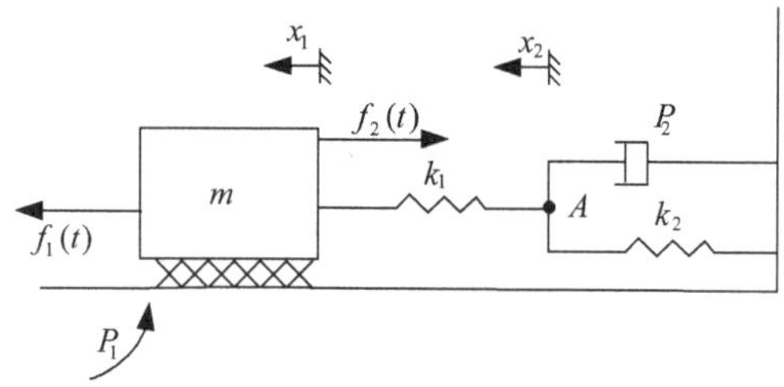

FIGURE P6.10 Mechanical system for Problem 6.10.

Problem 6.11

Consider the spring-mass system shown in Figure P6.11. The only friction in the system is fluid friction (with damping coefficient P) between block 1 and the surface. The external force $f_A(t)$ is constant. Reference positions for each block are their initial positions where the springs are neither stretched nor compressed. Develop a model that describes the position and velocity of each block.

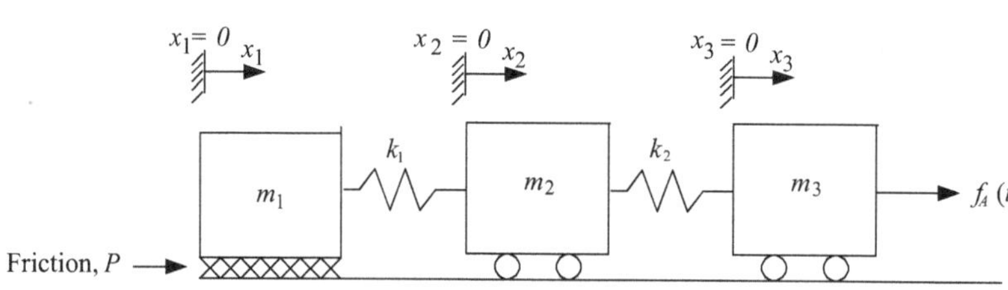

FIGURE P6.11 Mechanical system for Problem 6.11.

Problem 6.12

Consider the spring–mass system shown in Figure P6.12. There is no friction between the surface and mass 1, but there is fluid friction, with damping coefficient $P = 20\text{ N}\cdot\text{s/m}$ P, between mass 2 and the surface. The spring constant k_1 is 40 N/m, and $m_1 = m_2 = 10$ kg. Reference positions for each mass are their initial positions where the spring is neither stretched nor compressed. The force applied to mass 1 is $f_A(t) = 10u(t)$ N.

a. Develop a model that describes the position and velocity of each block.
b. Obtain the analytical solutions of the model describing the responses of the system.
c. Graph the responses.

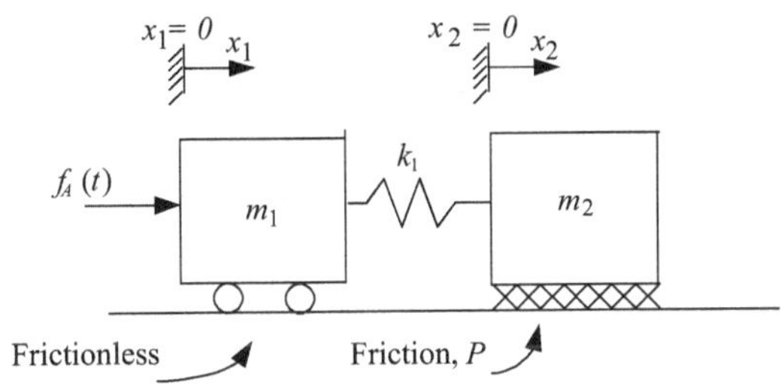

FIGURE P6.12 Mechanical system for Problem 6.12.

Problem 6.13

Consider the spring-mass system shown in Figure P6.13. The reference point $x = 0$ corresponds to the spring being under neither compression nor tension. The 2-kg block is resting on a belt that is initially motionless but begins moving with a constant velocity v of 2.5 m/s at time zero, or $v = 2.5u(t)$ m / s. At time zero, the block is motionless and at its reference position. There is fluid friction between the block and moving belt with a damping coefficient of $P = 8$ N·s / m and the spring constant $k = 40$ N / m.

Write the differential equations and initial conditions that model the velocity and the position of the block. Without solving the differential equation, calculate the ultimate position of the block.

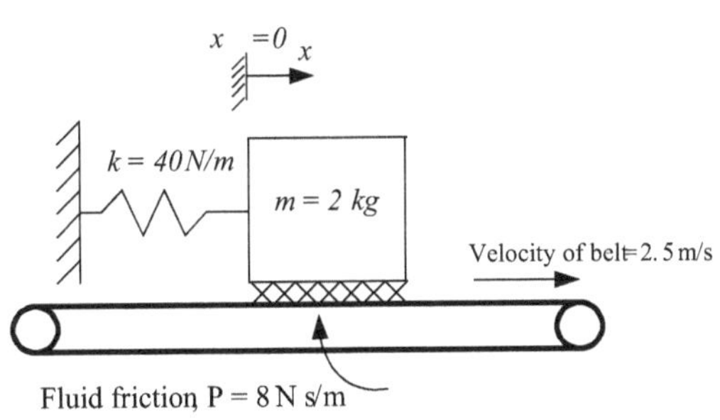

FIGURE P6.13 Mechanical system for Problem 6.13.

Problem 6.14

Consider the system shown in Figure P6.14. The bungee has spring constant k_1 for tension *only*. The wheels are frictionless and friction between blocks 1 and 2 can be modeled as fluid friction. Reference positions for each block correspond to the bungee being taut but not stretched and the spring being neither stretched nor compressed. Initial positions and velocities for both blocks are zero. Write the differential equations that represent the motion of blocks 1 and 2.

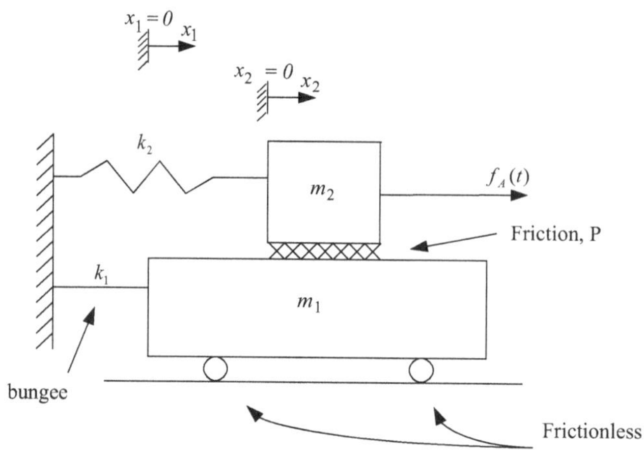

FIGURE P6.14 Mechanical system for Problem 6.14.

Problem 6.15

Consider the system shown in Figure P6.15. The values of the mass, damping factors, and spring constant are $m_1 = 5\ \text{kg}$, $P_1 = 35\ \text{N}\cdot\text{s/m}$, $P_2 = 5\ \text{N}\cdot\text{s/m}$, and $k = 200\ \text{N/m}$. The forcing function is $f_A(t) = 30u(t)$ N. Consider $y(0) = 0$ the position where the dashpot and spring are unstretched. Then attach the block and the external force at the same time.

a. Develop the model that describes the position of the block for $t > 0$.
b. Obtain the analytical solution of the model describing the response of the system.
c. Graph the response.

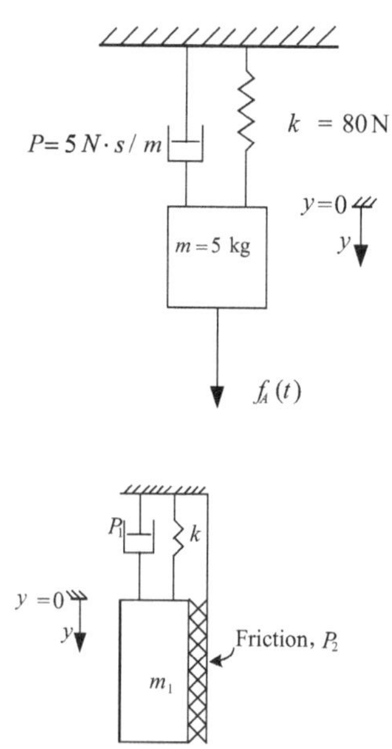

FIGURE P6.15 Mechanical system for Problem 6.15.

Problem 6.16

Consider the system shown in Figure P6.17. Originally, the block was not attached to the spring and dashpot; the end point of both devices is $y(0)=0$. Suppose that you attach the block.

a. Develop a model that describes the response of the system.
b. What is y_{final}?
c. Once the block reaches the new steady state, a force $f_A(t)=30u(t)$ is applied. Develop now the model that describes the response of the system due to only this force. Obtain the analytical solution of the model.

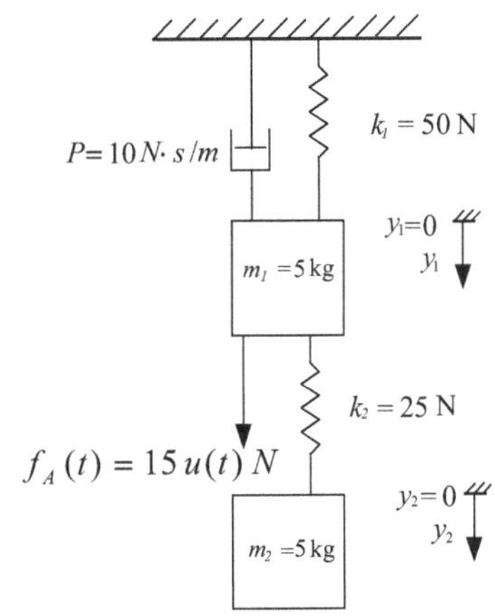

FIGURE P6.16 Mechanical system for Problem 6.16.

Problem 6.17

Consider the system shown in Figure P6.17. Develop the model that describes the response of both blocks due to *only* the applied force $f_A(t)$.

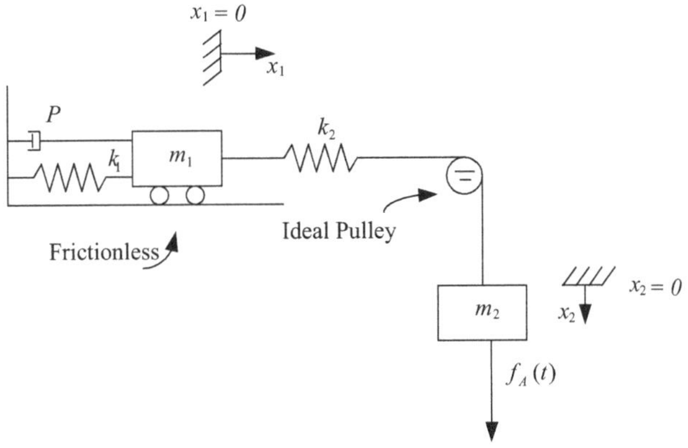

FIGURE P6.17 Mechanical system for Problem 6.17.

Problem 6.18

Consider the cart system shown in Figure P6.18. Assuming an ideal pulley, and with $m_1 = 15$ kg, $m_2 = 8$ kg, $k_1 = 30$ N / m, $k_2 = 100$ N / m, $P = 30$ N·s / m, and the applied force is $f_A(t) = 40u(t)$ N, determine the model that describes the displacement and position of the cart and of the block due to *only* the applied force. Use the initial conditions as shown in the figure, that is, having block 2 connected and the springs and dashpot extended.

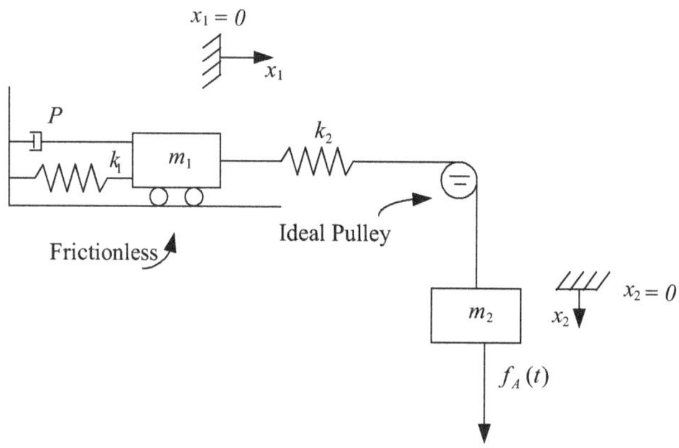

FIGURE P6.18 Cart system for Problem 6.18.

Problem 6.19

Consider the block system shown in Figure P6.19; the figure shows all the information. Develop the model that describes the displacement and velocity of each block after the external force is applied.

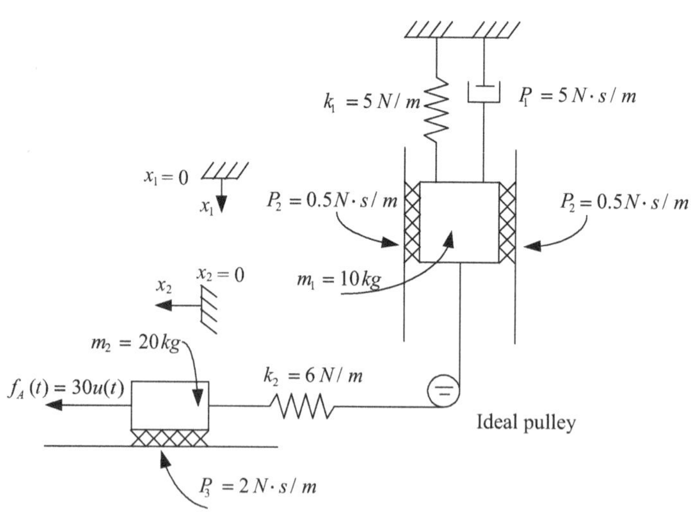

FIGURE P6.19 Block system for Problem 6.19.

Problem 6.20

Consider the system shown in Figure P6.20.

Develop a model that describes the displacement of the block, specify its initial conditions, obtain the analytical solution, and graph the response.

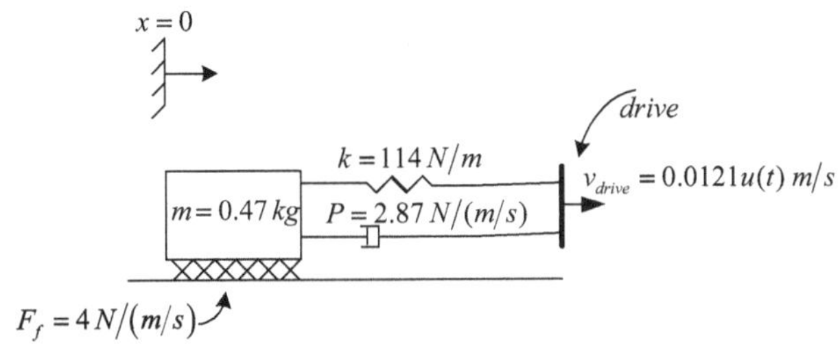

FIGURE P6.20 Block system for Problem 6.20.

Problem 6.21

Consider the mechanical system shown in Figure P6.21.

a. Obtain the model the system that provides the position x in meters, and velocity v in m/s of the block.
b. Obtain the frequency of the forcing function that will maximize the amplitude of the response. This frequency is what in Section 6.2.1 we called the practical, or pseudo-, resonance.

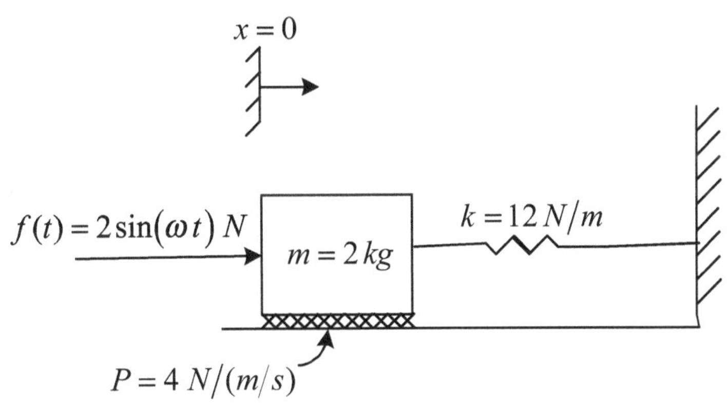

FIGURE P6.21 Block system for Problem 6.21.

Problem 6.22

Figure P6.22 shows a mechanical system to be designed. It is known that often the forcing function affecting the system will be $f(t) = 5\sin(10t)$ N.

a. What value of k will produce resonance?
b. Using twice that value, obtain the analytical solution providing the displacement of the block.

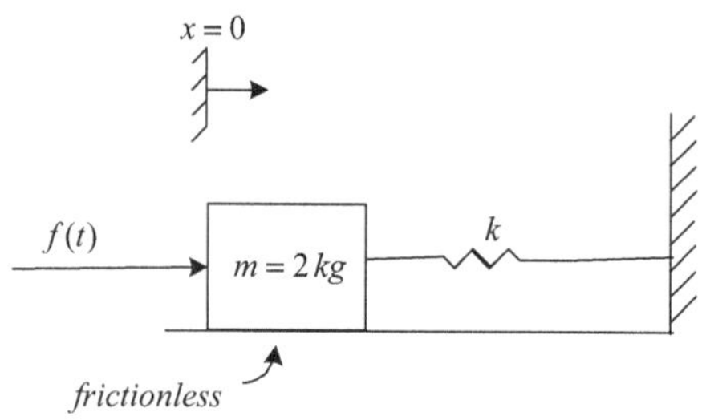

FIGURE P6.22 Block system for Problem 6.22.

Problem 6.23

Figure P6.23 shows a simplified suspension system for a vehicle. The objective of the system is to provide passengers comfort when the vehicle goes over disturbances (potholes, uneven roads, etc.). The effect of the disturbances on the response of the vehicle xv should dissipate quickly and with minimum oscillations. The figure shows the values of all the equipment, and the masses of the vehicle and suspension.

Model and simulate the entire system. Assuming that the road changes (a) $xr = 0.3u(t-3)$, m, and (b) $xr = 0.3\sin(10t)u(t-3)$, m. Graph the change of xr and the response xv for 15 seconds.

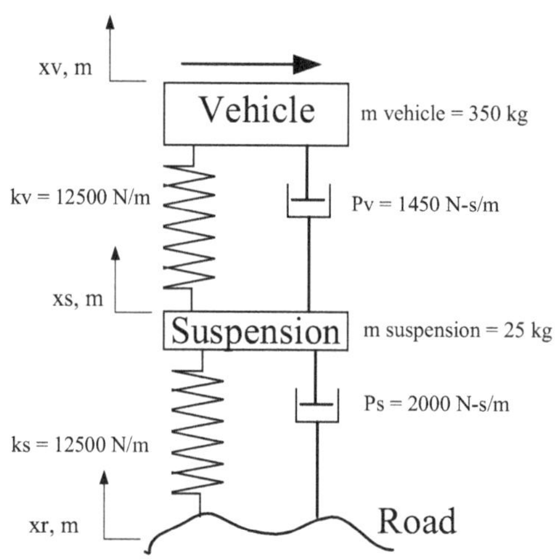

FIGURE P6.23 Block system for Problem 6.23.

Problem 6.24

Consider the mechanical system, shown in Figure P6.24, which is part of an agricultural machine. Model the system giving the displacement of all three masses $-x_1$, x_2, and x_3 –

Simulate the system and run the simulation using the values shown in the figure. Then make $P_2 = 400\,{}^{\mathrm{N}}\!/_{\mathrm{m/s}}$ *run again the simulation and compare and discuss the results.*

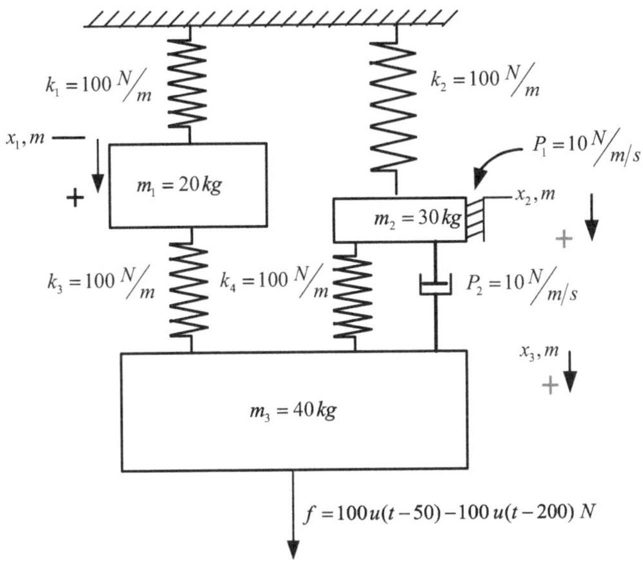

FIGURE P6.24 Block system for Problem 6.24.

Problem 6.25

Consider the mechanical system shown in Figure P6.25.

Model and simulate this system. Select the values for the second damper and second spring. Graph the responses of both blocks over 100 seconds. Discuss the responses including why you may like the values you chose for P_2 and k_2. Based on your results, what is the minimum required distance between the two blocks, and why?

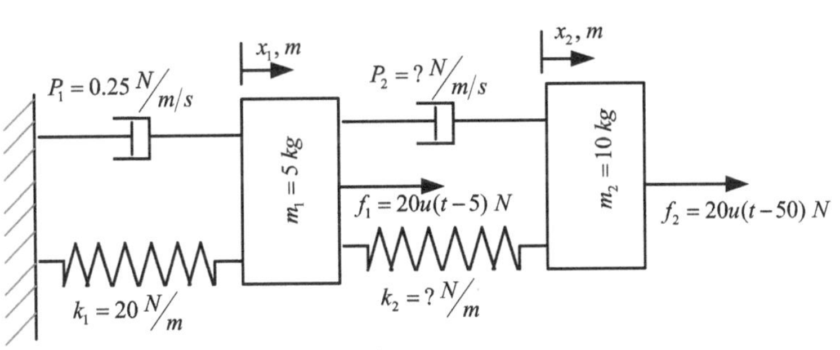

FIGURE P6.25 Block system for Problem 6.25.

7 Mechanical Systems
Rotational

The focus of Chapter 6 was on systems where the elements moved in pure translation. In this chapter, we extend the concepts from Chapter 6 to systems in which the elements rotate, such as those comprised of gears, wheels, levers, pendulums, or rotating shafts. We will confine ourselves to systems in which each rotating element is a rigid body that is rotating about a single fixed axis.

7.1 MECHANICAL LAW, MOMENT OF INERTIA, AND TORQUE

Before giving the mechanical law that governs the motion of rotational elements, we have to define the displacement, velocity, and acceleration variables that are convenient for describing rotational motion. The symbol θ is used to represent the angular position of an element with respect to a reference position as shown in Figure 7.1. Angles will be expressed in radians.

We define the angular velocity ω as the time rate of change of angular position

$$\omega = \frac{d\theta}{dt} \tag{7.1}$$

and the angular acceleration α as the time rate of change of angular velocity

$$\alpha = \frac{d\omega}{dt} = \frac{d^2\theta}{dt^2} \tag{7.2}$$

The positive directions for ω and α will be taken to be the same as that for θ.

Note that ω is in rad/s and α is in rad/s^2. In many applications, these quantities are expressed in terms of degrees or revolutions. If so, they should be converted to radians before using them in calculations. We note that radians are really dimensionless, but we will show "rad" as a unit when it adds clarity to do so, as in the case of angular position, velocity, and acceleration. In developing free-body diagrams (FBDs), we will assume that angular positions and velocities are positive.

There are many analogous relations between translational systems and rotational systems. We include several of them in Table 7.1 so that students can more easily make connections between the two.

We saw in Chapter 6 that, for a rigid body (a body for which any two points remain at a fixed distance from each other) in translation, the motion of a mass in the x direction was governed by Newton's second law

$$\sum F_i = ma \tag{7.3}$$

where the sum is over all external forces F_i acting on the body. For a rigid mass rotating around a fixed axis, the analogous law (known as Euler's second law) is

$$\sum \tau_i = J\alpha \tag{7.4}$$

where J is the mass moment of inertia of the body around the axis of rotation and where the left side of Equation 7.4 represents the sum of the *external* torques τ_i acting on the body. Torque (also

DOI: 10.1201/9781003619161-7

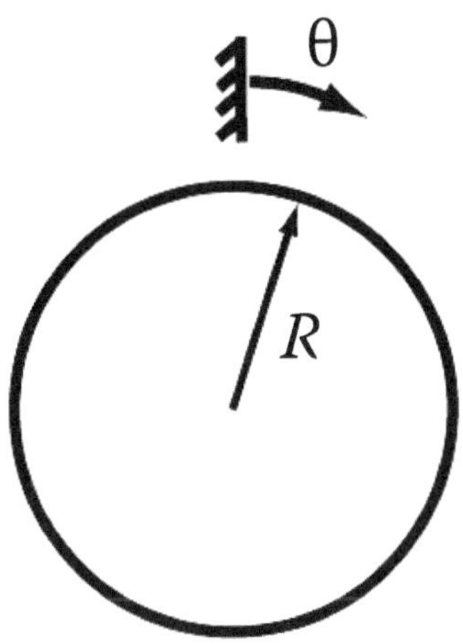

FIGURE 7.1 Angular position θ.

TABLE 7.1
Analogous Relations between Translational and Rotational Systems

Translational Quantity	Rotational Analog
Position x	Angular position θ
Velocity $v = dx/dt$	Angular velocity $\omega = d\theta/dt$
Acceleration $a = dv/dt$	Angular acceleration $\alpha = d\omega/dt$
Mass m	Mass moment of inertia J
Force $F = ma$	Torque $\tau = J\alpha$

known as moment) is the tendency of a force to induce rotation of a body around an axis. The mass moment of inertia J in rotational systems is analogous to mass in translational systems – both provide a measure of resistance to motion. We will discuss both torque and mass moment of inertia in more detail later. First, however, we will note that D'Alembert's principle, as applied to translational systems (see Equations 6.27a and 6.27b), can also be applied to rotational systems. Here, $-J\alpha$ in Equation 7.4 is interpreted as an *inertial torque*, and Equation 7.4 is rewritten as

$$\sum \tau_j = 0 \tag{7.5}$$

where the τ_j terms include the external torques τ_i and the inertial torque $-J\alpha$.

7.1.1 Mass Moment of Inertia

The mass moment of inertia J can be evaluated from

$$J = \int r^2 dm \tag{7.6}$$

where the integration is over the mass m of the body and r is the distance from the axis of rotation. So the moment of inertia depends on the size, shape, and density of the body but, most importantly, upon the location of the axis that it is being rotated around. In Example 7.1, we illustrate how the moment of inertia can be evaluated for a simple shape.

Example 7.1

Consider the long slender bar of mass m and length L shown in Figure 7.2. Evaluate its moment of inertia if the axis of rotation is (a) at the end of the bar and (b) at the middle of the bar.

By "slender bar," we mean that the thickness of the bar is very small relative to its length. In this case, we can apply the approximation that every point in the bar at a particular position x is the same distance from the y axis, or, in other words, $r = x$.

Let us define a differential element of the bar to have length dx and the cross-sectional area of the bar to be A. The volume of the differential element is then

$$dV = Adx \tag{7.7}$$

and the mass of the differential element is

$$dm = \rho dV = \rho Adx \tag{7.8}$$

where ρ is the density of the material making up the bar. We assume here that the density of the material making up the bar is uniform throughout.

For part (a), the bar extends from $x = 0$ to $x = L$. Substituting for dm and $r = x$ in Equation 7.6 and using the appropriate limits of integration:

$$J = \int_0^L x^2 \rho Adx = \rho A \frac{x^3}{3}\bigg|_0^L = \rho A \frac{L^3}{3} \tag{7.9}$$

However, the mass m of the entire bar is

$$m = \rho AL \tag{7.10}$$

So the moment of inertia can be expressed by

$$J = \frac{mL^2}{3} \tag{7.11}$$

For part (b), the bar extends from $x = -L/2$ to $x = L/2$. Changing the limits of integration in Equation 7.9, the moment of inertia about an axis through the midpoint of the bar is

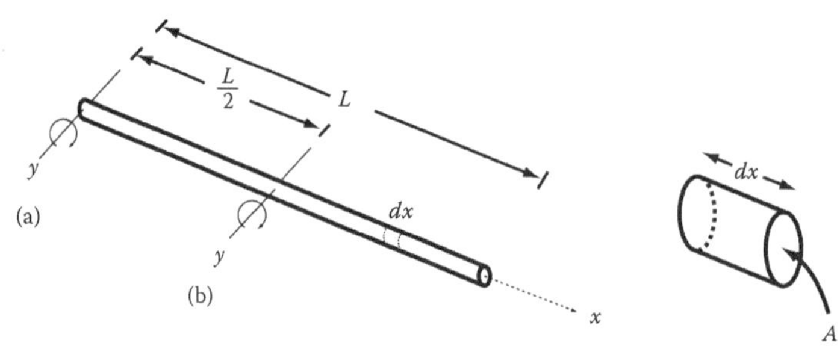

FIGURE 7.2 Rotation of a slender bar and enlarged view of volume element.

$$J = \int_{-L/2}^{L/2} x^2 \rho A dx = \rho A \frac{x^3}{3}\Bigg|_{-L/2}^{L/2} = \rho A\left(\frac{(L/2)^3}{3} - \frac{(-L/2)^3}{3}\right) = \rho A \frac{L^3}{12} \tag{7.12}$$

Finally, substitution of m for ρAL in Equation 7.12 yields

$$J = \frac{mL^2}{12} \tag{7.13}$$

7.1.2 Torque

When a force F acts upon a rigid body in such a way as to produce a tendency of the body to rotate, it is said to exert a torque (or moment) τ on the body. The magnitude of the torque is the magnitude of the force multiplied by the moment arm of the force, which is the *perpendicular* distance between the axis of rotation and the line of action of the force, as shown in Figure 7.3.

Specifically, the magnitude τ of the torque created by the force of magnitude F would be $\tau = Fd$, where d is the length of the moment arm. In this particular example, the force will produce a tendency for the body to rotate in the *clockwise* direction around an axis through point A.

The concept of torque is illustrated further in Example 7.2.

Example 7.2

The pendulum shown in Figure 7.4 is a long slender rod of length L and mass m. It is put into motion by rotating the rod to an angle θ_0 and releasing it from rest. Neglecting air resistance and friction around the pivot point of the pendulum, develop an expression for the position θ of the pendulum as a function of time.

We will begin by drawing an FBD for the pendulum. In the absence of friction and air resistance, the only force acting on the pendulum is that of gravity, which acts through the center of mass as shown in the left side of Figure 7.5.

In Figure 7.5, the displacement θ is taken as positive counterclockwise from the vertical axis. In the bar orientation shown in Figure 7.5a, it is clear that the gravity force mg will produce a clockwise torque τ_g on the rod. Trigonometry is used to determine that the length of its moment arm is $L/2 \sin \theta$. The FBD for the pendulum is shown in Figure 7.5b in which the inertial torque $-J\theta''$ is represented by a torque of magnitude $J\theta''$ drawn in the clockwise (negative) direction and where the clockwise torque τ_g due to the gravity force is also shown. We also show the reaction forces A_x and A_y at the pin. However, their contributions to the torques will be zero because the distance from their lines of action to the axis of rotation is zero. Because of this, we will suppress showing these reaction forces in subsequent examples.

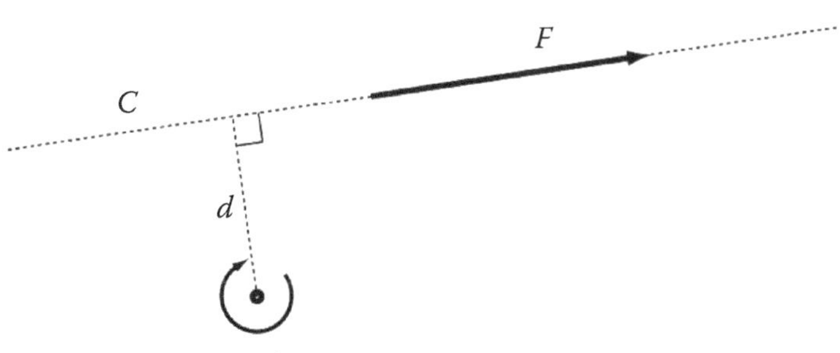

FIGURE 7.3 Moment arm d between axis of rotation A and the line of action C of a force F.

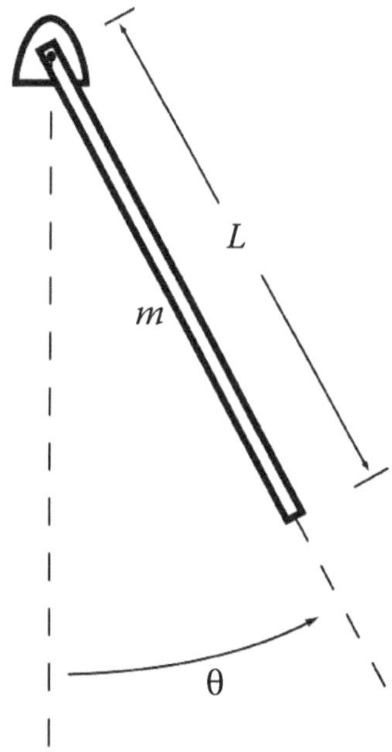

FIGURE 7.4 A pendulum composed of a slender rod.

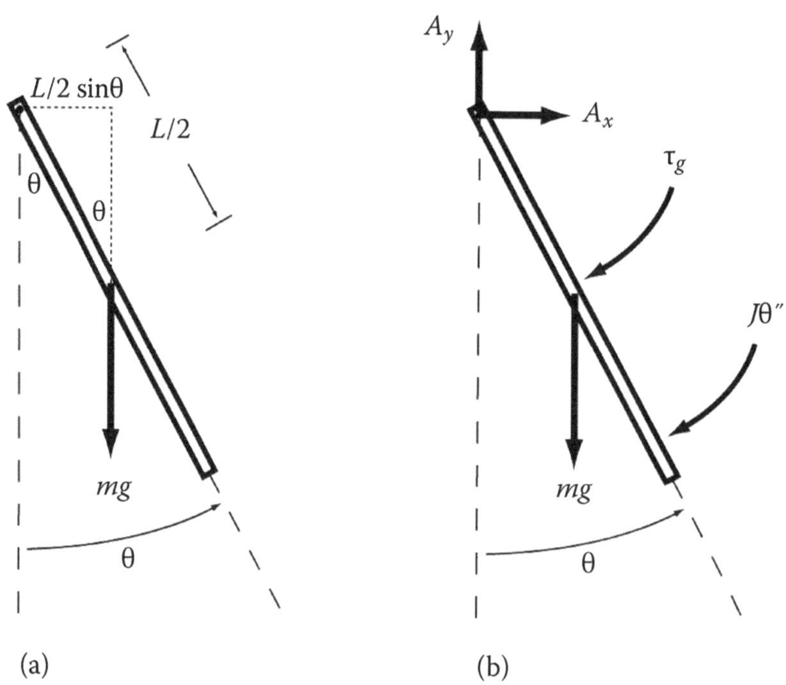

FIGURE 7.5 Moment arm evaluation (a) and free body diagram (b) for the pendulum of Example 7.2.

Applying Equation 7.5, treating torques acting in the positive θ direction as positive, yields

$$(+\text{CCW}) - \tau_g - J\theta'' = 0 \tag{7.14}$$

1 equation, 3 unknowns $[J, \theta, \tau_g]$.

The (+CCW) symbol is used to emphasize that the counterclockwise direction is taken as positive. We will also use (+CW), (+R), and (+L), when appropriate, to indicate that the clockwise, right, and left directions are positive.

Noting that torque is the product of force times the length of the moment arm

$$\tau_g = \text{mg}\frac{L}{2}\sin\theta \tag{7.15}$$

2 equations, 3 unknowns,

and that the moment of inertia for a slender bar rotating around an axis through its end (see Example 7.1) is

$$J = \text{m}\frac{L^2}{3} \tag{7.16}$$

3 equations, 3 unknowns,

results in a completely defined system. Combining Equations 7.14, 7.15, and 7.16 yields, after rearrangement,

$$\frac{d^2\theta}{dt^2} + \left(\frac{3\text{g}}{2\text{L}}\right)\sin\theta = 0 \tag{7.17}$$

The pendulum is put into motion by rotating it to an angle of θ_0 and releasing it from rest. The initial conditions are therefore

$$\theta(0) = \theta_0 \quad \theta'(0) = 0 \tag{7.18}$$

Equations 7.17 and 7.18 provide all the information needed to find $\theta(t)$. Note, however, that the differential equation is nonlinear because of the sin θ term and cannot be solved by the methods for linear equations described so far. It can be solved readily by simulation using the tools described in Chapter 11. Here, we will obtain an *approximate* analytical solution by linearizing the sin θ term. Expressing $f(\theta) = \sin\theta$ as a two-term Taylor series expanded about $\theta = 0$ yields

$$\sin\theta = f(0) + \theta f'(0) = \sin(0) + \theta\cos(0) = \theta \tag{7.19}$$

Replacing sin θ by θ in Equation 7.19 yields a linear second-order homogeneous differential equation with constant coefficients:

$$\frac{d^2\theta}{dt^2} + \left(\frac{3\text{g}}{2\text{L}}\right)\theta = 0 \tag{7.20}$$

Equation 7.20 can be solved by the methods of Section 3.4. Assuming a solution of the form $\theta = e^{rt}$ results in

$$\theta = C_1 \cos\left(\sqrt{\frac{3g}{2L}}t\right) + C_2 \sin\left(\sqrt{\frac{3g}{2L}}t\right) \quad (7.21)$$

Applying initial conditions (Equation 7.18) provides the final result

$$\theta = \theta_0 \cos\left(\sqrt{\frac{3g}{2L}}t\right) \quad (7.22)$$

It is of interest to know the effect of the linearization of the differential equation. In Figure 7.6 we compare the analytical solution (Equation 7.22) of the linearized equation (Equation 7.20) to a numerical solution of nonlinear Equation 7.17 for $L = 0.3$ m and two different values of θ_0 (30° and 90°). Figure 7.6a shows that for $\theta_0 = 30°$, the analytical solution to the linearized equation is in good agreement with the numerical solution to the nonlinear equation but that the agreement is poor when $\theta_0 = 90°$. This is consistent with the fact that the Taylor series expansion of Equation 7.19 was truncated after the second term. Thus, we would expect that its use would become progressively less accurate as θ_0 increases.

7.2 TORSION SPRINGS

In Section 6.1, we reviewed the force model for a spring, in which the force that a spring exerts on an attached mass is given by Hooke's law (Equation 6.3). In this section, we apply a similar concept to torsion springs, in which a torque arises not from stretching or compressing but by twist.

Torsion springs fall into two categories. Helical torsion springs, such as those found in a mousetrap or clothespin, exert torque through a wire shaped in a helix or coil. Torsion bars, such as those used to support automobile suspension components, are straight bars that are subjected to twist around their axis.

Consider the torsion bar connecting the two masses shown in Figure 7.7.

We neglect the mass of the spring (bar) so there is no inertial torque associated with it. In this case, Equation 7.5 predicts that the torque τ_S exerted on the two masses connected to the spring will be equal and in opposite directions – exactly analogous to the ordinary spring where the forces exerted by each end on the elements connected there are equal and opposite. The analog of Hooke's law for a torsion spring is

$$\tau_S = k(\theta_1 - \theta_2) \quad (7.23)$$

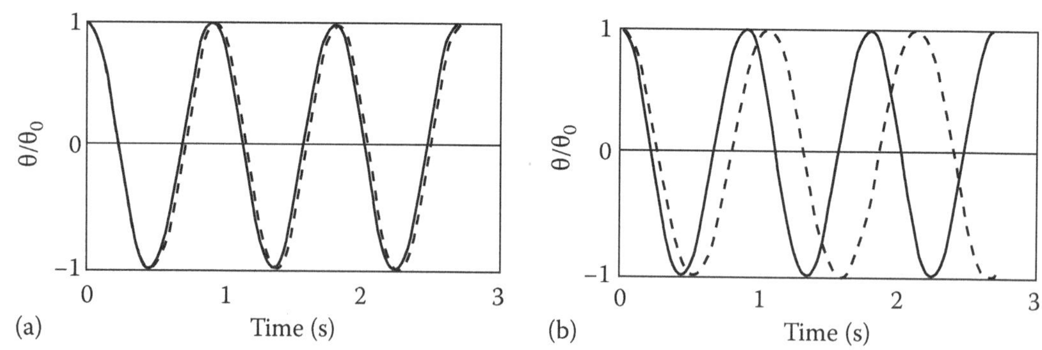

FIGURE 7.6 Comparison of analytical solution of linearized Eq. 7.20 to numerical solution of Equation 7.17 for (a) $\theta_0 = 30°$ and (b) $\theta_0 = 90°$.

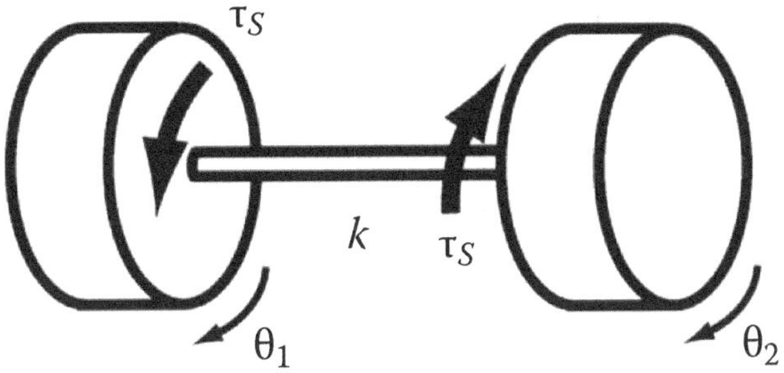

FIGURE 7.7 Torsion bar.

where τ_S is the torque exerted by the spring. Equation 7.23 is written for the case where both ends of the spring are attached to moving elements. Ordinarily, we will choose the reference positions of these elements to correspond to where there is no twist in the spring (i.e., $\theta_1 = \theta_2 = 0$). We illustrate the procedure for assigning directions of the torques in Example 7.3.

Example 7.3

Consider the system of two masses and two torsion springs shown in Figure 7.8. Initially, neither mass is in motion nor the spring twisted. A constant external torque T is applied to mass 2 at $t = 0$. Find the angular displacements of masses 1 and 2 as a function of time if $k_1 = k_2 = 10$ N · m, $T = 1$ N · m and $J_1 = J_2 = 1$ kg · m^2.

Let us look at the system lengthwise from the right side – so that the positive θ directions and the external torque T are both clockwise in sense. We then draw FBDs for each of the two masses as shown in Figure 7.9. The inertial torques $J_1\theta_1''$ and $J_2\theta_2''$ are then drawn counterclockwise. Now we have to assign directions to the spring torques. Let us assume that $\theta_2 > \theta_1 > 0$. The quantity $k_2(\theta_2 - \theta_1)$ is therefore positive. We expect that the torque of the rightmost spring will be counterclockwise (negative) on mass 2 and clockwise on mass 1. We also expect that the torque $k_1\theta_1$ of the leftmost spring will be counterclockwise on mass 1.

Applying Equation 7.5 to both masses with clockwise torques positive and counterclockwise terms negative:

$$(+\text{CW}) - J_1\theta_1'' - k_1\theta_1 + k_2(\theta_2 - \theta_1) = 0 \tag{7.24}$$

1 equation, 2 unknowns $[\theta_1, \theta_2]$;

$$(+\text{CW}) - J_2\theta_2'' + T - k_2(\theta_2 - \theta_1) = 0 \tag{7.25}$$

2 equations, 2 unknowns.

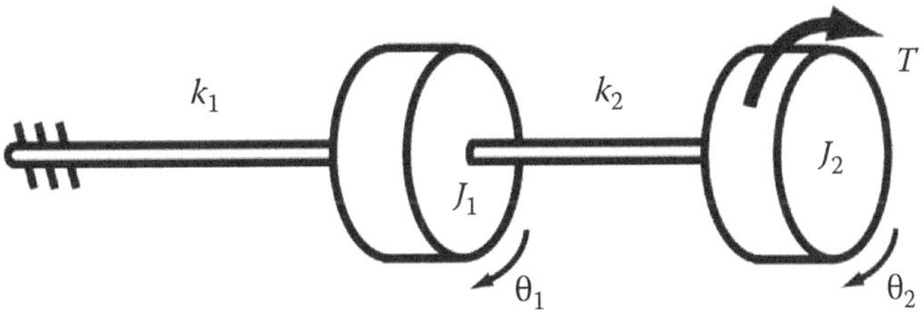

FIGURE 7.8 System of two rotating masses and two torsion springs.

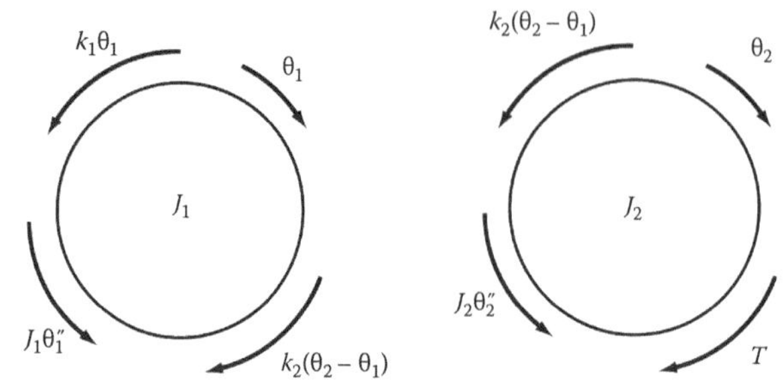

FIGURE 7.9 Free body diagrams of the masses in Example 7.3.

Equations 7.24 and 7.25, along with initial conditions $\theta_1(0) = \theta_2(0) = \theta_1'(0) = \theta_2'(0) = 0$ allow the determination of θ_1 and θ_2 as functions of time. Rearranging and substituting parameter values ($k_1 = k_2 = 10$ N · m, $T = 1$ N · m and $J_1 = J_2 = 1$ kg · m^2) yields

$$\theta_1'' = 10(\theta_2 - \theta_1) - 10\theta_1 \tag{7.26}$$

$$\theta_2'' = 1 - 10(\theta_2 - \theta_1) \tag{7.27}$$

Equations 7.26 and 7.27 can be easily solved using the simulation methods described in Chapter 11. With a bit more elbow grease, an analytical solution can be obtained using the method of Laplace transforms as illustrated in Section 4.5. The resulting solution is

$$\theta_1 = 0.1 + 0.01708\cos(5.1167t) - 0.11708\cos(1.9544t) \tag{7.28}$$

$$\theta_2 = 0.2 - 0.01056\cos(5.1167t) - 0.18944\cos(1.9544t) \tag{7.29}$$

The solution is oscillatory and undamped, as we would expect from the absence of friction in the system.

7.3 ROTATIONAL DAMPING

Damping occurs in rotational systems just as it does in translational systems. Two examples are shown in Figure 7.10, where damping coefficient P characterizes damping between two rotating elements and coefficient P_2 characterizes damping between a rotating element and a stationary surface.

As with damping for translational systems, we will employ a viscous fluid model here. In translational systems, the viscous drag force was proportional to the velocity difference between both ends of the damper. In rotational damping, drag exerts a *torque* that is proportional to the difference in *angular* velocity between both sides of the damper. For the frictional element between masses 1 and 2 in Figure 7.10, for instance, the torque caused by drag would be

$$\tau_D = P(\theta_1' - \theta_2') \tag{7.30}$$

and the torque caused by drag between mass 2 and the stationary surface would be

$$\tau_D = P_2(\theta_2') \tag{7.31}$$

Analogous to the case of damping in translational systems, the torque exerted on elements on both sides of a damper are equal and opposite. We illustrate motion in a rotational system with damping in Example 7.4.

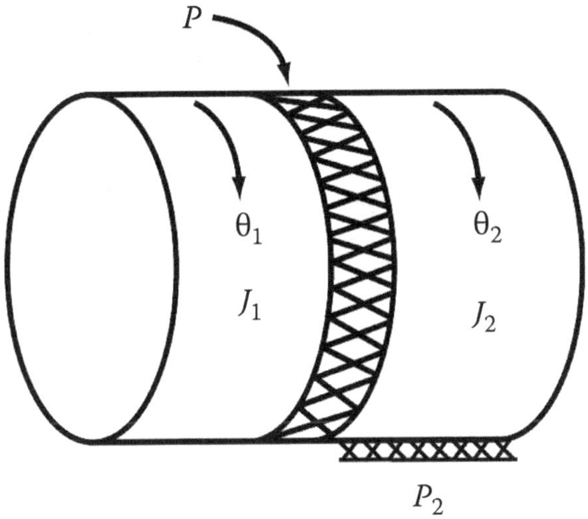

FIGURE 7.10 Damping in a rotational system.

Example 7.4

Consider the system shown in Figure 7.11. Initially, both masses 1 and 2 are stationary in a reference state where there is no twist in the spring. At $t = 0$, a constant external torque T is applied as shown:

a. Develop the differential equations that describe the motion of the two masses.
b. Given that $J_1 = 0.01$ kg · m², $J_2 = 0.08$ kg · m², $P = 0.08$ N · m · s, $k = 0.27$ N · m, and $T = 0.1$ N · m, calculate the angular displacement θ_1 of mass 1 and the angular velocity ω_2 of mass 2 after the external torque has been applied for a long time.

Part a

We imagine looking at the system vertically from the top down. The external torque T and the directions of positive angular displacement are then clockwise and the inertial torques $J_1\theta_1''$ and $J_2\theta_2''$ are then drawn counterclockwise. In addition to the external torque, a torque is exerted on mass 2 by the damper (characterized by damping coefficient P). Mass 1 has a torque exerted by the damper and another exerted by the torsion spring. We consider the case of $\theta_2' > \theta_1' > 0$ and conclude that the damping torque on mass 2 will be $P(\theta_2' - \theta_1')$ with a counterclockwise sense and that the damping torque on mass 1 will be the same but with a clockwise sense. We also note that the torque $k\theta_1$ exerted by the torsion spring on mass 1 will have a counterclockwise sense. FBDs for each mass are shown in Figure 7.12.

Applying Equation 7.5 to both masses with clockwise torques positive and counterclockwise terms negative:

$$(+CW) - J_1\theta_1'' - k\theta_1 + P(\theta_2' - \theta_1') = 0 \tag{7.32}$$

1 equation, 2 unknowns $[\theta_1, \theta_2]$;

$$(+CW) - J_2\theta_2'' + T - P(\theta_2' - \theta_1') = 0 \tag{7.33}$$

2 equations, 2 unknowns.

Equations 7.32 and 7.33, along with initial conditions $\theta_1(0) = \theta_2(0) = \theta_1'(0) = \theta_2'(0) = 0$ allow the determination of θ_1 and θ_2 as functions of time. They may be solved numerically using the methods

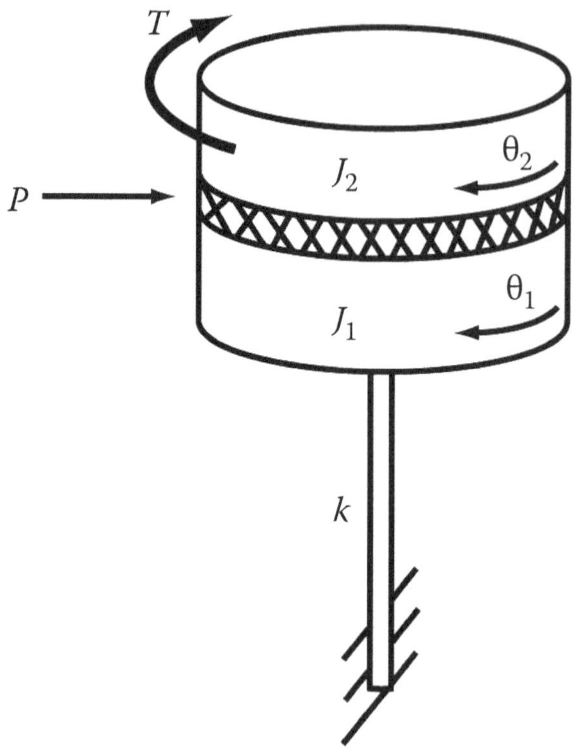

FIGURE 7.11 Rotational system of Example 7.4.

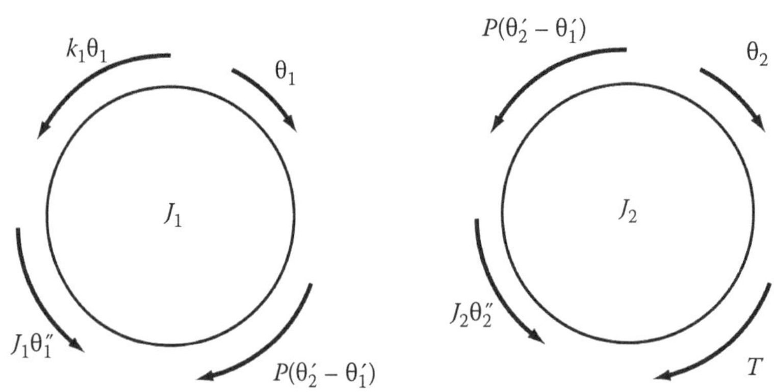

FIGURE 7.12 Free body diagrams for Example 7.4.

of Chapter 11 or analytically, most conveniently, by the method of Laplace transform. The details are left as a (lengthy) exercise. The results for θ_1 and $\omega_2 = \theta_2'$, using the parameter values given in part b, are shown in Figure 7.13. Note that there are no oscillations so the system is heavily damped.

Part b

We anticipate that, after a very long time, mass 2 will move with a constant angular velocity with the external torque balanced by the damping torque. The torsion spring cannot be twisted infinitely, and the presence of damping suggests that any oscillations will eventually die out. We anticipate

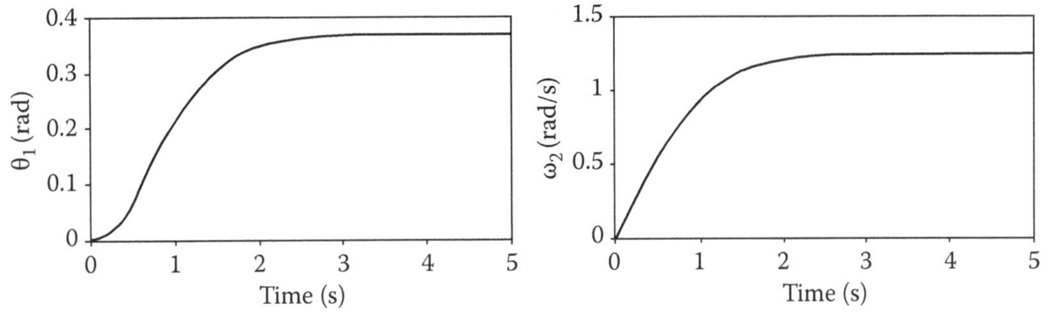

FIGURE 7.13 Solution to Equations 7.32 and 7.33 using parameters given in part (b).

therefore that mass 1 will have no angular velocity after a very long time. We apply Equations 7.32 and 7.33 with $\theta_1'' = \theta_2'' = \theta_1' = 0$ to obtain (noting $\theta_2' = \omega_2$)

$$\omega_2 = T/P \tag{7.34}$$

$$\theta_1 = P\omega_2/k = T/k \tag{7.35}$$

Substituting the given parameter values yields the ultimate values of ω_2 and θ_1:

$$\omega_2 = (0.1\,\mathrm{N\cdot m}/0.08\cdot\mathrm{N\cdot m\cdot s}) = 1.25(\mathrm{rad})/\mathrm{s} \tag{7.36}$$

$$\theta_1 = T/k = (0.1\,\mathrm{N\cdot m})/(0.27\,\mathrm{N\cdot m}) = 0.37(\mathrm{rad}) \tag{7.37}$$

Note that these values are in agreement with the full solution shown in Figure 7.13.

There are at least two reasons to examine the limiting case of the solution at long times. First, many systems are designed for steady-state operation, and it might therefore be a good idea to determine whether the system shows the appropriate steady-state behavior before investing the time to solve the differential equation(s) to obtain the transient behavior. Second, the limiting case supplies a check on both the analytical and numerical solutions to the model. As t approaches infinity, both the numerical and analytical models should provide the same displacements and/or velocities as the limiting case analysis or else an error was made somewhere.

7.4 GEARS

Rotational elements may be geared so that one can be used to rotate another. An example is shown in Figure 7.14a. We will assume here that there is *no slip* between gears. As a result, there is a fixed and simple relationship between the angular displacements of two connected gears. With no slip, the *tangential* displacement (the distance in meters that a point on the periphery of the gear would travel) must be the same for the two gears. It follows (for the gears shown in Figure 7.14) that the angular displacements are related by

$$R_1\theta_1 = R_2\theta_2 \tag{7.38}$$

Taking the time derivative once and then again of both sides of Equation 7.38 provides similar expressions relating to the angular velocities and angular accelerations of the two gears:

$$R_1\omega_1 = R_2\omega_2 \quad \text{or} \quad R_1\theta_1' = R_2\theta_2' \tag{7.39}$$

$$R_1\alpha_1 = R_2\alpha_2 \quad \text{or} \quad R_1\theta_1'' = R_2\theta_2'' \tag{7.40}$$

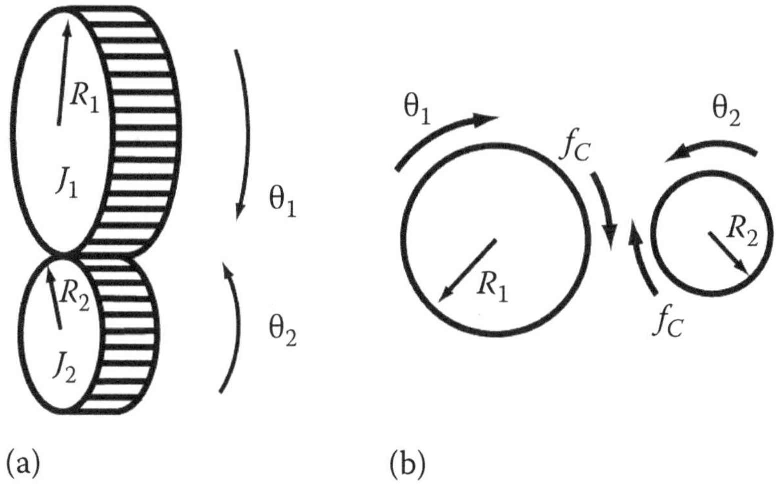

FIGURE 7.14 (a) Two gears and (b) partial free-body diagram showing the contact force f_C.

There will exist a contact force between two gears that acts on each in an opposed fashion as shown in Figure 7.14b. The directions of the contact force may be assigned arbitrarily. Here, gear 2 is interpreted as "driving" gear 1, which is placing a resistance on gear 2. It is interesting to note that the contact forces are opposed but that each acts on its gear in a clockwise sense. It should also be noted that it is the *magnitude* of this contact force that is the same on both gears, not the resulting *torques*.

Interconnection relations given by the contact force and equations such as Equations 7.38–7.40 often allow us to reduce the number of differential equations in a system model, resulting in an easier solution than we might expect. We illustrate the development of a system model for a system that includes gears with Example 7.5.

Example 7.5

Consider the system of two masses geared together as shown in Figure 7.15. Initially, both masses 1 and 2 are stationary in a reference state where there is no twist in the spring. At $t = 0$, a constant external torque T is applied as shown. The radii of the two masses are $R_1 = 0.2$ m and $R_2 = 0.1$ m. Given that $J_1 = 2\ \text{kg} \cdot \text{m}^2$, $J_2 = 1.5\ \text{kg} \cdot \text{m}^2$, $P = 4\ \text{N} \cdot \text{m} \cdot \text{s}$, $k = 10\ \text{N} \cdot \text{m}$, and $T = 4\ \text{N} \cdot \text{m}$, calculate the angular displacements θ_1 and θ_2 as a function of time.

Note that the positive directions for angular displacement have opposite senses for the two masses. Looking from the left side, positive displacement is clockwise for mass 1 and counterclockwise for mass 2. Mass 1 is subject to the external torque T and a torque generated by the contact force f_C between the two gears. Torques on mass 2 include those generated by the contact force f_C, the damping torque characterized by coefficient P and the torsion spring torque characterized by spring constant k. The FBD for each gear is shown in Figure 7.16, where the inertial torques have been drawn with opposite sense to the direction of positive displacement. For mass 1 moving clockwise and mass 2 moving counterclockwise, we expect the friction and spring torques to act clockwise on mass 2. We interpret mass 1 to be driving mass 2, so the torques exerted by the contact forces are drawn counterclockwise. Note that they are represented by the contact force multiplied by the moment arms and are not the same for both masses.

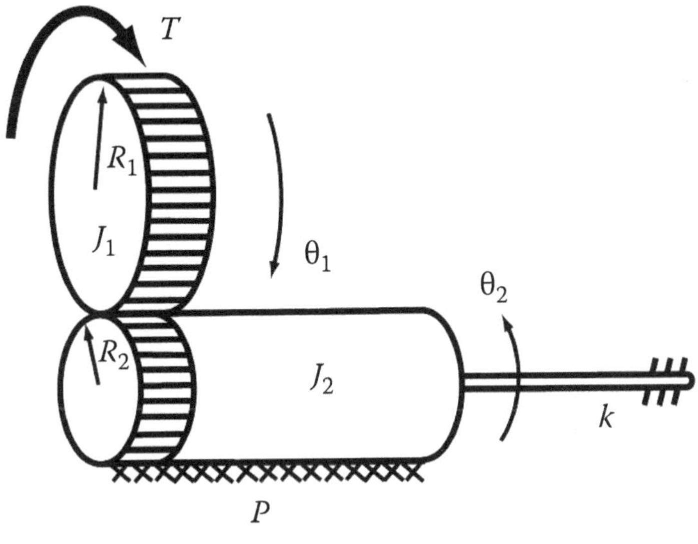

FIGURE 7.15 Geared elements of Example 7.5.

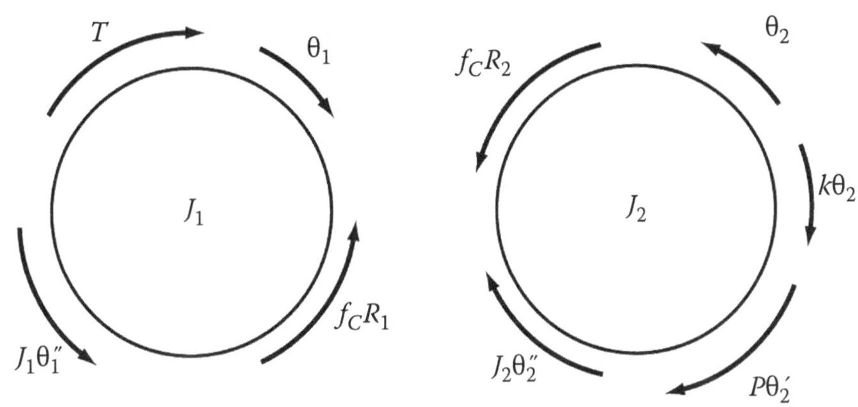

FIGURE 7.16 Free-body diagrams for the gears of Example 7.5.

Applying Equation 7.5 to each mass:

$$(+\text{CW}) - J_1\theta_1'' + T - f_C R_1 = 0 \tag{7.41}$$

1 equation, 2 unknowns $[\theta_1, f_C]$

$$(+\text{CCW}) - J_1\theta_1'' + f_C R_1 - k\theta_1 - P\theta_1' = 0 \tag{7.42}$$

2 equations, 3 unknowns $[\theta_2]$.

We note from Equation 7.40 that

$$\theta_1'' = R_2\theta_2''/R_1 \tag{7.43}$$

3 equations, 3 unknowns.

The system is therefore completely defined. Substituting for θ_1'' in Equation 7.41 using Equation 7.43 and solving for the contact force:

$$f_C = -J_1 R_2 \theta_2'' / R_1^2 + T / R_1 \tag{7.44}$$

Substituting for f_C in Equation 7.42 yields, after rearrangement,

$$\left[J_2 + \left(\frac{R_2}{R_1} \right)^2 J_1 \right] \theta_2'' + P\theta_2' + k\theta_2 = T\frac{R_2}{R_1} \tag{7.45}$$

Substituting for the parameter values given in the problem statement and dividing through by the coefficient of the θ_2'' term yields

$$\theta_2'' + 2\theta_2' + 5\theta_2 = 1 \tag{7.46}$$

Equation 7.46 is subject to the initial conditions $\theta_2 = \theta_2' = 0$. It is a nonhomogeneous second-order ordinary differential equation with constant coefficients that can be solved by the methods of Chapter 3, in which the homogeneous equation is solved followed by obtaining the particular solution. Alternatively, Laplace transforms may be used. The resulting solution is

$$\theta_2 = e^{-t}(-0.2\cos 2t - 0.1\sin 2t) + 0.2 \tag{7.47}$$

The expression for θ_1 is obtained as $\theta_1 = \theta_2/2$ by using Equation 7.38.

7.5 SYSTEMS WITH ROTATIONAL AND TRANSLATIONAL ELEMENTS

We will not consider here systems in which a given element may undergo *both* translation and rotation. However, there are practical situations in which a system may have some elements in pure rotation and others in pure translation that can be treated by the methods already described. Such a system is shown in Figure 7.17 and illustrated in Example 7.6.

Example 7.6

Consider the system shown in Figure 7.17. A block of mass m is connected by a cable to a rotating drum with moment of inertia J. The cable is very long and is wrapped around the drum, which

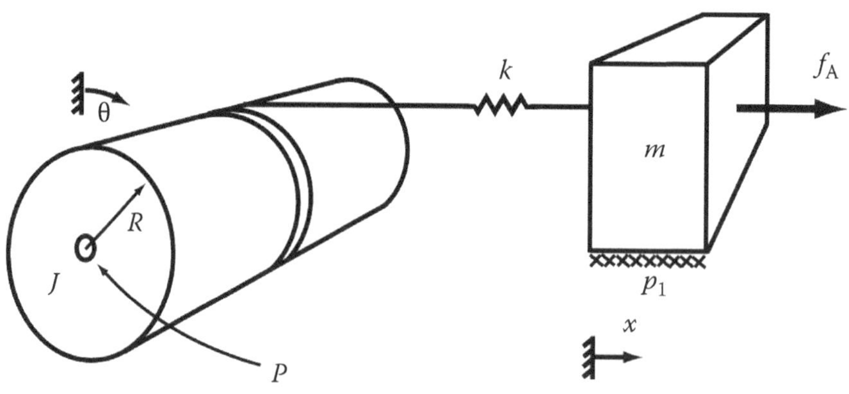

FIGURE 7.17 Drum and mass system of Example 7.6.

rotates about a fixed axis. Initially, both the block and drum are stationary in a reference state for which the spring is neither stretched nor compressed. At $t = 0$, a constant external force f_A is applied to the block as shown and the block moves to the right, unspooling the cable from the drum as it rotates. There is fluid friction between the block and surface and between the drum and its axis. There is no slip between the cable and drum.

a. Develop the differential equations that describe the motion of the block and drum.
b. Given that $J = 0.08$ kg · m², $m = 2$ kg, $P_1 = 20$ N · s/m, $P = 0.2$ N · m · s, $k = 100$ N/m, $R = 0.1$ m, and $f_A = 4$ N, calculate the ultimate (terminal) velocity v_t of the block and the ultimate stretch in the spring.

Part a

The directions of positive displacement are clockwise for the drum and to the right for the mass. If we consider the case where the spring is stretched, it will exert a force to the left on the block and a clockwise torque on the drum. The amount of stretch in the spring is the distance (x) that its right side has moved minus the distance ($R\theta$) that its left side has moved. Thus, the magnitude of the spring force on the block is $k(x - R\theta)$ and the magnitude of the torque exerted by the spring on the drum is this force times the moment arm or $kR(x - R\theta)$. Additionally, there is a counterclockwise torque of magnitude $P\theta'$ on the drum caused by friction between the drum and its axis and a force to the left of magnitude P_1x' on the block due to friction with the surface. The FBDs for the drum and block are shown in Figure 7.18 and include the inertial torque $-J\theta''$ of the drum and inertial force $-mx''$ of the block.

Applying Equation 7.5 to the drum and Equation 6.27b to the block result in

$$(+\text{CW}) - J\theta'' - P\theta' + kR(x - R\theta) = 0 \tag{7.48}$$

1 equation, 2 unknowns $[\theta, x]$;

$$(+R) - mx'' + f\text{A} - P_1x' - k(x - R\theta) = 0 \tag{7.49}$$

2 equations, 2 unknowns.

Equations 7.48 and 7.49, along with initial conditions $\theta(0) = x(0) = \theta'(0) = x'(0) = 0$ allow the determination of θ and x as functions of time. The system of equations is readily solved numerically by the methods of Chapter 11 and may be solved analytically by Laplace transforms. The solution for the parameter values given in part (b) is shown in Figure 7.19. Rather than present θ and x as functions of time, we show the block velocity $v = x'$ and the spring stretch $(x - R\theta)$. Note that the spring stretch is always positive. If it had not been, the solution would be non-physical because the cable cannot support compression. This could happen with other forcing functions f_A.

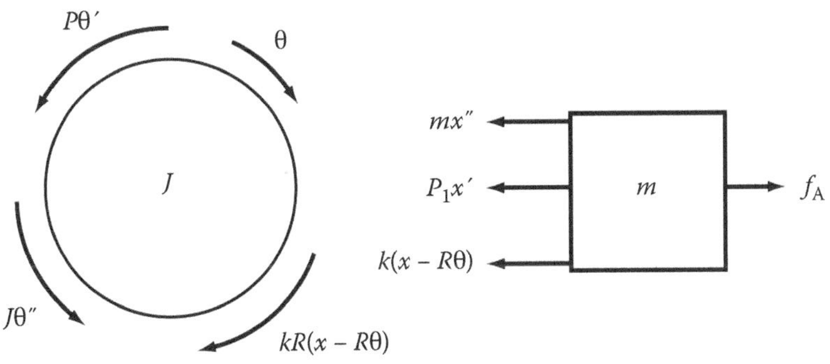

FIGURE 7.18 Free-body diagrams for Example 7.6.

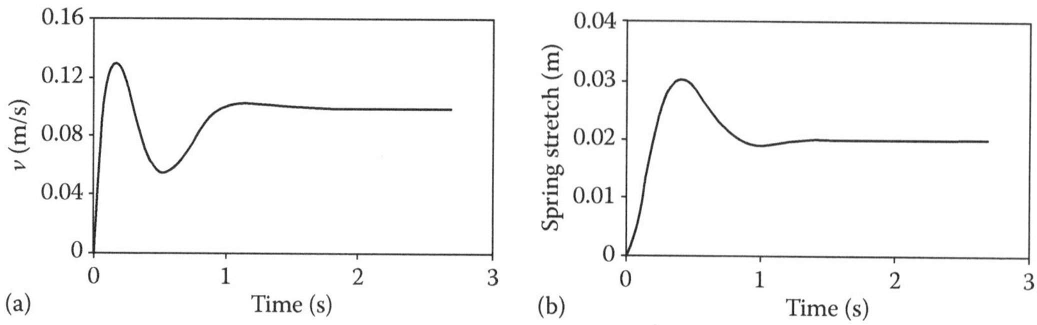

FIGURE 7.19 Velocity and spring stretch versus time for the system of Example 7.6.

Part b

When the block reaches its ultimate velocity, $\theta'' = 0$ and $x'' = 0$. The velocity x' will be the ultimate velocity v_t. Also, the spring cannot be stretched infinitely so it will reach a constant amount of stretch $S_t = (x - R\theta)_t$. Because the spring stretch reaches a constant value, the ultimate angular velocity ω_t of the drum will be simply related to the ultimate velocity v_t of the block by $R\omega_t = v_t$. Substituting $\theta'' = 0, x'' = 0, x' = v_t, \theta' = \omega_t = v_t/R$, and $x - R\theta = S_t$ into Equations 7.48 and 7.49 yields

$$-Pv_t/R + RkS_t = 0 \tag{7.50}$$

$$f_A - P_1 v_t - kS_t = 0 \tag{7.51}$$

which can be solved simultaneously for S_t and v_t:

$$v_t = \frac{f_A}{P_1 + P/R^2} = \frac{4N}{[20 + 0.2/(0.1)^2]\,\text{N s/m}} = 0.1\,\text{m/s} \tag{7.52}$$

$$S_t = \frac{Pv_t}{kR^2} = \frac{(0.2\,\text{N}\cdot\text{m}\cdot\text{s})(0.1\,\text{m/s})}{(100\,\text{N/m})(0.1)^2\,\text{m}^2} = 0.02\,\text{m} \tag{7.53}$$

Note that these values are in agreement with the solution shown in Figure 7.19.

7.6 SUMMARY

In this chapter, we applied Euler's second law (the rotational analog of Newton's second law) to systems in which rigid bodies rotate around a fixed axis. Specifically, we studied systems comprised of rotating masses, torsion springs, rotational dampers, and gears and extended the concepts of D'Alembert's principle and FBDs to rotational systems.

Problems

Problem 7.1

Refer to Figure P7.1. Evaluate the moment of inertia around the z-axis for (a) a solid cylinder of mass m, radius R, and length L and (b) a thin ring with mass m, radius R, and length L. Hints: Use the volume element shown in the figure. For part (b), assume that the disk of radius R has a thickness of ΔR and let ΔR approach zero as the final step.

Problem 7.2

Consider the pendulum of Example 7.2. Assume now that there is friction between the pendulum and pin, as shown in Figure P7.2. The pendulum is rotated to an angle θ_0 and then released from rest at $t = 0$.

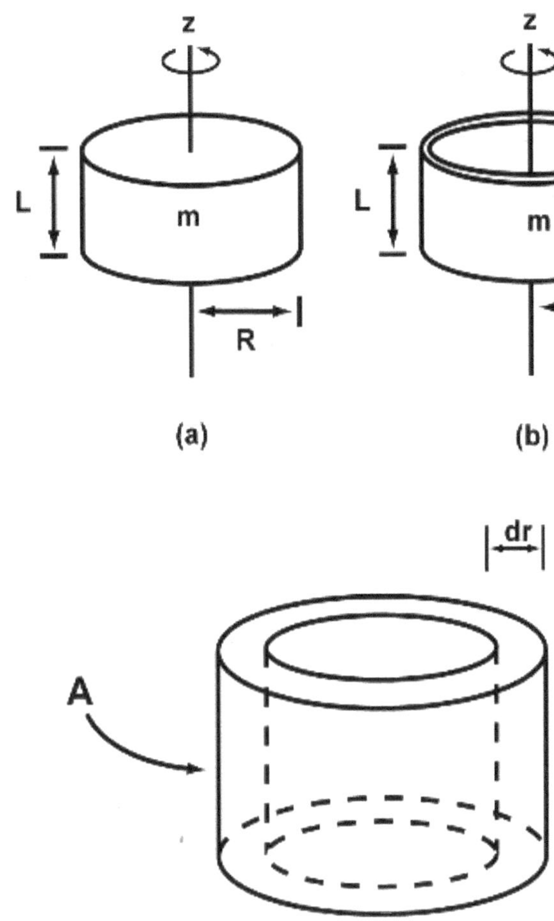

FIGURE P7.1 Disks of mass m, radius R, and length L for parts (a) and (b) and the suggested volume element.

a. Develop the model that represents the angular displacement θ of the pendulum as a function of time.
b. Let $P = 0.065$ N · m · s, $L = 0.3$ m, and $m = 1/3$ kg. Linearize the model developed in part a and obtain an analytical solution for $\theta_0 = 30°$.

Problem 7.3

Consider the torsion spring shown in Figure P7.3.

a. Develop the differential equation that represents θ as a function of time.
b. The spring is rotated by an angle of $\pi/2$ and released from rest. Solve the differential equation to obtain $\theta(t)$. Assume $J = 1$ kg·m^2, $P = 6$ N·m·s, and $k = 10$ N·m.

Problem 7.4

Consider the single torsion spring shown in Figure P7.4. Initially, masses 1 and 2 are in a state of rest with no twist in the spring. At $t = 0$, a constant external torque T_1 is applied to mass 1 and a constant external torque T_2 is applied to mass 2.

a. Develop the model that represents the angular displacements θ_1 and θ_2 of the two masses as functions of time.

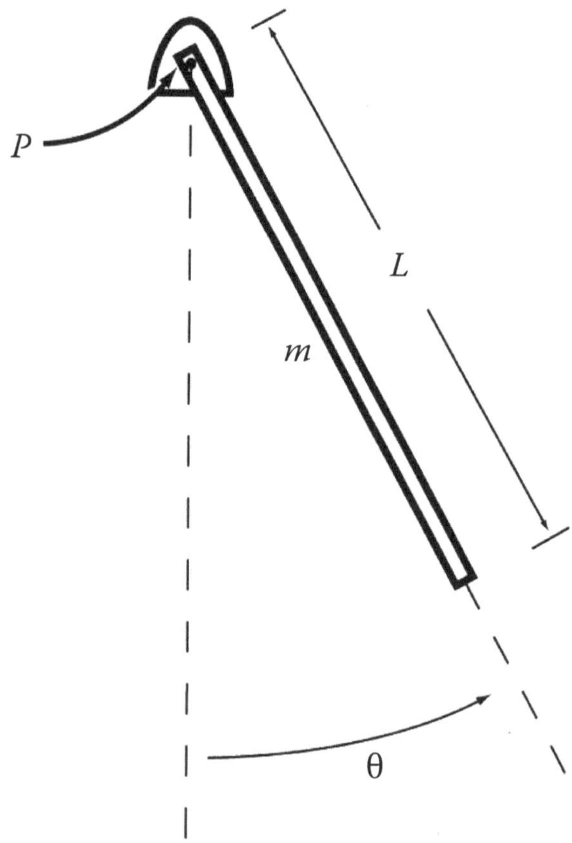

FIGURE P7.2 Pendulum made of a slender bar with friction at the pin.

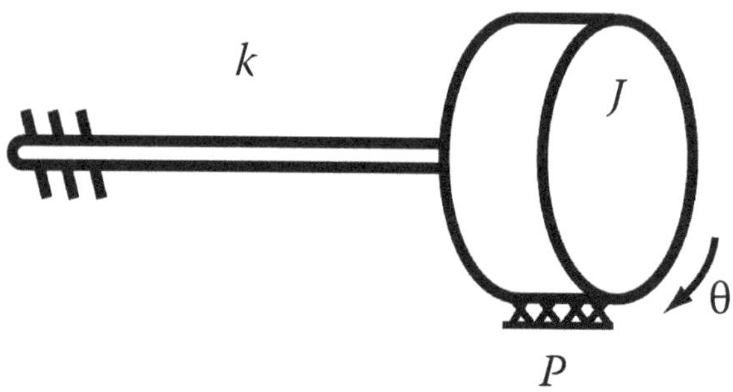

FIGURE P7.3 Torsion spring with damping.

b. Let $J_1 = J_2 = 1$ kg · m^2, $k = 8$ N · m and $T_1 = T_2 = 1$ N · m. Obtain the analytical solution for θ_1 and θ_2. Hint: Take advantage of the symmetry in the model parameters to simplify the model down to a single differential equation.

Problem 7.5

Consider the system shown in Figure P7.5. Mass 1 is a wheel, and mass 2 is a pulley connected to a belt, which exerts a torque T on the pulley. The two masses are connected by a rigid shaft. The

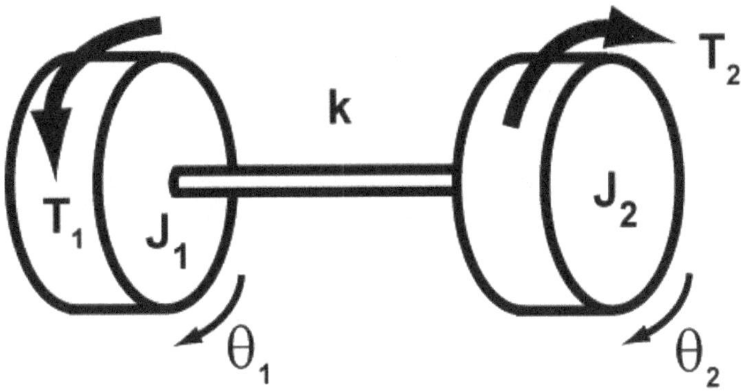

FIGURE P7.4 Single torsion spring of Problem 7.4.

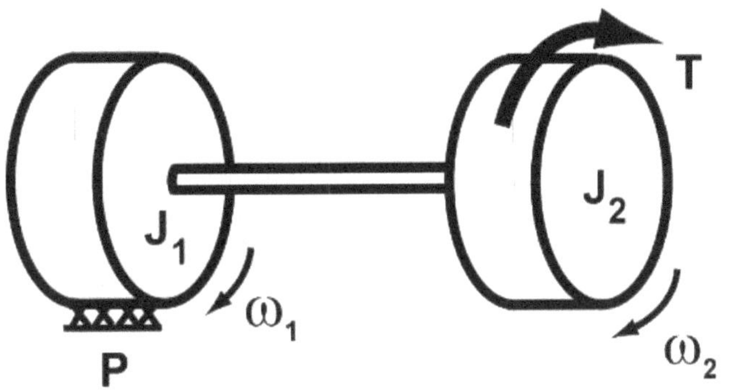

FIGURE P7.5 Wheel and pulley system of Problem 7.5.

mass and radius of the wheel are 2 kg and 0.1 m, respectively, and $J_2 = 0.005$ kg·m^2. Damping on the wheel is expressed by $P = 0.01$ N·m·s.

a. Develop the differential equation that represents ω_1 as a function of time. Solve the model to obtain $\omega_1(t)$. Assume that $\omega_1(0) = 0$.
b. It is desired that the angular velocity of the wheel at steady state be 2000 RPM. What torque should be applied? About how long will it take for the wheel to reach a steady state from rest?

Problem 7.6

Consider the viscous coupler shown in Figure P7.6. Mass 2 includes the shaft, and both turn as a rigid body. A constant torque T may be applied to mass 1, and a constant braking torque T_B may be applied to mass 2. Parameter values are $J_2 = 1.5$ kg·m^2, $P_1 = 5$ N·m·s, $P_2 = 0.5$ N·m·s.

a. Assuming that J_1 is negligible, develop the differential equation and algebraic equation that represent the general case where both T and T_B are applied.
b. Assume that $T_B = 0$ and the system is in a state of rest when a constant torque $T = 50$ N·m is applied at $t = 0$. Solve the model to obtain $\omega_1(t)$ and $\omega_2(t)$. What are their ultimate values?
c. Assume that the system reaches the steady state of part (a). At $t = 0$, T falls to zero and remains there. Solve the model to obtain $\omega_1(t)$ and $\omega_2(t)$. Approximately how long does it take for the system to come to rest?

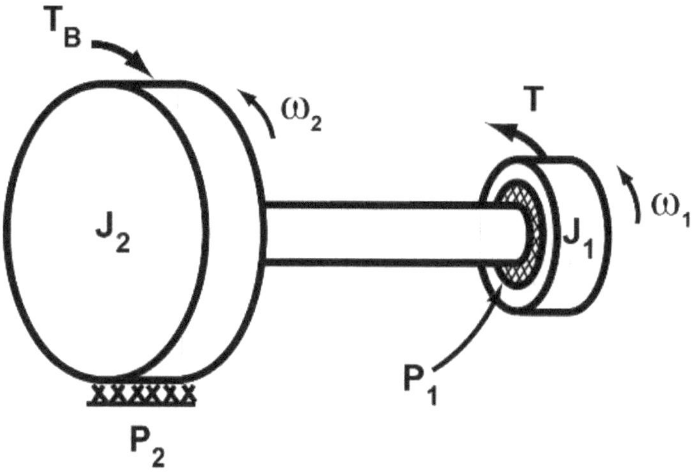

FIGURE P7.6 Viscous coupler.

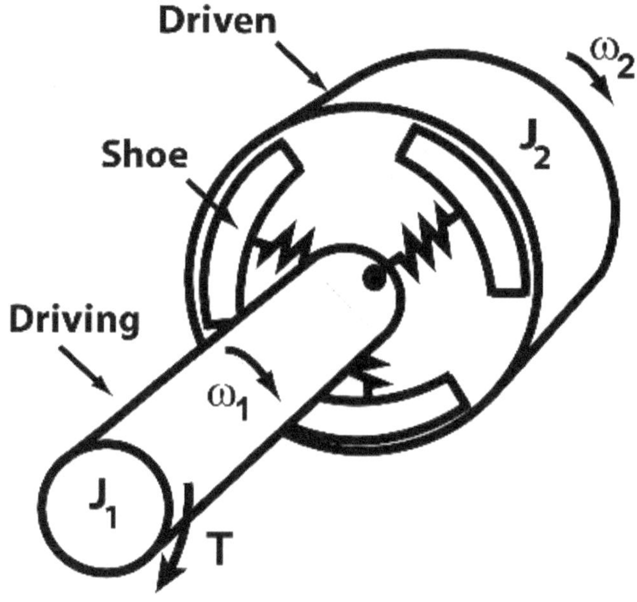

FIGURE P7.7 Centrifugal clutch.

d. Assume that the system reaches the steady state of part (a). At $t = 0$, T falls to zero and remains there, and T_B = constant = 10 N·m is applied to the mass 2. Solve the model to obtain $\omega_1(t)$ and $\omega_2(t)$. How long does it take for the system to come to rest?

Problem 7.7

A centrifugal clutch like that shown in Figure P7.7 is used in many mechanical systems including chainsaws, minibikes, weedwhackers and go-carts. It allows two rotating shafts to be matched at higher speeds and decoupled at lower ones. The clutch works as follows: At low engine speeds, the driving shaft (connected to an engine) and driven shafts (connected to a load) turn independently. As the RPMs increase, the shoes connected to the driving shaft move outwards and eventually contact the inside of the drum of the driven shaft. As the engine speed increases further, centrifugal force pushes the shoes even tighter to the drum and friction between the shoes and drum causes the driven shaft to rotate. Eventually, the rotational velocities of the driving and driven shafts will match. If the engine speed is decreased, the shoes are pulled back to the center by springs, and the shafts decouple. It is this decoupling that allows the engine to continue running if the driven shaft is braked.

For a centrifugal clutch, the frictional torque between the two shafts is given by

$$\tau_f = a(\omega_1^2 - \omega_{1,e}^2) \qquad \omega_1 > \omega_{1,e}$$

$$\tau_f = 0 \qquad \omega_1 < \omega_{1,e}$$

where a is a constant and $\omega_{1,e}$ is rotational velocity of the driving shaft at which the clutch is engaged (the shoes just reach the drum surface). The square dependence on ω_1 is due to centripetal acceleration of the shoes. There is no dependence on ω_2 because the friction is of the dry sliding type. It should be noted though that direction of the frictional torque would change if ω_2 exceeded ω_1.

In this example, the load is simply a flywheel with moment of inertia J_2. The moment of inertia J_1 is for the entire driving shaft assembly, including the springs and shoes. Any variation in J_1 due to movement of the shoes is neglected.

a. Assume $J_1 = 0.05$ kg·m², $J_2 = 0.10$ kg·m², $a = 0.0001$ N·m·s², and the engagement velocity $\omega_{1,e} = 150$ (radians)/s. The rotational velocities of both shafts are zero when a constant torque $T = 4$ N·m is applied to the driving shaft at $t = 0$. Develop the differential equations that express ω_1 and ω_2 as functions of time for the case in which the clutch is not yet engaged. Solve the equations to obtain $\omega_1(t)$ and $\omega_2(t)$. Determine the time at which the clutch will engage.
b. Develop the differential equations that express ω_1 and ω_2 as functions of time for the case where the clutch is engaged. Solve the equations to obtain $\omega_1(t)$ and $\omega_2(t)$. Find the time t_{lock} at which the clutch will lock ($\omega_1 = \omega_2$) and the corresponding value of ω_1.
c. Note that the model is not valid for $t > t_{lock}$. Why?

Problem 7.8

Consider the viscous coupler shown in Figure P7.8. Initially, both masses are at rest. At $t = 0$, a constant external torque T is applied to mass 1 as shown.

a. Develop the model that represents the angular velocities ω_1 and ω_2 of the two masses as functions of time.
b. Let $J_1 = J_2 = 0.1$ kg · m², $P = 0.03$ N · m · s, $P_2 = 0.08$ N · m · s, and $T = 1.2$ N · m. Calculate the ultimate angular velocities of the two masses.
c. Obtain the analytical solutions for ω_1 and ω_2. Verify that the solutions evaluated at infinite time match the answers to part (b).

Problem 7.9

Consider the system of torsion springs shown in Figure P7.9. Initially, masses 1 and 2 are at rest with no twist in either spring. At $t = 0$, a constant external torque T is applied to mass 2 as shown.

a. Develop the model that represents the angular displacements θ_1 and θ_2 of the two masses as functions of time.
b. Let $J_1 = 1$ kg · m², $J_2 = 0.725$ kg · m², $P = 6$ N · m · s, $k_1 = 15$ N · m, $k_2 = 29/3$ N · m, and $T = 1$ N · m. Calculate the ultimate angular displacements of the two masses.
c. Obtain the analytical solutions for θ_1 and θ_2. Verify that the solutions evaluated at infinite time match the answers to part (b).

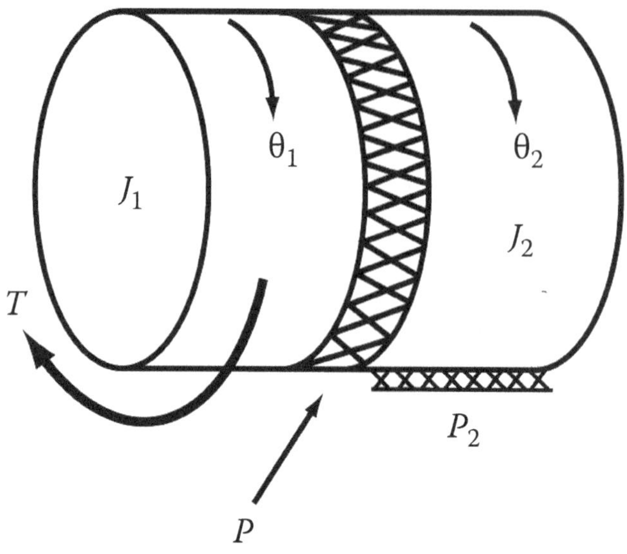

FIGURE P7.8 Viscous coupler of Problem 7.8.

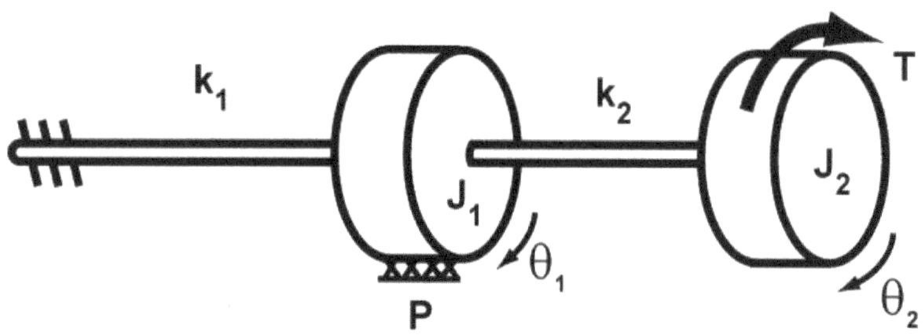

FIGURE P7.9 System of two masses and torsion springs for Problem 7.9.

Problem 7.10

Consider the system shown in Figure P7.10. An inelastic cable connects a mass to a rotating drum. Initially, the mass and drum are at rest. At $t = 0$, the mass m is released and allowed to lower, unspooling the cable from the drum as it does. There is no slip between the cable and drum.

a. Develop the model that represents the velocities v and ω of the mass and drum. Hint: Represent the tension in the cable explicitly by a force T.
b. Reduce the model of part a to a single differential equation for v by reasoning out the relationship between v and ω.
c. Let $J = 0.08\ \text{kg} \cdot \text{m}^2$, $R = 0.1\ \text{m}$, $P = 0.2\ \text{N} \cdot \text{m} \cdot \text{s}$, and $m = 2\ \text{kg}$. Calculate the ultimate (terminal) velocity of the mass.
d. Obtain the analytical solution for v. Verify that the solution evaluated at infinite time matches the ultimate velocity of part (b).

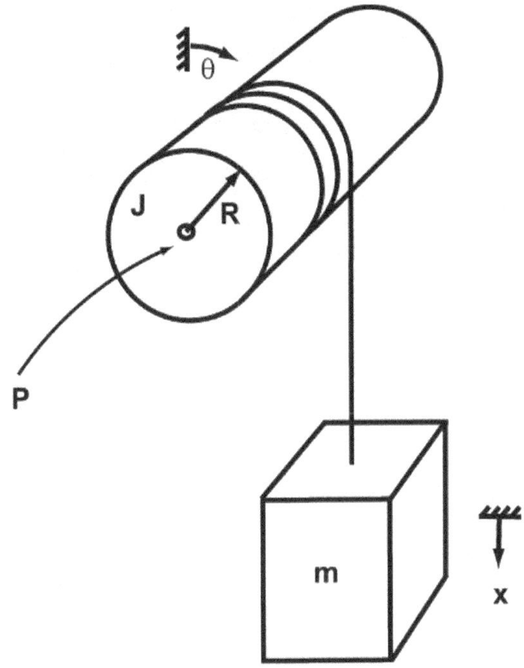

FIGURE P7.10 Mass and rotating drum of Problem 7.10.

Problem 7.11

Consider the system shown in Figure P7.11 in which a translational mass and rotational mass are geared together. Initially, elements m and J are stationary and there is no twist in the spring. At $t = 0$, a constant external force f_A is applied to mass m.

a. Develop the model that represents the displacements x and θ as a function of time. Reduce this model to a single differential equation for θ by reasoning out a relationship between x and θ.
b. Let $J = 2\ \text{kg} \cdot \text{m}^2$, $m = 100$ kg, $P = 60\ \text{N} \cdot \text{s/m}$, $k = 0.24\ \text{N} \cdot \text{m}$, $R = 0.2$ m, and $f_A = 0.6$ N. Calculate the ultimate angular displacement θ.
c. Obtain the analytical solution for θ. Compare its behavior at infinite time to the result of part (b).

Problem 7.12

Consider the system shown in Figure P7.12. Initially, the gear 1, drum 2, and mass are at rest with the spring neither stretched nor compressed. The cable between the drum and mass is inelastic and does not slip on the drum. At $t = 0$, an external torque T is applied to the gear as shown.

a. Develop the model that represents the displacements x, θ_1, and θ_2 as a function of time. Reduce this model to a single differential equation for x by using interconnection relations between the gear and drum and between the drum and mass.
b. Let $J_1 = J_2 = 0.2\ \text{kg·m}^2$, $m = 15$ kg, $P = 60$ N·s/m, $P_2 = 0.6$ N·m·s, $k = 80$ N/m, $R_1 = 0.2$ m, $R_2 = 0.1$ m, and $T = 0.8$ N·m. Calculate the ultimate displacement x_t of the mass.
c. Obtain the analytical solution for x. Compare its behavior at infinite time to the result of part (b).

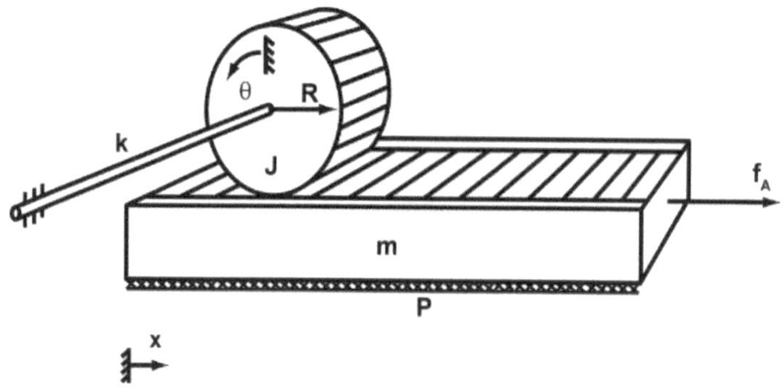

FIGURE P7.11 System of translating and rotating masses of Problem 7.11.

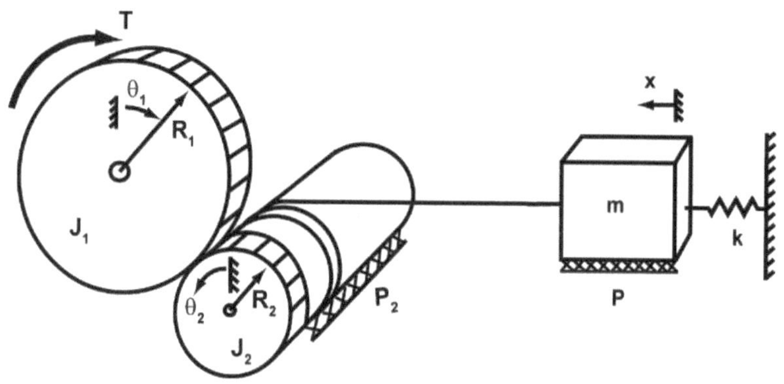

FIGURE P7.12 System of Problem 7.12.

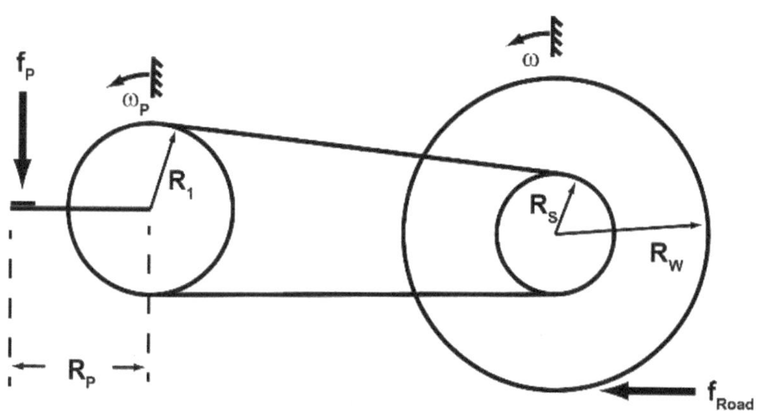

FIGURE P7.13 Single-speed bicycle.

Problem 7.13

We wish to determine a reasonable value of the sprocket ratio R_1/R_S for the single-speed bicycle shown in Figure P7.13.

Here, R_1 is the front sprocket radius, R_P is the pedal radius, f_P is the force exerted by the cyclist on the pedals, and f_C (not shown) is the force of the chain on the front sprocket. At the rear of the bicycle, R_W is the radius of the rear wheel, f_{Road} is the force of the road on the rear wheel, and R_S is the radius of the rear sprocket. The quantities ω and ω_P are the angular velocities (rad/s) of the rear wheel and front sprocket, respectively. There is also a drag force f_D (not shown) caused by air resistance on the bike and rider. We will neglect any other sources of friction in the system. In addition, we will neglect the moments of inertia of the sprockets and wheels.

On a level road, applying Newton's second law to the bicycle results in

$$\mathrm{m}\frac{dv}{dt} = f_{Road} - f_D$$

where v is the velocity of the bicycle and m is the *combined* mass of cyclist and cycle. We will assume here that the drag force is given by

$$f_D = -kv$$

where k is a drag coefficient.

a. Using the assumptions stated, show that

$$M\frac{dv}{dt} = \frac{f_P R_P}{R_W\left(R_1/R_S\right)} - kv$$

and that

$$\omega_P = \frac{v}{R_W\left(R_1/R_S\right)}$$

b. Assuming that the pedal force f_P is constant and that the bike starts from rest, solve the equations of part (a) to determine v and ω_P as functions of time. Also, develop expressions for the ultimate values of v and ω_P.
c. Assume $R_W = 0.3$ m, $R_P = 0.15$ m, $M = 90$ kg, and $k = 1.7$ kg/s. A serious rider can maintain a constant pedal force of $f_P = 100$ N as long as the cadence (ω_P) is less than 100 RPM. Therefore, we will choose the sprocket ratio R_1/R_S such that the ultimate value of ω_P is 100 RPM. What is the sprocket ratio and the corresponding ultimate velocity v (in mile/hr)? Is this reasonable for a single-speed bike?

Problem 7.14

Consider the Scotch yoke linkage shown in Figure P7.14. The Scotch yoke is one means of converting linear motion into rotational motion.

a. Assume that the force on the linkage is given by

$$F = f\sin\theta$$

Show that the model for $\theta(t)$ is given by

$$J\frac{d^2\theta}{dt^2} + P\frac{d\theta}{dt} = fR\sin^2\theta$$

b. If you have covered simulation, simulate the model and plot $\omega(t)$. Assume $J = 0.4$ kg·m^2, $P = 0.05$ N·m·s, $f = 2$ N, and $R = 0.2$ m. Use initial conditions that $\omega(0) = 0$ and $\theta(0) = \pi/2$. Run the model for 50 seconds.

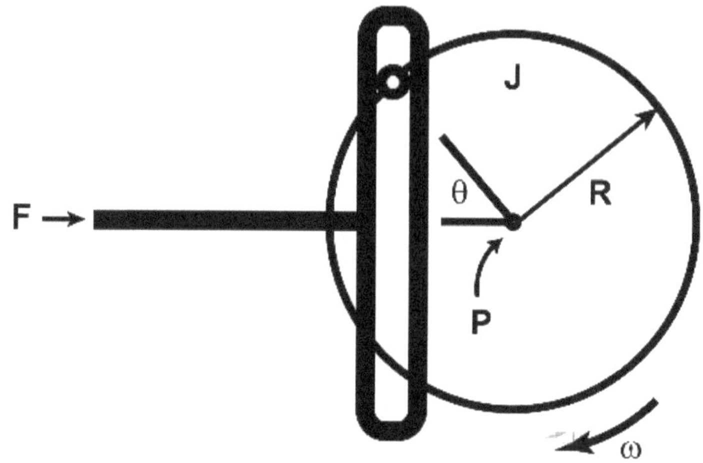

FIGURE P7.14 Scotch yoke linkage.

c. The differential equation of part (a) is nonlinear, which is why solving by numerical simulation is convenient. As discussed in the text, however, finding a limiting case of the model that can be solved analytically is often useful so that it may be used as a check of the numerical simulation. Here, let us consider a special case for which $P = 0$. Using the relation

$$\frac{d^2\theta}{dt^2} = \frac{d\omega}{dt} = \frac{dw}{d\theta}\frac{d\theta}{dt} = \omega\frac{d\omega}{d\theta}$$

along with $P = 0$ and the initial conditions of $\omega(0) = 0$ and $\theta(0) = \pi/2$, solve the model of part (a) analytically to obtain ω as a function of θ. This is not a complete solution for $\theta(t)$, which would require an additional integration, but it is sufficient to serve as a check on the simulation.

d. If you have done part (b), rerun your simulation with $P = 0$ and plot ω versus θ. Compare your simulation results to your analytical result from part (c). Do they agree?

Problem 7.15

The winch shown in Figure P7.15 is used to lift a mass m. The proportions of the various components (particularly to the mass m) are not to scale. The operator exerts a force F on the handle, rotating the gear from 0 to $\pi/2$ radians. At that point, a ratchet mechanism locks the spool and decouples the gear, allowing the handle to be rotated back to $\theta_1 = 0$ and the process to be repeated.

a. Develop the model (in the form of a single differential equation) that represents ω_1 as a function of time. Assume that the inertial effects of the gear and spool are negligible compared to that of the mass m and that the cable is inelastic and does not slip on the spool.
b. Solve the differential equation, using appropriate initial conditions, to obtain $\omega_1(t)$. Then set up and solve the differential equation for $\theta_1(t)$.
c. If $m = 100$ kg, $R_H = 0.5$ m, and $R_1 = 0.05$ m, what is the minimum force F needed to move the mass?
d. What force F would have to be exerted to rotate the gear $\pi/2$ radians in 1 second?

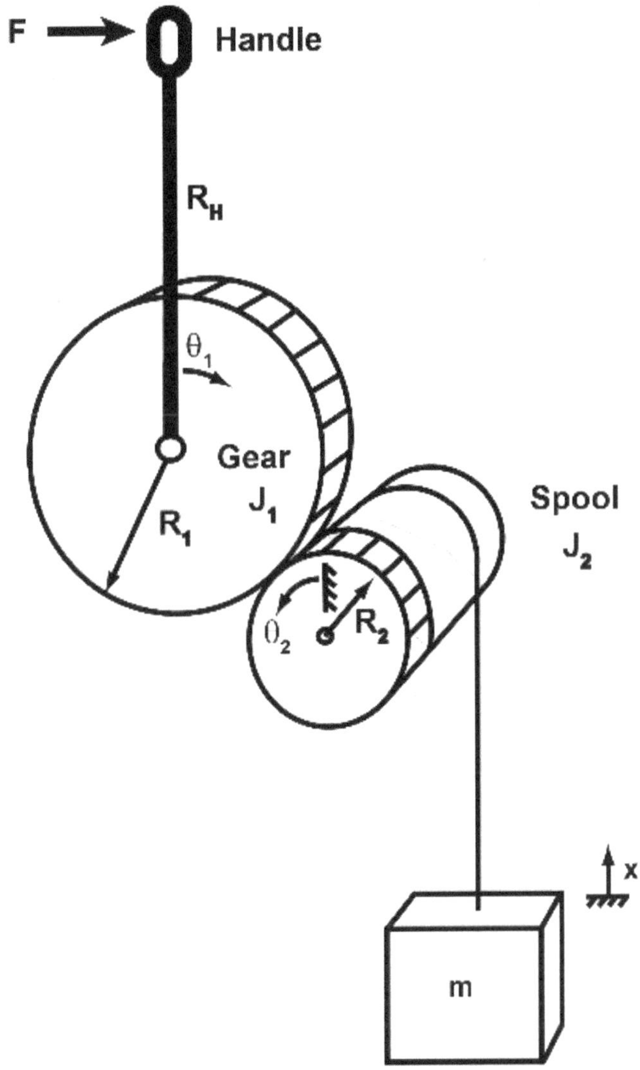

FIGURE P7.15 Winch of Problem 7.15.

Problem 7.16

The system shown in Figure P7.16 is used to transfer a thin cable onto a small product spool (right) from a supply spool (left). The process is started by applying a constant torque T to the product spool at $t = 0$. Initial conditions are $\omega_1(0) = \omega_2(0) = 0$.

a. Assuming no stretch in the cable and no slip of the cable on the spools, develop the differential equations that express ω_1 and ω_2 as functions of time.
b. It is desired that the tension in the cable never exceeds 6 N. What is the maximum torque T that should be used?
c. Use an interconnection relation to reduce the model of part (a) to a single differential equation for ω_2. Solve the model to obtain $\omega_2(t)$ and then $\theta_2(t)$. Use the value of T found in part (b) and assume $J_1 = 0.05$ kg·m^2, $R_1 = 0.1$ m, $J_2 = 80$ kg·m^2, $R_2 = 0.5$ m, and $P = 2$ N·m·s.
d. Determine the length of time needed to transfer 100 m of cable from the supply spool to the product spool.

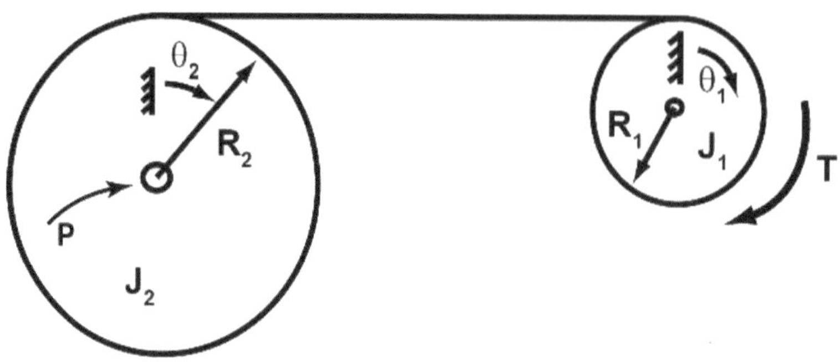

FIGURE P7.16 Spool transfer system of Problem 7.16.

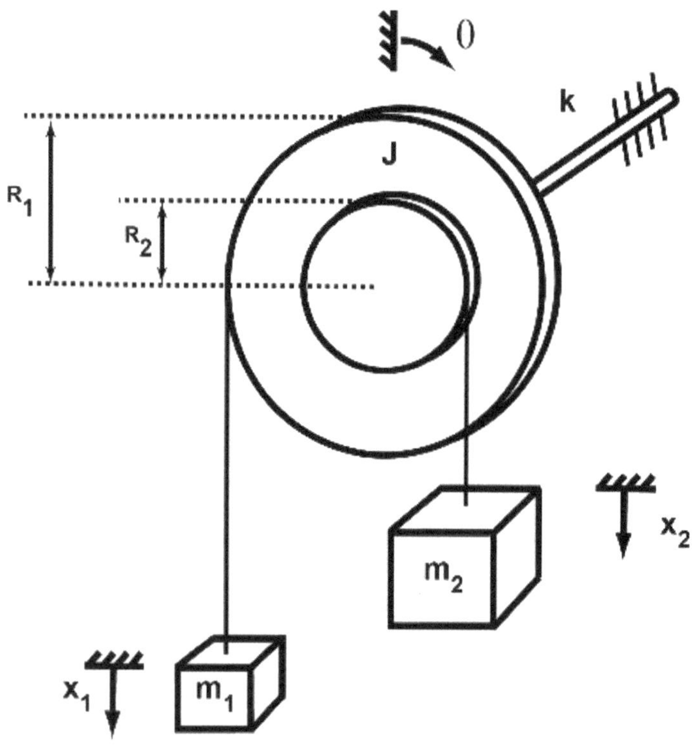

FIGURE P7.17 Disk and two masses of Problem 7.17.

Problem 7.17

Two masses are connected to a disk which is connected to a torsion spring as shown in Figure P7.17. The connections are inelastic cables that spool around the disk at different radii. Parameter values are $J = 0.1$ kg·m², $k = 8$ N·m, $m_1 = 5$ kg, $m_2 = 20$ kg, $R_1 = 0.2$ m, and $R_2 = 0.1$ m.

a. Develop the model, in the form of a single differential equation, that relates θ to t. Assume that there is always tension in both cables and that they do not slip on the spools.
b. Suppose the system is in static equilibrium. What is the equilibrium angle θ?
c. The system is in static equilibrium when, at time $t = 0$, mass m_2 is pulled downward by 0.0775 m and released from rest. Solve the model to obtain $\theta(t)$.
d. Because cables cannot support tension, the solution is nonphysical if the tension in either cable falls below zero. Determine whether this is the case.

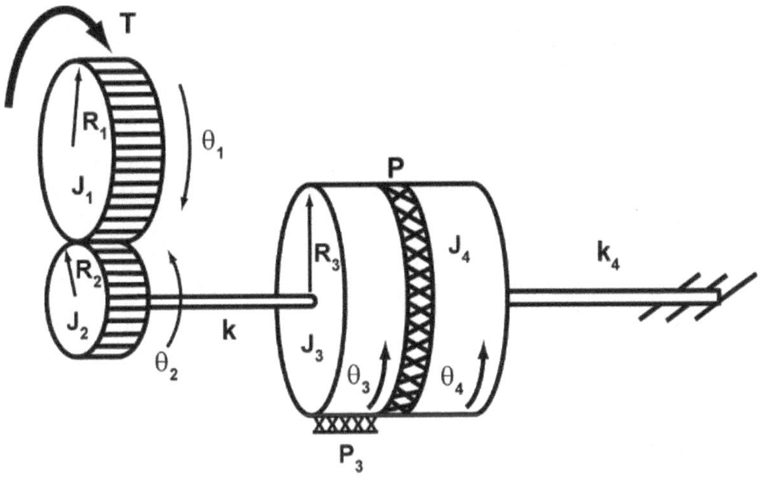

FIGURE P7.18 System of Problem 7.18.

Problem 7.18

Consider the system shown in Figure P7.18. Initially, all four mass elements are at rest with no twist in either spring. At $t = 0$, a constant external torque T is applied to mass 1.

a. Develop the differential equations that represent the angular displacements of all four masses. Use the connection relations between masses 1 and 2 to produce a model of three coupled equations in terms of θ_2, θ_3, and θ_4.
b. Let $J_1 = 0.25$ kg·m², $J_2 = 0.0625$ kg·m², $J_3 = J_4 = 1$ kg·m², $R_1 = 0.2$ m, $R_2 = 0.1$ m, $P = 2$ N·m·s, $P_3 = 5$ N·m·s, $k = k_4 = 5$ N·m, and $T = 7$ N·m. Calculate the ultimate angular displacement θ_4 and the ultimate angular velocity ω_3.

Problem 7.19

Consider Problem 7.11. Find the analytical solution for θ as a function of time if the constant force $f_A = 0.6$ N is replaced by a sinusoidal force $f_A = 0.6 \sin(0.4t)$.

Problem 7.20

Rework Problem 7.4 for $T_1 = 1\ \text{N} \cdot \text{m}$ and $T_2 = 2\ \text{N} \cdot \text{m}$.

8 Mass Balances

This chapter focuses on the application of mass balances to different processes. These balances are the most important tool a process engineer possesses for analysis, design, operation, and troubleshooting. They are also of prime importance for areas such as biomedical and biotechnology. The chapter begins by looking at the physical law that is used, as well as the equations that describe components and different phenomena occurring in processes, including those related to chemical reactions and separation processes, necessary to develop the models.

The terms *process(es)* and *system(s)* are used many times in the chapter and usually mean the same thing. They refer to industrial processes or systems (chemical, petrochemical, petroleum, environmental, etc.) as well as bioengineering processes or systems.

8.1 CONSERVATION OF MASS

The *law of conservation of mass* states that in any system *mass is neither created nor destroyed, only conserved* (Lavoisier, 1743–1794). The exception is atomic fission and fusion processes, where mass and energy can be interchanged as stated by Einstein's relation,

$$E = mc^2$$

where E = energy, m = mass, and c = velocity of light. Ignoring these cases, the law of conservation of mass always applies; that is, *mass is always conserved.*

Because mass is conserved, we can "track" the substances entering and exiting a system by performing a *mass balance*. The expression for a mass balance is

$$\begin{matrix} \textit{Rate of} & & \textit{Rate of} & & \textit{Rate of change of} \\ \textit{mass entering} & - & \textit{mass exiting} & = & \textit{mass accumulated} \\ \textit{system} & & \textit{system} & & \textit{in system} \end{matrix} \tag{8.1}$$

As an analogy, think of a money balance where the system is your checkbook. The rate of money into the checkbook (when one is paid, when a loan comes in, or when parents send money as in the case of some students) *minus* the rate of money out of checkbook (when you spend it – most every day) *is equal* to the rate of change of money accumulated in the checkbook (most times this rate is negative!).

Consider the "generic" system shown in Figure 8.1.

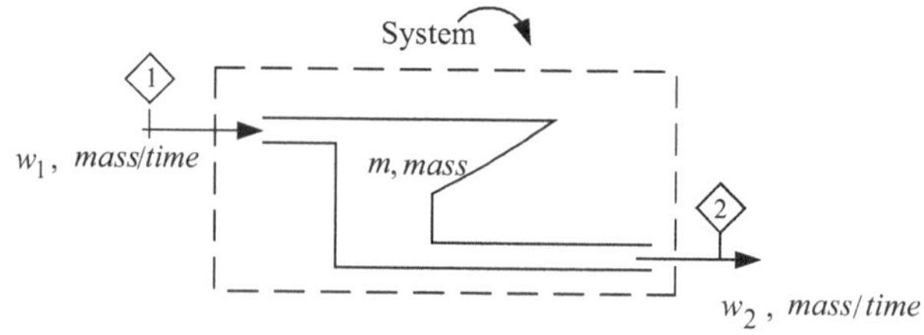

FIGURE 8.1 Generic system.

DOI: 10.1201/9781003619161-8

The dashed lines define the boundary of the system around which the mass balance will be written; to write this balance, we *only consider the streams that cross the boundary of the system.* The application of the mass balance equation, Equation 8.1, results in

$$\begin{array}{ccccc} \textit{Rate of} & & \textit{Rate of} & & \textit{Rate of} \\ \textit{mass entering} & - & \textit{mass exiting} & = & \textit{mass accumulated} \\ \textit{system} & & \textit{system} & & \textit{in system} \end{array}$$

$$w_1 - w_2 = \frac{dm}{dt}$$

where w_1 and w_2 are the mass flow rates of streams 1 and 2, respectively, and m is the mass accumulated in the system. Look at the units of each term shown in the figure, and make sure you understand that all the terms in the previous equation have the same units, mass/time. $\frac{dm}{dt}$ has units of mass/time.

For more than one input and output streams, the balance is written as

$$\sum_i w_i - \sum_o w_o = \frac{dm}{dt} \tag{8.2}$$

where the subscripts i and o stand for input and output streams, respectively.

Often, the assumption that the rate of change of the mass accumulated in the system is zero, $dm/dt = 0$, is made; this assumption is called the *steady-state assumption.* Note that this assumption says that there is no rate of change of mass in the system, meaning that the mass accumulated is constant. In this case, the mass balance statement is written as

$$\begin{array}{ccccc} \textit{Rate of} & & \textit{Rate of} & & \\ \textit{mass entering} & - & \textit{mass exiting} & = & 0 \\ \textit{system} & & \textit{system} & & \end{array} \tag{8.3}$$

and in equation form

$$w_1 - w_2 = 0$$

or

$$\sum_i w_i - \sum_o w_o = 0 \tag{8.4}$$

Recall from chemistry that the term *mole* is another way to express the quantity of a substance. *One mole is the amount of a substance numerically equal to the molecular mass of the substance.* (Molecular mass was formerly known as molecular weight.) The molecular mass is obtained by adding the mass of the atomic elements forming the substance. Using the definition of moles,

$$moles = \frac{mass}{molecular\ mass}$$

or

$$n = \frac{m}{MM} \tag{8.5}$$

We use the term n for moles and MM for molecular mass.

Because mole is another way to express the amount of a substance, we could write a mole balance instead of a mass balance to track the substances entering and exiting the system. A word of advice is important here. The following presentation on mole balance applies only in processes where there are no reactions occurring. Section 8.5.2 presents chemical reactions and discusses at that time the mole balances when reactions are taking place. The expression for a mole balance is similar to that of a mass balance:

$$\begin{matrix} \textit{Rate of} & & \textit{Rate of} & & \textit{Rate of change o} \\ \textit{moles entering} & - & \textit{moles exiting} & = & \textit{moles accumulate} \\ \textit{system} & & \textit{system} & & \textit{in system} \end{matrix} \tag{8.6}$$

Referring to Figure 8.1,

$$\frac{w_1}{MM_1} - \frac{w_2}{MM_2} = \frac{d\dfrac{m}{MM}}{dt}$$

or

$$n_1 - n_2 = \frac{dn}{dt}$$

and for multiple streams

$$\sum_i n_i - \sum_o n_o = \frac{dn}{dt} \tag{8.7}$$

where, as before, the subscripts i and o stand for input and output streams. Note that n_i refers to a flow in moles in stream i/time, but n by itself (no subscript) stands for only moles (in the previous case for moles accumulated in the system). Thus, the subscript indicates the flow of the stream numbered; without subscript n indicates the moles accumulated in the system.

Systems described by Equation 8.1, 8.2, 8.6, and 8.7 are referred to as *unsteady-state systems*; those described by Equation 8.3 and 8.4 are referred to as *steady-state systems*. What would be the equations for the steady state expression of Equations 8.6 and 8.7? For steady-state systems, the resulting equation from a mass balance is usually an algebraic equation. For unsteady-state systems, the resulting mathematical expression is a differential equation. This chapter is concerned with unsteady-state balances.

8.1.1 Multicomponent Systems

Very often streams are a mixture of several components, and thus, we need to consider different ways to express the composition of a mixture. Let us suppose that the streams in Figure 8.1 are a mixture of components A and B. In this case, two types of mass balances can be written: total mass balance, and component mass balance.

1. Mass Balance – Total

$$\begin{matrix} \textit{Rate of total} & & \textit{Rate of total} & & \textit{Rate of change of total} \\ \textit{mass entering} & - & \textit{mass exiting} & = & \textit{mass accumulated} \\ \textit{system} & & \textit{system} & & \textit{in system} \end{matrix}$$

or

$$w_1 - w_2 = \frac{dm}{dt} \tag{8.8}$$

2. Mass Balance – Component A

$$\begin{matrix} \textit{Rate of mass of} & & \textit{Rate of mass of} & & \textit{Rate of change of mass of} \\ \textit{component A} & - & \textit{component A} & = & \textit{component A accumulated} \\ \textit{entering system} & & \textit{exiting system} & & \textit{in system} \end{matrix}$$

or

$$w_1^A - w_2^A = \frac{dm^A}{dt} \tag{8.9}$$

and for component B,

$$w_1^B - w_2^B = \frac{dm^B}{dt} \tag{8.10}$$

where the notation used for the mass flow rate of a component is $w_{Stream}^{Component}$, and for the mass of a component accumulated in the system is $m^{Component}$.

In the mass balance expression of Equation 8.9, each term must have the units of *mass of A / time*; the terms in Equation 8.10 must have units of *mass of B / time*. What about the units of the terms in Equation 8.8?

When moles are used,

1. Mole Balance – Total

$$\begin{matrix} \textit{Rate of total} & & \textit{Rate of total} & & \textit{Rate of change of total} \\ \textit{moles entering} & - & \textit{moles exiting} & = & \textit{moles accumulated} \\ \textit{system} & & \textit{system} & & \textit{in system} \end{matrix}$$

or

$$n_1 - n_2 = \frac{dn}{dt} \tag{8.11}$$

2. Mole Balance – Component A

$$\begin{matrix} \textit{Rate of moles of} & & \textit{Rate of moles of} & & \textit{Rate of change of moles of} \\ \textit{component A} & - & \textit{component A} & = & \textit{component A accumulated} \\ \textit{entering system} & & \textit{exiting system} & & \textit{in system} \end{matrix}$$

or

$$n_1^A - n_2^A = \frac{dn^A}{dt} \tag{8.12}$$

and for component B

$$n_1^B - n_2^B = \frac{dn^B}{dt} \tag{8.13}$$

where the notation used for the mole flow rate of a component is $n_{Stream}^{Component}$, and for the moles of a component accumulated in the system are $n^{Component}$

In the mole balance expression of Equation 8.12, each term must have the units of *moles of A / time*; the terms in Equation 8.13 must have units of *moles of B / time*. What about the units of the terms in Equation 8.11?

Important note for modeling: As previously discussed in Chapter 2, there must be the same number of independent equations as unknowns to complete a model, that is, the degrees of freedom must be zero ($DoF = 0$). Considering a system consisting of two components A and B, as we just did in the previous paragraphs, there are three possible balances, a total balance, Equation 8.8, and two component balances, one for each component, Equations 8.9 and 8.10. However, only two of these balances, any two, are independent and can be used in the model. The third balance is not independent because it can be obtained by a mathematical manipulation of the two chosen balances. For example, if in writing the model we had used Equations 8.9 and 8.10, then Equation 8.8 could not be used because it could be obtained from the other two as

$$\text{Equation 8.9} + \text{Equation 8.10} = \text{Equation 8.8}$$

$$\begin{aligned} w^A{}_1 + w_2{}^A &= \frac{dm^A}{dt} \\ +w_1{}^B + w_2{}^B &= \frac{dm^B}{dt} \\ \hline (w_1{}^A + w_1{}^B) - (w_2{}^A + w_2{}^B) &= \frac{d(m^A + m^B)}{dt} \\ w_1 - w_2 &= \frac{dm}{dt} \end{aligned}$$

or mass balance on A plus mass balance on B gives the total mass balance.

Similarly, if we had chosen Equations 8.8 and 8.9 then Equation 8.10 is not independent because it can be obtained by subtracting Equation 8.9 from Equation 8.8.

Thus, in a process with R components there is a possibility of $R + 1$ balances (one for each component plus one total), but only R balances are independent and can be used – *any R*. This restriction equally applies for mass and mole balances.

8.1.2 Types of Processes

A comment about the type of processes is important before completing this section. Processes in general could be divided into two types, those where no reactions take place, and those where reactions take place. In processes without reactions, both mass and mole balances can be written without a problem. However, one must be careful when writing these balances in processes where reactions occur. Consider, for example, the reaction $SO_2 + 1/2O_2 \rightarrow SO_3$; a total of 1.5 moles of reactants enter the reactor (1 mole of SO_2 and 0.5 mole of O_2) and 1 mole of product exits. Thus, in this reaction the total moles are not conserved! However, considering the mass, 80 gm of reactants enter (64 gm of SO_2 plus 16 gm of O_2) and 80 gm of product SO_3 exits. Thus, total mass is conserved! The point is that in processes without reactions both mass and moles are always conserved, in processes in which reactions occur, total mass is always conserved, but total moles are not necessarily conserved. Furthermore, components are also not conserved – reactants disappear, and products appear! Sections 8.5, 8.6, and 8.7 show how to write the balances when reactions occur.

8.2 FLOW RATES AND CONCENTRATIONS

Most balances are concerned with stream *flow rates*, and ways to express the amount of each component in a mixture of several components, that is, the *concentration* of each component in a mixture. This section presents these terms.

Flow rate is the expression of a *quantity per unit time* of a stream. We consider the following three types of flow rates:

1. *Volumetric flow rate*, which is volume per unit time. Typical units are m^3/s, ft^3/h, and gal/min (gpm). f denotes volumetric flow rate. More specifically, $f_{Stream\ number}$ indicates the volumetric flow rate of the numbered stream.
2. *Mass flow rate,* which is mass per unit time. Typical units are *kg/min, lbm/hr* and *gm/s*. w denotes mass flow rate. More specifically, $w_{Stream\ number}$ indicates the mass flow rate of the numbered stream, and $w_{Stream\ number}^{Component}$ indicates the mass of the component in the numbered stream.
3. *Mole flow rate,* which is moles per unit time. Typical units are *mole/s* and *kmole/hr*. $n_{Stream\ number}$ indicates the mole flow rate of the numbered stream, and $n_{Stream\ number}^{Component}$ indicates the moles of the component in the numbered stream.

An important property of any fluid is its density (ρ). By definition the mass density is

$$\rho = \frac{m}{V} \text{ or mass per unit volume} \tag{8.14a}$$

and the molar density is

$$\bar{\rho} = \frac{n}{V} \text{ or mole per unit volume} \tag{8.14b}$$

A common use of density is to convert from volumetric flow rate to mass or mole flow rate or vice versa.

$$mass\ flow\ rate = \rho \times volumetric\ flow\ rate \tag{8.15a}$$

or

$$mole\ flow\ rate = \bar{\rho} \times volumetric\ flow\ rate \tag{8.15b}$$

Look at the units to make sure the equations are correct.

The term *mass fraction* is often used to express the concentration of a component in a solution; this term is defined as

$$Mass\ fraction\ of\ component\ i\ in\ a\ solution = \frac{mass\ of\ i\ in\ the\ solution}{total\ mass\ of\ mixture} = \frac{m^i}{\sum_{j=1}^{C} m^j} \tag{8.16}$$

where C is the number of components. Look at the units of these fractions. We tend to think that these terms are dimensionless, but this is not the case! The units are

$$Mass\ fraction\ of\ component\ i\ [=] \frac{mass\ of\ i}{total\ mass\ of\ mixture}; \frac{\text{kg}\ of\ i}{kg\ of\ mixture}, \frac{lb\ of\ i}{lb\ of\ mixture}$$

The symbol [=] stands for "has units of." The numerator and the denominator are both in masses, but they are different masses.

The term *mole fraction*, or *mole percentage*, is also often used to express the concentration of a component in a solution; this term is defined as

$$Mole\ fraction\ of\ component i = \frac{mole\ of\ i}{total\ moles\ of\ mixture} = \frac{n^i}{\sum_{j=1}^{C} n^j}$$

Often, books reserve the lowercase letters x, y, and z to designate mass or mole fractions, with x for liquids, y for gases, and z for either phase. Specifically, this book uses $x_{Stream\ number}^{Component}$ or $y_{Stream\ number}^{Component}$ or $z_{Stream\ number}^{Component}$ to specify the mass or mole fraction of a component in a mixture in the numbered stream.

Revisiting the mass balance on component A of Equation 8.9, and using the mass fraction definition we write the mass balance as

$$w_1^A - w_2^A = \frac{dm^A}{dt} \tag{8.9}$$

or

$$w_1 x_1{}^A - w_2 x_2{}^A = \frac{dmx^A}{dt} \tag{8.17}$$

As mentioned before, the units of every term in Equation 8.9 are *mass of A/time*. Obviously, that must also be the case in Equation 8.17, take a look,

$$w_1 x_1^A [=] \frac{total\ mass\ of\ stream\ 1}{time} \times \frac{mass\ of\ A\ in\ stream\ 1}{total\ mass\ of\ stream\ 1} = \frac{mass\ of\ A\ in\ stream\ 1}{time}$$

$$w_2 x_2^A [=] \frac{mass\ of\ A\ in\ stream\ 2}{time}$$

and

$$\frac{dmx^A}{dt} [=] \frac{(total\ mass\ accumulated) \times \dfrac{mass\ of\ A\ accumulated}{total\ mass\ accumulated}}{time} = \frac{mass\ of\ A\ accumulated}{time}$$

A common assumption taken in mixing systems is that the contents inside the system (tank in this case) are "well mixed." This means that the concentration – temperature in other cases – is the same all over the system including the exiting stream. If we had used this assumption in writing Equation 8.17, because the mass fraction exiting the system is the same as that in the system,

$$w_1 x_1^A - w_2 x_2^A = \frac{dmx_2^A}{dt} \tag{8.18}$$

As mentioned, this well-mixed assumption is commonly used, and we will remind you whenever is needed.

Another common concentration term is that of amount (mass or mole) of a component per unit volume or

$$C^i = \frac{\text{mass or moles of component } i}{\text{volume}}$$

This concentration term is quite used in chemistry courses, and when considering chemical reactions.

8.3 ELEMENTS AND EXPERIMENTAL FACTS

8.3.1 Flow Element

A very important flow element is a valve; Figure 8.2 shows a valve as we will represent it, where P_1 is the inlet fluid pressure, and P_2 is the exit fluid pressure with units in *kPa* or in lb_f/in^2 (*psi*). The flow through a valve is commonly expressed by

$$Flow = C_V \sqrt{P_1 - P_2} \tag{8.19}$$

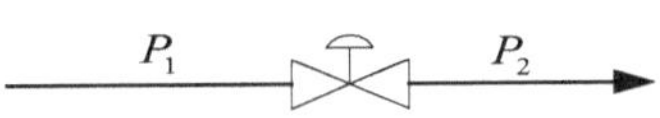

FIGURE 8.2 Process valve.

The flow could be in mass or volumetric units as presented in Section 8.2. C_V is called the *valve coefficient* and will have the necessary units to satisfy the equation – by the way, can you use the valve equation to obtain the units of C_V?

Manipulating flows is the principal use of valves. In these cases, a signal is transmitted to the valve to operate its opening; Figure 8.3 shows the schematic of the valve with its signal, where $s(t)$ is the signal to the valve. This signal varies between 0–100%, indicating the percentage of opening of the valve, when 0 % the valve is closed. Usually, we refer to the numerical value of signals in percentage.

The expression for the flow through the valve is adapted as follows to reflect this signal:

$$Flow = C_V \frac{s(t)}{100} \sqrt{P_1 - P_2} \tag{8.20}$$

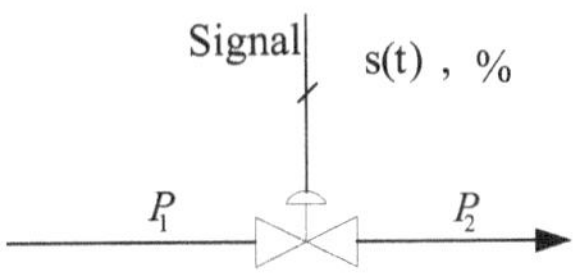

FIGURE 8.3 Control valve.

where the term $\frac{s(t)}{100}$ expresses the fraction open of the valve.

The valve equation indicates that the flow is proportional to the pressure drop (difference in pressure), or $Flow \sim \Delta P$ where $\Delta P = P_1 - P_2$. Thus, a pressure drop is the gradient necessary for material to flow.

8.3.2 Liquid Service

Consider the tank shown in Figure 8.4 where a liquid level exists; the tank is open to the atmosphere.

In this tank the inlet flow w_1, the discharge pressure P_2, and the signal to the valve s are known. Applying the mass balance of Equation 8.1 to the contents of the tank yields

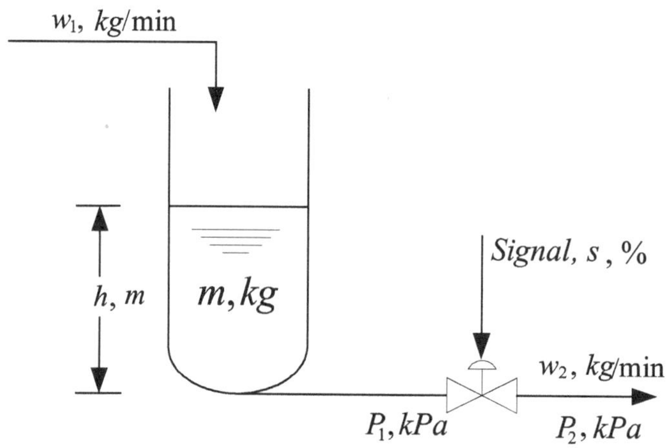

FIGURE 8.4 Liquid level

$$w_1 - w_2 = \frac{dm}{dt} \tag{8.21}$$

1 equation, 2 unknowns $[w_2, m]$,

where m is the mass accumulated in the tank.

The level in the tank is related to the mass accumulated. This mass m is equal to the volume V ((m^3)) the liquid occupies times its density ρ ((kg/m^3)). The liquid volume is equal to the cross-sectional area A ((m^2)) of the tank times the height h ((m)) of liquid; thus,

$$m = V\rho = \rho \text{Ah} \tag{8.22}$$

2 equations, 3 unknowns [h].

In solving Equation 8.21, the flow rate out of the tank, w_2, must either be known or calculated. A valve equation such as Equation 8.20 provides the necessary equation,

$$w_2 = C_V \frac{s}{100}\sqrt{P_1 - P_2} \tag{8.23}$$

3 equations, 4 unknowns $[P_1]$.

In this equation P_1 and P_2 are in kPa. The upstream pressure from the valve P_1 is a new unknown. Because the tank is opened to the atmosphere, P_1 is the summation of the atmospheric pressure plus the pressure generated by the liquid level, or what it is called the *hydrostatic pressure*,

$$P_1 = P_{atmospheric} + P_{hydrostatic} = 101.32 + \frac{\rho gh}{1000} \tag{8.24}$$

4 equations, 4 unknowns,

where assuming a sea-level location $P_{atmospheric} = 101.32\,\text{kPa}$, and g is the local acceleration due to gravity, *9.8* m/s^2 at sea level. The term 1000 in the denominator of the second term is needed for the units purpose (kPa = 1000 Pa). The model is now complete (0 *DoF*).

8.3.3 Gas Service

When dealing with gases, there is a very well-known expression that relates the pressure, temperature, volume, and moles of the gas. The expression, known as the *ideal gas law*, is

$$PV = nRT \tag{8.25}$$

where
P= pressure, Pa;
V= volume, m^3;
n= moles, moles.;
T= temperature, K; and
R= ideal gas law constant, $8.314\dfrac{\text{Pa}\cdot\text{m}^3}{\text{moles}\cdot\text{K}} = 8.314\dfrac{\text{kPa}\cdot\text{m}^3}{\text{kmoles}\cdot\text{K}}$.

The ideal gas law applies to gases at low density. Low density occurs when the gas is at either at a low pressure or a very high temperature. Gases that are not ideal are called real gases. For the purpose of the chapter, we assume all gases are ideal.

The ideal gas offers us a way for obtaining the density of the gas. From Equation 8.25,

$$\bar{\rho} = \frac{n}{V} = \frac{P}{RT} \tag{8.26}$$

To calculate the mass density of an ideal gas, we use the molecular mass

$$\rho = \bar{\rho} \times molecular\ mass = \frac{P \times molecular\ mass}{RT} \tag{8.27}$$

8.4 EXAMPLES

Example 8.1

Consider once more the tank shown in Figure 8.4. Suppose the liquid is water with a density of $1000\,\text{kg/m}^3$, the cross-sectional area of the tank is $1\,\text{m}^2$, the downstream pressure from the valve is $90\,\text{kPa}$, the steady-state flow into the tank is $10\,\text{m}^3/\text{min}$, and the valve equation is given by

$$f_2 = 1.5\sqrt{P_1 - P_2}\text{, m}^3/\text{min}$$

a. Find the steady-state liquid level.
b. Develop the model that describes how the level in the tank varies when the inlet volumetric flow changes by $2\,\text{m}^3/\text{min}$, or $f_1 = 10 + 2u(t)$.

a. At steady-state the mass flow into the tank must equal the mass flow out of the tank, Equation 8.4 applies.

$$w_1 - w_2 = 0$$
$$w_1 = w_2$$

or

$$w_1 = w_2 = f_1\rho = 10\frac{\text{m}^3}{\text{min}}\left(1000\frac{\text{kg}}{\text{m}^3}\right) = 1\times10^4\frac{\text{kg}}{\text{min}}$$

The outlet volumetric flow using Equation 8.15a is

$$f_2 = \frac{w_2}{\rho} = \frac{1\times 10^4 \dfrac{\text{kg}}{\text{min}}}{1000\dfrac{\text{kg}}{\text{m}^3}} = 10\frac{\text{m}^3}{\text{min}}$$

In this example $f_1 = f_2$ because the inlet and outlet densities are equal and ρ is a constant.

To find the pressure that produces this flow we make use of the valve equation,

$$f_2 = 10 = 1.5\sqrt{P_1 - P_2} = 1.5\sqrt{P_1 - 90}$$

or

$$P_1 = 134.44\text{kPa}.$$

Finally, to find the level to produce pressure P_1, make use of Equation 8.24:

$$P_1 = 134.44\,\text{kPa} = 101.32 + \frac{\rho g h}{1000} = 101.32 + \frac{\left(1000\dfrac{\text{kg}}{\text{m}^3}\right)\left(9.8\dfrac{\text{m}}{\text{s}^2}\right)\text{h}}{1000}$$

$$\text{h} = 3.38\,\text{m}$$

This is the steady-state liquid level.

b. The model that describes how the liquid level in the tank varies when the inlet flow varies was developed in the previous section, and it is repeated here for convenience. From a mass balance around the tank,

$$w_1 - w_2 = \frac{dm}{dt}$$

or

$$\rho f_1 - \rho f_2 = \frac{dm}{dt} \tag{8.28}$$

1 equation, 2 unknowns $[f_2, \text{m}]$.

Note that the input flow f_1 is not considered an unknown because it is a forcing function; thus, we set its value.

The relation between the mass accumulated in the tank and the level is

$$m = V\rho = \rho A h \tag{8.29}$$

2 equation, 3 unknowns $[\text{h}]$.

The flow through the exit valve is given by

$$w_2 = \rho f_2 = 1.5\rho\sqrt{P_1 - P_2} \tag{8.30}$$

3 equation, 4 unknowns $[P_1]$.

Finally, the upstream pressure from the valve is

$$P_1 = P_{atmospheric} + P_{hydrostatic} = 101.32 + \frac{\rho gh}{1000} \tag{8.31}$$

4 equations, 4 unknowns.

These four equations constitute the dynamic model of the tank system: zero degrees of freedom! Substituting Equations 8.29, 8.30, and 8.31 into Equation 8.28 yields

$$\rho f_1 - 1.5\rho\sqrt{101.32 + \frac{\rho gh}{1000} - P_2} = \rho A\frac{dh}{dt}$$

and substituting the numerical information

$$\rho f_1 - 1.5\rho\sqrt{11.32 + 9.8\text{h}} = \rho A\frac{dh}{dt}$$

$$\rho A\frac{dh}{dt} + 1.5\rho\sqrt{11.32 + 9.8\text{h}} = \rho f_1 = \rho[10 + 2u(t)] \tag{8.32}$$

Based on the definitions in Section 3.2, Equation 8.32 is a first-order *nonlinear* differential equation. An approximate analytical solution can be obtained using the linearization method presented in Chapter 3; otherwise, simulation can also be used to obtain a response; Chapter 11 presents the simulation of this system.

Example 8.2

Consider the mixing tank shown in Figure 8.5. In this tank a concentrated solution of $NaOH$ and H_2O (stream 1) is diluted using pure water (stream 2); the concentrated solution contains 0.75 mass fraction of $NaOH$. The figure shows the steady-state information. The exit stream flows out of the tank by overflow, the tank has a volume of $0.2845\,\text{m}^3$, and the density of the liquid accumulated in the tank can be assumed constant at $1200\,\text{kg/m}^3$. Also assume that the contents of the tank are mixed well (meaning that the concentration of $NaOH$ is the same in the entire volume, including the exiting stream). Develop the model and obtain the analytical solution that describes how the exit concentration of $NaOH$ varies when the concentration of stream 1 changes in a step change to 0.67 mass fraction of $NaOH$, or $x_1^{NaOH} = 0.75 - 0.08u(t)$.

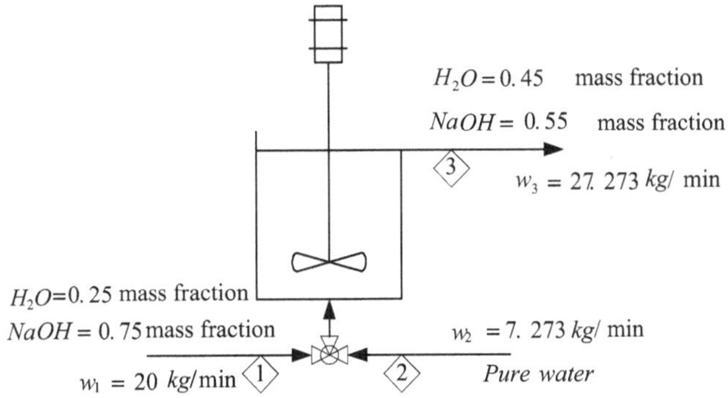

FIGURE 8.5 Mixing tank.

Defining the tank as the system, we start by using a mass balance on $NaOH$,

$$\begin{matrix} \textit{Rate of mass of} \\ \textit{NaOH entering} \end{matrix} - \begin{matrix} \textit{Rate of mass of} \\ \textit{NaOH exiting} \end{matrix} = \begin{matrix} \textit{Rate of change of mass of} \\ \textit{NaOH accumulated in system} \end{matrix}$$

$$w_1^{NaOH} - w_3^{NaOH} = \frac{dm^{NaOH}}{dt}$$

or

$$w_1 x_1^{NaOH} - w_3 x_3^{NaOH} = \frac{dm x_3^{NaOH}}{dt} \tag{8.33}$$

1 equation, 1 unknown $[x_3^{NaOH}]$,

where m is the total mass inside the tank, and because the volume and density are constant this mass is also constant at $m = \rho V = 1200 \times 0.2845 = 341.1\,\text{kg}$. Then,

$$w_1 x_1^{NaOH} - w_3 x_3^{NaOH} = m\frac{dx_3^{NaOH}}{dt}$$

$$\frac{w_1}{m} x_1^{NaOH} - \frac{w_3}{m} x_3^{NaOH} = \frac{dx_3^{NaOH}}{dt}$$

$$\frac{dx_3^{NaOH}}{dt} + 0.0799 x_3^{NaOH} = 0.0585 x_1^{NaOH}$$

or

$$12.5\frac{dx_3^{NaOH}}{dt} + x_3^{NaOH} = 0.73 x_1^{NaOH}$$

This first-order linear nonhomogeneous differential equation with constant coefficients can be easily solved by any of the methods shown Chapters 2, 3, 4, and 5. Actually, a general solution of a first-order differential equation when forced by a step change is given by Equation 5.10a, yielding

$$x_3^{NaOH} = 0.55 - 0.059\left(1 - e^{-\frac{t}{12.5}}\right)$$

Example 8.3

Consider the same mixing tank shown in Figure 8.5. This time develop the model that describes how the exit concentration of $NaOH$ varies when the concentration of stream 1 changes in a step change to 0.67 mass fraction of $NaOH$, or $x_1^{NaOH} = 0.75 - 0.08u(t)$, and at the same time, the flow of stream 1 changes in a step change to 15 kg/min, or $w_1 = 20 - 5u(t)$.

The model starts in the same manner as previously, that is, with a mass balance on $NaOH$ and actually, the resulting equation is the same Equation 8.33,

$$w_1 x_1^{NaOH} - w_3 x_3^{NaOH} = \frac{dm x_3^{NaOH}}{dt} \tag{8.33}$$

and using the value of the mass accumulated in the tank, rearranging gives

$$341.1\frac{dx_3^{NaOH}}{dt}+w_3x_3^{NaOH}=w_1x_1^{NaOH} \tag{8.34}$$

1 equation, 2 unknowns $\left[w_3, x_3^{NaOH}\right]$.

Writing a total mass balance,

$$w_1+w_2-w_3=0 \tag{8.35}$$

2 equations, 2 unknowns.

Equations 8.34 and 8.35 constitute the model for the mixing tank. Substituting Equation 8.35 into 8.34 yields

$$341.1\frac{dx_3^{NaOH}}{dt}+(w_1+w_2)x_3^{NaOH}=w_1x_1^{NaOH} \tag{8.36}$$

This time, obtaining the analytical solution is not as easy as before because the coefficient of the second term on the left-hand side, w_1+w_2, depends on one of the forcing functions, which is a function of time. However, in this particular case although it is a function of time, it is a step change that once it changes, it stays constant from then on. Using the separation of variable method, the analytical solution is

$$x_3^{NaOH}=0.444+0.106^{-\frac{t}{15.31}} \tag{8.37}$$

Obtaining the analytical solution, if the change in w_1 had been a different type, such as a ramp, would not have been this easy. In that case, Example 3.21 shows how to obtain an analytical solution, albeit an approximate one, using the method presented in Section 3.7. Obviously, using computer simulation, Chapter 11, is another option.

Before continuing to another example, consider Equation 8.35. This equation is a steady state total mass balance, but Equation 8.34 is an unsteady state mass balance on *NaOH*. Why is one equation a steady-state balance and the other an unsteady-state balance? That is, why is it that for the total mass balance we use an algebraic equation?

Example 8.4

Consider the gas tank shown in Figure 8.6. A fan blows air into a tank, and from the tank, the air flows out through a valve. For purposes of this example, let us suppose the air flow delivered by the fan is given by

$$f_i(t)=0.453\frac{s_i(t)}{100}$$

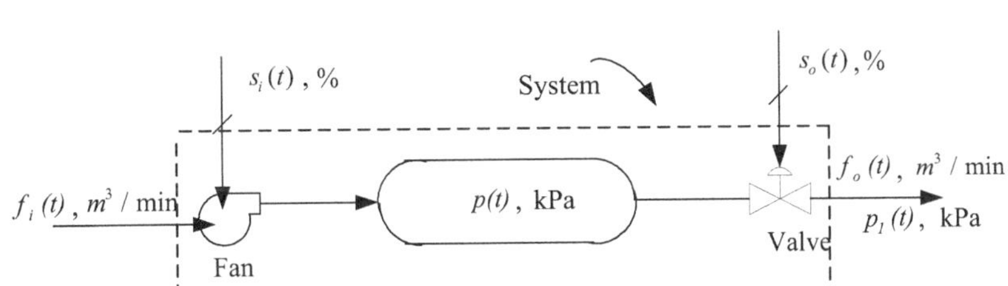

FIGURE 8.6 Gas system.

where
$f_i(t)$ = air flow in m^3/min at 23 °C and 101.32 kPa.
$s_i(t)$ = signal to fan, 0–100%.

The flow through the valve is expressed by

$$f_o(t) = 2.078 \times 10^{-3} \frac{s_o(t)}{100} \sqrt{p(t)[p(t) - p_1(t)]}$$

where
$f_o(t)$ = air flow in m^3/min also at 23 °C and at the pressure of the tank.
$s_o(t)$ = signal to valve, 0–100%.
$p(t)$ = pressure in tank, kPa.
$p_1(t)$ = downstream pressure from valve, kPa.

The volume of the tank is $0.569\,m^3$, and it can be assumed that the process occurs isothermally at 23°C. The initial steady-state conditions are the following:

$$f_i(0) = 0.2265\,\mathrm{m^3/min};\ f_o(0) = 0.08321\,\mathrm{m^3/min};\ p(0) = 275.788\,\mathrm{kPa};\ p_1(0) = 101.325\,\mathrm{kPa};$$
$$s_i(0) = 50\%;\ s_o(0) = 50\%$$

Develop the mathematical model that relates the pressure in the tank to changes in the signal to the fan, $s_i(t)$; in the signal to the valve, $s_o(t)$; and in the downstream pressure, $p_1(t)$.

An unsteady-state mass balance around the system, defined as the fan, tank, and outlet valve, provides the starting relation. That is,

$$w_i(t) - w_o(t) = \frac{dm(t)}{dt}$$

or

$$\rho_i\, f_i(t) - \rho_o\, f_o(t) = \frac{dm(t)}{dt} \tag{8.38}$$

1 equation, 4 unknowns $[f_i(t), \rho_o, f_o(t), m(t)]$,

where $m(t)$= mass of air in tank, kg.

Because the pressure and temperature of the inlet air are known, the entering density is also known and given by Equations 8.26 as (the molecular mass of air is $\left(28.9 \frac{\text{kgm}}{\text{kmole}}\right)$:

$$\rho_i = \bar{\rho}_i \times molecular\ mass = \frac{p_i \times molecular\ mass}{RT_i} = \frac{(101.325\,\text{kPa})\left(28.9\frac{\text{kgm}}{\text{kmole}}\right)}{\left(8.314\frac{\text{kPa}\cdot\text{m}^3}{\text{kmole}\cdot\text{K}}\right)(296\,\text{K})} = 1.189\frac{\text{kgm}}{\text{m}^3}$$

Equation 8.27 also provides the exit density:

$$\rho_0 = \bar{\rho}_0 \times molecular\ mass = \frac{p(t) \times molecular\ mass}{RT} \tag{8.39}$$

2 equations, 5 unknowns $[p(t)]$.

The fan provides another equation:

$$f_i(t) = 0.453\frac{s_i(t)}{100} \tag{8.40}$$

3 equations, 5 unknowns.

The valve provides still another equation:

$$f_o(t) = 7.58\times10^{-4}\frac{s_o(t)}{100}\sqrt{p(t)[p(t)-p_1(t)]} \tag{8.41}$$

4 equations, 5 unknowns.

There is still 1 degree of freedom; thus, one more relation is still needed. Because the pressure in the tank is low, the ideal gas equation of state can be used to relate the moles in the tank to the pressure.

$$p(t)\,V = n(t)\,R\,T \tag{8.42}$$

5 equations, 6 unknowns $[n(t)]$,

and using Equation 8.5,

$$n(t) = \frac{m(t)}{molecular\,mass} \tag{8.5}$$

6 equations, 6 unknowns.

This set of equations constitutes the mathematical model for this process. The solution describes how the pressure in the tank responds to changes in $s_i(t)$, $s_o(t)$, and $p_1(t)$. Substituting the last five equations into Equation 8.38, a single equation results in a first-order ordinary nonlinear differential equation. Either the nonlinear equation(s) can be linearized to obtain an approximate solution, or computer simulation can be used to obtain the solution.

8.5 EXPRESSIONS FOR MASS TRANSPORT AND CHEMICAL REACTIONS

8.5.1 Mass Transport

Separation processes are very common in process industries. These processes consist in the transfer of a certain component from one phase to a different phase. There are many different reasons for this transfer, but certainly, the most common one is to purify one of the phases, and many examples are found in chemical, petrochemical, petroleum, environmental, biotechnology, and biomedical. To help us visualize the mathematical description of mass transfer, consider the process shown in Figure 8.7.

Stream 1 is composed of components B and C. Because component B can "poison" further processing of component C, it is necessary to separate it from C. A possible process to do so is to feed stream 1 to a volume of a tank that is separated from another volume by a membrane; this is shown in Figure 8.7. Volume I is composed of components B and C. A liquid solvent, i.e., water, is fed to Volume II. The membrane is permeable to component B, but not to C. Thus, the water receives component B. In the figure N^B represents the *mass transfer rate* of component B from Volume I to Volume II. The units of this rate are either mass/time or mole/time.

The discussion at the beginning of Section 8.3 showed that for a fluid to flow, a difference in pressure – a pressure drop, a ΔP – must exist. That is, the pressure drop is the gradient necessary for material to flow. For mass transfer to occur – that is, for N^B to happen –a *concentration gradient*

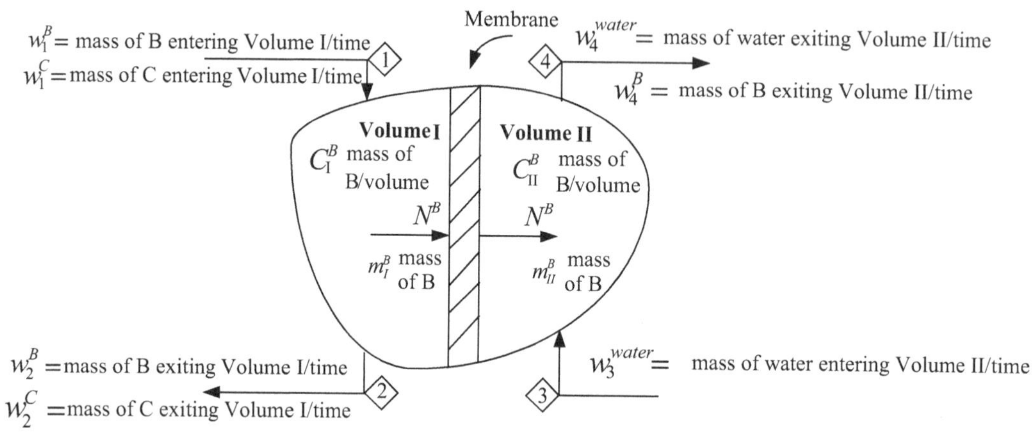

FIGURE 8.7 Mass transfer of component A from Phase I to Phase II.

of component B must exist between the two phases, a ΔC. We may say that the mass transfer rate $-N^B$ – is proportional to the difference in concentration of component B between the phases or

$$N^B \sim C_I^B - C_{II}^B \tag{8.43}$$

where the symbol ~ stands for "proportional to." We can also write this expression as

$$N^B = k\left(C_I^B - C_{II}^B\right) \tag{8.44}$$

where k is the proportionality constant and often called a mass transfer coefficient. This coefficient is empirically obtained, and it depends on the substances involved and process operating conditions. Actually, referring to Figure 8.7 we may realize that the surface area A separating the two phases also affects the transfer rate – the larger the area, the more component B can transfer. Thus, to show explicitly this effect Equation 8.44 is better written as

$$N^B = kA\left(C_I^B - C_{II}^B\right) \tag{8.45a}$$

Suppose that Volume II is very big and that the amount of component B being transferred from Volume I hardly affects its concentration and in fact, this concentration is so low that the assumption $C_{II}^B \approx 0$ is acceptable. In this case, we write

$$N^B = kAC_I^B \tag{8.45b}$$

By the way, what are the units of k in this expression?

Equations 8.45a and 8.45b are very commonly used to describe the transfer of mass of a component from one phase to another phase.

For the process of Figure 8.7, realizing that N^B acts as another stream exiting Volume I, a mass balance on component B in that volume is written as

$$w_1^B - w_2^B - N^B = \frac{dm_I^B}{dt} \tag{8.46a}$$

For Volume II, N^B acts as another entering stream, and the mass balance on component B is

$$N^B - w_4{}^B - = \frac{dm_{II}{}^B}{dt} \tag{8.46b}$$

It behooves the reader to make sure they understand these balances, and to think about the balances in component C and water, we will provide more examples in this chapter.

8.5.2 Chemical Reactions

Chemical reactions are common occurrences in industrial processes and the human body. Consider the reactor shown in Figure 8.8 where the reaction $A \rightarrow P$ takes place. We may desire to know what happens to the exit concentration C_2^A if the inlet moles of A n_1^A change.

Consider the same reaction, $A \rightarrow P$. A general way of writing any equation is

$$aA \rightarrow pP \tag{8.47}$$

where the lowercase letters are called stoichiometric coefficients and indicate the number of moles of each component taking part in the reaction. If $a = p = 1$ as previously used, this means that when one mole of A reacts, one mole of P is produced. So, if r^A is the rate at which one mole of A reacts, it is also the rate at which one mole of P is produced. If $a = 1$ and $p = 2$ means that for every one mole of A that reacts, 2 moles of P are produced. In this case, r^A indicates the rate at which 2 moles of product P are produced. Thus, the stoichiometric coefficients indicate the relative number of moles reacted/produced by the reaction and must be considered in the mass/mole balances.

As we learned in Section 8.1, the law of conservation of mass states that mass is always conserved. In processes in which reactions occur the total mass is obviously conserved, *but the total moles may not be conserved, and certainly, the mass and moles of individual reactants and products are not conserved* (reactants disappear and products are formed!). Consider the following reaction:

$$CO + \frac{1}{2}O_2 \rightarrow CO_2$$

In this case, CO is disappearing; its mass and moles are not conserved. The total moles are not conserved either; that is, there is 1 mole of CO and 1/2 mole of O_2 for a total of 1.5 moles of reactants, but only 1 mole of product CO_2. However, the total mass is conserved, that is, there are 28 gm of CO and 16 gm of O_2 for a total of 44 gm of reactants, and there are also 44 gm of the product CO_2. As explained next, we need to consider the moles being reacted away and those formed in the balances.

There is a need for an equation that describes how fast reactions occur, or what is called *reaction rate*. These equations are empirical in nature (there is hardly any theory – go into the lab, get data, and analyze it). In dealing with reactions, we use moles instead of mass (because moles are the ones satisfying the reaction expressions); this is why we show the flows in moles instead of mass in Figure 8.8.

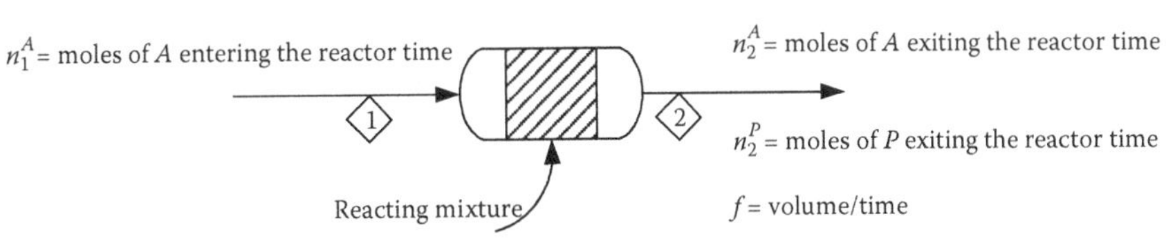

FIGURE 8.8 Chemical reactor.

As just mentioned, the reaction rate expression is obtained empirically however, experience shows that a possible expression for the aforementioned reaction is

$$r^A = kC^A \tag{8.48}$$

where r^A = reaction rate, with units of $\dfrac{moles\ of\ A\ reacted}{volume\ of\ reacting\ mixture \cdot time}$;

k = reaction rate constant; and

C^A = concentration of A in the reacting mixture with units of

$$\frac{moles\ of\ A}{volume\ of\ reacting\ mixture}$$

The reaction rate constant k is usually a strong function of temperature. In this chapter, we consider processes that occur at constant temperature (isothermal), and thus, we treat k as a constant. Most times, as in this case, the reaction rate r^A is given in *per* unit of volume of reacting mixture. By the way, what are the units of k in Equation 8.48?

In the reactor shown in Figure 8.8, the reacting mixture occupies volume V and happens that the exit volumetric flow f is known. Note that in addition to exiting the reactor as part of the exiting stream, reactant A is also "exiting," or disappearing, due to the reaction. The reaction rate times the volume of the reacting mixture gives an expression for the disappearance of A. Considering the reacting mixture to be mixed well, a mole balance on component A around the reactor yields,

Rate of moles of A exiting in stream2

Rate of moles of A entering in stream 1

Rate of moles of A dissappearing or "exiting" due to the reaction

$$n_1^A - n_2^A - Vr^A = \frac{dn^A}{dt} \tag{8.49}$$

1 equation, 3 unknowns $[n_2^A, r^A, n^A]$,

where n^A are the moles of component A accumulated in the reacting mixture. In this case, n_1^A is the forcing function.

The reaction rate expression provides another equation:

$$r^A = kC^A \tag{8.48}$$

2 equations, 4 unknowns $[C^A]$.

C^A is the concentration of component A in the reacting mixture. Because we are assuming a well-mixed reacting mixture, C^A is also the exit concentration; it is related to the moles accumulated by

$$C^A = \frac{n^A}{V} \tag{8.50}$$

3 equations, 4 unknowns.

Because it happens that the exiting volumetric flow f is known,

$$n_2^A = fC^A \tag{8.51}$$

4 equations, 4 unknowns.

Great, zero degrees of freedom (0 *DoF*), and we can now solve this model.

8.5.2.1 Half-Life

The term *half-life* is quite used in chemistry, physics, medical sciences, and other natural sciences. Simply put, it indicates the amount of time it takes to reduce by 50% of the concentration, radioactivity, or any other activity of interest of a substance. So, in one half-life, the activity reduces by 50% of the original activity; in two half-lives the activity reduces by 50% of whatever is left after the first half-life, or has reduced by 75% of the original activity; in three half-lives, the activity reduces by 50% of whatever is left after two half-lives, or has reduced by 87.5% of the original activity, and so on. We will use this half-life concept in the following sections.

8.5.3 Batch Processes

Figures 8.7 and 8.8 show streams entering and exiting the processes; we often refer to these processes as open systems. That is, an *open system* is one that has streams entering and/or exiting. We refer to a *closed system* if there are no streams entering and exiting the process; see Figure 8.9.

A *batch process* is an obvious example of a closed system. In a batch process, a material is loaded and some process starts; after a certain time has elapsed, referred to as *batch time*, the process is stopped, and the material unloaded. For example, consider the process shown in Figure 8.9, which may be considered a batch separation process. Pure water is loaded into Volume II, and a concentrated solution of components B and C into Volume I. Mass transfer starts, the transfer of B from Volume I to Volume II, and at the end of the batch time, the two volumes are unloaded, and the equipment prepared to start another batch.

Another common example is a batch reactor. In this reactor, the reactants are loaded into the reactor and the reaction starts. At the end of the batch time, the reactants and products are unloaded.

In any batch process, the important term is the batch time. The process engineer needs to decide on this time before running the process. In the first example mentioned of the batch separation process, the batch time is the time needed to reach a desired separation. In a batch reactor, the batch time is the amount of time required to reach a desired conversion of the reactant(s).

8.5.3.1 Batch Separation

Referring to the process shown in Figure 8.9, it is required to obtain the batch time necessary to reduce by 90% the initial mass of B in Volume I. All physical measurements – volumes V_I and V_{II}, and surface area A -, property k, and initial mass of B in volume I, $m_I^B(t=0)$, are known.

To develop the balances, note that there is no stream entering or exiting either volume; however, there is mass of component B exiting Volume I and entering Volume II given by N^B in mass/time. Thus, a mass balance on component B in Volume I is

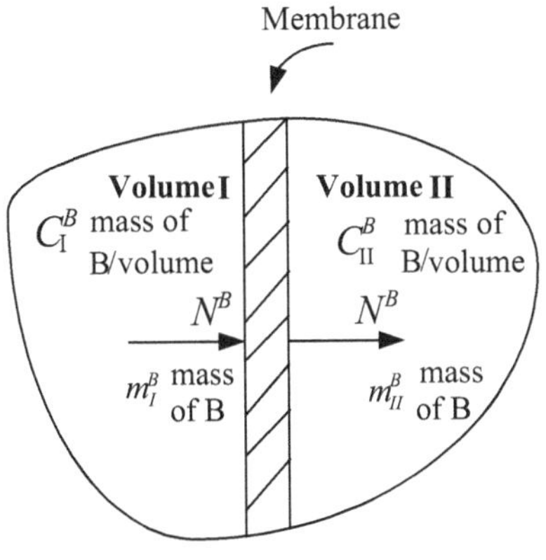

FIGURE 8.9 Closed system.

$$-N^B = \frac{dm_I^B}{dt} \tag{8.52}$$

1 equation, 3 unknowns $[N^B, m_I^B, t]$.

Note that time t is an unknown because it is what we are searching for. That is, in batch processes, we often need to decide for how long to run the process to reach a desired outcome, in this case, for how long we need to integrate Equation 8.52 to reach the desired value of m_I^B.

The mass transfer rate expression provides another equation. Let us assume that the mass of B transferring to Volume II hardly affects its composition; thus, Equation 8.45b is used:

$$N^B = kAC_I^B \tag{8.45b}$$

2 equations, 4 unknowns $[C_I^B]$.

From the definition of concentration,

$$C_I^B = \frac{m_I^B}{V_I} \tag{8.53}$$

3 equations, 4 unknowns.

We are still lacking one equation to fully describe the amount of B in Volume I that is, there is still 1 degree of freedom. The three equations so far do a good job in describing this amount; however, the important question is for how long is necessary to integrate Equation 8.52? The process specification states that the final condition is when 90% of component B is transferred to Volume II. At this condition 10% of B is left in Volume I; this specification is written as

$$m_I^B(t = t_f) = 0.1 m_I^B(t = 0) \tag{8.54}$$

4 equations, 4 unknowns.

Therefore, we start integrating at some time $t = 0$ when m_I^B is at its initial value, and the time it takes to be at 10% of this value is t_f.

Substituting Equations 8.45b and 8.53 into Equation 8.52 and rearranging,

$$V_I \frac{dC_I^B}{dt} + kAC_I^B = 0 \tag{8.55}$$

This is a first-order homogeneous linear differential equation with constant coefficients; thus, it is easily solved by any of the methods learned in previous chapters. Here, we use the separation of variables on purpose to show using the limits of integration. From Equation 8.55,

$$\int_{\frac{m_I^B(t=0)}{V_I} = C_I^B(0)}^{\frac{0.1 m_I^B(t=0)}{V_I} = 0.1 C_I^B(0)} \frac{dC_I^B}{C_I^B} = -\frac{kA}{V_I} \int_0^{t_f} dt$$

or

$$t_f = -\frac{V_I}{kA} \int_{\frac{m_I^B(t=0)}{V_I} = C_I^B(0)}^{\frac{0.1 m_I^B(t=0)}{V_I} = 0.1 C_I^B(0)} \frac{dC_I^B}{C_I^B} = \frac{V_I}{kA} \ln\left(\frac{C_I^B(0)}{0.1 C_I^B(0)}\right)$$

Note that Equation 8.54 is used as the upper limit of integration indicating the ending of the batch time (indicating the ending of integration).

In obtaining this solution we assumed that the transfer of B does not affect its concentration in Volume II and, therefore, the transfer rate is given by Equation 8.45b, or $N_B = kAC_I^B$. Let us think about this. What if indeed the transfer of B does affect its concentration in Volume II? In this case, the transfer rate would be expressed by $N^B = kA\left(C_I^B - C_{II}^B\right)$. How would this change affect the process model (set of equations)? Could you obtain an analytical equation? If so, how would you do it? Would you be expecting the new resulting batch time to be longer or shorter than the previous time, and if so, why? Think about it, this is a problem at the end of the chapter.

8.5.3.2 Batch Reactor

As already mentioned, another common batch process is a batch reactor; Figure 8.10 shows a simple schematic. Essentially, the reactor is a tank where reactants are loaded and kept well mixed. Therefore, the assumption in the analysis and design of these reactors is that their composition and temperature are uniform at any instant of time (well mixed). Suppose that the same reaction $A \rightarrow P$ in Section 8.5.2 occurs in this reactor and that the reaction rate is given by $r^A = kC^A$. Remember, the units of r^A are

$$\frac{moles\ of\ A\ reacted}{volume\ of\ reacting\ mixture \cdot time}$$

The design and analysis question is, *How long must the reaction last to obtain a desired conversion (say, 90%) – the batch time?*

The only difference between this process and the one presented in Section 8.5.2 is that there are no entering and exiting streams; component A only "exits" by reaction. Thus, the mole balance on reactant A gives

$$-Vr^A = \frac{dn^A}{dt} \tag{8.56}$$

1 equation, 3 unknowns [r^A, n^A, t].

Similar to the previous example of batch separation, time t is an unknown because it is what we are searching for. That is, for how much time we need to integrate Equation 8.56 to reach the desired value of n^A.

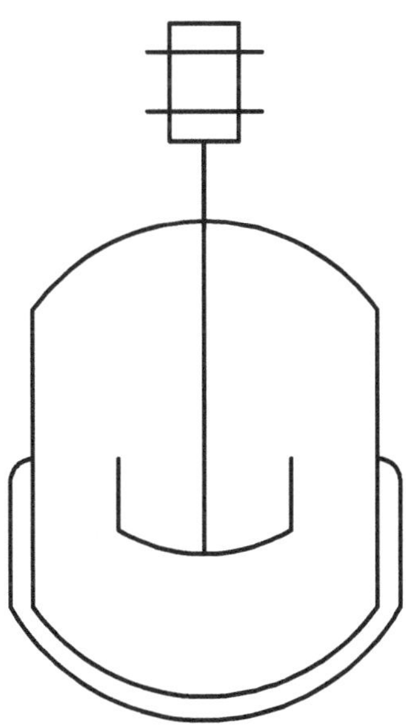

FIGURE 8.10 Batch reactor

The reaction rate expression provides another equation:

$$r^A = kC^A \tag{8.57}$$

2 equations, 4 unknowns $[C^A]$.

From the definition of concentration,

$$C^A = \frac{n^A}{V} \tag{8.58}$$

3 equations, 4 unknowns.

We still need one more equation. Once again, similar to the previous example, we can use the fact that at the end of the batch time 90% of the reactant A must have been converted, or reacted, to P; thus, only 10% is left as A:

$$n^A(t = t_f) = 0.1 \times n^A(t = 0) \tag{8.59}$$

4 equations, 4 unknowns.

Following exactly the same procedure as in the batch separation example for obtaining the batch time t yields

$$\int_{\frac{n^A(t=0)}{V}=C^A(0)}^{\frac{0.1n^A(t=0)}{V}=0.1C^A(0)} \frac{dC^A}{C^A} = -k\int_0^{t_f} dt$$

or

$$t_f = -\frac{1}{k}\int_{\frac{n^A(t=0)}{V}=C^A(0)}^{\frac{0.1n^A(t=0)}{V}=0.1C^A(0)} \frac{dC^A}{C^A} = \frac{1}{k}\ln\left(\frac{C^A(0)}{0.1C^A(0)}\right)$$

We again use one of the model equations, Equation 8.59 this time, as the upper limit of the integration, indicating when to stop the batch process.

8.6 ADDITIONAL EXAMPLES

Example 8.5

A water treatment plant uses large mixed basins in series to remove volatile pollutants from inlet water; Figure 8.11 shows the process.

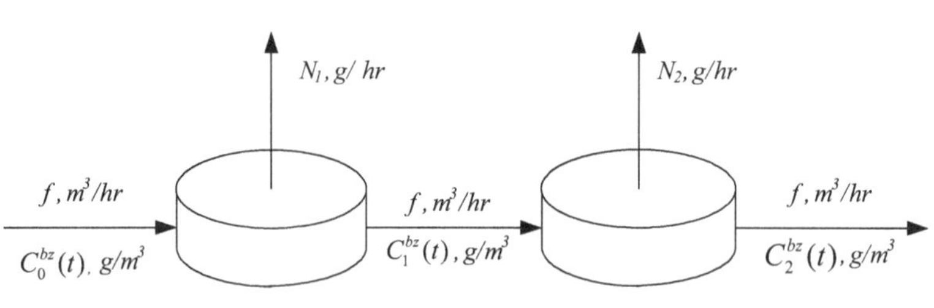

FIGURE 8.11 Water treatment basins.

The flow rate of water entering the basin system is f = 100 m³/h containing the volatile organic pollutant (benzene in this case) at a concentration of C_0^{bz}, g/m^3. Each basin is assumed to be mixed well. The feed enters the first basin where benzene evaporates at a rate of $N_1(t)$(in g/h) and then flows to the second basin where benzene evaporates at a rate $N_2(t)$. The rate $N_i(t)$ (in g/h) of benzene evaporation from basin i is given by the expression

$$N_i(t) = AKC_i^{bz}(t)$$

where A is the surface area (m^2) of the basin (both basins have the same diameter), K is a mass transfer coefficient and $C_i^{bz}(t)$ is the concentration of benzene in basin i. The mass transfer coefficient for benzene in this situation is estimated to be 15 m/h. The diameter of each basin is 10 m, and each has a depth of 0.5 m. You may assume that the benzene evaporated hardly affects the volumetric flow rate f.

Suppose that initially there is no benzene in the inlet stream or in the basins. At $t = 0$, the entering concentration C_0^{bz} spikes to 5 gr/m^3 and remains there for 0.5 hours, at which point it drops back to zero.

Determine the concentration of benzene as a function of time in each basin for the period from 0 to 1 hour. Plot the results for both basins in the form of $C_i^{bz}(t)$ versus t. What maximum concentration will be reached in the second basin, and when will it occur? We would like concentration $C_2^{bz}(t)$ to never exceed 1 g/m^3. Will that happen here? If so, what might be done to avoid the problem in the future?

This is the first time we encounter more than one unit (basins in this case). As in Chapter 6 when we encountered more than one mass in a mechanical system, we consider one basin at a time.

Basin 1

We start by writing an unsteady state mass balance on benzene,

$$f(t)C_0^{bz} - f(t)C_1^{bz}(t) - N_1(t) = \frac{dm_1^{bz}(t)}{dt} = V\frac{dC_1^{bz}(t)}{dt} \tag{8.60}$$

1 equation, 2 unknowns $[C_1^{bz}(t), N_1(t)]$.

The evaporation rate is given by

$$N_1(t) = AKC_1^{bz}(t) \tag{8.61}$$

2 equations, 2 unknowns.

Basin 2

An unsteady-state mass balance on benzene gives

$$f(t)C_1^{bz} - f(t)C_2^{bz}(t) - N_2(t) = \frac{dm_2^{bz}(t)}{dt} = V\frac{dC_2^{bz}(t)}{dt} \tag{8.62}$$

3 equations, 4 unknowns $[C_2^{bz}(t), N_2(t)]$.

The evaporation rate is given by

$$N_2(t) = AKC_2^{bz}(t) \tag{8.63}$$

4 equations, 4 unknowns.

Equations 8.60–8.63 describe the response of both basins. These equations are not completely interactive; Basin 1 affects Basin 2 – Equation 8.62 depends on the response of Equation 8.60 – but Basin 2 does not affect Basin 1 – Equation 8.62 does not affect Equation 8.60. Thus, we can first solve Equations 8.60 and 8.61 and use the solution to solve Equations 8.62 and 8.63.

Substituting Equation 8.61 into Equation 8.60, using the numerical values of the terms, and rearranging yields

$$39.27\frac{dC_1^{bz}(t)}{dt}+1278.1C_1^{bz}(t)=100_0^{bz} \quad (8.64)$$

This is a first-order linear nonhomogeneous differential equation with constant coefficients. The analytical solution can be obtained by any of the methods studied in previous chapters. For a pulse forcing function of 5 g/m³ units of magnitude and 0.5 hour of duration or,

$$C_0^{bc}=\begin{cases}5 & 0\le t\le 0.5\\ 0 & t\ge 0.5\end{cases}$$

or simply

$$C_0^{bz}=5u(0.5-t).$$

Because of the assumption that originally, there is no benzene in the inlet stream, $C_1^{bz}(0)=0\,\text{gm/cm}^3$. Using the integrating factor method, the solution for Basin 1 is

$$C_1^{bz}(t)=0.3912\left(1-e^{-32.546t}\right)\text{ for }0\le t\le 0.5 \quad (8.65a)$$

and

$$C_1^{bz}(t)=0.3912e^{-32.546(t-0.5)}\text{ for }t\ge 0.5 \quad (8.65b)$$

Showing that Equations 8.65a and 8.65b are the analytical solutions of Equation 8.64 is a problem at the end of Chapter 3; you may want to review this solution.

For Basin 2, the model is

$$39.27\frac{dC_2^{bz}(t)}{dt}+1278.1C_2^{bz}(t)=100C_1^{bz} \quad (8.66)$$

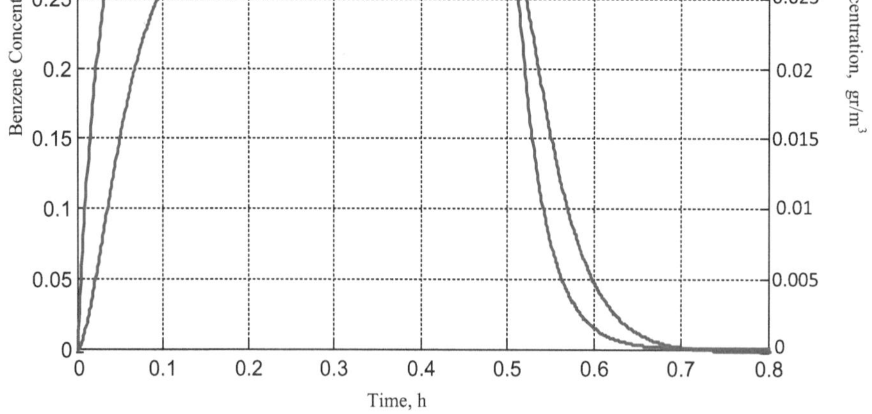

FIGURE 8.12 Benzene concentration in basins.

and the solution using the integrating factor method again is

$$C_2^{bz}(t) = 0.0306\left(1 - e^{-32.546t}\right) - 0.996te^{-32.546t} \text{ for } 0 \le t \le 0.5 \quad (8.67a)$$

and

$$C_2^{bz}(t) = \left(0.996(t-0.5) + 0.0306\right)e^{-32.546(t-0.5)} \text{ for } t \ge 0.5 \quad (8.67b)$$

Figure 8.12 shows the concentration response for both basins. Equations 8.65a, 8.65b, 8.67a, and 8.67b were used to obtain these graphs. The figure shows the highest concentration attained in each basin and that the concentration in Basin 2 never reaches 1 g/m^3.

Example 8.6

Consider a batch reactor where the gas reaction $2NOBr \rightarrow 2NO + Br_2$ takes place. The reaction rate is described by $r^{NOBr} = 0.1(C^{NOBr})^2$ with units of *mole of NOBr reacted* / m$^3 \cdot$s. The reactor volume is 2 m^3, the reaction occurs at 10 °C, and the initial charge of NOBr is 0.5 mole. What is the batch time to reach 90% conversion?

A mole balance on $NOBr$ gives,

$$-Vr^{NOBr} = \frac{dn^{NOBr}}{dt} \quad (8.68)$$

1 equation, 3 unknowns $[r^{NOBr}, n^{NOBr} t]$.

The rate of reaction is another equation:

$$r^{NOBr} = 0.1(C^{NOBr})^2 \quad (8.69)$$

2 equations, 4 unknowns $[C^{NOBr}]$.

From the definition of concentration,

$$C^{NOBr} = \frac{n^{NOBr}}{V} \quad (8.70)$$

3 equations, 4 unknowns.

The last equation needed is related to the batch time. This is the time required for 90% conversion of NOB$_r$. When this conversion is reached 10% of NOB$_r$ is left unreacted so,

$$C^{NOBr}(t = t_f) = 0.1C^{NOBr}(0) \quad (8.71)$$

4 equations, 4 unknowns.

From these last four equations, using the separation of variables method,

$$\int_{C^{NOBr}(0)}^{C^{NOBr}} \frac{dC^{NOBr}}{(C^{NOBr})^2} = -0.1\int_0^t dt$$

$$t = \frac{1}{0.1}\left(\frac{1}{C^{NOBr}} - \frac{1}{C^{NOBr}(0)}\right) \quad (8.72)$$

The initial concentration of NOBr is $C^{NOBr}(0) = \dfrac{n^{NOBr}(0)}{V} = \dfrac{0.5}{2} = 0.25\dfrac{\text{mole}}{\text{m}^3}$ and for a 90% conversion $C^{NOBr}(t = t) = \dfrac{0.1n^{NOBr}(0)}{V} = \dfrac{0.5}{2} = 0.25\dfrac{\text{mole}}{\text{m}^3}$. Thus, the batch time is

$$t_f = \frac{1}{0.1}\left(\frac{1}{0.025} - \frac{1}{0.25}\right) = 360 \text{ seconds.}$$

Example 8.7

Consider a reactor where the reaction $S \rightarrow P$ takes place and P is the desired product. It happens that once P is produced, it may react, giving back the reactants; that is, the actual reaction is reversible and is shown as

$$S \Leftrightarrow P$$

It is known that the forward reaction rate producing product P is given by $r^S = 0.085C^S$ in *moles of S reacted* / $\text{m}^3 \cdot \text{h}$. For the backward reaction, breaking P back into S again, the reaction rate is given by $r^P = 0.025C^P$ in *moles of P reacted* / $\text{m}^3 \cdot \text{h}$.

The reaction takes place in a batch reactor, shown in Figure 8.13, and it is desired to obtain a conversion of 90% of reactant S. The reactor has a volume of $2.5\,\text{m}^3$, and 12.5 moles of reactant S are initially loaded. It is suspected that because of the backward reaction, breaking P back into its reactants, the 90% conversion level may not be reached. Thus, the design department proposes to remove P from the reactor as soon as it is produced to minimize this undesired (backward) reaction. It is proposed to design the reactor such that its walls are formed by a membrane permeable to P and not to S. On the other side of the membrane, water will be flowing at a very high rate; thus, the transferred P will have no effect in its concentration $((C^P_{water} = 0))$. The transfer of P to the water stream can be described by Equation 8.45b, $N^P = kAC^P = 1.25AC^P$, with units of *moles of P/h*; A is the membrane's surface area $((\text{m}^2))$ between the water and the reactor's content.

a. Assuming no mass transfer of product P to the water, model the system, obtain the analytical solution, and calculate how long (batch time) it takes to reach a 90% conversion.
b. Assuming mass transfer of product P to the water, model the system, obtain the analytical solutions, and calculate how long (batch time) it takes to reach a 90% conversion. How would you calculate the batch time this time? What would be the considerations?
   a. We start with a mole balance on component S in the reacting mass. Note that, although there are no streams in and out of the reactor, the reactions from S to P and from P back to S act as if they were streams – by means of the reactions is how the components exit and enter the reacting mass.

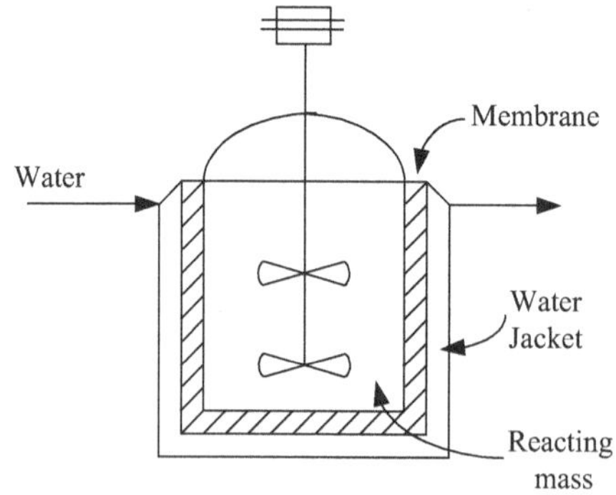

FIGURE 8.13 Batch reactor for Example 8.7.

Mole Balance – S

$$-Vr^S + Vr^P = \frac{dn^S}{dt}$$

or, dividing both sides by V,

$$-r^S + r^P = \frac{dC^S}{dt} \tag{8.73}$$

1 equation, 3 unknowns $[r^S, r^P, C^S]$.

Reactions

$$r^S = 0.085C^S \tag{8.74}$$

2 equations, 3 unknowns

$$r^P = 0.025C^P \tag{8.75}$$

3 equations, 4 unknowns $[C^P]$

Mole Balance – P

$$Vr^S + Vr^P = \frac{dn^P}{dt}$$

or, dividing both sides by V,

$$r^S + r^P = \frac{dC^P}{dt} \tag{8.76}$$

4 equations, 4 unknowns.

These four equations constitute the model for the reactor: zero degrees of freedom. Substituting the reaction rate equations into the mole balance equations yields two coupled ordinary linear differential equations with constant coefficients with two unknowns, C^S and C^P. The initial conditions are $C^S(0) = \frac{12.5\,\text{moles}}{2.5\,\text{m}^3} = 5\frac{\text{moles}}{\text{m}^3}$ and $C^P(0) = 0\frac{\text{mole}}{\text{m}^3}$

Using the Laplace transforms presented in Chapter 4 gives as solutions

$$C^S(t) = 1.136 + 3.864e^{-0.11t} \tag{8.77}$$

and

$$C^P(t) = 3.864(1 - e^{-0.11t}) \tag{8.78}$$

These solutions indicate that the concentrations change in a monotonic stable response.

A 90% conversion of S means that 10% (or 1.25 moles) of the original quantity loaded into the reactor has not reacted. At this moment the concentration is $C^S = 0.5\,\text{moles/m}^3$. Equation 8.77 indicates that at no time the concentration reaches the desired value. Actually, the equation shows that the lower value (at $t - \infty$), the concentration will reach is $1.136\,\text{mole/m}^3$. What happens is that at this concentration the forward reaction rate becomes equal to the backward rate,

$$r^S = 0.085(1.136) = 0.0966$$

and

$$r^P = 0.025(3.864) = 0.0966,$$

meaning that the rate at which product P is formed is equal to the rate at which P goes back to S. At this moment equilibrium is reached and no more changes occur.

b. Let us consider the mass transfer of P to the water side and its effect on the reaction of S. Equations 8.73, 8.74, and 8.75 remain the same with no change however, the mole balance on product P must now consider the mass transfer out of the reacting mass,

$$r^s - r^P - \frac{1.25AC^P}{V} = \frac{dC^P}{dt} \tag{8.79}$$

4 equations, 5 unknowns [A].

The new unknown is the surface area of the membrane.

At this time, we need one more equation to solve this set of equations; that is, we still have 1 degree of freedom. It is up to the engineer to make one extra specification so that set can be solved. For the time being, let us suppose for simplicity that

$$A = 1\text{m}^2 \tag{8.80}$$

5 equations, 5 unknowns.

This model can again be easily solved by Laplace transforms yielding

$$C^S(t) = 4.94e^{-0.08t} + 0.06e^{-0.53t} \tag{8.81}$$

and

$$C^P(t) = e^{-0.08t} - e^{-0.53t} \tag{8.82}$$

This time the concentrations again change in a stable monotonic response. Equation 8.81 shows that as $t = \infty$, $C^S \to 0$ meaning that more than 90% conversion is achieved; actually, 100% conversion is achieved!

The time for 90% conversion, $C^s = 0.5\dfrac{\text{mole}}{\text{m}^3}$, is

$$0.5 = 4.94e^{-0.08t} + 0.06e^{-0.53t}$$

and

$$t = 28.63\text{h}$$

To obtain this batch time, the engineer specified a surface area for the membrane of 1 m^2; that is, this is the batch time based on the specification of $A = 1\text{m}^2$. Another possible specification, for making the degrees of freedom equal to zero, is the required batch time t. In this case, the engineer would specify the batch time and solve the set of equations to find the membrane surface area required to satisfy the specification.

Example 8.8

You are a brand-new engineer working for an environmental company. As your first job, your supervisor mentions that it is desired to design a clay filter for filtering water for drinking in

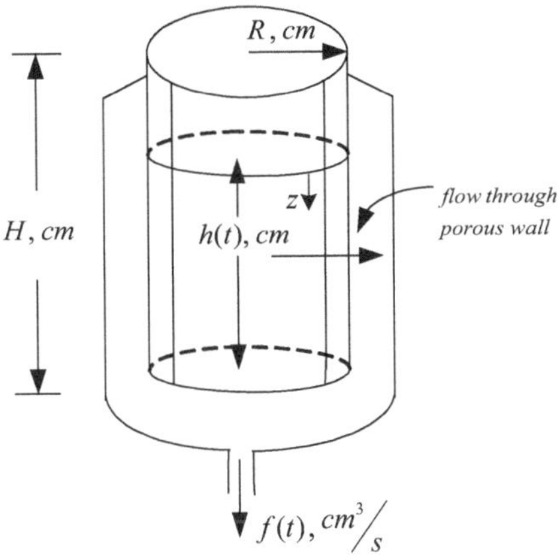

FIGURE 8.14 Water filter.

underdeveloped countries. He gives you the simple schematic shown in Figure 8.14. The cylindrical side of the filter is porous so water will drain out through any of the area that is wetted. The base of the filter is impermeable to water. The water is collected in a jacket and flows out to the users. In the drawing, H is the total height of the filter, $h(t)$ is the instantaneous liquid level, R is the radius of the filter, and $f(t)$ is the volumetric flow out of the filter.

The average velocity of water through the wetted surface is dictated by Darcy's law (you will learn about this law in your fluid courses),

$$v(t) = C\rho g h(t)$$

where $v(t)$ is the water velocity in *cm/s*, C is a coefficient characteristic of the filter, $\rho = 1\,\text{gm}/\text{cm}^3$ is the density of the water, $g = 980$ cm/s² is the acceleration due to gravity. For the clay being used, $C = 9\times10^{-8}\ \text{cm}^2\cdot\text{s/g}$. The filter is filled to height H at $t = 0$. Once the liquid level reaches 15% of the total height, the operation of the filter ceases.

Your supervisor asks you to model this filter system and simulate it to study the proposed design. Specifically, consider two filters with the same total height of 200 inches, one of them with an 11-inch radius, and the other one with a 6-inch radius. Compare the performance of these two filters by providing

a. a graph showing the total water output in gallons, as a function of time, of each filter. Make sure the graphs show the output for a bit of time after the liquid level reaches 15%.
b. a graph showing the instantaneous water output in gallons per minute (gpm), as a function of time, of each filter. Make sure the graphs show the output for a bit of time after the liquid level reaches 15%.
c. the liquid level in inches, as a function of time, of each filter. Make sure the graphs show the level for a bit of time after the liquid level reaches 15%.

As a well-educated engineer, you decide that the first equation to write in this case is a mass balance on the water (the law of conservation of mass) inside the cylinder,

$$-w_{out}(t) = \frac{dm(t)}{dt} \tag{8.83}$$

1 equation, 2 unknowns $[m(t), w_{out}(t)]$,

where $m(t)$ is the mass accumulated in the cylinder, and $w_{out}(t)$ is the mass flow rate out of the cylinder, which is equal to the flow of water to the users; $w_{out}(t)$ is obtained from the volumetric flow as

$$w_{out}(t) = \rho f(t) \tag{8.84}$$

2 equations, 3 unknowns $[f(t)]$.

The volumetric flow rate out of the tank depends on the filter's total wetted surface area at any moment. The surface area is $2\pi R h(t)$, or

$$f(t) = v(t) 2\pi R h(t) \tag{8.85}$$

3 equations, 5 unknowns $[v(t), R, h(t)]$.

Using Darcy's equation,

$$v(t) = C\rho g h(t) \tag{8.86}$$

4 equations, 5 unknowns.

The mass $m(t)$ accumulated in the tank is obtained by

$$m(t) = \rho A_{\substack{cross \\ sectional}} h(t) = \rho(\pi R^2) h(t) \tag{8.87}$$

5 equations, 5 unknowns.

This example shows once more the importance of having a method to keep up with the information needed. The solution seems complex but actually, it is not. We discuss the modeling and the solution in more detail in Chapter 11.

8.7 APPLICATION TO BIOENGINEERING PROCESSES

For purposes of this chapter, the term bioengineering processes refers to both biomedical and biotechnology processes. Both have tremendous interdisciplinary challenges and contributions for the future. There is a multitude of reactions and separation processes in the different processes. Consequently, the concepts presented in the previous sections equally apply in these areas.

8.7.1 Compartmental Modeling

It is interesting to note the similarity between reaction engineering modeling (part of process engineering) and pharmacokinetic and metabolic modeling (part of biomedical engineering). In pharmacokinetics and metabolic modeling, the body is divided into different "compartments" as shown in Figure 8.15. In this modeling, several familiar assumptions are taken such as that each compartment is assumed to be well mixed and that its volume is constant.

A pharmacokinetic model describes the movement of drugs, also sometimes referred to as solutes, through different sites of the body. The most common way a drug exits the body is either by excretion in the urine or by bio-dissipation (metabolic reaction). From a compartment point of view, the drug will exit either as transfer to another compartment or as bio-dissipation; bio-dissipation may be thought of as a rate of loss or disappearance of the drug (similar to a chemical reaction).

Commonly a first-order model describes the bio-dissipation of a solute, and it is given by,

$$r_{dissipation} = \mathrm{K m}^{solute} \tag{8.90a}$$

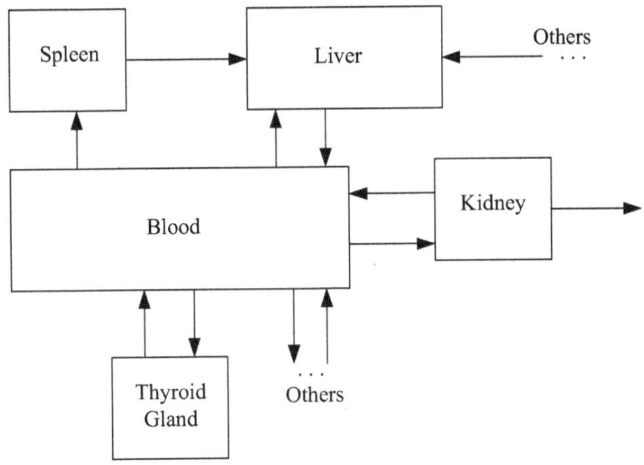

FIGURE 8.15 Body divided into compartments for modeling purposes.

in *mass of solute / time* or

$$r_{dissipation} = \mathrm{K}C^{solute} \tag{8.90b}$$

in *mass of solute / volume · time*. We refer to the constant K as the dissipation rate constant with units of time^{-1}.

Consider a one-compartment system, for example one that represents the blood, with a volume of 5 liters. Assume that 10 mg of a protein (solute) is injected into the department and realize the physics of the process (injection) that just happened. The injection is rather fast, and we assume the compartment is well mixed. Thus, we could state that at the moment the disappearance starts, the initial condition is $m^{solute}(0) = 10\,\text{mg}$ or $C^{solute}(0) = \frac{m^{solute}(0)}{V} = 2\,\text{mg/L}$. Assuming a first-order disappearance,

$$-Km^{solute} = \frac{dm^{solute}}{dt}$$

using the separation of variables,

$$\int_{m^{solute}(0)=10}^{m^{solute}} \frac{dm^{solute}}{m^{solute}} = -K\int_0^t dt$$

or

$$m^{solute}(t) = m^{solute}(0)e^{-Kt} = 10e^{-Kt} \tag{8.91}$$

It is necessary to have information to obtain the value for K. This required information is usually empirical and should show the disappearance of the solute in a system as the one being modeled; Figure 8.16 shows a typical disappearing curve. Using the concept of half-life discussed earlier, $\frac{1}{2}m\ solute(0) = 5\,\text{mg}$, and this occurs roughly at $t_{\frac{1}{2}} = 3.5\,\text{min}$. From Equation 8.91 after some algebra,

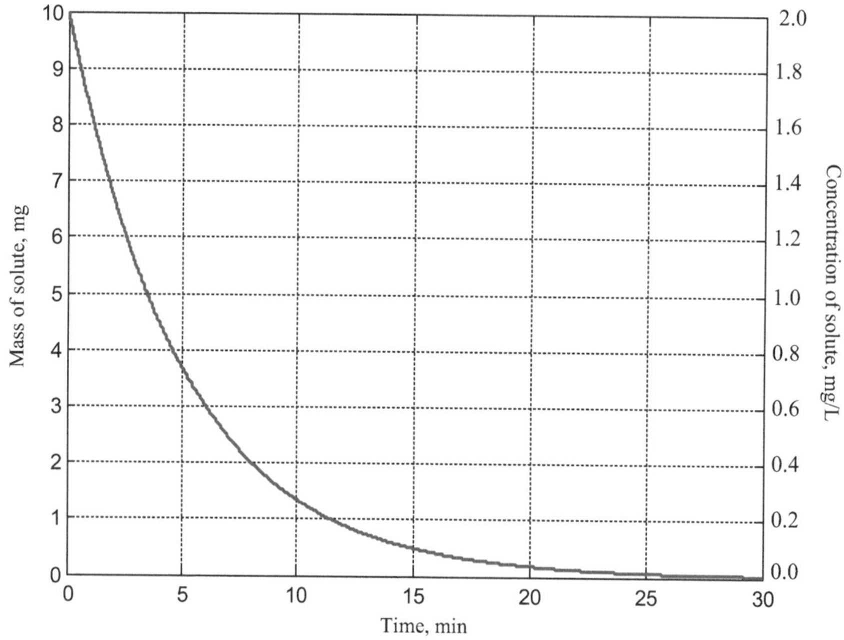

FIGURE 8.16 Disappearance of a protein (solute) in blood

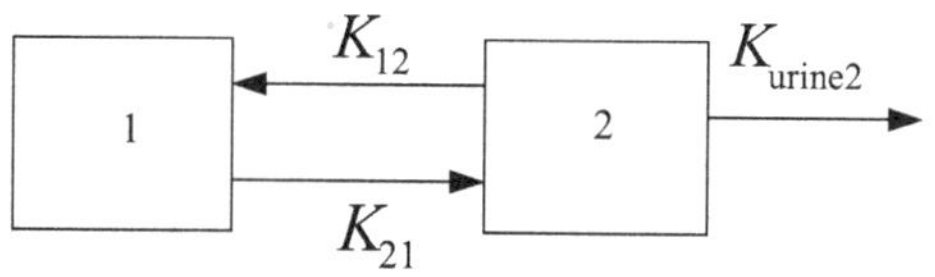

FIGURE 8.17 Transfer of a solute between two compartments.

$$K = \frac{\ln 2}{t_{\frac{1}{2}}} = \frac{0.69}{t_{\frac{1}{2}}} \tag{8.92}$$

In this case,

$$K = \frac{\ln 2}{t_{\frac{1}{2}}} = \frac{0.69}{t_{\frac{1}{2}}} = \frac{0.69}{3.5} = 0.197\,\text{min}^{-1}$$

Therefore, $m^{solute}(t) = 10e^{-0.197t}$.

The reader is encouraged to graph the resulting equation and compare it to Figure 8.16.

The transfer of solutes across compartments is commonly expressed as $K_{ij}m_j^{solute}$, in mass/time, where the subscript *i* refer to the receiving compartment and the subscript *j* for the sending compartment.m_j^{solute} stands for the mass of solute accumulated in the sending compartment.K_{ij} is the transfer rate coefficient.

Let us look at this transfer of solute in more detail. Figure 8.17 shows the transfer between two compartments. The transfer of solute is given by the following:

Mass of solute from 1 to 2 $= K_{21} m_1^{solute}$ in mass/time

Mass of solute from 2 to 1 $= K_{12} m_2^{solute}$ in mass/time

Mass of solute from 2 to urine $= K_{urine2} m_2^{solute}$ in mass/time

Therefore, the unit of K_{ij} is time^{-1}, and it represents the fraction of the mass of solute accumulated in the compartment that is transferred out per unit time.

8.7.2 Biological Reactions

Most biological reactions are catalyzed by enzymes (a catalyst helps increase the rate of the reaction, but it is not part of the reaction – it is not altered by the reaction). Thus, a biological reaction is shown as

$$\text{Enzyme} + \text{Reactant} \rightarrow \text{Enzyme} + \text{Product}$$

The term *substrate* is often used to refer to the reactant, and the product is still referred as "product." Using E for enzyme, S for substrate and P for product we write,

$$E + S \rightarrow E + P \tag{8.93}$$

As with any other reaction, the description of the kinetics of the reaction is of prime importance. Section 8.5.2 discussed chemical reactions and presented Equation 8.48 as a way to express the rate of reaction, which is still commonly used in biological reactions. Laboratory experiments have also shown that for many enzymes catalyzed reactions the Michaelis–Menten equation describes quite well the reaction kinetics:

$$r^S = \frac{r_{\max} C^S}{K_m + C^S} \tag{8.94}$$

where r^S is the rate of reaction of the substrate in $\frac{\textit{mass or moles}}{\textit{volume time}}$; C^S is the concentration of the substrate in $\frac{\textit{mass or moles}}{\textit{volume}}$; K_m is the Michaelis constant with, of course, the same units as C^S; and $r_{\max}$ is the maximum rate of reaction that occurs when C^S is infinite – note that as $C^S \rightarrow \infty$, the term $\frac{C^S}{K_m + C^S} \rightarrow 1$ and $r^S \rightarrow r_{\max}$. The constants K_m and $r_{\max}$ are empirically obtained.

Sometimes, the enzymes used to catalyze reactions experience deactivation, and of course, this affects the rate of reaction. Example 8.11 shows a way to handle this effect.

Example 8.9

Carbohydrates are broken down into glucose when food is digested. As glucose levels in the blood rise, the body releases insulin, which signals the body to store the glucose as glycogen in the liver and fatty tissues. Diabetes occurs when the body cannot produce enough insulin to regulate glucose in the blood, with the result that glucose levels can become very high and cause damage to the body through various mechanisms. The following data for glucose concentration in blood were obtained by a person who monitored their glucose levels after ingesting a meal. Time = 0 corresponds to when the individual started to eat.

t (minutes)	C^G (mg/dL)	t (minutes)	C^G (mg/dL)
0	90	120	105
30	151	180	94
60	137	240	92
90	118	320	91

A crude model for the processing of glucose in the human body might be $C \rightarrow G \rightarrow P$, where C represents the undigested food (primarily carbohydrates) in the stomach that ultimately produces glucose, G represents the glucose in the blood, and P represents the glycogen in the liver and fatty tissues. Along with this reaction scheme, it is proposed that the kinetics of the reactions of C to G and G to P are first orders of the form

$$r^c = k_1 C^c, \frac{\text{mg of carbohydrate digested}}{\text{dL} \cdot \text{min}} \tag{8.95}$$

$$r^G = k_2 (C^G - G_0), \frac{\text{mg of glucose converted}}{\text{dL} \cdot \text{min}} \tag{8.96}$$

r^C is the rate at which the carbohydrate disappears forming glucose, and r^G is the rate at which glucose disappears forming glycogen.C^C (mg/dL) represents the concentration of carbohydrates in the stomach and C^G (mg/dL) represents the concentration of glucose in the blood. The quantity $G_0 = 90\,\text{mg/dL}$ is the baseline value of C^G, and it is the value at t = 0. The rate constant $k_1 = 0.0453\,\text{min}^{-1}$ is related to how quickly the food is digested while the rate constant $k_2 = 0.0224\,\text{min}^{-1}$ is related to how quickly insulin is released and reacts with the glucose to produce glycogen.

Consider the stomach, blood, liver, and fatty tissues as a single compartment, as shown in Figure 8.18, with a volume of 1,000 mL. Assuming that the initial concentration of G is $C^G(0) = 90\,\text{mg/dL}$ and assuming that 1,217 mg of carbohydrates are ingested (this is the initial amount of carbohydrates in the compartment), let us examine the ability of a model to represent blood glucose data.

a. Write a model that describes the concentration of carbohydrates and obtain the analytical solution. Be careful in calculating the initial condition; look at the units.
b. Write a model that describes the concentration of glucose and with the help of part (a) obtain the analytical solution.
c. Compare the glucose response given by the analytical solution with that of the previous table.

a. Writing a carbohydrate mass balance in the compartment gives

$$-V r^C = \frac{dm^C}{dt}$$

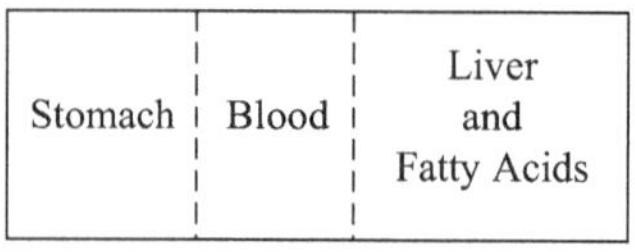

FIGURE 8.18 Single compartment for Example 8.9.

using the rate of reaction and dividing both sides by the volume

$$-k_1C^C = \frac{dC^C}{dt} \tag{8.97}$$

1 equation, 1 unknown $[C^C]$.

The initial condition is

$$C^C(0) = \frac{m_C}{V} = \frac{m_{injected}}{V} = \frac{1{,}217\,\text{mg}}{1{,}000\,\text{mL}} = 1.217\frac{\text{mg}}{\text{mL}}\left(\frac{1{,}000\,\text{mL}}{10\,\text{dL}}\right) = 121.7\frac{\text{mg}}{\text{dL}}.$$

Using the separation of variables,

$$\int_{121.7}^{C^C} \frac{dC^C}{C^C} = -k\int_0^t dt$$

$$C^C(t) = C^C(0)\,e^{-k_1 t} = 121.7e^{-0.0453t} \tag{8.98}$$

b. Writing a glucose mass balance in the compartment gives

$$Vr^C - Vr^G = \frac{dm^G}{dt}$$

using the rates of reaction and dividing both sides by the volume

$$k_1C^C - k_2(C^G - G_0) = \frac{dC^G}{dt} \tag{8.99}$$

1 equation, 1 unknown $[C^G]$.

Rearranging this equation with the help of Equation 8.98 gives

$$\frac{dC^G}{dt} + 0.0224C^G = 2.016 + 5.513e^{-0.0453t} \tag{8.100}$$

and using the integrating factor method of Chapter 3 we obtain the analytical solution

$$C^G(t) = 90 + 240.74\left(e^{-0.0224t} - e^{-0.0453t}\right) \tag{8.101}$$

c. The following table shows the comparison of the data from the person and the result from analytical solution.

t (minutes)	C^G person	C^G equation	t(minutes)	C^G person	C^G equation
0	90	90	120	105	105.3
30	151	151.3	180	94	94.2
60	137	136.9	240	92	91.11
90	118	118	320	91	90.19

The table shows an excellent match between the real data and the predictive model.

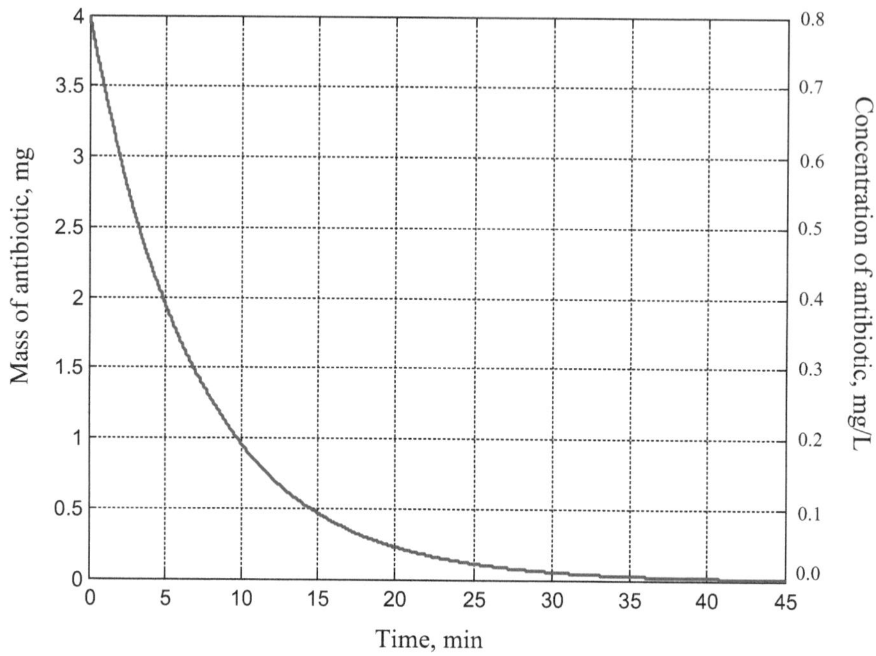

FIGURE 8.19 Response of an antibiotic.

Example 8.10

For a person with a blood volume of 5 L, 4 mg of an antibiotic is injected. Based on previous experience the disappearance of the antibiotic in this person is as shown in Figure 8.19.

Assuming a one compartment system, how often must the antibiotic be injected to maintain a minimum of 20% of the injected amount (0.8 mg) in the compartment?

Writing an antibiotic mass balance in the single compartment,

$$-r_{dissipation} = -\mathrm{Km}^{antibiotic} = \frac{dm^{antibiotic}}{dt} \tag{8.102}$$

1 equation, 3 unknowns $[K, m^{antibiotic}, t]$.

The variable t is considered an unknown because it is the time taken in going from the initial condition $m^{antibiotic}(0) = 4\,\text{mg}$ to 20% of this value. Thus,

$$m^{antibiotic}(t_{final}) = 0.8\,\text{mg} \tag{8.103}$$

2 equation, 3 unknowns.

To obtain the rate constant K we follow the meaning of half-life discussed earlier in this section, applying Equation 8.92,

$$K = \frac{\ln 2}{t_{\frac{1}{2}}} = \frac{0.69}{t_{\frac{1}{2}}} \tag{8.92}$$

3 equations, 4 unknowns $\left[t_{\frac{1}{2}}\right]$.

From Figure 8.19,

$$t_{\frac{1}{2}} \approx 5\,\text{min} \tag{8.104}$$

4 equations, 4 unknowns.

These last four equations constitute the model to solve and obtain the time interval to maintain 0.8 mg in the compartment. From Equations 8.92 and 8.104,

$$K = \frac{\ln 2}{t_{\frac{1}{2}}} = \frac{0.69}{t_{\frac{1}{2}}} = \frac{0.69}{5} = 0.138\,\text{min}^{-1}$$

From the other equations using the separation of variables,

$$\int_0^{t_{final}} dt = -\frac{1}{K}\int_{m_{initial}^{antibiotic}}^{m_{final}^{antibiotic}} \frac{dm^{antibiotic}}{m^{antibiotic}} = \frac{1}{K}\ln\left(\frac{m_{initial}^{antibiotic}}{m_{final}^{antibiotic}}\right) \tag{8.105}$$

$$t_{final} = \frac{1}{0.138}\ln\left(\frac{4}{0.8}\right) = 11.66\,\text{min}$$

This is the time it takes for the mass of the antibiotic in the blood compartment to reach 0.8 mg.

Once we inject 4 mg again, the mass of the antibiotic in the compartment becomes 4.8 mg; thus, it will take longer to reach the lower limit of 0.8 mg. Applying Equation 8.105, the new final time is

$$t_{final} = \frac{1}{0.138}\ln\left(\frac{4.8}{0.8}\right) = 12.98\,\text{min}$$

From then on, it will take the same amount of time to lose the antibiotic down to its minimum limit. To summarize, 11.66 min after the first injection the second injection must be administered, and from then on, every 12.98 min a new injection must be administered (up until the doctor decides to stop!). Figure 8.20 shows the amount of antibiotics in the blood as injections (doses) are administered.

Example 8.11

An enzyme is used to produce a high-cost product. Suppose a 3 L reactor is used and 40 mmoles of the substrate is initially charged to the reactor.

a. Assuming that the Michaelis–Menten equation describes the enzymatic reaction with $r_{max} = 10\,\text{mmole}/\text{L}\cdot\text{h}$ and $K_m = 12\,\text{mmole}/\text{L}$, plot the concentration of the substrate as a function of time.
b. It has been noticed that the enzyme used suffers deactivation with a half-life of 3 hours. Plot the concentration of the substrate as a function of time and compare the plot with the one obtained in part (a).
   a. The initial concentration of the substrate is

$$C^s(t=0) = C^s(0) = \frac{40\,\text{mmoles}}{3\text{L}} = 13.33\,\text{mmoles/L}$$

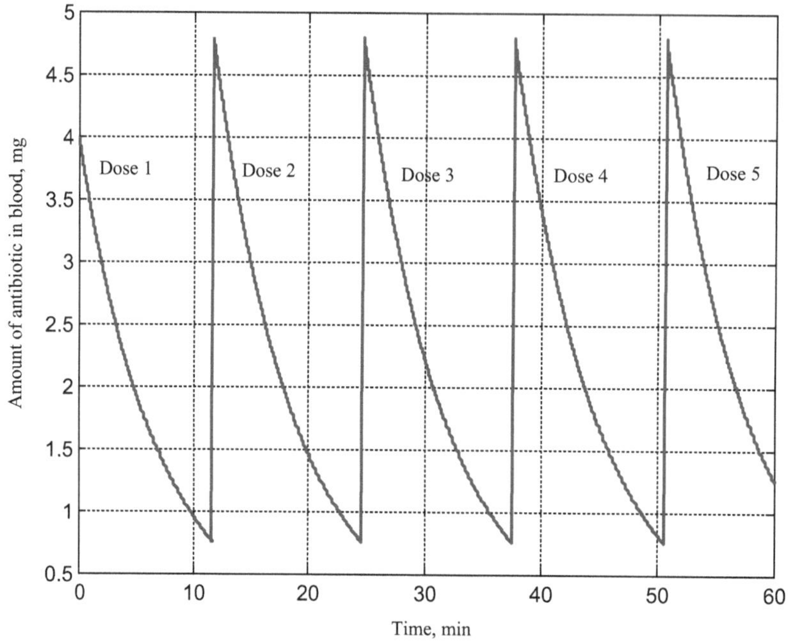

FIGURE 8.20 Antibiotic in blood.

Writing a mole balance on the substrate,

$$-Vr^s = \frac{dn^s}{dt}$$

Dividing both sides by V and using the definition of C^S,

$$r^s = \frac{r_{\max} C^s}{K_m + C^s} = \frac{dC^s}{dt} \tag{8.106}$$

1 equation, 1 unknown $[C^S]$.

Using the separation of variables,

$$t = \frac{1}{r_{\max}}[K_m \ln C^s(0) + C^s(0) - K_m \ln C^s - C^s]$$

or

$$t = 4.44 - 1.2\ln(C^S) - 0.1C^S \tag{8.107}$$

b. When the enzyme deactivates, it affects the kinetics in that it does not have the same activity to aid the reaction. An analysis of the Michaelis–Menten equation indicates that the term that should be affected is $r_{\max}$. Actually, this term should become a function of time. As we have previously discussed, it is common practice that these losses be described by first-order equations. Similar to Equation 8.91, we can write for the deactivation of the enzyme

$$r_{\max}(t) = r_{\max}(0)e^{-Kt} = 10e^{-Kt} \tag{8.108}$$

and the value of K is obtained using the half-life information:

$$K = \frac{\ln 2}{t_{\frac{1}{2}}} = \frac{0.69}{3} = 0.23\,\mathrm{h}^{-1}$$

Thus, the expression for r^S is

$$r^s = -\frac{r_{\max} C^s}{K_m + C^s} = \frac{10e^{-0.23t} C^s}{K_m + C^s} \tag{8.109}$$

The fact that the enzyme deactivates does not affect the mole balance:

$$-r^s = -\frac{10e^{-0.23t} C^s}{K_m + C^s} = \frac{dC^s}{dt} \tag{8.110}$$

The analytical expression is a bit more complex but obtaining the solution is not. Using the separation of variables again,

$$t = -4.35\ln\left[\frac{1}{43.48}(12\ln C^s + C^s - 44.41) + 1\right] \tag{8.111}$$

Figure 8.21 shows the substrate concentration vs. time for both cases. The graph makes sense because when there is no deactivation, meaning that the enzyme does not disappear, the reaction is faster, and the amount of substrate also drops faster.

8.7.3 Fermentation

Fermentation is among the oldest biotechnology processes. Briefly, fermentation is a reaction that converts sugars to alcohol, acids, and gases. This reaction is aided by yeast and bacteria (microorganisms), and often the term *biomass* refers to this combination. In aiding the reaction, the yeast also grows. The reader is strongly encouraged to search the vast information on fermentation processes, yeasts, and bacteria.

In modeling these fermentation processes, there is a need to describe the reactions, and other interactions between the biomass, reactants (substrates) and products, including the growth of yeast. These are quite complex interactions; their descriptions are obtained empirically and solely depend on the biomass (yeast and bacteria) and chemicals involved in the fermentation. Thus, the literature shows a number of models developed for this purpose. Next, we show a couple of these models.

When microorganisms are used to aid reactions and these microorganisms grow, the Monod model is often used to describe this growth:

$$\mu = \frac{1}{X}\frac{dX}{dt} = \mu_{\max}\frac{S}{K_S + S} \tag{8.112}$$

where μ is the specific growth rate of the microorganism usually in $time^{-1}$; X is the concentration of the biomass in $amount/volume$; $\mu_{\max}$ is the maximum rate of growth in $time^{-1}$; S is the concentration of the limiting substrate in amount (mass or moles) per volume (($amount/volume$)); and K_S is a constant in $amount/volume$. $\mu_{\max}$ and K_S are empirically obtained and depend on the species

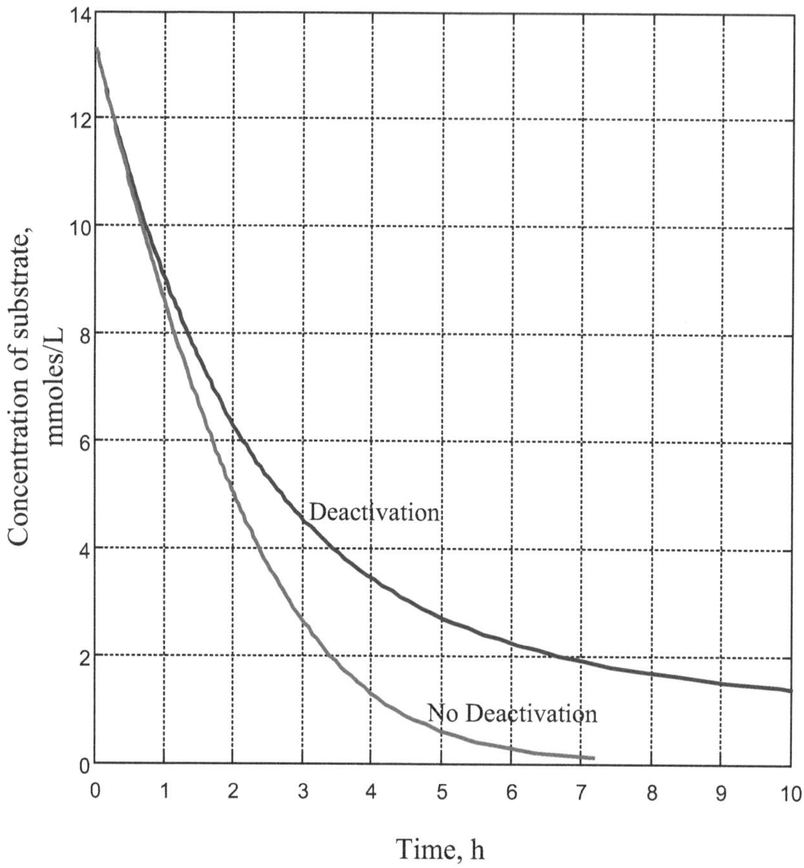

FIGURE 8.21 Substrate concentration versus time.

involved in the reaction and operating conditions. Many articles show how to obtain these constants. The growth rate of the biomass is obtained from Equation 8.112 as

$$\frac{dX}{dt} = \mu_{max} X \frac{S}{K_S + S} \tag{8.113}$$

The Monod model is quite used in environmental engineering applications to describe activated sludge processes for water treatment.

Another commonly used model for describing the rate of growth of the biomass is the "logistic equation":

$$\frac{dX}{dt} = \mu_{max} X \left[1 - \frac{X}{X_m}\right] \tag{8.114}$$

μ_{max} and X_m are constants empirically obtained.

A possible model for the reaction rate of the substrate (rate of disappearance of the reactant) in $amount/volume \cdot time$ is related to the growth rate of the biomass and given by

$$r^{Substrate} = Y \frac{dX}{dt} \tag{8.115}$$

where Y is referred to as a "yield coefficient," and it is related to the amount of substrate reacted per amount of biomass growth; this numerical value must also be empirically obtained.

Researchers have also noted that for some conditions, the reaction rate of the substrate (disappearance of the substrate) and the reaction rate of the product (formation of the product) are often related to both the rate of growth of the microorganism *and* the concentration of the biomass; this reaction rate is given by the Luedeking–Piret model:

$$r = \alpha\frac{dX}{dt} + \beta X \tag{8.116}$$

Where r is in $mass\,/\,volume \cdot time$; α, often referred to as the growth associated coefficient; and β, often referred to as the non-growth associated coefficient, are constants evaluated empirically.

Example 8.12

Consider the fermentation of glucose ($C_6H_{12}O_6$), producing ethanol (C_2H_5OH) and carbon dioxide (CO_2):

$$C_6H_{12}O_6 + Yeast\ (Biomass) \rightarrow 2C_2H_5OH + 2CO_2 + Yeast\ (Biomass).$$

This reaction occurs in a fermenter of volume V in cubic meters. Based on the biomass being used and operating conditions it is known that the growth rate of biomass can be described by the following Monod expression:

$$\frac{dX}{dt} = X\frac{0.0476C^G}{0.031 + C^G}$$

where C^G is the concentration of glucose and X is the concentration of biomass. The initial concentration of glucose in the fermenter is $500\,\text{moles/m}^3$, that of the biomass is $4.2\,\text{moles/m}^3$, and there is no ethanol. The yield coefficient is $Y = 7.46$ moles of glucose reacted/mole of biomass grow. Obtain the model that describes the concentration of ethanol once the reaction starts.

We start by writing a mole balance on glucose (substrate),

$$-Vr^G = \frac{dn^G}{dt}$$

where r_G is the reaction rate of glucose in $\text{moles/m}^3 \cdot \text{h}$ and n_G is the moles of glucose. Dividing both sides by the volume and using the definition of concentration,

$$-r^G = \frac{dC^G}{dt} \tag{8.117}$$

1 equation, 2 unknowns $[r^G, C^G]$.

The reaction rate is given by

$$r^G = Y\frac{dX}{dt} = 7.46X\frac{0.0476C^G}{0.031 + C^G} \tag{8.118}$$

2 equations, 3 unknowns $[X]$.

Next, we write a mole balance on ethanol (product),

$$2Vr^G = \frac{dn^E}{dt}$$

and dividing both sides by the volume yields

$$2r^G = \frac{dC^E}{dt} \tag{8.119}$$

3 equations, 4 unknowns $[[C^E]]$,

where C^E is the concentration of ethanol in moles/m^3. Note from the reaction that 2 moles of ethanol are formed for every 1 mole of glucose reacted. This is why we write 2 times the rate of reaction.

We still need one equation (there is 1 degree of freedom). Analyzing the equations, we note that there is no expression for the concentration of biomass, *X*. The rate of growth of biomass yields *X* or

$$\frac{dX}{dt} = X\frac{0.0476C^G}{0.031 + C^G} \tag{8.120}$$

4 equations, 4 unknowns.

Equations 8.117–8.120 are the model for this fermentation process.

An analytical solution of this model is rather difficult to obtain mainly because of the nonlinearity developing in Equations 8.118 and 8.120. Although the linearization procedure shown in Chapter 3 is possible, simulation (Chapter 11) is probably the easiest and best way to obtain a response curve.

Example 8.13

The production of gluconic acid by fermentation of glucose using an *Asperigillus niger* culture is reported in References [1, 2]. The fermentation reaction may be represented as

$$\underset{\text{Glucose}}{C_6H_{12}O_6} + \frac{1}{2}O_2 + Biomass \rightarrow \underset{\text{Gluconic acid}}{C_6H_{12}O_7} + Biomass$$

The reader is referred to Reference 2 for the details of the mechanism. Laboratory studies show that the following logistic equation describes the growth rate of the biomass at the operating conditions:

$$\frac{dX}{dt} = \mu_{\max} X\left(1 - \frac{X}{X_m}\right) = 0.3768X\left(1 - \frac{X}{18.91}\right)$$

where *X* is the biomass concentration in gm/L and the time unit is hour (h). Laboratory studies also showed that the reaction rate, in gm/L·h, producing the acid is given by the Luedeking–Piret model,

$$r^A = \alpha\frac{dX}{dt} + \beta X = 0.0068\frac{dX}{dt} + 0.0001X$$

The initial concentration of the biomass is 9.52 gm/L, that of glucose is 151.9 gm/L, and there is no gluconic acid. Develop a model that describes the concentration of biomass and glucose for this fermentation.

We begin by writing a mass balance on glucose,

$$-Vr^G = \frac{dm^G}{dt}$$

and dividing by volume V,

$$-r^G = \frac{dC^G}{dt} \tag{8.121}$$

1 equation, 2 unknowns $[r^G, C^G]$

The reaction rate of glucose is

$$r^G = 9.85\frac{dX}{dt} + 0.101X \tag{8.122}$$

2 equations, 3 unknowns $[X]$.

As indicated by the laboratory experiments, the model for the biomass concentration is given by

$$0.3768X\left(1 - \frac{X}{18.91}\right) = \frac{dX}{dt} \tag{8.123}$$

3 equation, 3 unknowns.

At this moment we have zero degrees of freedom, so we could solve this set of equations.

Equations 8.121–8.123 constitute the model for this fermentation reaction. Without a doubt, in this case, obtaining a solution using simulation is easier. The reaction rate expressions make the solution of the differential equations quite complex.

8.8 FINAL COMMENTS

All the modeling shown in this chapter has started using the law of conservation of mass; mass or mole balances have always been the starting point. The equations that follow depend on the phenomenon occurring such as mass transfer, chemical reactions, flow-through elements, and so on. These phenomenological equations are usually based on experimental information; the form of the equations and the terms in them are obtained empirically. You will be learning about these phenomena in other courses. The important fact is that you should first select the system, and then think about the basic law that applies, in this case the law of conservation of mass.

8.9 SUMMARY

This chapter considered mass balances which are based on the law of conservation of mass. The chapter presented details to model systems of liquids and gases, as well pure components or mixture of components and simple separation processes and chemical reactors in industrial-type processes, as well as in biomedical and biotechnology processes. The next chapter presents energy balances that are based on the law of conservation of energy.

REFERENCES

1. Fatmawati, A., et al., "Kinetic Study of Gluconic Acid Batch Fermentation by Aspergillus Niger," *World Academy of Science, Engineering and Technology*, vol. 3, 2009.
2. Znad, H., J. Markos, V. Bales, "Production of Gluconic Acid by Aspergillus Niger: Growth and Non-Growth Conditions," *Process Biochemistry*, vol. 39, no. 5, pp. 527–534, 1998.

Problems

Problem 8.1

The tank shown in Figure P8.1 has a diameter of 6.15 m and a height of 3.05 m. The output volumetric flow from this tank is given by $f_{out}(t) = 2h(t)$, where $h(t)$ is the height of liquid in the tank. At a particular time, the tank is at a steady state with an input flow of $2.8\,m^3/min$.

a. What are the units of the number 2 in the expression of the output volumetric flow?
b. What is the steady-state liquid height in the tank?
c. Develop a model that describes how the level in the tank varies if the input flow changes.
d. Obtain the expression that relates the level in the tank to the input flow. Assume the input flow becomes $7.1\,m^3/min$ that is, $f_{in}(t) = 2.8 + 4.3u(t)$.
e. Will the tank overflow? If it overflows how long does it take? If it does not overflow, what is the new steady-state height?

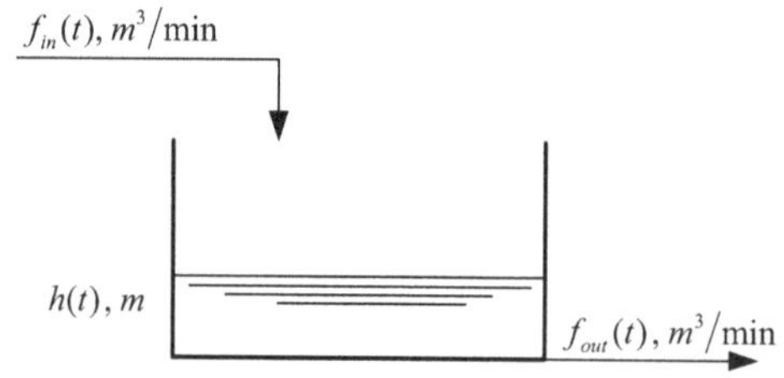

FIGURE P8.1 Tank for Problem 8.1.

Problem 8.2

In a small town in Florida, it hardly rains during mid-March, April, May, and mid-June, a dry season of roughly 90 days. The forecast for this coming year is that there will not be any rain during this time. The town has a water reservoir to be used during the dry season. The forecast also states that during this period, the town will use approximately $3.5 \times 10^5 + 6.7 \times 10^5 e^{-t/18}$ gal/day, where t is the day starting the dry season. Assume no rain and that on the first day of the season, the volume of water in the reservoir is 5.4×10^7 gallons.

a. Write a model that describes the volume of the tank as the dry season advances.
b. How many days of available water are left in the reservoir at the end of the dry season? Graph the volume of water left as time goes on. Solve this part by obtaining the analytical solution of the model *and, if you have studied simulation, by straight simulation. Make sure you obtain the same answer.*

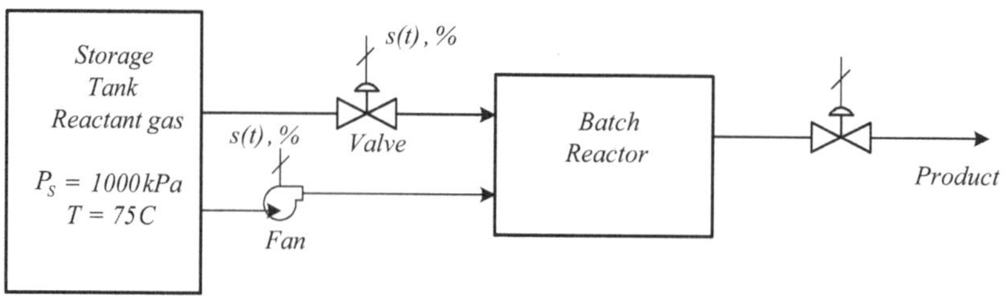

FIGURE P8.2 Process for Problem 8.3.

Problem 8.3

Figure P8.2 shows a batch chemical reactor where a reactant gas produces an expensive product gas. The reactant gas is loaded; the reaction occurs; at the end, the product gas is unloaded; and a new batch starts. The figure shows two feeds to the reactor – however, only one is used at a time; this design is used for redundancy purposes. The reactant gas is loaded into the reactor until the pressure reaches 500 kPa at which time the input flow stops. After unloading the reactor, the pressure is 100 kPa; the reactor has a volume of 78 m^3. The reactant gas is flammable, and therefore, it is recommended to be careful when loading the reactor; its molecular mass is 40. Based on the safety procedures, as the reactant is loaded, the pressure in the reactor must not take less than 15 minutes to go from the initial pressure of 100 kPa to the final pressure of 500 kPa. The reactant gas flows from a storage tank that is maintained at 1000 kPa and 75 °C at all times.

It is necessary to size the valve and the fan that will deliver the reactant gas to the reactor. The flow in m^3/min provided by the fan is given by $f(t) = k_{fan}\dfrac{s(t)}{100}$, where $s(t)$ is a control signal to the fan; when 0%, the fan is not running, and when 100%, the fan runs delivering the gas. For safety reasons, $s(t)$ should be 50% (fan at half-speed). The flow in m^3/min provided by the valve is given by $f(t) = C_V\dfrac{s(t)}{100}\sqrt{P_s - P_{(t)}}$, where C_V is the valve coefficient, $s(t)$ is the same previous signal, P_S is the pressure at the storage tank, and $P(t)$ is the pressure in the reactor in kPa. For safety reasons, $s(t)$ should be 50% (valve half-opened). In sizing the valve, the value of C_V must be known; once C_V is known, the valve size can be selected. In sizing the fan, the value of k_{fan} must be known; once k_{fan} is known, the fan size can be selected.

a. Model the reactor system for the instance when the fan is used. Using the analytical solution obtain k_{fan}. Show the units of k_{fan}.
b. Model the reactor system for the instance when the valve is used. Using the analytical solution obtain C_V. Show the units of C_V.

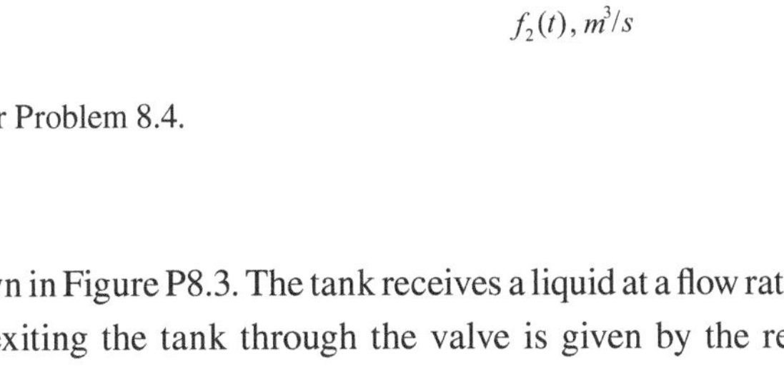

FIGURE P8.3 Tank for Problem 8.4.

Problem 8.4

Consider the tank shown in Figure P8.3. The tank receives a liquid at a flow rate $f_1(t) = 6.0\times10^{-3}\,\mathrm{m^3/s}$. The liquid flow rate exiting the tank through the valve is given by the relation $f_2 = 1.565\times10^{-3}\sqrt{P_1(t) - P_a}$, where $P_1(t)$ denotes the pressure at the bottom of the tank. $P_1(t)$ is given by $P_a + \dfrac{\rho g h(t)}{1000}$ and $g = 9.8\,\mathrm{m/s^2}$ (acceleration due to gravity). The cross-sectional area of the tank is $A = 2\,\mathrm{m^2}$, and the liquid density is $\rho = 1000\,\mathrm{kg/m^3}$.

a. What is the steady-state level in the tank?
b. The forcing function is $f_1(t)$. Derive a model that describes how the level in the tank varies as the forcing function varies; provide the initial condition(s).

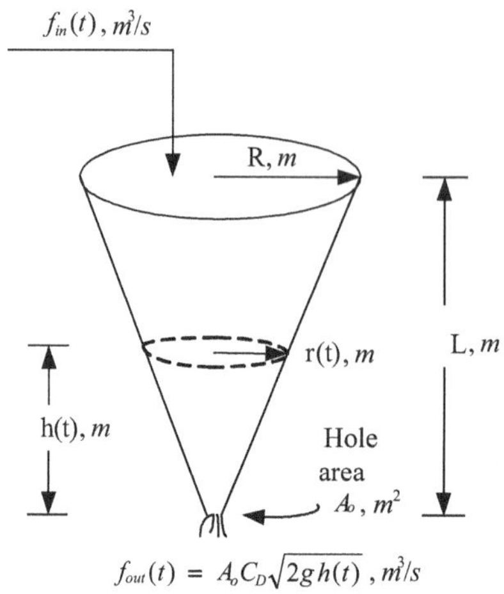

FIGURE P8.4 Tank for Problem 8.5.

Problem 8.5

Consider the conical tank, shown in Figure P8.4, with length L and radius R. The volume of liquid in the tank is $V = \frac{\pi}{3}\left(\frac{R}{L}\right)^2 h^3(t)$. Develop a model that describes the level in the tank; the forcing function is the inlet volumetric flow.

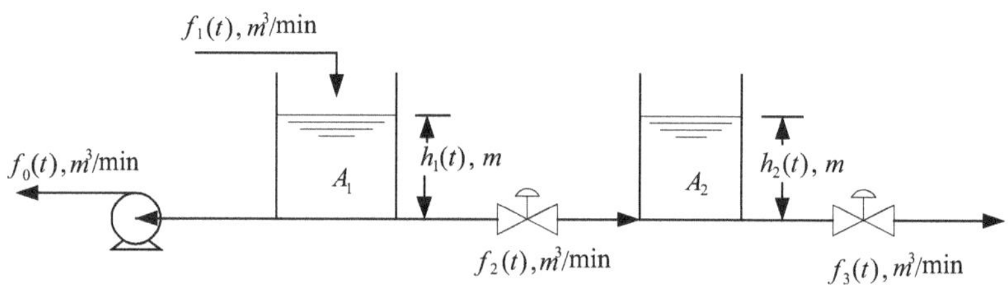

FIGURE P8.5 Tanks for Problem 8.6.

Problem 8.6

Consider the tank system shown in Figure P8.5. Assuming that both tanks are open to the atmosphere and that the exit flow from the second tank also discharges to the atmosphere, the flow through the valves are given by

$$f_2(t) = C_{V1}\sqrt{h_1(t) - h_2(t)} \text{ and } f_3(t) = C_{V2}\sqrt{h_2(t)}$$

The possible forcing functions are $f_1(t)$ and $f_0(t)$. Develop a model describing the levels in both tanks for changes in forcing functions. What are the units of C_{V1} and C_{V2}?

Problem 8.7

Consider the tank shown in Example 8.2.

a. Develop a model that describes the outlet N_aOH concentration, x_3^{NaOH}, when the two input flows change by +10%, and the inlet concentration x_1^{NaOH} changes by −20%.
b. Linearize the model obtained in part (a).
c. *If you are familiar with simulation, simulate the real model of part (a) and the linearized model of part (b), and compare the results. Comments?*

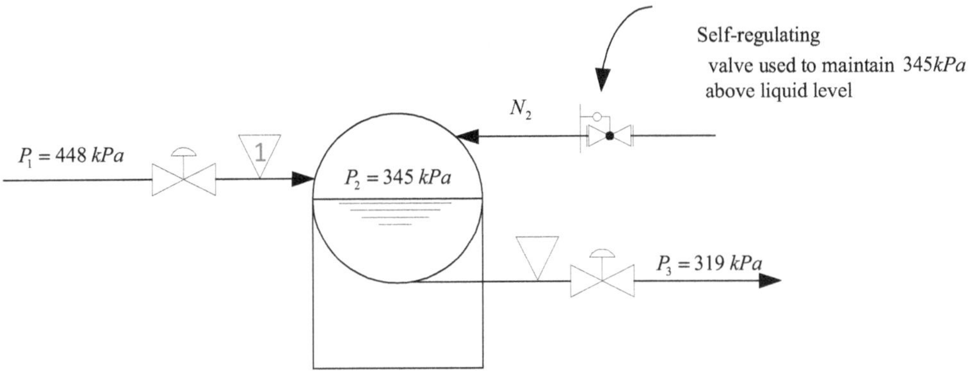

FIGURE P8.6 Tank for Problem 8.10.

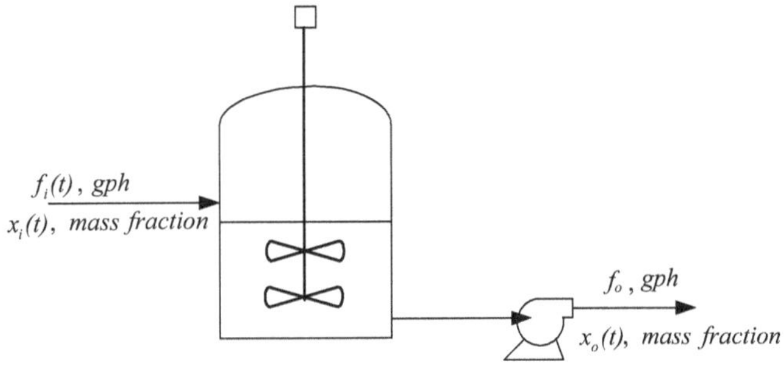

FIGURE P8.7 Mixing tank for Problem 8.11.

Problem 8.8

Process wastewater (density = 1,000 kg/m^3) flows at 500,000 kg/h into a holding pond with a volume of 5,000 m^3 and then flows from the pond to a river. Initially, the pond is at a steady state with a negligible concentration of pollutants, $x(0) = 0$. Because of a malfunction in the wastewater treating process, the concentration of pollutants in the inlet stream suddenly increases to 500 mass ppm (kg pollutant per million kg of liquid) and stays constant at that value (step change).

Assuming a perfectly mixed pond, obtain the transfer function of the pollutant concentration in the outlet stream to the concentration of the inlet stream, and determine for how long the process malfunction can go undetected before the outlet concentration of pollutants exceeds the regulated maximum value of 350 ppm.

Problem 8.9

Water is poured at a rate $f_i(t)$, cm^3/s, into a cup measuring 6.5 cm in diameter and 10 cm high. The cup has a circular hole in the bottom measuring 0.2 cm in diameter. The velocity of the water through the hole (in cm/s) is given, from Bernoulli's equation, by

$$v = \sqrt{2gh}$$

where g is the local acceleration of gravity, 980 cm/s^2, and h cm is the level of the water in the cup.

a. Obtain the model that describes how the level of the water in the cup varies as the inlet flow changes.

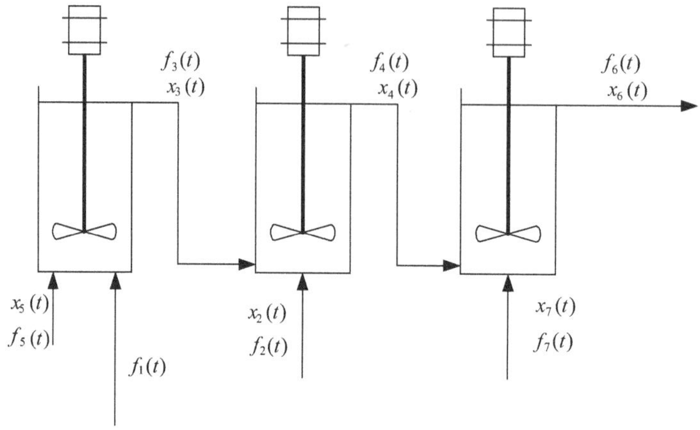

FIGURE P8.8 Tank system for Problem 8.12.

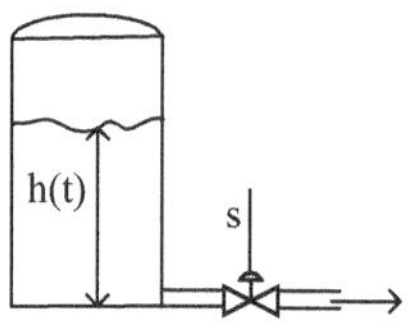

FIGURE P8.9 Storage tank for Problem 8.15.

b. Suppose the cup is originally full, that there is no flow in, and that the hole is capped. At $t = 0$ uncap the hole, how long does it take to drain the cup? Draw the level of liquid versus time.

Problem 8.10

Consider the process shown in Figure P8.6. The tank is spherical with a radius of 4 m. The nominal mass flow into and out of the tank is 14,000 kg/h, the density of the liquid is 1,120 kg/m³, and the steady-state level is 1.52 m. Stream 1 enters the tank 5 meters from the bottom. The volume of a sphere is given by $4\pi r^3/3$. The relation between volume and height is given by

$$V(t) = V_T\left[\frac{h^2(t)[3r - h(t)]}{4r^3}\right]$$

and the flows through the valves by

$$w(t) = 500C_v vp(t)\sqrt{G_f \Delta P(t)}$$

where r = radius of sphere, m;
$V(t)$ = volume of liquid in tank, m³;
V_T = total volume of tank, m³;
$h(t)$ = height of liquid in tank, m;
$w(t)$ = mass flow rate, kg/h;
C_v = valve coefficient, $m^3/(kPa^{1/2})$;
$C_{v1} = 7.3\ m^3/(kPa^{1/2})$ and $C_{v2} = 7.28\ m^3/(kPa^{1/2})$.
$\Delta P(t)$ = pressure drop across valve, kPa;

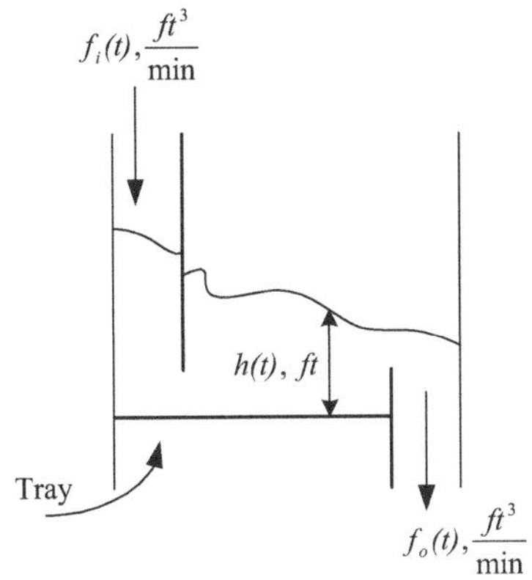

FIGURE P8.10 Tray of a distillation column for Problem 8.16.

G_f = specific gravity of fluid, 1.12; and
$vp(t)$ = valve position, a fraction (0–1) of the valve opening.

The pressure above the liquid level is maintained constant at a value of 345 kPa. The upstream pressure from valve 1 is 448 kPa, and the downstream pressure from valve 2 is 258.8 kPa.

The entrance liquid port is 5 m above the bottom of the tank.

a. Obtain the steady-state valve positions vp_1 and vp_2.
b. Assuming the liquid level in the tank is never above the entrance port of the stream 1, develop the model that relates the liquid level in the tank to changes in valve positions of valves 1 and 2.
c. Remove the assumption in part (b), allowing the liquid level going above the port of the stream 1. Develop a new model that relates the liquid level in the tank to changes in valve positions of valves 1 and 2, indicating the information you may need and may not have been indicated in the figure.

   If you are familiar with simulation, work the next two parts. These are challenging simulations.
d. *Simulate part (c). Increasing the opening of the inlet valve, vp_1, the inlet flow will increase. What is the minimum opening that will drive the liquid level up to the inlet liquid port?*
e. *Back to the initial steady state. After 10 units of time, increase the outlet valve opening by 10% of its present value and obtain how much time it takes for the liquid level to reach its new steady value. What is this new value?*

Problem 8.11

Consider the tank shown in Figure P8.7. A 10% (±0.2 %) by mass NaOH solution is being used for a caustic washing process. To smooth variations in flow rate and concentration, an 8,000-gallon tank is being used as surge tank.

The steady-state conditions are as follows:

$$V = 4000\text{ gal};\ f_i = f_o = 2500\text{ gph};\ x_i = x_o = 0.10\ \textit{mass fraction}$$

The tank contents are well mixed, and the density of all streams is 4 kg/gal.

a. An alarm will sound when the outlet concentration drops to 9.8 mass% (or rises to 10.2 mass%). Assume that the flows are constant. Because of an upset, the inlet

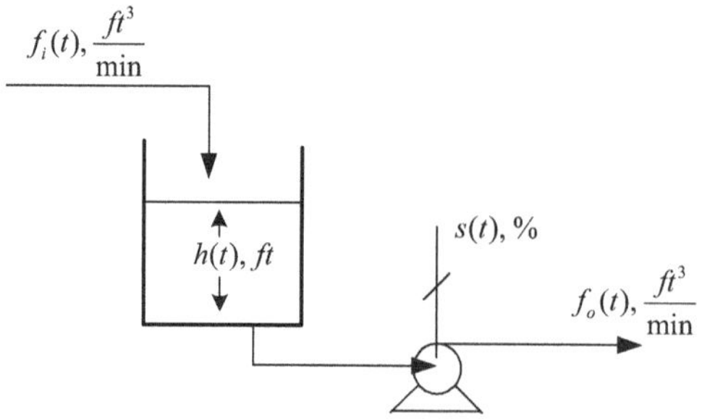

FIGURE P8.11 Process tank with a pump manipulating the exit stream.

concentration, x_i, drops to 8% NaOH instantaneously; that is, $x_i(t) = 0.1 - 0.02u(t)$. It is necessary to determine how long it will take before the alarm sounds.

i. Develop a model that relates the concentration of NaOH in the tank to inlet concentration.
ii. Obtain the analytical solution of the model to determine how long it will take before the alarm sounds.

b. Consider now that the inlet flow, $f_i(t)$, can vary whereas the outlet flow is maintained constant at 2500 gph. Therefore, the volume in the tank can also vary.

i. Develop a model that relates the volume in the tank to the flows in and out.
ii. Suppose that the inlet flow to the tank drops to 1000 gph that is, $f_i(t) = 2500 - 1500u(t)$. Determine how long it takes to empty the tank.

Problem 8.12

Several streams are mixed in the tank system shown in Figure P8.8. Streams 5, 2, and 7 are solutions of water and component A; stream 1 is pure water. The steady-state value and other important process information are given next.

$$\text{Tank Volumes: } V_1 = V_2 = V_3 = 7{,}000 \text{ gal}$$

The density of all streams can be considered similar and constant at 4 kg/gal.

Steady-State Values

Stream	Flow, gpm	Mass Fraction
1	1,900	0.000
2	1,000	0.990
5	500	0.800
7	500	0.900

We are interested in studying the start-up and behavior of this process. To do so, you may assume that the tanks are originally empty.

a. Develop a model that describes the level of liquid and the mass fraction of component A in the first tank when streams 1 and 5 start flowing.
b. Develop a model that describes the level of liquid and the mass fraction of component A in the second tank when streams 1, 5, and 2 start flowing.
c. Develop a model that describes the level of liquid and the mass fraction of component A in the third tank when streams 1, 5, 2, and 7 start flowing.

Note: This is somewhat of a challenging modeling system, namely, to model the flow out of each tank.

d. *If you are familiar with simulation, let us study a slightly modified start-up of the process. The procedure calls for first to enter stream 1 (water) to build levels in every tank. Once the levels are established, all three process streams (streams 5, 2, and 7) enter the process. Simulate this start-up procedure. In a graph, show how the tanks levels build up. In another graph, show the response of the mass fraction in each tank. Make sure to identify the level and mass fraction in each tank. What do the graphs show you? Do they make sense? Why?*

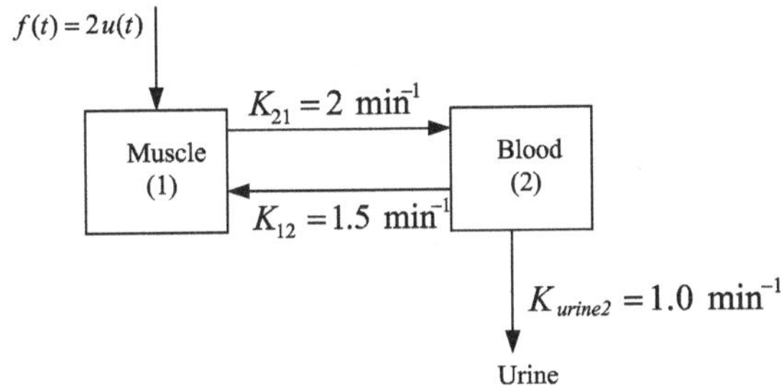

FIGURE P8.12 Compartments representing the muscle and the blood.

Problem 8.13

Consider a tank originally containing 300 kg of a magnesium hydroxide–water ($Mg(OH)_2 - H_2O$) solution with 0.45 mass fraction of $Mg(OH)_2$. At time $t = 0$, a stream of water at 25 kg/min enters the tank, and at the same time, another stream of equal amount exits the tank.

a. Assuming the contents of the tank are well mixed, develop a model that provides the mass fraction of $Mg(OH)_2$ inside the tank after $t = 0$.
b. Solve the model to obtain the analytical expression relating the mass fraction in the tank to time. How long will a take to reduce the mass fraction of $Mg(OH)_2$ from the initial value of 0.45 to 0.15 and to 0.03?
c. Draw the shape of the response curve showing the mass fraction of $Mg(OH)_2$ in the tank. The corresponding curve if the streams entering and exiting the tank have a flow of 75 kg/min instead of 25 kg/min. What would be the curve if they were 125 kg/min? What is the difference between the three curves? Explain from a physical point of view and how this difference is reflected in the mathematical model.

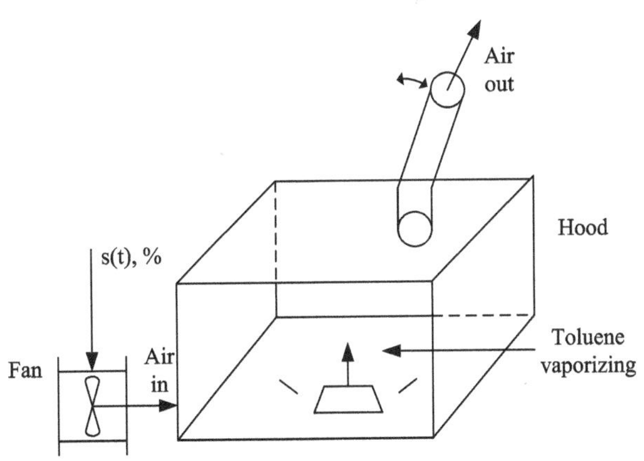

FIGURE P8.13 Chemical hood for Problem 8.26.

Problem 8.14

Consider the following three liquid reactions in series,

$$A \rightarrow B \rightarrow C$$

Component B is the desired product, but unfortunately, once formed, it reacts fairly quickly to component C. The rate of reaction of component A is given by $r_A = k_1 C^A$ and that of component B is $r_B = k_2 C^B$, with $k_1 = 0.5\,\text{min}^{-1}$ and $k_2 = 0.4\,\text{min}^{-1}$. The rates of reaction are given in $\text{mole/m}^3 \cdot \text{min}$.

Assuming that the contents of a batch reactor have an initial concentration $C^A(0) = 1\,\text{mole/m}^3$ and neither B nor C. Develop a model that describes the concentration of each component as a function of time. How much time will it take to reach the maximum concentration of B? It is important to know this time to stop the reaction for obtaining the most B possible. *Do this by obtaining the analytical solution of the model, and by simulation. Check that both methods provide the same solution.*

Problem 8.15

A pharmaceutical plant needs to design a cylindrical tank, Figure P8.9, to store and deliver liquid to its process.

The tank radius is 1 m. The valve equation is given by $f = C_V(s)\sqrt{h}$ where $C_V = 0.0005657$ is the valve coefficient, f is the outlet flow in m^3/sec, s is the signal to the valve, and h is the level (height) of water in the tank in *m*. The valve is such that if the signal it receives increases, the valve opens increasing the flow out. To indicate this behavior, we give the signal a value of 0 to 1.0; 0 gives a value of no flow and 1 provides maximum flow through the valve. At a value of s = 0.5, the valve is half-open.

a. The most common delivery condition is the one that provides a flow such that at the end of 4 hours the liquid delivered to the process be 11.0 m^3; the tank is initially fully filled. The manufacturing department does not want to deal with the valve during this delivery, that is, they want to completely open the valve (signal $s = 1.0$) and not touch it. So they asked to find the height H of the tank that satisfies this operation. Model this system to obtain the height.
b. Operating conditions have changed, and it is now necessary to deliver 7.5 m^3 of liquid in 4 hours. Because we cannot change the height of the tank to satisfy the new condition, we must use the signal to the valve to manipulate the flow. Once the liquid is delivered, the valve closes. Develop a model to obtain the necessary valve opening for this new requirement.
c. *If you are familiar with simulation, simulate part (a) and obtain the minimum height required for the job. Provide a graph showing the liquid level in the tank as a function of time, and another graph providing the total flow delivered as a function of time.*
d. *Simulate part (b) and obtain the minimum signal to the valve to do the job. Provide similar graphs as those suggested in part (c).*

Problem 8.16

The figure shown in Figure P8.10 shows a tray of a distillation column. The liquid flow from the tray is given by the Francis weir formula (adapted from Perry's *Chemical Engineering Handbook*, 6th edition, 1984, p. 5–19)

$$(c)\; f_o(t) = 24.9\,\text{w}\text{h}^{1.5}(t)\sqrt{2\text{g}},$$

where

$h(t)$ = the liquid level on the tray above the top of the weir, ft.

w = width of the weir over which the liquid overflows, ft; and

g = local acceleration of gravity (32.2 ft/s^2).

The density of the liquid is $62.4\,lb/ft^3$. The steady-state inlet flow and process parameters are as follows: tray cross-sectional area = 11.2 ft^2; w = 3.0 ft; $f_i(0)$ = 30 ft^3/min.

a. Model and simulate this tray (Note: you may treat the tray as a tank)
b. Comment on the solution of the model using analytical methods (easy, not easy, and why?)

Problem 8.17

Figure P8.11 shows a tank with one inlet stream and one exit stream. The exit stream is manipulated by a variable speed pump that receives a remote signal provided by an operator or a control system.

The exit flow depends on the pump speed, and this speed depends on the signal the pump receives. You may assume the following relates the exit flow to the signal:

$$f_o(t) = 125\frac{s(t)}{100} = 1.25s(t)$$

At the steady state, the conditions are $f_i(0) = 50\,ft^3/\text{min}; s(0) = 40\%; h(0) = 3ft$. The tank has a diameter of 3 ft and a maximum height of 10 ft. The liquid density is 62.4 lb/ft^3.

a. Develop a model describing how the liquid level in the tank is affected by the inlet flow and signal to the pump.
b. Assuming the inlet flow changes from 50 to 55 ft^3/min, and the signal to the pump remains constant, how long does it take the liquid level to overflow?
c. Let us assume now that there is no liquid level, with no inlet and outlet flows, $s(t) = 0$. The inlet flow now changes as a ramp from 0 to 25 ft^3/min in 5 min, at which time it goes back to 0 ft^3/min. What level does the liquid reach?

Problem 8.18

Dissolved oxygen is very important to support fish populations in rivers, lakes, and streams. In particular, a fish population can become at risk if the dissolved oxygen content falls below 4 mg/l in the water. The oxygen deficit D is defined as

$$D = C^{sat} - C$$

where C is the concentration of oxygen in the water and C^{sat} is the concentration of oxygen in the water at saturation. C^{sat} is about 8 mg/l at ambient conditions so we would like to maintain D at less than 8 – 4 = 4 mg/l.

Oxygen depletion can occur in highly polluted rivers when microorganisms in the water use up the available oxygen in breaking down organic matter. The *biological oxygen demand L* is a measure of how much oxygen is required to break down the organic material and is therefore an indirect measure of the degree of contamination of the river.

The rate $\text{mg}/\text{l}\bullet\text{h}$ at which the organic material will use up the dissolved oxygen in the water is given by $k_d L$. The rate at which oxygen is transferred from the atmosphere to water is given by $K_R D$. Reasonable values of K_R and k_d are 0.2/h and 0.04/h, respectively.

A river saturated with oxygen, that is $D(0) = 0\,\text{mg/l}$, is suddenly exposed to organic waste such that $L(0) = 35\,\text{mg/l}$.

a. Develop a model and obtain the analytical solution that describes the oxygen concentration in the water. (Hint: Consider a volume of liquid, for example 1 liter, and describe what happens to the oxygen in that volume.) (Another hint: As the microorganism's breakdown the organic material, the biological oxygen demand, L, will reduce thus, you may need to write a model of how L changes with time.)
b. If the stream flows at 1 m/s, for what range of distances (in km) downstream will the oxygen level in the river be unacceptable?

Problem 8.19

Section 8.5.3 discussed the separation of component B from component C. Component B transfers across a membrane to a volume containing just water, and it was assumed that the transfer of B to the water would not affect the concentration of B in the water. We assume that B is so diluted in the water that its concentration is essentially zero at all times, $C_{II}^{B} \approx 0\,\text{mass/volume}$. Using this assumption, the transfer rate of B was expressed as $N^{B} = kAC_{I}^{B}$. Assume now that the transfer of B indeed affects its concentration in the water volume. In this case, the transfer rate would be expressed by $N^{B} = kA\left(C_{I}^{B} - C_{II}^{B}\right)$. How would this change affect the process model (set of equations)? Could you obtain an analytical equation? If so, how would you do it? Would you expect the resulting batch time to be longer or shorter than the batch time with the original assumption, and why?

Problem 8.20

Creatine is a waste metabolically produced in the muscles. From the muscle, the creatine transfers to the blood plasma, and from there, it is excreted to urine. Figure P8.12 shows two compartments illustrating the muscles and the plasma. The system is originally at a steady state with zero amount of creatine. Suppose now that the muscle starts producing creatine at 2 mg/min; this is indicated in the figure as $f(t) = 2u(t)$. Model the system and obtain the analytical expressions that describe the amount of creatine in each compartment.

Problem 8.21

Theophylline is a drug used to treat respiratory diseases. Suppose that this drug is given intravenously to a patient at a rate of 10 mg/min for 30 min; after 30 min, the dose stops. The elimination rate constant is 0.08 h^{-1}. Assuming a single compartment, develop the model that describes the amount of theophylline in the compartment and obtain the analytical solution. Hint: It takes two models: one for the first 30 minutes and one for after the dose stops.

Problem 8.22

Reference 3 presents the kinetics and modeling fermentation reaction of glucose to L-glutamic acid in a batch bioreactor:

$$\text{Glucose} + \text{Biomass} \rightarrow \text{L-glutamic acid} + \text{Biomass}.$$

The article presents that the logistic equation describes the biomass growth,

$$\frac{dX}{dt} = \mu_{max} X\left(1-\frac{X}{X_m}\right) = 0.21X\left(1-\frac{X}{3.88}\right)$$

where X is the biomass concentration in kg/m^3, and the time unit is in hour (h). The initial biomass concentration is 0.164 kg/m^3. The article also presents that the rate of reaction of the glucose is given by the Leudeking–Piret model,

$$r_G = \alpha\frac{dX}{dt} + \beta X = 8.33\frac{dX}{dt} + 0.07X$$

and that for the production of L-glutamic acid,

$$r_A = \alpha\frac{dX}{dt} + \beta X = 3.23\frac{dX}{dt} + 0X$$

The initial concentration of glucose is 49.87 kg/m^3, and there is no L-glutamic acid.

Develop a model that describes the glucose and the acid produced. *If you have gone over simulation, simulate the model and compare the results with the following experimental results.*

Time, h	Biomass, kg/m³	Glucose, kg/m³	Acid, kg/m³
0	0.164	49.87	0
5	0.5	48.1	1.01
10	1.01	43.1	3.02
15	1.6	34.8	5.15
20	2.7	29.1	9.1
25	3.0	25.1	10.9
30	3.5	25.0	11.8
35	3.8	25.0	12.1

Problem 8.23

The reaction of A

$$2A + B \rightarrow 2C$$

is a third-order reaction of the form $r_A = kC_A^2 C_B$ with units of *kmoles of A reacted* / liter · hr. The rate constant at 30°C is $k = 85.28$ liter^2/mole^2 hr. A reactor is charged with 250 moles of a gas containing 18.42 mole% A, 10.53 mole% B, and 71.05 mole% of an inert that does not take part in the reaction. The volume of the reactor is 25 liters. An isothermal operation at 30°C is desired. How long a time is required for a 90% conversion of A?

Problem 8.24

Consider the following fermentation reaction:

Substrate (S) + Biomass (X) – > Product (P) + Biomass (X).

The growth rate of the biomass is given by the following Monod expression,

$$r^{Biomass} = 1.5X\left[\frac{C^S}{25 + C^S}\right], \frac{g}{L \cdot h}$$

The following Leudeking-Piret model provides the product formation,

$$r^P = 2.987 r^{Biomass} + 1.1319X, \frac{g}{L \cdot h}$$

and for the reaction of the substrate,

$$r^S = 0.232 r^{Biomass} + 0.2789X, \frac{g}{L \cdot h}$$

Assume $X(0) = 1\frac{g}{L}$ and $C^S(0) = 40\frac{g}{L}$.

Using the information given,

a. develop a model that describes the concentrations of substrate C^S, biomass X, and product C^P as the batch time progresses.
b. *if you are familiar with simulation, simulate the process and find out how much time it takes to deplete the substrate.*

Problem 8.25

Example 8.8 provides a model for water filter. Using the model and information provided in the example obtain the following:

a. The required radius. R in cm, and total height, H in cm
b. The amount of time it would take to filter the water and for the remainder liquid to reach 10% of the total height

Problem 8.26

Figure P8.13 shows a chemical hood with a fan that blows air through. In the middle of the hood stands a big beaker with liquid toluene. Based on the temperature, the toluene vaporizes at an estimated rate of 0.1 lb/min. The environmental regulations state that the maximum toluene concentration in the exit air flow be 0.00002336 lb/ft^3. The flow of air, f_{air} in ft^3/min, provided by the fan is related to the signal it receives by a first-order differential equation with a time constant of 0.1 min and a gain of 100 ft^3/min/%, or

$$0.1\frac{df_{air}}{dt} + f_{air} = 100\,s(t)$$

The volume of the hood is 10 ft^3.

a. Calculate the minimum required air flow to dilute the toluene that vaporizes so that the toluene concentration in the air exiting the hood is 10% below the environmental regulation limit.
b. Calculate the required signal to the fan, $s(t)$ in %, that provides this flow.

c. Suppose now that, by mistake, the fan stops, $s(t) = 0\ \%$, and no air flows into the hood (major mistake). After a while, the process operator realizes this problem, but by then, the toluene concentration inside the hood is 0.002 lb/ft^3. At this moment, the individual turns on the fan by setting $s(t)$ to its desired value. Assuming that the air–toluene mixture inside the hood is mixed well, obtain the model describing the toluene concentration in the air exiting the hood. *Using simulation, obtain the response of this system.*

9 Thermal Systems

In this chapter, we focus on thermal systems, or systems that undergo changes in energy content as a result of the transfer of heat to or from them. The governing physical principle that will guide the development of mathematical models for these systems is the concept of conservation of energy. In the first part of the chapter, we will study dynamic systems where the independent variable is time and the mathematical models are initial value problems. In the last part of the chapter, we will study systems in steady state where the independent variable is a position coordinate – which results in boundary value problems.

9.1 CONSERVATION OF ENERGY

If we restrict ourselves to processes that do not involve atomic fission or fusion then energy, like mass, is conserved. Writing Equation 8.1 in terms of energy, rather than mass, yields

$$\begin{matrix} \textit{RATE OF} & & \textit{RATE OF} & & \textit{RATE OF CHANGE OF} \\ \textit{ENERGY ENTERING} & - & \textit{ENERGY EXITING} & = & \textit{ENERGY ACCUMULATED} \\ \textit{SYSTEM} & & \textit{SYSTEM} & & \textit{IN SYSTEM} \end{matrix} \tag{9.1}$$

Energy exists in many different forms and this equation has a tremendous number of applications. In fact, its use comprises about one half of the subject of thermodynamics. We will not examine a large number of applications here but rather focus on systems for which the energy flows into and out of the system are in the form of heat and the amount of energy accumulated in the system is proportional to its temperature. Thus, we will not consider systems in which energy transfers to or from the system because of mechanical motion (work). Similarly, we will not consider systems where the stored energy is in the form of macroscopic kinetic or potential energy or systems in which phase changes (e.g., solid to liquid, liquid to vapor) occur. We *will*, however, consider systems where there is a transfer of thermal energy to or from a system due to energy stored in mass that is entering or leaving the system. In addition, we will extend the coverage of chemical reactions given in Chapter 8.

We will begin our progress towards developing mathematical models for thermal systems by considering the two sides of Equation 9.1 separately. First, it is important to discuss some vocabulary because terms such as *heat* take on meanings in common usage that are sometimes different from how an engineer or scientist would use them. Actually, the term *heat* applies only to the left side of Equation 9.1 because heat refers only to energy that is moving because of a temperature gradient – in the direction of hot to cold. Once heat transfer to a system has taken place, we do not use the word heat to describe the energy that has accumulated in the system. This energy is no longer moving but is stored, and we call it internal energy. So, while it would be correct to say, "That bucket of hot water has a large amount of energy stored in it," it would not be correct to claim that it has a lot of heat stored in it.

9.2 MODES OF HEAT TRANSFER

We will begin our development of mathematical models for thermal systems by looking at the "heat flow" terms on the left side of Equation 9.1. It turns out that there are three different modes by which the transport of heat can occur: conduction, convection and radiation.

Conduction is heat transfer that occurs because of atomic and molecular motion on a microscopic scale. In gases or liquids, for instance, heat is conducted by the collision of faster moving

DOI: 10.1201/9781003619161-9

molecules with slower moving molecules with a resulting exchange of energy and momentum. In solids, heat is conducted both by the movement of free electrons and by vibrational energy propagated between neighboring atoms in a lattice.

If you leave a metal spoon in a pan that you are using to heat soup, and you return to stir the soup only to find that the handle of the spoon is uncomfortably hot, it is because of conduction of heat through the spoon.

Convection is heat transfer that occurs because of bulk, or macroscopic, movement of a fluid (liquid or gas) through which the heat is moving. Because it occurs on a macroscopic, rather than microscopic, scale, it is a more rapid process than conduction. Convection may either be forced, if it is caused by externally applied mechanical motion, or natural, if it is caused by density gradients within the fluid.

Imagine that you have poured your soup into a bowl and begin to eat it with a spoon. You know it is hot, so you blow across the top of the spoon. This is an example of forced convection in which an external agent (you) has created mechanical motion in the medium (air) through which heat is being transferred.

What if you just held the spoon up in front of you – would the soup on it cool? Yes, it would, only now the primary heat transfer mode would be natural convection. Here the air closest to the spoon would increase in temperature first. As its temperature increases, its density decreases and that parcel of air will begin to rise and will be replaced with a parcel of cooler air, thus setting up a circulation.

Does conduction occur in gases and liquids? It does, but since convection is a more rapid process, the heat transfer by conduction can usually be neglected.

Radiation is the transfer of thermal energy in the form of electromagnetic waves. Unlike conduction and convection, it does not require a medium for its transport. Good thing, or else we would not receive energy from the sun through the vacuum of space. While radiation has many important applications in engineering, this introductory exposure to heat transfer will focus on conduction and convection.

9.3 CONDUCTION

We have an appreciation of what conduction heat transfer is, but we need to be able to represent it quantitatively if we wish to write a mathematical model for a system in which it occurs. Let us consider a simple example of heat conduction through a plate as shown schematically in Figure 9.1. Assume that temperatures T_1 and T_2 on the left and right surfaces of the plate are known. Because conduction will occur whenever there is a temperature gradient, we know that there will be conduction of heat from left to right (higher temperature to lower temperature).

Let us define the rate of heat transfer (in J/s) through the plate by $\dot{q}$. What do we think that this rate of heat transfer will depend on? Certainly it will depend on what the plate is made out of – our experience tells us that a metal plate will conduct heat much faster than one made out of glass. We also have the sense that a larger difference between temperatures T_1 and T_2, as well as a smaller plate thickness L, would increase the rate of heat conduction. Finally, the rate of heat transfer per unit area (in J/s m^2) might be similar for a small and large plate. But the total rate of heat transfer (in J/s) will be higher for the plate with the largest area.

Fourier put these ideas together to propose his law of heat conduction, which is

$$\dot{q} = -kA\frac{dT}{dx} \tag{9.2}$$

where A is the cross-sectional area of the plate (in this case, the area normal to the direction of heat transfer), dT/dx is the temperature gradient in the x-direction, and k is the thermal conductivity. It is in k that the dependence of the rate of heat transfer on the plate material shows up. For example, the

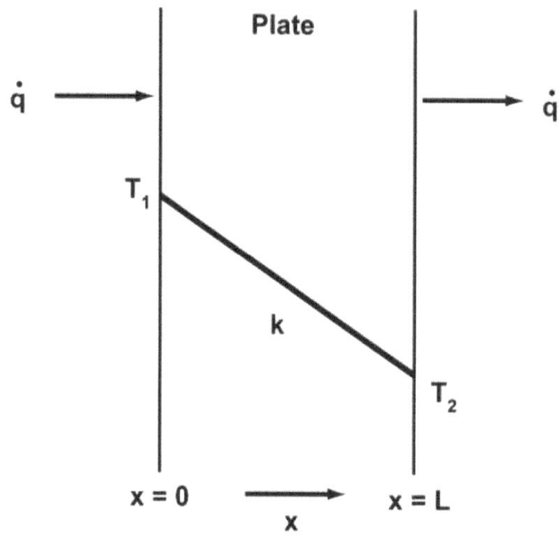

FIGURE 9.1 Heat conduction through a plate.

thermal conductivity of water is about 30 times higher than that of air and a few hundred times lower than that of copper. Thermal conductivity *does* depend on temperature, but it is generally not a strong dependence and can be neglected if the temperature differences in the material are not too large.

Note the minus sign in Equation 9.2. This is needed because conduction occurs down a temperature gradient. The temperature gradient in Figure 9.1 is negative and heat flows in the positive-x direction, making it a positive quantity. Likewise, if T_2 were greater than T_1, the temperature gradient would be positive but the heat flow would be negative (right to left) – thus the requirement of the minus sign.

Equation 9.2 is a differential equation that relates the rate of heat conduction to the temperature gradient. In general, both may change with position x. Let us solve a simple problem of steady state conduction through the plate shown in Figure 9.1. If the plate is at a steady state, then the temperature may change with position in the plate but not with time. It follows, then, that at steady state, $\dot{q}$ must be constant along position x. If $\dot{q}$ changed with x, it would be because some interior slice of the plate was cooling or heating – which means it could not be in a steady state.

Assuming the thermal conductivity k may be treated as constant, Equation 9.2 can be separated and integrated:

$$-kA\int_{T_1}^{T_2} dT = \dot{q}\int_0^L dx \tag{9.3}$$

or

$$\dot{q} = \frac{kA}{L}(T_1 - T_2) \tag{9.4}$$

Note that the heat transfer rate $\dot{q}$ will increase with increasing area, increasing temperature difference and increasing thermal conductivity and will decrease with increasing plate thickness – just as we expected.

9.4 CONVECTION

Expressing convection in a quantitative way is a more difficult problem than for conduction because of the motion of the fluid medium. Consequently, modeling of convection is still highly empirical.

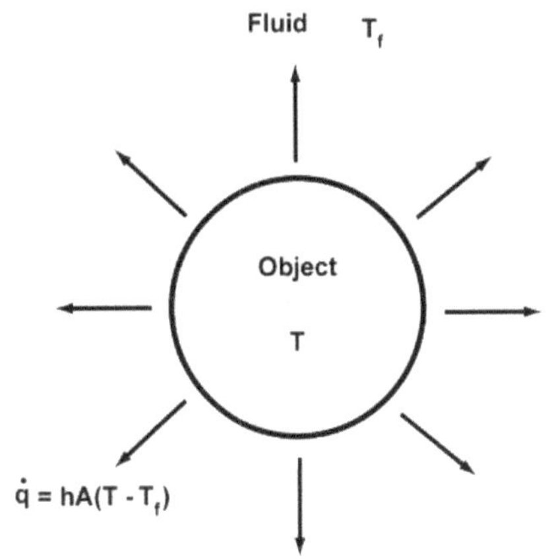

FIGURE 9.2 Convection of heat from an object of surface temperature T and surface area A to a fluid at temperature T_f.

To fix ideas, consider Figure 9.2. An object with a surface area A and a surface temperature T is transferring heat at a rate $\dot{q}$ to a fluid, which has a bulk temperature of T_f.

As in conduction, we expect the heat transfer rate to depend on the difference between T and T_f and on the surface area A. We also expect that the rate of heat transfer will depend on the fluid, knowing from experience that this heat transfer would occur much more quickly if the fluid was water than if it was air. In addition, the motion of the fluid is important – we have the feeling that heat will transfer more rapidly if the fluid is being swept past the object at a high speed as opposed to if the fluid was quiescent.

These features are captured in an empirical expression known as Newton's law of cooling:

$$\dot{q} = hA(T - T_f) \tag{9.5}$$

where A, T, and T_f are as defined before. The quantity h is called the heat transfer coefficient and depends on the fluid and its flow profile. Because the fluid velocity is not always easy to characterize, h can be a difficult quantity to determine. Often, one has no choice but to measure it. For simpler situations, empirical correlations that relate h to the object shape, flow velocity, and fluid material are available in handbooks and heat transfer texts.

We should note that Newton's law of cooling is also applicable to situations in which the object is being *heated* by the fluid. In this case, one would simply replace $T - T_f$ with $T_f - T$ in Equation 9.5. In other words, whether used for heating or cooling, write Equation 9.5 with the higher temperature minus the lower temperature in parentheses.

9.5 CONDUCTION AND CONVECTION IN SERIES

Often, the system we wish to model cannot be represented as simply conduction or convection but consists of several different modes in series. Consider Figure 9.3, which represents schematically the transfer of heat through a wall. Temperatures T_1 and T_2 represent the temperatures on the inside and outside surfaces of the wall, and if we knew these, we could use Equation 9.4 to calculate the heat flow $\dot{q}$ through the wall. Unfortunately, we seldom know the temperatures on the wall surface; instead, we know the inside and outside air temperatures T_{in} and T_{out}. However, the transfer of heat from T_{in} to T_1 and from T_{out} to T_2 is by convection, because it is through a fluid. Thus, because we know only the inside and outside air temperatures, we have to consider a system in which there is transfer through two convective layers and one conductive layer.

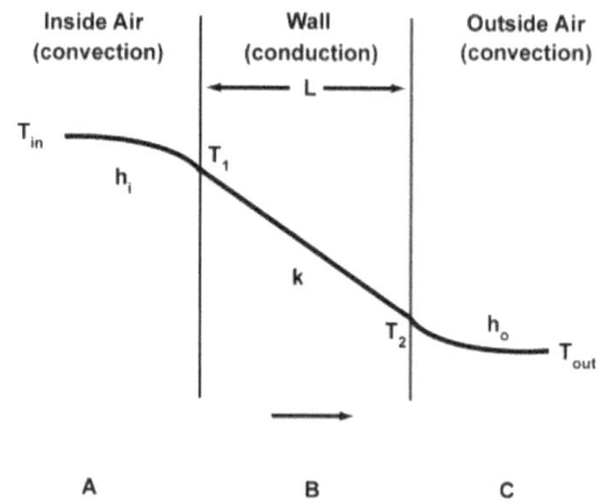

FIGURE 9.3 Heat transfer by convection and conduction.

Let us find the heat flow $\dot{q}$ through the wall for the *steady-state* case. Let $\dot{q}_A$, $\dot{q}_B$, and $\dot{q}_C$ represent the heat transfer rates in the inside air, wall, and outside air, respectively. Let h_i and h_o represent the inside and outside heat transfer coefficients and k represent the thermal conductivity of the wall material. For each layer, A, B and C, we may write Equation 9.4 or 9.5, depending on whether the heat transfer is convective or conductive:

$$\dot{q}_A = h_i A(T_{in} - T_1) \tag{9.6}$$

$$\dot{q}_B = \frac{kA}{L}(T_1 - T_2) \tag{9.7}$$

$$\dot{q}_C = h_o A(T_2 - T_{out}) \tag{9.8}$$

However, if this system is at steady state, then $\dot{q}_A$, $\dot{q}_B$, and $\dot{q}_C$ must all be equal; otherwise, there would be an accumulation or loss of energy somewhere, and the temperature would change, invalidating the steady-state assumption. Replacing $\dot{q}_A$, $\dot{q}_B$, and $\dot{q}_C$ with the symbol $\dot{q}$ and rearranging Equation 9.6 to 9.8, we obtain

$$\dot{q}/h_i A = (T_{in} - T_1) \tag{9.9}$$

$$\dot{q}L/kA = (T_1 - T_2) \tag{9.10}$$

$$\dot{q}/h_o A = (T_2 - T_{out}) \tag{9.11}$$

By adding these equations, we can drop out the unknown temperatures T_1 and T_2 to obtain

$$\left(\frac{1}{h_i} + \frac{L}{k} + \frac{1}{h_o}\right)\frac{\dot{q}}{A} = (T_{in} - T_{out}) \tag{9.12}$$

Usually, Equation 9.12 is rearranged further into the form

$$\dot{q} = UA(T_{in} - T_{out}) \tag{9.13}$$

where U is called the *overall heat transfer coefficient* and, for this particular case, has the form

$$U = \frac{1}{\left(\frac{1}{h_o} + \frac{L}{k} + \frac{1}{h_i}\right)} \tag{9.14}$$

Actually, Equation 9.13's similarity to 9.5 is striking. Equation 9.5 is used when there is a single convective heat transfer layer, and Equation 9.13 is used when there are different layers in series. Extending these equations to systems is easy when there are additional layers by redefining U in Equation 9.14. For instance, work out how the definition of U in Equation 9.14 would change if there was an additional solid layer of a different wall material with thickness L_1 and thermal conductivity k_1. You should be able to convince yourself that U for this case would be

$$U = \frac{1}{\left(\frac{1}{h_o} + \frac{L}{k} + \frac{L_1}{k_1} + \frac{1}{h_i}\right)} \tag{9.15}$$

9.6 ACCUMULATED OR STORED ENERGY

In Sections 9.2–9.5, we discussed topics related to the transport of energy, or the left side of Equation 9.1. Let us now address the right-hand side of this equation. Neglecting stored kinetic and potential energy, all the energy transported into a system must become internal energy. Often, but not always, this change in internal energy will be proportional to the change in the temperature of the system. Exceptions occur if there is a chemical reaction or a change of phase within the system. For instance, adding heat to a system composed of water without changing its temperature is entirely possible if the conditions are such that all the added energy goes into converting liquid water to water vapor. Likewise, the internal energy of a reacting system may change with time at constant temperature if heat is added or removed from the system to compensate for the energy required or liberated by the chemical reaction. Systems in which a change in phase occurs will be addressed in your thermodynamics course and we will discuss systems with a chemical reaction in more detail in Section 9.9. For the present, we will address systems where chemical reactions or phase changes are absent – which means that changes in internal energy will show up as a change in temperature.

We expect that, for nonreactive materials of a single phase, the internal energy E will be proportional to the temperature. But what is the proportionality constant? If we measured the temperature change resulting from adding 1 J of heat to 1 kg of water, we would find it to be different than if we added 1 J of heat to 1 kg of air. Different materials must therefore have different capacities to store energy. Furthermore, we can anticipate that if we add 1 J of heat to 1 kg of water, we will observe a different temperature change than if we added 1 J of heat to 10 kg of water – so the mass of the system must be important too.

It happens that the stored internal energy E of a nonreactive, single phase, solid material may be represented by

$$E = mC\left(T - T_{ref}\right) \tag{9.16}$$

where T_{ref} is a reference temperature, m is the mass of material and C is its constant volume heat capacity. Like thermal conductivity, heat capacity varies from material to material and depends on temperature. However, its temperature dependence may be neglected when temperature changes are not too large. As written, Equation 9.16 assumes that C is independent of temperature.

It was stated that Equation 9.16 expresses the internal energy for a solid material. It is valid for liquids and gases at low pressure as well. The internal energy of gases and liquids at *high* pressures depends on pressure as well as temperature, but this is another topic for your thermodynamics course.

Equation 9.16 shows that internal energy is proportional to the difference between the temperature T and a reference temperature T_{ref}. In fact, none of the reported values for internal energy are absolute – they are always relative to the value at an arbitrary reference temperature. This is because measurements can detect only changes in energy – not absolute values. Fortunately, this does not cause any difficulty because it is only changes in energy that we ever need to calculate. In this sense, internal energy is like potential and kinetic energy in that they are not absolute but are relative to a reference value. For instance, when we calculate the potential energy mgz of an object in a gravitational field, it is usually with the understanding that the reference (where $z = 0$) is the surface of the Earth.

Now let us address the right-hand side of Equation 9.1. Let us assume that the system is comprised of a nonreactive material that remains in a single phase. Let us also assume that the system is such that the properties within it are uniform (not necessarily constant). In other words, we will assume that although the temperature of the system is changing with time, its value at *any instant of time* is the same throughout the system. We call a system that follows this assumption a *lumped parameter* system. Later, we will discuss the circumstances for which the lumped parameter assumption may be expected to apply.

If we let E represent the stored energy of the system at any time, its rate of change is given by dE/dt. Applying the assumptions in the previous paragraph means that we can represent E by Equation 9.16. The reference temperature is a constant and, assuming constant mass and heat capacity, we find

$$\text{Rate of change of energy accumulated in system} = \frac{dE}{dt} = \text{mC}\frac{dT}{dt} \tag{9.17}$$

Equation 9.17 can be used to represent the rate of change of stored energy for a nonreactive single-phase constant mass system, where the lumped parameter assumption is reasonable.

Next, we will work some example problems that illustrate the concepts discussed and developed in Sections 9.1–9.6.

9.7 SOME EXAMPLES

Example 9.1: Convective Heat Transfer

Spherical steel pellets at a high initial temperature T_0 are quenched by dropping them into a very large cooling bath containing a fluid which is at temperature T_f. Develop an expression for the temperature T of a pellet as a function of time after it is dropped into the cooling bath.

We will make two assumptions in solving this problem:

1. The bath temperature T_f remains constant.
2. The temperature within the pellet is uniform – it depends on time but not on radial position within the pellet. This is equivalent to treating the pellet as a lumped parameter system.

We can justify the first assumption by noting that the bath is very large. Obviously, if we drop a hot pellet into a thimble of water, the temperature of the water will increase. However, if we dropped the pellet into a swimming pool, we would not detect the small temperature change of the water.

The second assumption is more problematic. We can imagine that the temperature in the pellet will be a function of both time and radial position. We know that the temperature within the pellet will decrease with time, but there is no reason to expect that it will do so uniformly. In fact, it seems very likely that the outer layers of the pellet will cool more quickly than the inner layers. For this example, let us solve the problem using assumption 2 and then ask under what circumstances the solution might be valid.

We begin by choosing a pellet as our system and applying the energy conservation law given by Equation 9.1:

$$\begin{matrix} \textit{RATE OF CHANGE OF} & & \textit{RATE OF} & & \textit{RATE OF} \\ \textit{ENERGY ACCUMULATED} & = & \textit{ENERGY ENTERING} & - & \textit{ENERGY EXITING} \\ \textit{IN PELLET} & & \textit{PELLET} & & \textit{PELLET} \end{matrix} \tag{9.18}$$

Because the pellet is a lumped parameter system and is undergoing neither a phase change nor chemical reaction, the rate of change of energy accumulated in the system is given by Equation 9.17, where m, C, and T refer to the mass, heat capacity, and temperature of the pellet. Because the pellet is transferring heat to the fluid, there is no energy flow into the pellet, and the first term on the right side of Equation 9.18 is therefore zero. The heat leaving the pellet is by convection through the bath fluid. Because we are assuming the pellet to be a lumped parameter system, the surface temperature of the pellet and the pellet temperature are identical. Thus, the rate of energy leaving the pellet is given by Equation 9.5, where A is the surface area of the pellet and h is the heat transfer coefficient. Substituting Equation 9.17 and 9.5 into Equation 9.1 yields

$$mC\frac{dT}{dt} = 0 - hA(T - T_f) \tag{9.19}$$

which may be written more simply as

$$\frac{dT}{dt} = -\alpha(T - T_f) \tag{9.20}$$

where $\alpha = hA/mC$. Equation 9.20 is a first-order equation so we need one initial condition. This is

$$\mathrm{T}(0) = \mathrm{T}_0 \tag{9.21}$$

Equation 9.20 can be solved by separating variables and integrating (shown in Chapter 2):

$$\int \frac{dT}{T - T_f} = -\int \alpha\, dt \tag{9.22}$$

Integrating yields

$$\ln(T - T_f) = -\alpha t + C_1 \tag{9.23}$$

where C_1 is an integration constant. Applying the initial condition $T(0) = T_0$,

$$\ln(T_0 - T_f) = C_1 \tag{9.24}$$

Substituting for C_1 in Equation 9.23, followed by rearranging, yields

$$\ln\left(\frac{T - T_f}{T_0 - T_f}\right) = -\alpha t \tag{9.25}$$

Taking the exponential of both sides, and rearranging, yields the expression for pellet temperature T as a function of time:

$$T = T_f + (T_0 - T_f)\exp(-\alpha t) \tag{9.26}$$

where $\alpha = hA/mC$. This solution is called a *decaying exponential* because of the exponential to a negative power. It comes up frequently in engineering and the sciences. Note that if we rearrange Equation 9.20 into $(1/\alpha)T' + T = T_f$ and compare it to Equation 5.3, we see that the quantity α is simply the reciprocal of the time constant τ. As we saw in Chapter 5, the time constant provides information about how long it takes for a first-order process to occur.

Let us see what this solution looks like when plotted. A couple of useful points to consider are the values of T at zero time and at infinite time. At $t = 0$, the exponential goes to unity and Equation 9.26 simply becomes

$$T(0) = T_f + T_0 - T_f = T_0 \tag{9.27}$$

Hopefully, this was no surprise because it was our initial condition. However, it does not hurt to check.

As t approaches infinity, the exponential goes to zero and Equation 9.26 becomes

$$T(\infty) = T_f \tag{9.28}$$

This should come as no large surprise either. Eventually, the pellet will lose energy until it reaches thermal equilibrium with the bath fluid. Once the pellet temperature becomes that of the bath, heat transfer ceases.

The exact graph of T versus t depends on the value of $\alpha = hA/mC$. However, it will have the general characteristic of starting at T_0 and decaying to T_f, approaching it asymptotically, as shown in Figure 9.4 for two different values of α. Increasing the value of α (decreasing the time constant) will simply make the decay happen faster. Remember from Chapter 5 that, practically, the decay will be complete after $t = 5\tau$.

Now that we have a solution, how do we know if it is valid? Remember, this analysis assumes that the temperature of the pellet changes uniformly. We begin by asking under what circumstances the assumption of uniform (not constant) temperature in the pellet is reasonable. In this situation, we have heat conduction inside the pellet from the center (where the temperature will be highest) to the pellet surface. Heat that reaches the surface is then transferred to the bath fluid by convection. It seems reasonable that the assumption of uniform temperature will be valid if the rate of conduction in the pellet is much faster than the rate of convection from the pellet. In this situation, energy would be able to redistribute very quickly within the pellet and maintain an almost uniform profile. It also seems reasonable that the lumped parameter assumption would work better for small pellets, because there is less volume for the internal energy to redistribute over.

In fact, there is a dimensionless grouping called the Biot number that is used to determine whether the lumped parameter analysis is valid. The Biot number N_{Bi} for a sphere is defined by

$$N_{Bi} = hD / k \tag{9.29}$$

where h is the convective heat transfer coefficient, D is the diameter of the sphere, and k is the thermal conductivity of the material composing the sphere. Note that a small diameter and a high

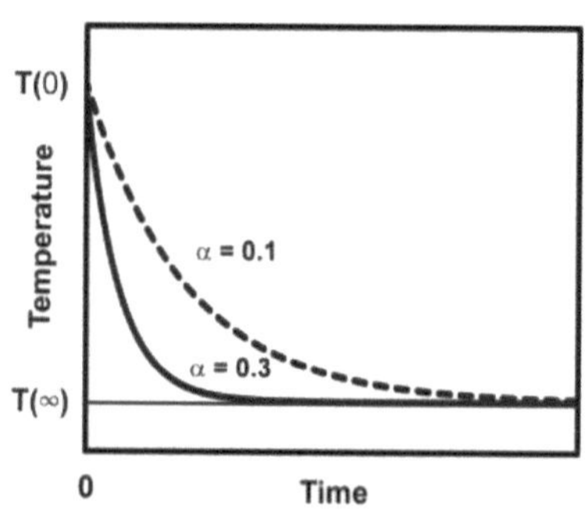

FIGURE 9.4 Graph of Equation 9.26 with two different values of α.

conduction to convection ratio (as discussed earlier) correspond to small values of the Biot number. In practice, a lumped parameter analysis is usually used when the Biot number is less than 0.1.

The next example illustrates the use of an overall heat transfer coefficient.

Example 9.2: Heating a Liquid in a Jacketed, Stirred Vessel

A stirred tank (see Figure 9.5) contains 30 m^3 of water, originally at 25°C. The water is heated by saturated steam condensing at 100°C in the jacket surrounding the vessel. The heat transfer area of the jacket is 80 m^2 and the overall heat transfer coefficient between the condensing steam and the water in the vessel is 500 $J/m^2 \cdot s \cdot °C$. The density and the heat capacity of water are 1000 kg/m^3 and 4200 J/kg ·°C, respectively. Determine how long it will take for the water in the vessel to reach a temperature of 50°C.

We begin our development of a mathematical model by writing an energy balance for the water in the vessel:

$$\begin{matrix} \textit{RATE OF CHANGE OF} & & \textit{RATE OF} & & \textit{RATE OF} \\ \textit{ENERGY ACCUMULATED} & = & \textit{ENERGY INTO} & - & \textit{ENERGY OUT} \\ \textit{IN THE WATER} & & \textit{THE WATER} & & \textit{OF THE WATER} \end{matrix} \quad (9.30)$$

The water is agitated so we will assume it is well-mixed and can therefore be treated as a lumped parameter system. Thus, Equation 9.17 can be used to represent the left side of Equation 9.30 where m, C, and T represent the mass, heat capacity, and temperature of the water in the vessel.

Neglecting any heat loss from the water to the surroundings through the top of the vessel, there will be no rate of energy leaving the water, and the second term on the right side of Equation 9.30 will be zero. The rate of energy *into* the water represents the heat transferred from the condensing steam to the water in the vessel (we are neglecting here any energy input into the water from the mechanical stirrer). The heat transfer from the steam to the vessel wall is convective, as is the heat transfer from the vessel wall to the water. Heat transfer through the vessel wall is by conduction. As a result, we must express the first term on the right of Equation 9.30 in terms of the overall heat transfer coefficient U. Following Equation 9.13, writing the temperature difference so that it is positive, gives us

$$\dot{q}_{in} = UA(T_S - T) \quad (9.31)$$

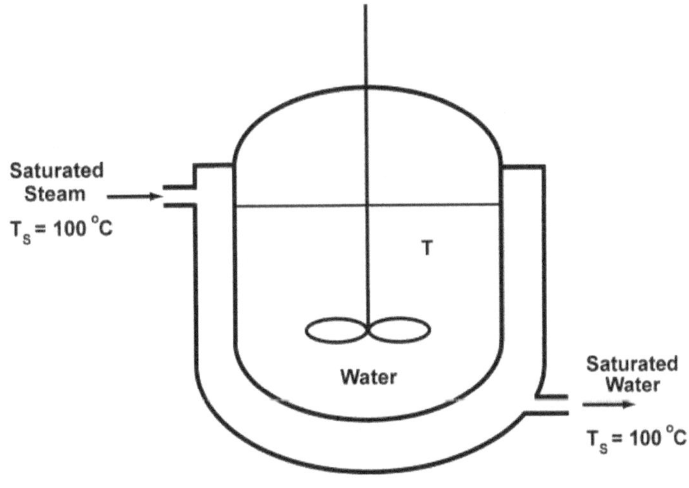

FIGURE 9.5 Heating of water in a jacketed, stirred vessel.

where A is the area of the jacket, T_S is the temperature of the condensing steam, and T is the temperature of the water in the vessel.

Substituting for the left and right sides of Equation 9.30 using Equations 9.17 and 9.31, we obtain

$$mC\frac{dT}{dt} = UA(T_S - T) \tag{9.32}$$

The initial condition is $T(0) = T_0 = 25°C$.

Defining $\alpha = UA/mC$, we find that Equation 9.32 can be placed into the form

$$\frac{dT}{dt} = -\alpha(T - T_S) \tag{9.33}$$

which is identical to Equation 9.20 if T_f is replaced by T_S. Following the steps in the solution to that equation would yield Equation 9.25 (again with T_f replaced by T_S):

$$\ln\left(\frac{T - T_S}{T_0 - T_S}\right) = -\alpha t \tag{9.34}$$

We note that the mass m is equal to the product of density ρ and volume V and evaluate α by

$$\alpha = \frac{UA}{\rho VC} = \left(\frac{500\,\text{J}}{\text{s}\cdot\text{m}^2\cdot{}^\circ\text{C}}\right)\left(\frac{80\,\text{m}^2}{1}\right)\left(\frac{\text{m}^3}{1000\,\text{kg}}\right)\left(\frac{1}{30\,\text{m}^3}\right)\left(\frac{{}^\circ\text{C}\cdot\text{kg}}{4200\,\text{J}}\right) = 0.00032\,\text{s}^{-1} \tag{9.35}$$

Noting that the initial temperature T_0 is 25°C, that the steam temperature T_S is 100°C, and that we wish to find the time needed for the temperature T to reach 50°C, we substitute into Equation 9.34

$$\ln\left(\frac{50 - 100}{25 - 100}\right) = -0.00032\,\text{t} \tag{9.36}$$

Solving for t indicates that 1270 seconds (or approximately 21 minutes) are required for the water in the vessel to reach a temperature of 50°C.

In the next example, we solve a problem in which there is internal heat generation within the system.

Example 9.3: Convective Heat Transfer with Internal Generation

A clothes iron (Figure 9.6) has a sole plate weighing 1.75 kg with an exposed area of 0.05 m^2. The sole plate is made of steel, which has a heat capacity of 450 J/kg·°C, and the heat transfer coefficient for convection from the iron to the surrounding air (which is at 25°C) is 20 J/s·m^2·°C. The iron is rated at 150 W and is initially at the temperature of the air.

a. If the iron is plugged in, what temperature will the sole plate ultimately reach?
b. How long will it take the sole plate to reach a temperature of 100°C?

Before building a mathematical model for this problem let us analyze what is going on physically. The sole plate is heated by resistance wires. When we plug the iron into an electrical socket, 150 W of electrical power is converted to heat in this resistance. Where does the heat go? The sole plate temperature increases so some of it becomes internal energy stored in the iron. However, as soon as the sole plate becomes warmer than the surrounding air, there is a driving force for the convective

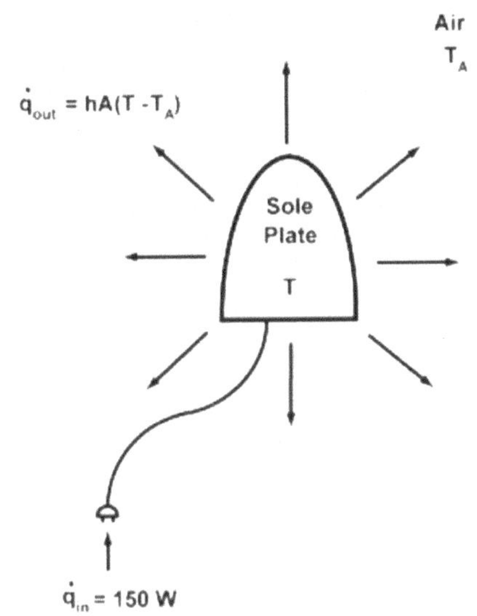

FIGURE 9.6 Heat transfer from the sole plate of an electric iron.

transfer of heat from the sole plate to the air. This problem is somewhat analogous to the fluid storage tank discussed in Chapter 8. A constant inflow of energy (or mass) to a system results in an increase in the system temperature (or liquid level height) which then drives an outflow of energy (or mass) from the system.

Now let us build a mathematical model to solve this problem. Choosing the steel sole plate as the system and applying the energy conservation law, Equation 9.1, yields

$$\begin{matrix} \textit{RATE OF CHANGE OF} & & \textit{RATE OF} & & \textit{RATE OF} \\ \textit{ENERGY ACCUMULATED} & = & \textit{ENERGY ENTERING} & - & \textit{ENERGY EXITING} \\ \textit{IN SOLE PLATE} & & \textit{SOLE PLATE} & & \textit{OF SOLE PLATE} \end{matrix} \tag{9.37}$$

Actually, one might wonder why we are choosing the sole plate as the system. After all, is the entire iron not being warmed? Yes, it is, but the handles, electronics, and gadgets are thermally insulated from the sole plate. Furthermore, most of the iron mass is in the sole plate. So while the other parts of the iron will increase some in temperature, the vast majority of stored energy will reside in the sole plate.

If we treat the sole plate as a lumped parameter system, the rate of change of energy accumulated in it will be given by Equation 9.17, where m, C, and T represent the mass, heat capacity, and temperature of the sole plate, respectively. The heat transferred *from* (out of) the sole plate will be by convection to the air and is represented by $hA(T - T_A)$, where A is the surface area of the sole plate, h is the heat transfer coefficient, T is the temperature of the sole plate and T_A is the air temperature.

Energy is being generated in the sole plate by the conversion of electrical power into heat. Because the conversion of electrical power is total, we can simply interpret this energy as an input to the system and equal to the power drawn by the iron. We will represent this energy inflow by the symbol $\dot{q}_{in}$. In this example, $\dot{q}_{in}$ is constant and equal to 150 W.

Substituting for the terms in Equation 9.37, the energy balance for this system becomes

$$mC\frac{dT}{dt} = \dot{q}_{in} - hA(T - T_A) \tag{9.38}$$

with the initial condition $T(0) = T_0 = 25°C$.

We are now in a position to answer questions (a) and (b). Actually, we do not even have to solve Equation 9.38 to answer question (a). Think about what is happening here. There is a constant inflow of energy to the iron. As the sole plate begins to warm it is accumulating energy, but it is also transferring energy to the surrounding air. The higher the temperature of the sole plate becomes the faster heat is transferred to the surrounding air because this heat transfer is equal to $hA(T - T_A)$. So when will the temperature of the sole plate stop changing (when will it reach its ultimate temperature)? This will happen when the outflow of energy by convection becomes equal to the inflow. In other words, the ultimate temperature will be reached when the sole plate reaches a steady state. Another way of saying this is that the sole plate will reach its ultimate temperature $T = T_{ss}$ when $dT/dt = 0$. Setting $dT/dt = 0$ and $T = T_{ss}$ in Equation 9.38 yields, after rearrangement,

$$T_{ss} = \frac{\dot{q}_{in}}{hA} + T_A \tag{9.39}$$

Using the values given in the problem statement (and noting that 1 W = 1 J/s):

$$T_{ss} = \left(\frac{150\,\text{J}}{\text{s}}\right)\left(\frac{\text{s}\cdot\text{m}^2\cdot{}^\circ\text{C}}{20\,\text{J}}\right)\left(\frac{1}{0.05\,\text{m}^2}\right) + 25^\circ\text{C} = 175^\circ\text{C} \tag{9.40}$$

The sole plate will therefore eventually reach a temperature of 175°C. Take note that, depending on the question you have to answer, it might not be necessary to solve a differential equation at all!

In part b, we are asked to determine the temperature of the sole plate at a particular time – so we have no choice but to solve the differential equation. Equation 9.38 looks more complicated than those in the previous examples, but it is still separable. Sometimes, it is more convenient to solve an equation by collecting constants together and defining the collection by a new symbol. For instance, if we write Equation 9.38 as

$$\frac{mC}{hA}\frac{dT}{dt} = \frac{\dot{q}_{in}}{hA} + T_A - T \tag{9.41}$$

and if, as before, we define $\alpha = hA/mC$ and a new quantity β by

$$\beta = \frac{\dot{q}_{in}}{hA} + T_A \tag{9.42}$$

then the differential equation (Equation 9.41) can be written

$$\frac{1}{\alpha}\frac{dT}{dt} = \beta - T \tag{9.43}$$

which can be separated:

$$\int \frac{dT}{T-\beta} = \int -\alpha dT \tag{9.44}$$

Equation 9.44 looks just like Equation 9.22 except β appears rather than T_f. Following the same steps as those used in going from Equation 9.22 to Equation 9.25, we find that

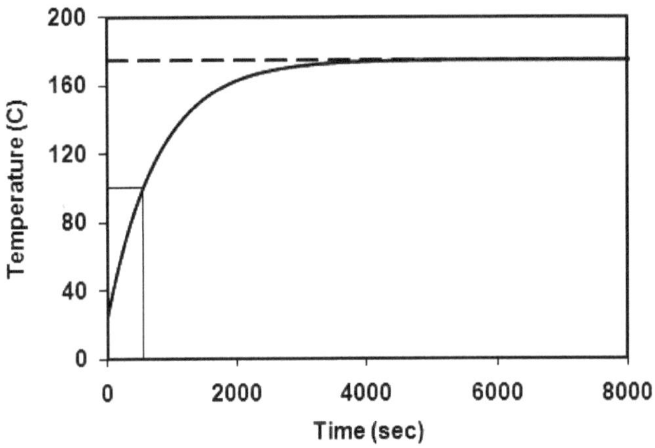

FIGURE 9.7 Temperature profile for the sole plate showing ultimate temperature and time to reach 100°C.

$$\ln\left(\frac{T-\beta}{T_0-\beta}\right) = -\alpha t \tag{9.45}$$

We wish to find the time t needed for the temperature T to reach 100°C. The initial temperature T_0 is equal to the air temperature T_A and is 25°C. The quantity α is

$$\alpha = \frac{hA}{mC} = \left(\frac{20\,\mathrm{J}}{\mathrm{s}\cdot\mathrm{m}^2\cdot{}^\circ\mathrm{C}}\right)\left(\frac{0.05\,\mathrm{m}^2}{1.75\,\mathrm{kg}}\right)\left(\frac{{}^\circ\mathrm{C}\cdot\mathrm{kg}}{450\,\mathrm{J}}\right) = 0.00127\,\mathrm{s}^{-1} \tag{9.46}$$

Comparing the definition of β (Equation 9.42) to Equation 9.39, we see that β is nothing other than the steady-state temperature, which we found to be 175°C.

Substituting for α, β, and T_0 in Equation 9.45 and setting $T = 100$°C, yields, after rearrangement,

$$t = \frac{-1}{\alpha}\ln\left(\frac{T-\beta}{T_0-\beta}\right) = \frac{-1}{0.00127}\ln\left(\frac{100-175}{25-175}\right) = 546\,\mathrm{s} \tag{9.47}$$

In conclusion, the sole plate starts out at 25°C, takes 546 seconds to reach 100°C and eventually levels out at a temperature of 175°C. The temperature profile is illustrated in Figure 9.7.

Note that the time constant for this process is $\tau = 1/\alpha = 787$ seconds. The rule of thumb is that the response will be complete (actually 99.3% complete) after $t = 5\tau$ suggests that the iron will essentially be at its ultimate temperature after 3,935 seconds. The reader may verify from Figure 9.7 that this is the case.

9.8 HEAT TRANSFER IN A FLOW SYSTEM

One might wonder how the concepts discussed here apply to systems where there are mass flows in and out of the system. Consider, for example, the system shown in Figure 9.8. Here, the fluid in the vessel is being heated while fresh fluid flows in and heated fluid flows out. The quantities m_{system} and T represent the mass and temperature of the fluid in the system at any instant of time while w_1 and T_1 represent the mass flow rate and temperature of the inlet stream (with similar designation for the outlet stream). The quantity $\dot{q}_{in}$ is the flow rate of heat to the vessel.

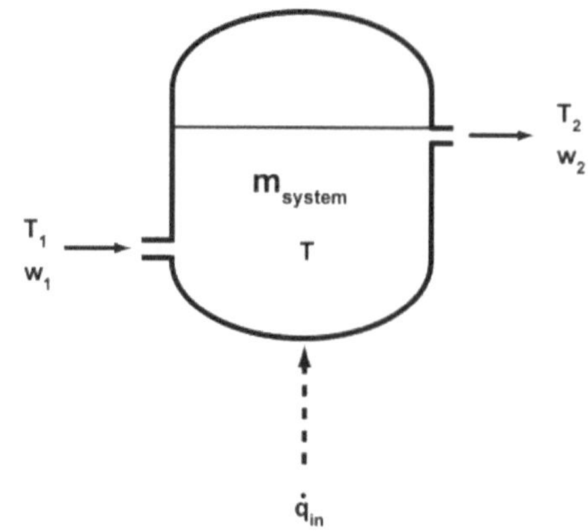

FIGURE 9.8 Heating of a vessel with inlet and outlet flows.

We sense that the amount of thermal energy stored in the flowing fluid should contribute to the energy balance because, for instance, if the entering fluid is already warm, it will require less heat to bring it to the desired temperature T_2.

To apply the conservation of energy equation, Equation 9.1, to this process we must define the system. Conventionally, for this type of problem, the system is defined as the vessel along with the recognition that, because there are mass flows across the system boundary, it is an *open* system. If we assume that the fluid in the vessel is well mixed, the rate of accumulation of energy in the system can be expressed by Equation 9.17 with one modification – we cannot pull m_{system} out of the derivative because the mass in the tank might be changing with time. Therefore,

$$\text{Rate of accumulation} = \frac{dE}{dt} = \frac{Cd\left(m_{system}\, T\right)}{dt} \tag{9.48}$$

Obviously, $\dot{q}_{in}$ will contribute to the rate of energy *into* the system. However, it is not the *only* contribution because there is thermal energy stored in the fluid entering the system. It is beyond the scope of this text to go into details but you will learn in your thermodynamics course that the stored energy in a stream flowing in or out of a system is represented by a quantity called the specific enthalpy h. Enthalpy actually consists of two terms – a stored energy term and a term for the so-called flow work that the fluid does (or has done on it) to push into (or be pushed out of) the system. The rate of energy entering a system due to a mass flow is therefore given by

$$\text{Rate of energy entering with stream } i = w_i h_i \tag{9.49}$$

where w_i is the mass flow rate of stream i and h_i is the specific enthalpy of stream i. Specific enthalpy has dimensions of energy per unit mass so note that the product $w_i h_i$ has units of energy/time. Applying the same concepts to the outlet stream of the system shown in Figure 9.8 allows us to write the energy balance, Equation 9.1 as

$$C\frac{dm_{system}T}{dt} = \dot{q}_{in} + w_1 h_1 - w_2 h_2 \tag{9.50}$$

For a system with an arbitrary number of inflows and outflows, we would write

$$C\frac{dm_{system}T}{dt} = \dot{q}_{in} + \sum_i w_i h_i - \sum_o w_o h_o \tag{9.51}$$

where the first sum is over all i inlets and the second is over all o outlets.

Experimentally, it is found for gases at low pressure, and for solids and liquids, that the difference in specific enthalpy between the inlet and outlet states depends only on temperature and is given by

$$h_i - h_o = C_p\left(T_i - T_o\right) \tag{9.52}$$

where C_p is the constant pressure heat capacity. For liquids and solids, it is observed that the constant volume heat capacity C and the constant pressure heat capacity C_p are essentially indistinguishable.

In solving problems, it will be necessary to couple the energy balance given by Equation 9.51 with the mass balance, given by Equation 8.2:

$$\frac{dm_{system}}{dt} = \sum_i w_i - \sum_o w_o \tag{9.53}$$

We will illustrate the application of these mass and energy balances with an example.

Example 9.4: Heating of a Vessel with Inflows and Outflows

Consider the system shown in Figure 9.8. A steady flow of water enters and leaves the vessel at a flow rate of 2 kg/min. Initially, the entering water, the water in the vessel, and the exiting water are at 20°C. It is desired to heat the water in the vessel so that the water exits the vessel at 30°C. The hold-up of water in the tank is 20 kg and the heat capacity of water is 4.184 kJ/kg°C. Assuming that the fluid in the vessel is well mixed and that the mass flows remain constant at 2 kg/min after the heater is activated:

a. What heat flow $\dot{q}_{in}$ would be required to reach the desired result at a steady state?
b. For the heat flow calculated in part (a), determine how long it will take the water to reach 25 °C from the initial state.

Applying the mass and energy balances, Equation 9.51 and Equation 9.53 to this system with one inflow and one outflow results in

$$\frac{dm_{system}}{dt} = w_1 - w_2 \tag{9.54}$$

and

$$C\frac{dm_{system}T}{dt} = \dot{q}_{in} + w_1h_1 - w_2h_2 \tag{9.55}$$

We begin by noting that, while the temperature of the water in the vessel will change with time after the heater is activated, the inlet and exiting mass flows remain constant. Thus, $w_1 = w_2 \equiv w$ and, from Equation 9.54, $dm_{system}/dt = 0$ or $m_{\text{system}} =$ constant. This allows simplification to both sides of Equation 9.55:

$$m_{system}C\frac{dT}{dt} = \dot{q}_{in} + w(h_1 - h_2) \tag{9.56}$$

Also, we can use Equation 9.52 to represent the enthalpy difference $h_1 - h_2$. Noting that $C_p \approx C$ for a liquid,

$$m_{system}C\frac{dT}{dt} = \dot{q}_{in} + wC(T_1 - T_2) \tag{9.57}$$

Now, we note that if the fluid in the vessel is well mixed, the temperature T_2 of the outlet stream will, at any instant, be the same as the temperature T of the fluid in the vessel, since the outlet stream is merely a drain for the vessel. We arrive then at

$$m_{system} C \frac{dT}{dt} = \dot{q}_{in} + wC(T_1 - T) \tag{9.58}$$

In part (a), we wish to find the heat flow required for the system at steady state. At steady state, $dT/dt = 0$, so Equation 9.58 becomes

$$\dot{q}_{in} = wC(T - T_1) \tag{9.59}$$

At steady state, $T_1 = 20°C$ and $T = T_2 = 30°C$. The required heat flow is therefore

$$\dot{q}_{in} = \left(\frac{2\,\text{kg}}{\text{min}}\right)\left(\frac{4.184\,\text{kJ}}{\text{kg}\cdot{}^{0}\text{C}}\right)\frac{(30-20)^{\circ}\text{C}}{1} = 83.7\ \text{kJ/min} \tag{9.60}$$

For part (b), we use $q_{in} = 83.7$ kJ/min, an initial condition of $T(0) = 20°C$, and wish to find the time required for T to reach 25°C. To do this, we need to solve the differential equation (Equation 9.58). Rearranging Equation 9.58 yields

$$\frac{m_{system}}{w}\frac{dT}{dt} = \frac{\dot{q}_{in}}{wC} + T_1 - T \tag{9.61}$$

If we define

$$\alpha = \frac{w}{m_{system}} \tag{9.62}$$

and

$$\beta = \frac{\dot{q}_{in}}{wC} + T_1 \tag{9.63}$$

then Equation 9.61 becomes

$$\frac{1}{\alpha}\frac{dT}{dt} = \beta - T \tag{9.64}$$

The initial condition for T is $T(0) = T_1$ because, initially, the water in the system is the same temperature as that of the inlet stream. You should be able to show by now that the solution to this equation is

$$\ln\left(\frac{T-\beta}{T_1-\beta}\right) = -\alpha t \tag{9.65}$$

Evaluating α from Equation 9.62 and β from Equation 9.63,

$$\alpha = \frac{2\,\text{kg/min}}{20\,\text{kg}} = 0.1/\text{min} \tag{9.66}$$

$$\beta = \frac{83.7 \text{ kJ/min}}{(2 \text{ kg/min})(4.184 \text{ kJ/kg}\cdot{}^{\circ}\text{C})} + 20 = 30^{\circ}\text{C} \tag{9.67}$$

That β should equal the steady-state outlet temperature is not surprising – compare its definition, Equation 9.63, to the steady state, Equation 9.59.

We wish to find the time required for the vessel (and outlet) temperature to reach 25°C. Rearranging Equation 9.65 for T, followed by substitution of numerical values, yields

$$t = \frac{-1}{\alpha}\ln\left(\frac{T-\beta}{T_1-\beta}\right) = \frac{-1}{0.1/\text{min}}\ln\left(\frac{25-30}{20-30}\right) = 6.9\,\text{min} \tag{9.68}$$

9.9 THERMAL EFFECTS IN A REACTIVE SYSTEM

Modeling of systems in which a chemical reaction occurs was discussed in Section 8.5.2, where it was noted that the reaction rate constant k is a (generally strong) function of temperature. Because thermal effects had not yet been discussed, the examples given in Chapter 8 were limited to isothermal systems. Now that we have a background in applying the energy balance, we can extend the discussion of chemical reactions to systems where the temperature can change.

Specifically, we will consider here the case of an adiabatic batch reactor. As discussed in Section 8.5.3.2, a *batch* reactor is one in which the reactant or reactants are charged to the reactor and the reaction is allowed to occur with no further addition of reactants or removal of products until the process is ended. An *adiabatic* reactor is one in which the reactor is insulated so that there is no heat exchange with the surroundings. If a reaction is exothermic in an adiabatic reactor, the heat generated by the reaction does not transfer to the surroundings – consequently, all this energy goes into heating the reacting mixture and the temperature of the mixture will increase with time. As the temperature changes, the rate constant does as well, which affects the reaction rate, and, in turn, the rate at which energy is released. We expect, then, that the mole balance and the energy balance will be coupled.

To illustrate, we will simulate the hydrolysis of acetic anhydride $(CH_3CO)_2O$ in an adiabatic batch reactor. The reaction is

$$(CH_3CO)_2O + H_2O \rightarrow 2CH_3COOH$$

This reaction is second order, but it is to be run with H_2O in large excess, which makes it pseudo–first order. Let the symbol A represent $(CH_3CO)_2O$. From section 8.5.3.2, the rate of disappearance of A per unit volume r^A (in mol/s·m^3) is

$$r^A = kC^A \tag{9.69}$$

1 equation, 3 unknowns [r^A, k, C^A],

where k is the reaction rate constant and C^A is the concentration of acetic anhydride. The temperature of the rate constant is generally expressed in the *Arrhenius* form:

$$k = k_0 e^{-a/T} \tag{9.70}$$

2 equations, 4 unknowns [T],

where k_0 and a are constants that have to be obtained from experimental data. Here, we assume they are known.

From Section 8.5.3.2, the mole balance on A is

$$\frac{dC^A}{dt} = -r^A \tag{9.71}$$

3 equations, 4 unknowns.

Now we will apply the energy balance. In the case of an adiabatic reaction, there is no energy transferred between the system and the surroundings so the energy balance (Equation 9.1) requires that the rate of accumulation of energy in the system is zero. This is not a trivial result, however, because there are two contributions to the internal energy change for a reactive system and they act in such a way that the overall change in internal energy (for an adiabatic reaction) is zero.

Specifically, the internal energies of the products of the reaction are generally different than those of the reactants, resulting either in a generation or absorption of energy as the reaction proceeds. Secondly, there will be a change in internal energy due to the fact that temperature of the reacting mixture is changing with time. Equation 9.1 can therefore be interpreted as follows:

$$0 = \begin{matrix} \textit{RATE OF CHANGE OF} \\ \textit{ENERGY ACCUMULATED} \\ \textit{IN THE REACTING} \\ \textit{MIXTURE} \end{matrix} = \begin{matrix} \textit{RATE OF INTERNAL ENERGY} \\ \textit{ACCUMULATION DUE} \\ \textit{TO CHEMICAL} \\ \textit{REACTION} \end{matrix} + \begin{matrix} \textit{RATE OF INTERNAL} \\ \textit{ENERGY ACCUMULATION} \\ \textit{DUE TO TEMPERATURE} \\ \textit{CHANGE} \end{matrix} \tag{9.72}$$

The rate of change of internal energy due to the chemical reaction is expressed in terms of the heat of reaction L (in J/mol), which represents the energy liberated or absorbed for each mole of A that reacts. Its value must be determined from experiment. Specifically, the rate of change of internal energy due to the chemical reaction is $L\,(r^A V)$, where V is the volume of fluid in the reactor. The rate of change of internal energy due to temperature change is given by the familiar Equation 9.17. Substituting for both terms in Equation 9.72 results in

$$0 = Lr^A V + mC\frac{dT}{dt} \tag{9.73}$$

The reader should verify that the product $Lr^A V$ has units of J/s. Noting that $m = \rho V$, substitution and rearrangement yields

$$\frac{dT}{dt} = \frac{-Lr^A}{\rho C} \tag{9.74}$$

4 equations, 4 unknowns.

Now that the energy balance is obtained it can be interpreted as follows. If no heat flow into or out of the reactor is allowed, then any energy generated by an exothermic ($L < 0$) chemical reaction will be used to raise the temperature of the reacting mixture. If the reaction is endothermic ($L > 0$), the temperature of the reacting mixture will decrease with time.

The mathematical model for the adiabatic batch reactor (with a first order reaction) is represented by Equations 9.69, 9.70, 9.71, and 9.74. The model can be expressed with two equations by eliminating r^A and k with the other two:

$$\frac{dC^A}{dt} = -k_0 e^{-a/T} C^A \tag{9.75}$$

$$\frac{dT}{dt} = \frac{-Lk_0}{\rho C} e^{-a/T} C^A \tag{9.76}$$

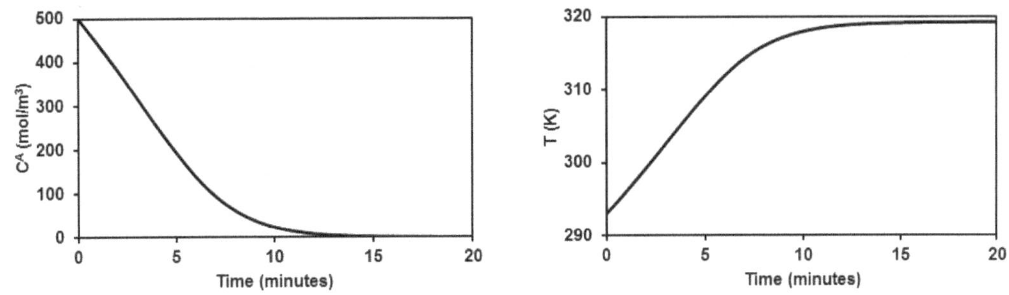

FIGURE 9.9 Solution to Equations 9.75 and 9.76 for an adiabatic batch reactor using given parameter values.

One issue that has been glossed over is that the heat capacity C and density ρ of the reacting mixture (even if the reaction was isothermal) would change with time because both are functions of composition and the composition changes with time. We will assume here that both quantities can be treated as constants. This is a reasonable assumption because water is present in large excess and the heat capacity and density will very nearly be those of water.

To simulate the reaction, two initial conditions are required. These are the initial concentration $C^A(0)$ of acetic anhydride and the initial temperature $T(0)$.

Note that Equations 9.75 and 9.76 are coupled in C^A and T. In addition, because of the exponential in a/T, they are nonlinear as well. As a result, obtaining an analytic solution is difficult (but not impossible). Alternatively, Equations 9.75 and 9.76 can be linearized as shown in Section 3.7, which leads to a reasonably straightforward analytical solution, or solved numerically, using the tools provided in Chapter 11. The latter approach was used to generate the temperature and concentration profiles shown in Figure 9.9. The following parameter values were used: $L = -209{,}000$ J/mol, $\rho = 1050$ kg/m³, $C = 3800$ J/kg·K, $T(0) = 293$ K, $C^A(0) = 500$ mol/m³, $a = 5627$ K, and $k_0 = 24.15 \times 10^6$ min⁻¹. Note that the temperature increases as the reactant disappears and eventually levels off after all the acetic anhydride has reacted, which occurs at about 15 minutes. One job of the design engineer is to ensure that the system does not allow the temperature to reach unacceptable levels, where equipment damage or degradation of the chemical products might occur. This was not a danger here because the initial mixture was dilute in acetic anhydride

9.10 BOUNDARY VALUE PROBLEMS IN HEAT TRANSFER

Throughout this chapter (and book) the emphasis has been on dynamic systems, in which the independent variable of the differential equations is time t. Not all models result in time-dependent differential equations however – often the independent variable is a position variable. Rather than being called *initial* value problems, which consist of differential equation(s) and a set of *initial* conditions, models with an independent *position* variable are called *boundary* value problems, consisting of differential equation(s) and a set of *boundary* conditions. In this section, we will expose the reader to boundary value problems in the context of steady-state heat transfer models.

Because temperature will usually vary with position in the examples considered in this section, we will not often take a lumped parameter approach in setting up the solutions. We will still apply Equation 9.1 – but it will be to a *differential* volume element *within* a body rather than to the body as a *whole*. Because the focus is on steady state problems, the accumulation term in Equation 9.1 will be zero.

Example 9.5: Temperature Profile and Heat Flow in Wire Insulation

Consider the insulated wire shown in Figure 9.10. The wire is made of platinum, has a length of $L = 1$ m, and has a radius $R_i = 0.001$ m. It is surrounded by insulation with an outer radius

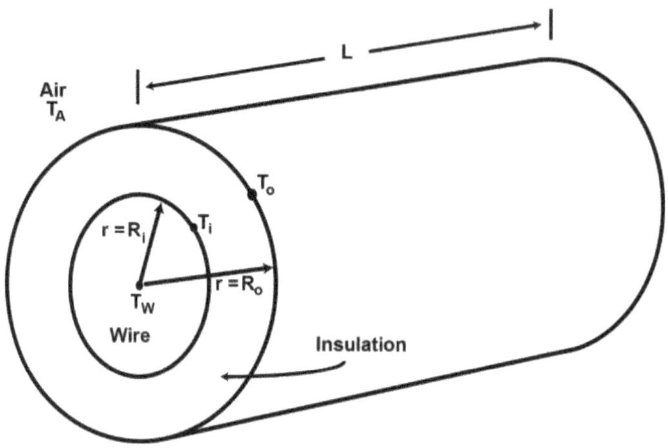

FIGURE 9.10 An insulated wire.

$R_o = 0.002$ m. The thermal conductivity of the insulation is 0.1 J/s•m•°C. At steady state, the wire carries a current of 10 A, which, for a wire of this radius and length, results in resistive heat generation of 3.5 J/s.

a. Determine the steady state temperature T in the insulation as a function of radial position r. Use boundary conditions that $T(R_i) = T_i$ and $T(R_o) = T_o$.
b. Most times, the temperatures T_i and T_o are not known. Express the temperature profile from part (a) in terms of the wire temperature T_W (also unknown) and the air temperature T_A. T_A is 25°C, and the convective heat transfer coefficient h_o between the insulation surface and air is 60 J/s•m²•°C. Develop an expression for the rate of heat flow through the insulation and use it to find the wire temperature T_W.

The wire is made of platinum and has a high thermal conductivity (about 70 J/s•m•°C), so we will treat the wire on a lumped parameter basis. However, the insulation has a much lower thermal conductivity so we will not make the lumped parameter assumption for it. Support for these assumptions is obtained by considering the Biot number defined in Equation 9.29 as $N_{Bi} = hD/k$. It was stated in Section 9.7 that a lumped parameter analysis is usually applied when $N_{Bi} < 0.1$.

Let us suppose the wire was not insulated. With a diameter of 0.002 m, the Biot number for this case would be (60)(0.002)/70 = 0.0017 which is substantially lower than 0.1 and would suggest applying a lumped parameter analysis. The wire in this problem is insulated, but that actually makes the argument for treating the wire on a lumped parameter basis stronger in that it applies an additional external resistance to the wire, making the effective value of h smaller.

For the insulation, which is annular in shape, the effective diameter is the difference between the outer and inner diameters, or $D = 2R_o - 2R_i = 0.002$ m. The Biot number for the insulation is therefore (60)(0.002)/0.1 = 1.2, suggesting that a lumped parameter assumption should *not* be applied to it.

We begin the solution to part a by focusing on the insulation. Consider the volume element between radius r and $r + \Delta r$ as shown in Figure 9.11. The element is a cylindrical shell of insulation with length L, inner radius r, and outer radius $r + \Delta r$. The rate of energy entering the volume element is the conductive heat flow rate $\dot{q}|_r$ and the rate leaving is the conductive flow rate $q|_r+_{\Delta r}$. Because we are assuming steady state, there is no accumulation of energy in the volume element.

Applying Equation 9.1 to the volume element,

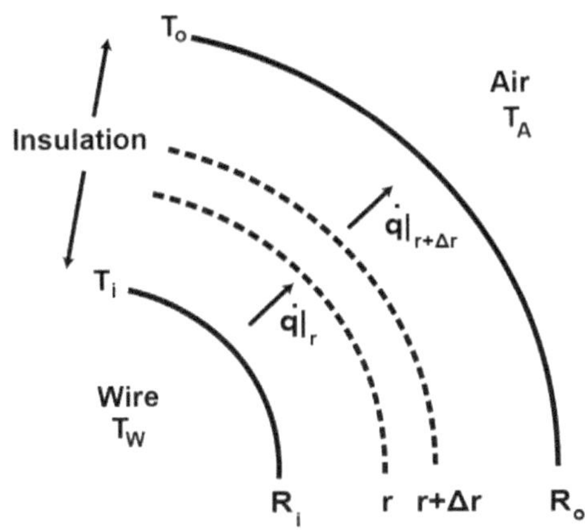

FIGURE 9.11 Expanded view of the insulation showing the chosen volume element.

$$\begin{matrix} \textit{RATE OF ENERGY} & & \textit{RATE OF ENERGY} & \\ \textit{ENTERING VOLUME} & - & \textit{EXITING VOLUME} & = 0 \\ \textit{ELEMENT AT } r & & \textit{ELEMENT AT } r+\Delta r & \end{matrix} \tag{9.77}$$

or

$$\dot{q}\big|_r - \dot{q}\big|_{r+\Delta r} = 0 \tag{9.78}$$

The vertical bars in Equation 9.78 can be read as "at." Equation 9.78 suggests that the heat transfer rate is independent of radius r or that

$$\dot{q} = C_1 \tag{9.79}$$

where C_1 is a constant.

The mode of heat transfer in insulation is conduction, so we use Fourier's law (Equation 9.2) to represent the heat flow:

$$\dot{q} = -kA\frac{dT}{dr} = C_1 \tag{9.80}$$

Note that we have replaced the distance coordinate x in Equation 9.2 with radial position r, because the flow of heat here is in the radial direction. The area A in Equation 9.80 is the area normal to the direction of heat flow. In this case, it represents the area of a cylinder with radius r and length L. Substituting for A by $2\pi rL$ in Equation 9.80 and rearranging yields

$$2\pi kL\frac{dT}{dr} = \frac{-C_1}{r} \tag{9.81}$$

Separating variables and integrating yields

$$2\pi kL\int dT = \int \frac{-C_1}{r}\,dr \tag{9.82}$$

or

$$2\pi kLT = -C_1 \ln(r) + C_2 \tag{9.83}$$

We can now apply the two *boundary* conditions $T(R_i) = T_i$ and $T(R_o) = T_o$:

$$2\pi kLT_i = -C_1 \ln(R_i) + C_2 \quad \text{and} \quad 2\pi kLT_o = -C_1 \ln(R_o) + C_2 \tag{9.84}$$

which may be solved simultaneously for constants C_1 and C_2:

$$C_1 = \dot{q} = \frac{2\pi kL(T_i - T_o)}{\ln\left(\frac{R_o}{R_i}\right)} \tag{9.85}$$

and

$$C_2 = 2\pi kLT_i + \frac{2\pi kL(T_i - T_o)}{\ln\left(\frac{R_o}{R_i}\right)} \ln(R_i) \tag{9.86}$$

where we have explicitly noted using Equation 9.79 that $C_1 = \dot{q}$.

Substituting Equations 9.85 and 9.86 into Equation 9.83 yields, after simplification,

$$T = T_i - (T_i - T_o)\frac{\ln\left(\frac{r}{R_i}\right)}{\ln\left(\frac{R_o}{R_i}\right)} \tag{9.87}$$

Equations 9.85 and 9.87 provide expressions for the heat flow rate $\dot{q}$ through the insulation and the temperature T in the insulation, as requested in part (a) of the problem statement. However, they contain unknown temperatures T_i and T_o. For part (b), we are asked to express these quantities in terms of the air and wire temperatures T_A and T_W. To do this, we must consider both the wire and the convective heat transfer to the air.

Let us consider the wire first. At steady state, the rate of heat generation in the wire must equal the rate of heat conducted through the insulation; otherwise, the temperature of the wire would change. As a result, $\dot{q} = 3.5$ J/s. Also, we are treating the wire on a lumped parameter basis, so the temperature T_W is assumed uniform throughout the wire. Because temperature is continuous at the boundary between the wire and insulation, it follows that $T_i = T_W$. Rewriting Equations 9.85 and 9.87 to reflect these observations yields

$$\dot{q} = \frac{2\pi kL(T_W - T_o)}{\ln\left(\frac{R_o}{R_i}\right)} = 3.5\,\text{J/s} \tag{9.88}$$

and

$$T = T_W - (T_W - T_o)\frac{\ln\left(\frac{r}{R_i}\right)}{\ln\left(\frac{R_o}{R_i}\right)} \tag{9.89}$$

Now let us consider the convective heat transfer to the air. As we noted in Section 9.5, it is the temperature of the fluid surrounding the surface, and not the surface temperature, that is usually known. We can proceed using the approach taken for heat transfer through a flat wall illustrated by Equations 9.9–9.11. Using Equation 9.88 to obtain the conductive term and Newton's law of cooling (Equation 9.5) to obtain the convective term yields:

$$\dot{q}\ln\left(\frac{R_o}{R_i}\right)/2\pi kL = T_W - T_o \tag{9.90}$$

$$\dot{q}/h_o A_o = T_o - T_A \tag{9.91}$$

where $A_o = 2\pi R_o L$. Adding Equations 9.90 and 9.91 yields

$$\dot{q}\left[\frac{\ln(R_o/R_i)}{2\pi kL} + \frac{1}{A_o h_o}\right] = T_W - T_A \tag{9.92}$$

Equation 9.92 can be arranged into the form

$$\dot{q} = U_o A_o (T_W - T_A) \tag{9.93}$$

where the overall heat transfer coefficient U_o is defined as

$$U_o = \frac{1}{\left[\frac{R_o \ln(R_o/R_i)}{k} + \frac{1}{h_o}\right]} \tag{9.94}$$

Equations 9.93 and 9.94 provide the requested result for the heat flow $\dot{q}$. Note the similarity between the overall heat transfer coefficient of Equation 9.94 and that given for the case of a flat wall in Equation 9.14.

We can now evaluate the wire temperature T_W. Using the given parameter values to evaluate the overall heat transfer coefficient from Equation 9.94:

$$U_o = \frac{1}{\left[\ln(2)(0.002\,\text{m})\left(\frac{\text{s}\cdot\text{m}\cdot{}^\circ\text{C}}{0.1\,\text{J}}\right) + \left(\frac{\text{s}\cdot\text{m}^2\cdot{}^\circ\text{C}}{60\,\text{J}}\right)\right]} = \frac{32.75\,\text{J}}{\text{s}\cdot\text{m}^2\cdot{}^\circ\text{C}} \tag{9.95}$$

The wire temperature T_W is then obtained by rearranging Equation 9.93

$$T_W = T_A + \frac{\dot{q}}{U_o A_o} = 25 + \left(\frac{3.5\,\text{J}}{\text{s}}\right)\left(\frac{\text{s}\cdot\text{m}^2\cdot{}^\circ\text{C}}{32.75\,\text{J}}\right)\left(\frac{1}{2\pi(1\,\text{m})(0.002\,\text{m})}\right) = 33.5\,{}^\circ\text{C} \tag{9.96}$$

The only issue remaining is that we cannot calculate the temperature profile in the insulation from Equation 9.89 without knowing temperature T_o. We can evaluate T_o by rearranging Equation 9.91, substituting $A_o = 2\pi R_o L$, and substituting known values:

$$T_o = \dot{q}/h_o A_o + T_A = \left(\frac{3.5\,\text{J}}{\text{s}}\right)\left(\frac{1}{2\pi(1)(0.002)\text{m}^2}\right)\left(\frac{\text{s}\cdot\text{m}^2\cdot{}^\circ\text{C}}{60\ \text{J}}\right) + 25 = 29.6\,{}^\circ\text{C} \tag{9.97}$$

The temperature profile of Equation 9.89 becomes, after substituting for T_W, T_o, R_i, and R_o:

$$T = 33.5 - (33.5 - 29.6)\frac{\ln\left(\frac{r}{0.001}\right)}{\ln(2)} = 33.5 - 5.6\ln\left(\frac{r}{0.001}\right) \quad (9.98)$$

where r is in m and T is in°C.

Example 9.6: Steady-State Temperature Profile in a Sphere with Internal Heat Generation

A spherical nuclear fuel pellet of radius R generates heat per unit volume uniformly at a constant rate of $\dot{s}$ J/s•m^3. Develop an expression for the steady-state temperature T inside the pellet as a function of radial position r. Also, determine the steady-state temperature T_o at the pellet center if the pellet radius R is 0.01 m, its rate of heat generation per unit volume is $\dot{s} = 18 \times 10^6$ J/s•m^3, its thermal conductivity k is 2 J/s•m•°C, the air temperature T_A is 25°C, and the convective heat transfer coefficient h between the pellet surface and surrounding fluid is 100 J/s•m^2•°C.

The sphere is shown in the upper right corner of Figure 9.12. The Biot number hD/k for the sphere is (100)(0.02)/2 = 1.0, so a lumped parameter analysis is not appropriate.

We will begin by focusing on the sphere and then turn our attention to the convection from the sphere surface. We consider a volume element comprised of a spherical shell between radius r and $r + \Delta r$ as shown in Figure 9.12. Applying Equation 9.1 to this element, noting that there is no accumulation of energy in the element at the steady state, yields

$$\begin{matrix} \textit{RATE OF ENERGY} & & \textit{RATE OF ENERGY} & \\ \textit{ENTERING VOLUME} & - & \textit{EXITING VOLUME} & = \\ \textit{ELEMENT AT } r & & \textit{ELEMENT AT } r+\Delta r & \end{matrix} \quad (9.99)$$

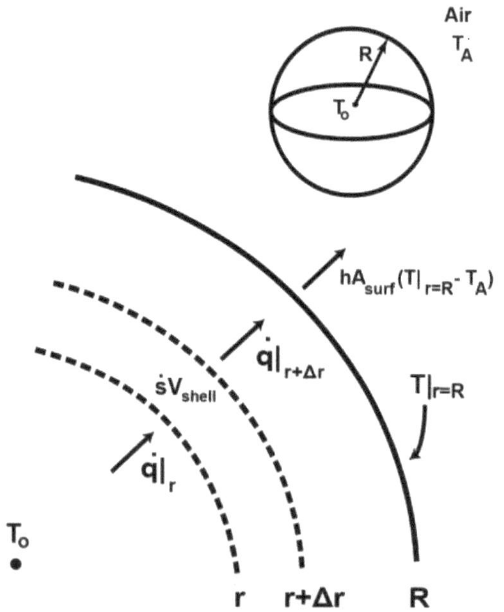

FIGURE 9.12 Spherical pellet and selected volume element.

As in Example 9.3 (the iron sole plate), we can interpret the internal heat generation as an energy input ($\dot{s}\ V_{shell}$ since $\dot{s}$ is the rate of energy generation per unit volume) to the volume element. Therefore, the energy inputs to the volume element are the rate of heat conducted into the element at radial position r plus the rate of heat generation in the element, and the energy output is the rate of heat conducted out of the volume element at radial position $r + \Delta r$. Note that the volume of the shell V_{shell} is the difference between the volume of a sphere of radius $r + \Delta r$ and one of radius r. Equation 9.99 thus becomes

$$\dot{q}\big|_r + \dot{s}\left(4\pi(r+\Delta r)^3/3 - 4\pi r^3/3\right) - \dot{q}\big|_{r+\Delta r} = 0 \tag{9.100}$$

Rearranging, and expanding the volume term, we obtain

$$\dot{q}\big|_{r+\Delta r} - \dot{q}\big|_r = 4\pi\dot{s}\left(r^3 + 3r^2\Delta r + 3r\Delta r^2 + \Delta r^3 - r^3\right)/3 \tag{9.101}$$

Simplifying the term in parentheses and dividing both sides by Δr results in

$$\frac{\dot{q}\big|_{r+\Delta r} - \dot{q}\big|_r}{\Delta r} = 4\pi\dot{s}\left(3r^2 + 3r\Delta r + \Delta r^2\right)/3 \tag{9.102}$$

Now, the limit as Δr approaches zero is taken of both sides, recognizing that, in this limit, the left-hand side becomes the derivative $d\dot{q}/dr$:

$$\frac{d\dot{q}}{dr} = 4\pi\dot{s}r^2 \tag{9.103}$$

Note that, unlike the previous example, the conductive heat flow rate depends on radial position.

For heat conduction in the radial direction, the appropriate form of Fourier's law is

$$\dot{q} = -kA\frac{dT}{dr} \tag{9.104}$$

Remember that A is the area normal to the direction of heat flow. For a sphere, this area is $4\pi r^2$. Equation 9.104 can be substituted into Equation 9.103 to obtain

$$\frac{d\dot{q}}{dr} = \frac{d}{dr}\left(-kA\frac{dT}{dr}\right) = -k\frac{d}{dr}\left(4\pi r^2\frac{dT}{dr}\right) = 4\pi\dot{s}r^2 \tag{9.105}$$

or

$$\frac{d}{dr}\left(r^2\frac{dT}{dr}\right) = \frac{-\dot{s}r^2}{k} \tag{9.106}$$

Separating variables and integrating

$$\int d\left(r^2\frac{dT}{dr}\right) = \int \frac{-\dot{s}r^2}{k}dr \tag{9.107}$$

or:

$$\left(r^2\frac{dT}{dr}\right) = \frac{-\dot{s}r^3}{3k} + C_1 \tag{9.108}$$

Dividing through by r^2, we obtain

$$\frac{dT}{dr} = \frac{-\dot{s}r}{3k} + \frac{C_1}{r^2} \tag{9.109}$$

Separating variables again and integrating

$$\int dT = \int \left(\frac{-\dot{s}r}{3k} + \frac{C_1}{r^2} \right) dr \tag{9.110}$$

or

$$T = \frac{-\dot{s}r^2}{6k} - \frac{C_1}{r} + C_2 \tag{9.111}$$

To evaluate the integration constants C_1 and C_2, we need two boundary conditions. For one of them, we will make use of the fact that the conduction heat transfer rate $\dot{q}$ at $r = R$ must equal the rate of convection to the surrounding fluid:

$$\dot{q}\big|_R = hA_{surf}(T\big|_R - T_A) \tag{9.112}$$

where A_{surf} is the surface area of the sphere $= 4\pi R^2$ and $T|_R$ is the temperature at the sphere surface. What about the other boundary condition? A notion that is often used in polar coordinate systems for which the region of interest includes $r = 0$ is to note that it is physically impossible for a quantity (temperature in this case) to obtain an infinite value. Thus, our second boundary condition is

$$T(0) = \textit{finite} \tag{9.113}$$

Applying the boundary condition in Equation 9.113 to Equation 9.111 results in $C_1 = 0$; otherwise, the temperature would approach infinity at $r = 0$. Thus,

$$T = C_2 - \frac{\dot{s}r^2}{6k} \tag{9.114}$$

We will now apply the boundary condition in Equation 9.112. The rate of heat leaving the pellet by conduction can be obtained by evaluating the conduction heat flow rate $\dot{q}$ at $r = R$, using Fourier's law and Equation 9.114 to evaluate the derivative dT/dr:

$$\dot{q}\big|_R = -k\left(A\frac{dT}{dr} \right)_{r=R} = -k\left(4\pi r^2 \frac{(-2r\dot{s})}{6k} \right)_{r=R} = \frac{4\pi R^3}{3}\dot{s} \tag{9.115}$$

We could have simply asserted this result, rather than derive it, because $\dot{s}$ is the rate of heat generation in the sphere per unit volume and $4\pi R^3/3$ is the volume of the sphere. Thus, the product of the two is the total rate of energy generation in the sphere and it follows that, at steady state, this must equal the rate of conduction at the sphere surface. Nevertheless, deriving the correct result using our obtained temperature profile provides a check on our work so far.

Using Equation 9.114 to substitute for $T|_R$ in Equation 9.112 results in

$$\dot{q}\big|_R = hA_{surf}(T\big|_R - T_A) = 4\pi R^2 h\left(C_2 - \frac{\dot{s}R^2}{6k} - T_A \right) \tag{9.116}$$

Equating the conduction heat flow rate at $r = R$ from Equation 9.115 to the convective transfer Equation 9.116:

$$\frac{4\pi R^3}{3}\dot{s} = 4\pi R^2 h\left(C_2 - \frac{\dot{s}R^2}{6k} - T_A\right) \tag{9.117}$$

and solving for C_2,

$$C_2 = T_A + \frac{\dot{s}R}{3h} + \frac{\dot{s}R^2}{6k} \tag{9.118}$$

Finally, substitution of C_2 from Equation 9.118 into Equation 9.114 provides the temperature profile in the pellet:

$$T = T_A + \frac{\dot{s}R}{3h} + \frac{\dot{s}}{6k}\left(R^2 - r^2\right) \tag{9.119}$$

The steady-state temperature T_o in the center of the pellet is simply T evaluated at $r = 0$:

$$T_o = T_A + \frac{\dot{s}R^2}{6k} + \frac{\dot{s}R}{3h} \tag{9.120}$$

or

$$T_o = 25\,^{\circ}\mathrm{C} + \left(\frac{18\times10^6\,\mathrm{J}}{\mathrm{s\cdot m^3}}\right)\left(\frac{0.01^2\,\mathrm{m}^2}{6}\right)\left(\frac{\mathrm{s\cdot m\cdot {}^{\circ}C}}{2\,\mathrm{J}}\right) + \left(\frac{18\times10^6\,\mathrm{J}}{\mathrm{s\cdot m^3}}\right)\left(\frac{0.01\,\mathrm{m}}{3}\right)\left(\frac{\mathrm{s\cdot m^2\cdot {}^{\circ}C}}{100\,\mathrm{J}}\right) = 775\,^{\circ}\mathrm{C} \tag{9.121}$$

Now that the solution is complete, it is worth spending a little time reconsidering our conclusion that this problem should not be solved by a lumped parameter approach. Suppose we had made the assumption that a lumped parameter approach could be used. This would correspond to assuming that, at steady state, the sphere temperature T was uniform. The total rate of heat generation in the sphere would be $4\pi R^3\,\dot{s}/3$ and the rate of convection heat transfer would be $4\pi R^2 h(T - T_A)$. Equating these two and solving for the sphere temperature results in

$$T = T_A + \frac{\dot{s}R}{3\mathrm{h}} = 25 + \left(\frac{18\times10^6\,\mathrm{J}}{\mathrm{s\cdot m^3}}\right)\left(\frac{0.01\mathrm{m}}{3}\right)\left(\frac{\mathrm{s\cdot m^2\cdot {}^{\circ}C}}{100\,\mathrm{J}}\right) = 625\,^{\circ}\mathrm{C} \tag{9.122}$$

To summarize, a lumped parameter approximation results in a uniform sphere temperature of 625°C and the full solution results in a nonuniform sphere temperature with a temperature in the center of 775°C and the temperature profile of Equation 9.119. These are quite different, but it does not rule out the possibility of using a lumped parameter approach in this problem. In particular, different nuclear fuels can have widely different thermal conductivities, and some may have thermal conductivities large enough to result in small Biot numbers. To demonstrate, we use Equation 9.119 to plot several different sphere temperature profiles in Figure 9.13. The uppermost dashed curve corresponds to the present example ($N_{Bi} = 1$), and the solid line represents the lumped parameter solution. The middle two dashed curves correspond to $N_{Bi} = 0.1$ and $N_{Bi} = 0.02$, representing fuels with higher thermal conductivities (20 and 100 J/s • m • °C) than in the present example. It can be observed from this figure that, as we would expect, the agreement between the full solution and one obtained using a lumped parameter approximation becomes better as the Biot number becomes smaller.

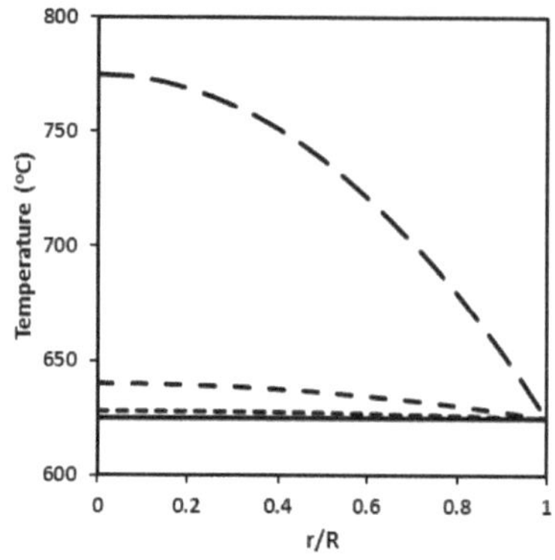

FIGURE 9.13 Temperature versus radial position for the sphere of Example 9.6. The lumped parameter approximation is represented by the solid line and full solutions by the dashed curves. The long dashes, medium dashes, and short dashes correspond to Biot numbers of 1.0, 0.1, and 0.02, respectively.

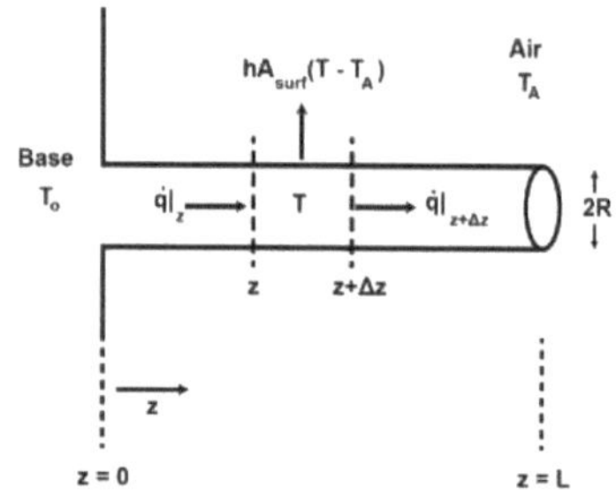

FIGURE 9.14 Heat transfer from an extended surface.

Example 9.7: Steady-State Heat Transfer from an Extended Surface

Consider the extended surface shown in Figure 9.14. Such surfaces (actually many such surfaces in bundles) are often used in air cooled systems (such as lawnmower engines or computers) as a means to increase the heat transfer area between the object being cooled (the base) and the surrounding air, thereby increasing the heat transfer rate between the two. The extended surface shown in Figure 9.14 is a rod that is circular in cross section, but it could be of any shape, including that of a long rectangular bar or, commonly, a thin fin.

There are two important modes of heat transfer in extended surfaces: conduction along the rod or bar from the base temperature T_o to increasingly cooler temperatures along the length and convection from the surface of the bar or rod to the cooler surrounding fluid at T_A. For the rod shown in Figure 9.14,

a. develop an expression for the steady state temperature of the rod as a function of axial position z.
b. If the bar radius R, bar thermal conductivity k, and convective heat transfer coefficient h are 0.01 m, 200 J/s • m • °C and 10 J/s • m² • °C, respectively, find the total rate of heat transfer from the rod in J/s. Assume that the base temperature and air temperature are 100°C and 25°C, respectively.

Extended surfaces are generally made from metals or other highly conductive materials, so it seems that a lumped parameter treatment should be reasonable. However, they are generally long enough that the temperature gradient in the z-direction cannot be neglected. However, they are often thin enough that temperature gradients in the x- and y- (or, for a circular rod, r-) directions *can* be neglected. Consequently, we make the approximation here that temperature in the rod is a function only of z. We begin the analysis by applying Equation 9.1 to a volume element of the rod between positions z and $z + \Delta z$, as shown in Figure 9.14:

$$\begin{matrix} \textit{RATE OF ENERGY} & & \textit{RATE OF ENERGY} & \\ \textit{ENTERING VOLUME} & - & \textit{EXTING VOLUME} & = 0 \\ \textit{ELEMENT} & & \textit{ELEMENT} & \end{matrix} \tag{9.123}$$

The rate of energy entering the volume element is the heat conduction rate $\dot{q}$ at position z. The rate of energy leaving the element is the heat conduction rate $\dot{q}$ at position $z + \Delta z$ and the rate of convection to the surroundings from the rod surface area between z and $z + \Delta z$. Using Newton's law of cooling (Equation 9.5) to represent the convective heat transfer to the surroundings:

$$\dot{q}\big|_z - hA_{surf}\left(T - T_A\right) - \dot{q}\big|_{z+\Delta z} = 0 \tag{9.124}$$

The surface area of the rod between z and $z + \Delta z$ is $A_{\text{surf}} = 2\pi R \Delta z$. Substituting for A_{surf}, rearranging, and dividing both sides by Δz yields

$$\frac{\dot{q}\big|_{z+\Delta z} - \dot{q}\big|_z}{\Delta z} = -2\pi Rh\left(T - T_A\right) \tag{9.125}$$

Taking the limit of both sides as Δz approaches zero results in

$$\frac{d\dot{q}}{dz} = -2\pi Rh\left(T - T_A\right) \tag{9.126}$$

where it is recognized that the left side of Equation 9.125 becomes the derivative $d\dot{q}/dz$ in the limit. It might appear that nothing happened to the *right* side of Equation 9.125 when the limit has taken – but it has. We know that the temperature T in the rod varies with position z, so the temperature T on the right side of Equation 9.125 represents the *average* temperature between z and $z + \Delta z$. When the limit is taken as Δz approaches zero, T (in Equation 9.126) becomes the temperature *at* position z.

Next, we substitute for $d\dot{q}/dz$ in Equation 9.126 using Fourier's law:

$$\frac{d\dot{q}}{dz} = \frac{d}{dz}\left(-kA_{CS}\frac{dT}{dz}\right) = -kA_{CS}\frac{d^2T}{dz^2} = -\pi R^2 k\frac{d^2T}{dz^2} = -2\pi Rh(T - T_A) \tag{9.127}$$

where we have labeled the area A as A_{CS} (for cross section) to emphasize that it is the area (πR^2) normal to the direction of heat conduction. Because the cross-sectional area of the rod is constant here, it may be pulled from the derivative. Rearranging yields

$$\frac{d^2T}{dz^2} - \frac{2h}{kR}T = -\frac{2h}{kR}T_A \tag{9.128}$$

For simplification, we define the constant m by $m^2 = 2h/kR$. The reason for squaring m will become evident shortly:

$$\frac{d^2T}{dz^2} - m^2T = -m^2T_A \tag{9.129}$$

Equation 9.129 is a second-order nonhomogeneous ordinary differential equation with constant coefficients. As such, it can be easily solved by the methods of Sections 3.4 and 3.5. Before doing so, however, we should discuss boundary conditions. Because the equation is second order, we need two boundary conditions. One of them can be obtained by noting that the temperature of the rod at $z = 0$ is equal to the base temperature:

$$T(0) = T_o \tag{9.130}$$

For the other boundary condition, several different approaches are taken, depending on the problem. One common one is to neglect heat transfer through the *end* of the rod. Assuming the length of the rod in the z direction is L, this results in

$$\dot{q}\Big|_{z=L} = -kA_{CS}\frac{dT}{dz}\Big|_{z=L} = 0 \quad or \quad \frac{dT}{dz}\Big|_{z=L} = 0 \tag{9.131}$$

If the rod is fairly long, it is reasonable to expect that the temperature at the end of the rod might approach the surroundings temperature T_A. An alternative boundary condition to Equation 9.131 is therefore obtained by assuming that the rod is infinitely long and its temperature at $z = \infty$ is T_A:

$$T(\infty) = T_A \tag{9.132}$$

In this example, we will employ the infinite length assumption, and the two boundary conditions will be Equations 9.130 and 9.132.

To solve Equation 9.129, we begin by solving the homogeneous equation:

$$\frac{d^2T_H}{dz^2} - m^2T_H = 0 \tag{9.133}$$

Assuming a solution of the form $T_H = e^{rz}$, and following the methods of Section 3.4, results in a characteristic equation of

$$r^2 - m^2 = 0 \tag{9.134}$$

from which we determine that $r = +/-m$. The solution to the homogeneous equation is therefore

$$T_H = C_1e^{-mz} + C_2e^{mz} \tag{9.135}$$

(Our reason for squaring m in its definition earlier can now be seen – it simply results in exponents having "m" in them rather than $\sqrt{m}$). For the particular solution, we note that the right-hand side of Equation 9.129 is a constant, suggesting that we use for a trial particular solution:

$$T_P = A = \text{constant} \tag{9.136}$$

Substituting Equation 9.136 into Equation 9.129 results in

$$-m^2A = -m^2T_A \quad or \quad A = T_A \quad \text{so} \quad T_P = T_A \tag{9.137}$$

The general solution is the sum of the homogeneous solution and particular solution:

$$T = T_H + T_P = T_A + C_1e^{-mz} + C_2e^{mz} \tag{9.138}$$

Applying boundary condition 9.132 to Equation 9.138,

$$T_A = T_A + C_1 e^{-m\infty} + C_2 e^{m\infty} = T_A + C_2 e^{m\infty} \tag{9.139}$$

we see that C_2 must equal zero. Applying boundary condition 9.130 to Equation 9.138 (with $C_2 = 0$) results in

$$T_0 = T_A + C_1 e^0 \quad \text{or} \quad C_1 = T_0 - T_A \tag{9.140}$$

Finally, substituting for C_1 and C_2 in Equation 9.138 provides the desired result for the temperature profile in the rod:

$$T = T_A + (T_0 - T_A)e^{-mz} \tag{9.141}$$

where $m^2 = 2h/kR$.

For part (b), we are asked to calculate the total rate of heat transfer from the rod. The easiest way to obtain this is to note that, at steady state, the rate of heat transferred from the rod surface must equal the rate of heat conducted *into* the rod $\dot{q}|_{z=0}$. Using Fourier's law and Equation 9.141 to evaluate the derivative dT/dz:

$$\dot{q} = -kA_{CS}\frac{dT}{dz} = kA_{CS}m(T_o - T_A)e^{-mz} = k\pi R^2\sqrt{\frac{2h}{kR}}(T_o - T_A)e^{-mz} \tag{9.142}$$

Evaluating the conduction heat rate at $z = 0$ yields

$$\dot{q}\big|_{z=0} = k\pi R^2\sqrt{\frac{2h}{kR}}(T_o - T_A) = \pi R\sqrt{2Rhk}(T_o - T_A) \tag{9.143}$$

Finally, substituting for R, h, k, T_o and T_A by the values provided yields the total rate of heat transfer from the rod:

$$\dot{q}\big|_{z=0} = \pi(0.01\,\text{m})\sqrt{2(0.01\,\text{m})\left(\frac{10\,\text{J}}{\text{s}\cdot\text{m}^2\cdot{}^\circ\text{C}}\right)\left(\frac{200\,\text{J}}{\text{s}\cdot\text{m}\cdot{}^\circ\text{C}}\right)}(100-25)\,{}^\circ\text{C} = 14.9\ \text{J/s} \tag{9.144}$$

To obtain this solution, we argued that we could neglect the dependence of temperature in the rod on radial position r, and therefore, it would be a function only of z. Our logic was that the radial dimension of the rod was much smaller than the axial dimension. Allowing temperature to depend on both r and z is possible of course, but it involves solving a partial differential equation:

$$\frac{1}{r}\frac{\partial}{\partial r}r\frac{\partial T}{\partial r} + \frac{\partial^2 T}{\partial z^2} = 0 \tag{9.145}$$

A discussion of how to solve a partial differential equation like Equation 9.145 is beyond the scope of this book. However, comparing the total heat transfer from the rod obtained from Equation 9.145 to that from Equation 9.143 is useful. In Figure 9.15, we plot the percent difference between heat transferred from the rod calculated from the one-dimensional approach taken here and that calculated using the two-dimensional approach suggested by Equation 9.145. The percentage difference between the two calculated heat flow rates is plotted as a function of the Biot number hD/k. Note that the solutions are within 1% of each other for Biot numbers less than 0.1 and that the Biot

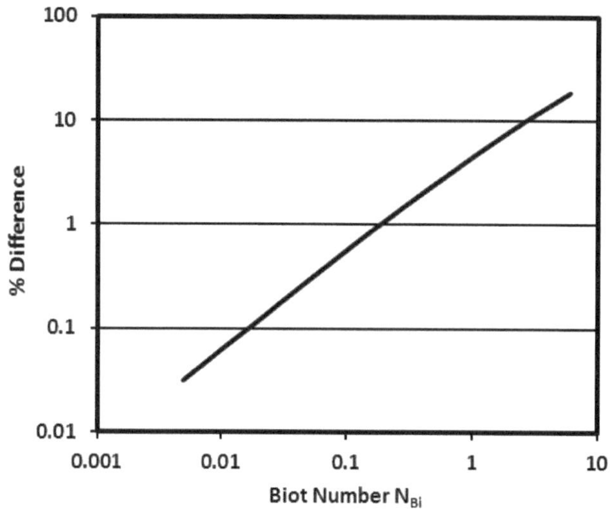

FIGURE 9.15 Percentage difference between the heat flow from an extended, infinitely long circular rod calculated assuming two-dimensional conduction and that calculated assuming one-dimensional conduction.

number of the rod in this example is 0.001, suggesting that our assumption of one-dimensional conduction was justified.

9.11 SUMMARY

In this chapter, we applied the concept of conservation of energy to systems that have heat transferred to or from them. We began with initial value problems for dynamic systems for which changes in stored internal energy are reflected only in a change in temperature and which could be reasonably solved with a lumped parameter approach. In the last part of the chapter, we focused on steady state systems in which temperature varied with a position variable, thus providing an introduction to boundary value problems. Conservation of energy (Equation 9.1) was just as important to these solutions, but it was applied to *differential* volume elements within a body rather than to the body as a whole. The concepts described here provide a reasonable introduction to the subjects of thermodynamics and heat transfer and are sufficient for solving a considerable number of practical problems.

Problems

Problem 9.1

Consider Example 9.3 involving the clothes iron (m = 1.75 kg, A = 0.05 m^2, C = 450 J/kg·°C, $\dot{q}_{in}$ = 150 W, and h = 20 J/s·m^2·°C). The air temperature is 25°C. Assume that the iron is plugged in and has reached its steady state temperature. Now, at $t = 0$, the iron is unplugged.

a. Develop the mathematical model for the temperature T of the iron.
b. Evaluate the time required for the temperature of the iron to fall to 30°C.

Problem 9.2

A metal thermocouple is used to measure the temperature of a water bath. We are interested in finding the response time of the thermocouple. The water bath is at a constant temperature of 40°C. The thermocouple is originally at 25°C, and then it is immersed in the water bath at $t = 0$. The diameter of the thermocouple is 0.001 m, and its immersed length is 0.1 m. The density and heat capacity of the metal comprising the thermocouple are 8,930 kg/m^3 and 383 J/kg · °C, respectively. The area of the end of the thermocouple may be neglected. The heat transfer coefficient between the thermocouple and the water is 80 J/s·m^2 · °C.

a. Develop the mathematical model for the temperature T of the thermocouple.
b. What is the time constant (in seconds) of the thermocouple?
c. Obtain the analytical solution for $T(t)$. How long does it take for the thermocouple to reach 39.9 °C? What does the time constant suggest for the response time? Why might we expect these to agree?

Problem 9.3

Metal bearings have a surface area of 0.001 m^2, a mass of 0.04 kg, and a heat capacity of 400 J/kg·°C. They are to be hardened by heating them in an oven maintained at 800 °C. They are removed from the oven after they have been at or above a temperature of 600 °C for 15 minutes. The heat transfer coefficient between the bearings and the oven air is 30 J/s·m^2·°C. The bearings enter the oven at 25 °C.

a. Develop the mathematical model for the temperature T of the bearing and obtain an analytical solution for $T(t)$.
b. How long will it take the bearings to reach 600 °C?
c. What will a bearing temperature be when it is removed from the oven?

Problem 9.4

Metal bearings have a surface area of 0.002 m^2, a mass of 0.02 kg, and a heat capacity of 800 J/kg °C. They exit a hardening process at a temperature of 200 °C.

a. The bearings are to be cooled in air which is at a constant temperature of 25 °C. The heat transfer coefficient between the bearings and air is 10 J/s m^2 °C. How long (in seconds) will it take to cool the bearings to 30 °C?
b. An alternative cooling process is to cool the bearings in air from 200 to 100 °C and then cool them to 30 °C in water. The heat transfer coefficient between the bearings and water is 200 J/s m^2 °C. What is the total time required (air + water) for the bearings to reach a temperature of 30 °C? The water temperature is maintained constant at 25 °C.

Problem 9.5

An electrical resistor in a direct current generates energy internally at a rate of 0.01 W. It has a mass of 5×10^{-5} kg, a surface area of 4×10^{-5} m^2, and a heat capacity of 300 J/kg·°C. Before the current starts flowing, it is at the temperature of the surroundings (25 °C). The heat transfer coefficient between the resistor and the surrounding air is 20 J/s·m^2 · °C.

a. Develop the mathematical model for the temperature T of the resistor.
b. What temperature will the resistor ultimately reach? Using the notion that the resistor will reach a steady state after 5 time constants, about how long will it take for the resistor to reach its ultimate temperature?
c. Obtain an analytical solution for $T(t)$ of the resistor. When will the temperature of the resistor reach 30 °C?

Problem 9.6

An electrical resistor with a mass of 5×10^{-5} kg and a heat capacity of 300 J/kg·°C is initially at a temperature of $T_0 = 25$ °C. The resistor is perfectly insulated so it exchanges no heat with its surroundings. The discharge of a capacitor creates a short-lived current through the resistor, resulting in an internal heat generation given by

$$\dot{q}_{in} = A\ \exp(-bt),\ \text{where } A \text{ is } 0.06\ \text{J/s and } b \text{ is } 0.5\ \text{s}^{-1}$$

a. Develop the differential equation that represents the temperature T of the resistor.
b. Solve the differential equation to provide an expression for the temperature of the resistor as a function of time.
c. What temperature will the resistor ultimately reach? About how long will it take to get there?

Problem 9.7

A hand warmer works by generating heat internally according to

$$\dot{q}_{in} = be^{-at}$$

where a is 0.001 s^{-1} and b is 30 J/s. The hand warmer has a mass of 0.1 kg, a heat capacity of 2000 J/kg°C, and a surface area of 0.01 m^2. The convective heat transfer coefficient between the hand warmer and the surroundings (which are at 10°C) is 40 J/s m^2°C. When the hand warmer is activated at $t = 0$, its temperature is that of the surroundings. Determine the temperature of the hand warmer at 500 s.

Problem 9.8

Consider the water bath shown in Figure P9.1. The objective is to keep the water in the bath at a temperature below that of the surrounding air T_A (= 25°C). This is done by passing a refrigerant (at temperature $T_C = -10$°C) through a cooling coil. The refrigerant temperature is colder than that of the bath, which allows the refrigerant to remove heat from the water. Heat enters the water because the surrounding air is at a temperature higher than that of the bath.

The mass m and heat capacity C of water are 8 kg and 4200 J/kg°C, and the external area A of the water bath is 0.2 m^2. The heat transfer coefficients U (between the surroundings and the bath) and U_C (between the bath and the refrigerant) are 10 and 100 J/s· m^2·°C, respectively. Air temperature

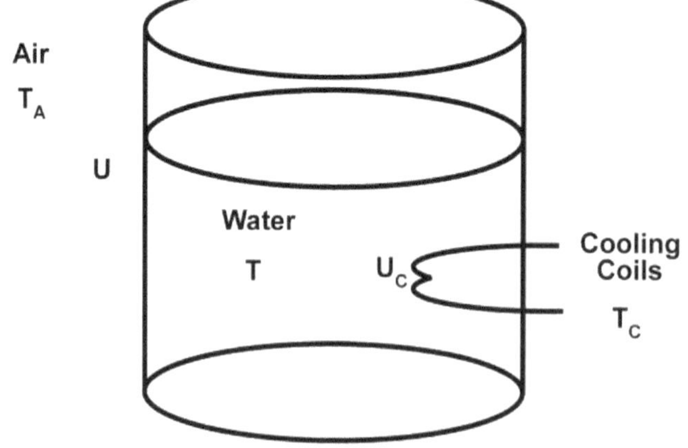

FIGURE P9.1 Cooled water bath of Problem 9.8.

T_A and refrigerant temperature T_C may be assumed to be constant, and the water temperature is initially at the temperature of the air.

a. Let the temperature of the water in the bath at any instant be T. Develop the mathematical model that describes the dependence of T on time t. Include the initial condition.
b. From part (a), develop an expression that gives the steady-state temperature of the water in the bath in terms of the parameters described earlier. Suppose we design the coil for the coldest bath we would want to achieve, say, 5 °C at the steady state. What should the area A_C of the cooling coils be?
c. Obtain an analytical solution for $T(t)$. Using the A_C calculated in part (b), how long will it take the bath to reach a temperature of 20 °C?

Problem 9.9

A resistance heater is used to heat a nonvolatile fluid in the bath shown in Figure P9.2. Initially, the temperatures T_E of the heating element and T of the fluid are at the air temperature T_A (which is constant). The heating element has a mass m_E, heat capacity C_E, and surface area A_E. It draws electrical power at a constant rate $\dot{q}_{in}$, and the heat transfer between the heating element and the fluid is convective with a heat transfer coefficient h. The fluid has mass m and heat capacity C, and the total surface area of the fluid is A. Heat is transferred from fluid to air directly at the top surface and through the container walls on the other surfaces. An overall heat transfer coefficient U includes both effects. The heating element is activated at $t = 0$.

a. Write the mathematical model (differential equations) that represents the temperatures T_E and T of the heating element and fluid. Show the initial conditions as well.
b. The air temperature is constant at 25°C. If $h = 50$ J/m^2•s•°C, $A_E = 0.02$ m^2, $U = 10$ J/m^2•s•°C, $A = 0.08$ m^2, and $\dot{q}_{in} = 40$ W, find the temperatures of both the heating element and fluid at the steady state.

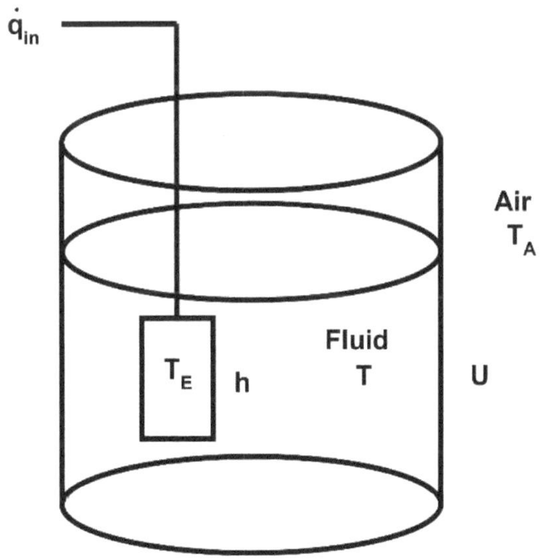

FIGURE P9.2 Heated bath of Problem 9.9.

Problem 9.10

A nichrome wire ($hA = 0.005$ W/°C, $mC = 0.04$ W·s/°C) is initially at 25°C. At $t = 0$, a surge of electrical power of $20 \cdot \delta(t)$ watts occurs and the wire breaks, stopping the current. Here, $\delta(t)$ is the Dirac delta function discussed in Chapter 4.

a. Write the mathematical model for the temperature T of the wire.
b. Obtain the analytical solution for $T(t)$ What is the highest temperature that the wire reaches?

Problem 9.11

Consider Example 9.4. Suppose the tank is operating in the steady state of part (a) with all flow rates and amounts as indicated in the problem statement, $T_1 = 20$°C and $T = T_2 = 30$°C. Now the heater is shut off. Assuming the tank is perfectly insulated from the surroundings, write and solve the differential equation for the tank temperature $T(t)$. How many minutes will it take for the water in the tank to fall to 25°C?

Problem 9.12

Consider Problem 9.11. Suppose now that the tank is not insulated from the surroundings but transfers heat to the surrounding air (at 20°C), with an overall heat transfer coefficient $U = 10$ kJ/min·m^2·°C. The heat transfer area is 2 m^2. Write and solve the differential equation for the tank temperature $T(t)$. How many minutes will it take for the water in the tank to fall to 25°C?

Problem 9.13

Consider the adiabatic batch reactor described in Section 9.9. Equations 9.75 and 9.76 are not easily solved analytically for C^A and T as functions of time. However, it is possible to obtain a direct relation between C^A and T rather easily. Divide Equation 9.76 by Equation 9.75 and simplify. Then integrate to obtain the relation between C^A and T. Using the parameter values given in Section 9.9 for the acetic anhydride hydrolysis, and assuming that the acetic anhydride completely reacts, calculate the final temperature of the reacting mixture. Does this agree with the simulation results shown in Figure 9.9? Can you think of a reason why this calculation would be useful?

Problem 9.14

Consider the adiabatic batch reactor described in Section 9.9. Suppose it is desired to operate the reactor isothermally rather than adiabatically. This could be approached, for instance, using a temperature controller to control the flow rate of a coolant through a jacket on the reactor.

a. It was stated in the text that, for the adiabatic batch hydrolysis of acetic anhydride, the reaction went to completion in about 15 minutes. More precisely, the simulation indicates that 99% of the acetic anhydride reacts in exactly 13.0 minutes. Using the parameter values given in Section 9.9 and assuming the reaction occurs at a constant temperature of 293 K (which was the initial temperature for the adiabatic case), determine the time needed for 99% of the acetic anhydride to react. Why is this time longer than for the adiabatic case?
b. Find the temperature at which an isothermal reaction would reach 99% completion at the same time as the adiabatic case. Does your result make sense?

Problem 9.15

Consider heat transfer to an object where there is no internal energy generation so that $dT/dt = (hA/mC)(T_A - T)$. Suppose that $(hA/mC) = 0.1\ \text{s}^{-1}$, $T(0) = 10°\text{C}$, and the ambient temperature T_A is a linear function of time, given by $T_A = 0.01t + 10$, where t is in seconds and T_A in degrees Celsius. Solve this initial value problem for $T(t)$ using the integrating factor method. Hint: You may have to apply integration by parts.

Problem 9.16

In Example 9.1, we considered a problem in which a spherical metal pellet was dropped into a temperature bath in which the bath fluid was assumed to remain at a constant temperature T_f. Let us reexamine that problem by relaxing this assumption. Now, assume that there is no heat exchange between the bath and the surroundings but that the bath temperature can change due to heat released from the pellet. Now both the pellet temperature $T(t)$ and the bath temperature $T_f(t)$ will vary with time. Take the initial pellet and bath temperatures to be 100°C and 25°C, respectively; the mass m, surface area A, and heat capacity C of the pellet to be 0.01 kg, 0.001 m^2, and 800 J/kg°C; and the heat transfer coefficient h to be 200 J/s m^2°C. The mass m_f and heat capacity C_f of the bath are 0.1 kg and 4000 J/kg°C.

a. Write the differential equations that represent the changes in T and T_f.
b. Obtain analytical solutions for $T(t)$ and $T_f(t)$. The method of Laplace transform is convenient, although somewhat tedious. Alternatively, the simulation methods of Chapter 11 may be applied.

Problem 9.17

Consider the extended rod of Example 9.7 (shown in Figure 9.14). Suppose that the rod has length $L = 1$ m and that at $z = L$, the end of the rod is connected to a separate base that is maintained at temperature T_0.

a. Develop an expression for the steady-state temperature of the rod as a function of axial position z.
b. If the bar radius R, bar thermal conductivity k, and convective heat transfer coefficient h are 0.01 m, 200 J/s•m•°C, and 10 J/s•m^2•°C, respectively, find the total rate of heat transfer from the rod in J/s. Assume that the base temperatures and air temperature are 100°C and 25°C, respectively.
c. Determine the minimum temperature in the rod.

Problem 9.18

Consider the extended rod of Example 9.7 (shown in Figure 9.14). Suppose that the rod is square in cross section with dimensions of 0.02 m × 0.02 m. The rod has length $L = 1$ m, and its end is insulated.

a. Develop an expression for the steady-state temperature of the rod as a function of axial position z.
b. If the bar thermal conductivity k and the convective heat transfer coefficient h are 200 J/s•m•°C and 10 J/s•m^2•°C, respectively, find the total rate of heat transfer from the rod in J/s. Assume that the base temperature and air temperature are 100°C and 25°C, respectively.

c. To estimate the effectiveness of the rod, calculate the rate of heat transfer from a 0.02 m × 0.02 m section of the base. Assume the convective heat transfer coefficient is 10 J/s • m^2 • °C. By what factor is the heat transfer to the surroundings increased by using the extended surface?

Problem 9.19

Consider the nuclear fuel problem of Example 9.6. Suppose that, instead of spherical pellets, the fuel is in the form of cylindrical rods of radius R and length L. The fuel rods are long enough that end effects can be neglected and, therefore, that the temperature inside the rod can be considered to be a function only of radial position r.

a. Assuming a uniform and constant heat generation rate per unit volume $\dot{s}$ (J/s • m^3) and that the air temperature T_A, the rod thermal conductivity k, and convective heat transfer coefficient h between the rod and surrounding air are known, develop an expression for the steady-state temperature in the rod as a function of radial position r.
b. Find the steady state temperature in the center of the rod if $k = 2$ J/s • m • °C, $h = 100$ J/s • m^2 • °C, $\dot{s} = 18 \times 10^6$ J/s • m^3, $R = 0.01$ m, and $T_A = 25$°C.

Problem 9.20

Because of continued radioactive decay, spent nuclear fuel pellets still produce energy after their useful life has ended. To isolate the pellets before waste storage, it is proposed to coat them with a boron-impregnated ceramic. Consider that the spent spherical nuclear fuel pellets have a radius R and generate heat per unit volume uniformly at a constant rate of $\dot{s}$. The radius of the pellet with the added ceramic coating is R_0, the thermal conductivity of the ceramic coating is k_c, and the convective heat transfer coefficient between the surface of the coating and the surrounding air is h.

a. Assume that the thermal conductivity of the fuel pellet is high enough that (unlike Example 9.6), a lumped parameter approach for the fuel pellet (not the ceramic) can be applied. Develop an expression for the steady-state temperature T inside the ceramic coating as a function of radial position r, fuel pellet radius R, coated pellet radius R_0, the fuel temperature T_f, and the temperature T_s of the ceramic surface.
b. Using the given heat generation rate and the fact that the value of the heat flow is continuous between fuel and ceramic and between ceramic and air, develop expressions for the steady-state temperatures T_f and T_s.
c. It is desired that the temperature T_s of the ceramic surface be only about 5 °C higher than the surroundings. How thick should the ceramic coating be? What would be the corresponding temperature T_f of the fuel pellet? Assume R is 0.01 m, the rate of heat generation per unit volume in the fuel is $\dot{s} = 20 \times 10^4$ J/s • m^3, the thermal conductivity k_c of the ceramic is 0.2 J/s • m • °C, the air temperature T_A is 25°C, and the convective heat transfer coefficient h between the ceramic surface and surrounding air is 20 J/s • m^2 • °C.

Problem 9.21

A person in a cold environment loses heat more rapidly when the wind is blowing – a phenomenon that is described by the term *wind chill factor*. Here, our goal is to quantify this effect for the specific case of heat loss through the cheeks of a person whose face is exposed to a low ambient temperature. In this event, there is heat *conduction* from the inside of the mouth to the outer surface of the cheek and heat *convection* from the outer surface of the cheek to the ambient temperature. At a steady state, which is the case examined here, these two heat flows must equal each other.

Heat conduction through the cheek is modeled by Fourier's law:

$$\dot{q} = -kA\frac{dT}{dz}$$

where k is the thermal conductivity of the tissue in the cheek (equal to 0.2 W/m K), z is the distance coordinate ($z = 0$ at inside cheek surface and $z = L = 0.01$ m at the outside skin surface), and A is the surface area of the cheek. Here, we will model the cheek as a flat plate. The inside temperature of the cheek is $T_i = 34°C$, and the temperature T_O of the outer surface will vary with both ambient temperature and wind speed.

The heat transfer from the cheek outer surface (at temperature T_O) to the ambient temperature T_A is given by Newton's law of cooling:

$$\dot{q} = hA(T_O - T_A)$$

where h is a convective heat transfer coefficient, which depends on the wind speed. For heat loss from a human head, researchers have used the following correlation to estimate the heat transfer coefficient:

$$\frac{2R_S h}{k_{air}} = 0.945\,\mathrm{Re}^{0.5}\,\mathrm{Pr}^{0.4}$$

where k_{air} is the thermal conductivity of air (0.024 W/m K), Pr is the Prandtl number for air (Pr = 0.7), R_S is the radius of the head (taken here to be 0.07 m), and Re is the Reynolds number given by

$$\mathrm{Re} = \frac{2R_S V \rho_{air}}{\mu_{air}}$$

where ρ_{air} and μ_{air} are the density and viscosity of air (which can be taken to be 1.3 kg/m^3 and 1.7×10^{-5} kg/m•s, respectively) and V is the wind speed in m/s.

Research shows that heat loss from the face becomes painful when the outer cheek temperature falls to 10°C. The goal here is to compute the ambient temperatures T_A at which the outer cheek temperature will reach 10°C for different wind speeds. Examine wind speeds ranging from 2 to 20 m/s and evaluate T_A for each case.

10 Electrical Systems

This chapter focuses on electrical systems. First, some terms, variables, and units are defined and discussed. The chapter then follows the same organization as in previous chapters in that it looks at the laws used, as well as the electrical components necessary to develop the models.

The reader is encouraged to read and study Chapter 11, where several simulations of examples of this chapter are presented.

10.1 SOME DEFINITIONS AND CONVENTIONS

The letter q denotes an *electrical charge*, and its unit is the **c**oulomb, C; 6.25×10^{18} electrons form a charge of 1 C. Current is the rate of flow of electrical charges, $i = \frac{dq}{dt}$; letter i denotes current, and its unit is the **a**mpere, A, so $1\ A = 1\ C/s$. By convention, current is positive if it flows from positive to negative polarity. (Note: Current is defined as a flow of electrons – negative charges – in reality, electrons move from negative to positive polarity. It is only *by convention* that current is defined as moving from positive to negative polarity; we refer to this current as *conventional current.*)

We have seen in previous chapters that the potential, or gradient, necessary for fluid flow is a difference in pressure, a ΔP, and for heat transfer, it is a difference in temperature, a ΔT. That is, a ΔP is what makes a fluid flow, and a ΔT is what makes heat flow. A difference in electrical potential or *voltage drop* is what makes electrical charges flow (current). The letter E denotes electrical potential at a point in a circuit with respect to a reference point (E comes from EMF, which stands for **e**lectro**m**otive **f**orce). The letter v denotes a difference in electrical potential (voltage drop) across an electrical component, that is $v = \Delta E = E_{in} - E_{out}$. The unit of electrical potential is the *Volt*, V, which is potential energy per unit charge, or 1 V = 1 *joule/coulomb.*

In the sections that follow, two different electrical sources are used. The *ideal voltage source* generates a constant voltage, v_s, regardless of the load attached at the terminals – a fresh, strong battery approximates an ideal voltage source. The *ideal current source* delivers the same current, i_s, regardless of the load attached at the terminals – a lightning strike approximates an ideal current source. Figure 10.1 shows how these two sources will be shown in electrical drawings.

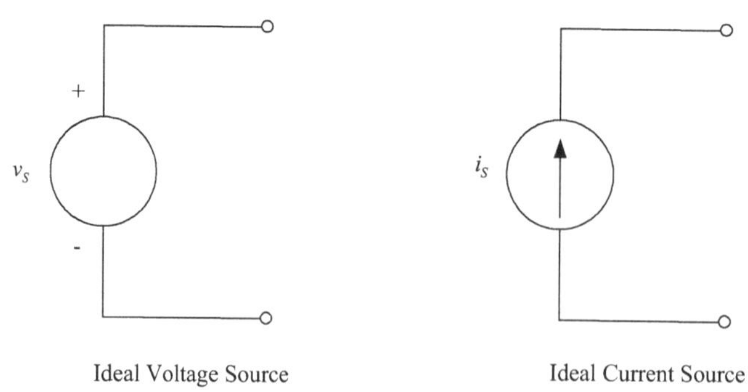

FIGURE 10.1 Schematic of ideal voltage and current sources.

DOI: 10.1201/9781003619161-10

10.2 ELECTRICAL LAWS, COMPONENTS, AND INITIAL CONDITIONS

10.2.1 Electrical Laws

Two of the most fundamental electrical laws are *Kirchhoff's voltage law* (*KVL*) and *Kirchhoff's current law* (*KCL*).

KVL: The sum of voltage drops around any closed loop is zero.

$$\sum v_i = 0 \quad \text{closed loop} \tag{10.1}$$

Consider the electrical system shown in Figure 10.2; the boxes represent different electrical components.

The solid lines connecting the different electrical components, including the voltage source, represent *electrical conductors*. In practice, there is some drop of electrical potential as current flows through a conductor due to its resistance, $R = \rho\frac{L}{A}$, where L is the length of the conductor, ρ is the resistivity of the conductor, and A is the cross-sectional area of the conductor. This chapter, unless otherwise noted, assumes ideal conductors ($\rho = 0$), meaning that the drop in electrical potential in conductors is zero. Thus, in this case, the electrical potential at the exit of component 1 is the same as entering component 2 and so on. The difference in electrical potential across a component is referred to as *voltage drop*, and as mentioned earlier, the letter v denotes this drop; for example, $v_1 = E_1 - E_2$. Note that we have indicated the polarities (+ and −) across each component; positive current flows from the highest potential (+) to the lowest potential (−), $E_1 > E_2$.

As the previous section indicated, a *volt* is the potential energy per unit charge or joule/coulomb. Therefore, a voltage drop means that some amount of energy (joules) has been lost. However, as discussed in Chapter 9, the first law of thermodynamics states that energy is always conserved; that is, energy cannot disappear! So where did this lost energy go? The answer is that this energy is dissipated by electrical components, and the effect is a rise in temperature within the circuit. This is why all electrical circuits must have some cooling mechanisms, usually in the form of air flow, to avoid excessive temperatures that would permanently harm the circuit components. More on this energy dissipation in Section 10.5.

Applying KVL, Equation 10.1, to the circuit in Figure 10.2,

$$\sum_{i=1}^{n} v_i = v_1 + v_2 + v_3 - v_s = 0 \tag{10.2}$$

Note the negative sign in front of v_S. The previous section mentioned that in going from positive polarity to negative polarity the electrical potential would decrease, indicating a positive voltage

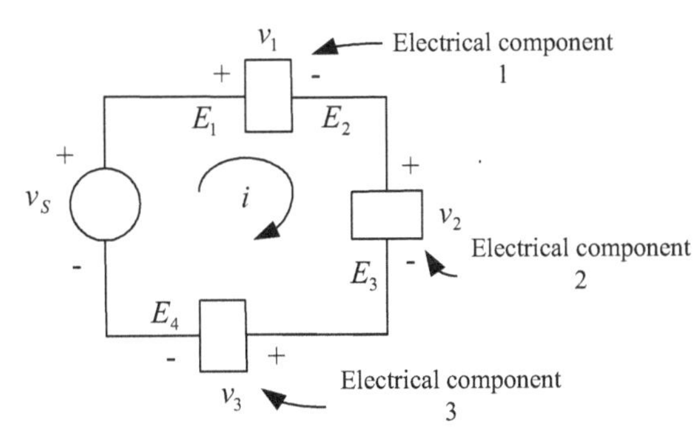

FIGURE 10.2 Closed-loop electrical circuit.

drop. A voltage source is actually a voltage rise, or "negative voltage drop." This is the reason for the negative sign.

KCL: The sum of currents entering a node is equal to the sum of currents exiting the node (in its most basic definition, a node is a place at which two or more electrical components are joined together).

$$\sum currents\ in = \sum currents\ out$$

or

$$\sum i_{in} = \sum i_{out} \text{ @node} \tag{10.3}$$

Consider Figure 10.3a showing a node; applying Equation 10.3,

$$i_1 = i_2 + i_3 + i_4 \tag{10.4a}$$

or

$$i_1 - i_2 - i_3 - i_4 = 0 \tag{10.4b}$$

For Figure 10.3b, applying Equation 10.3,

$$i_5 = i_6 \tag{10.4c}$$

or

$$i_5 - i_6 = 0 \tag{10.4d}$$

Equations 10.4b and 10.4d illustrate another way to state KCL: *The algebraic summation of all currents entering (+) and exiting (−) a node is zero.*

A brief interesting comment is about the analogy between KCL and the steady-state form of the law of conservation of mass presented in Chapter 8. We may say that KCL is a "conservation of charges," and write it as

$$\sum \begin{matrix} Rate\ of\ charges \\ (current)\ entering\ a\ node \end{matrix} - \begin{matrix} Rate\ of\ charges \\ (current)\ exiting\ a\ node \end{matrix} = 0$$

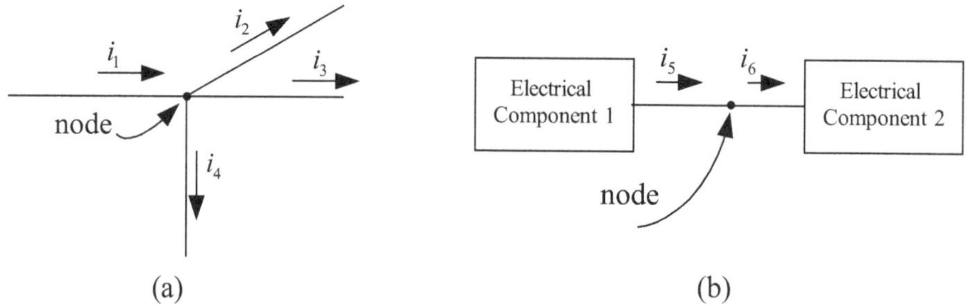

FIGURE 10.3 Electrical node.

10.2.2 Electrical Components

The following are the electrical components used in this chapter.

- The *resistor* is shown graphically in Figure 10.4.

The figure shows the direction of the current i, the electrical potentials, E_1 and E_2, at both ends, and the voltage drop, $v_R = E_1 - E_2$, across the resistor. The current through the resistor is proportional to the voltage drop across the resistor, $i \sim v_R$($\sim$ stands for "proportional to"), or,

$$i = \frac{v_R}{R} \text{ or } v_R = i\,R \tag{10.5}$$

where R is the proportionality constant. Equation 10.5 is called *Ohm's Law*; R is called the *resistance* of the resistor with units of *ohms*, Ω, or $1\Omega = 1V / A$.

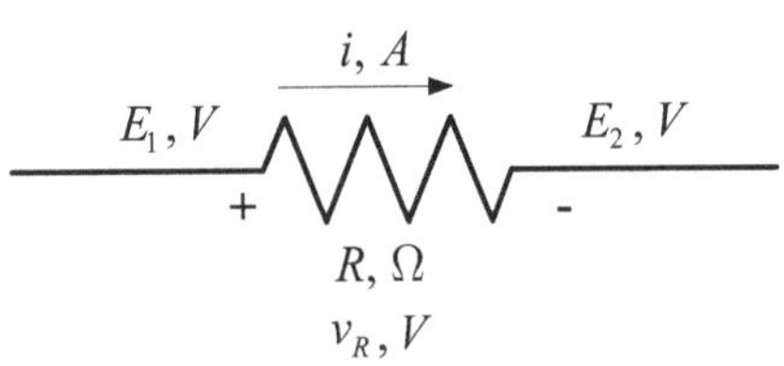

FIGURE 10.4 Resistor.

- A *capacitor* is depicted in Figure 10.5.

The capacitor is constructed from two parallel conducting plates separated by an insulating material, called a *dielectric*. The figure shows that current flows from left to right, from the positive to the negative polarity. As the current flows to the right electrons move left and start to collect on the right plate because it cannot cross the insulator. These negative charges on the right plate produce an electric field in the insulator (dielectric) forcing the electrons on the left plate to leave the plate (move to the left) and positive charges start to collect on the left plate. When this happens, electricity *seems to flow*, although note that no current (electrons) has crossed the dielectric! This process continues until no more electrons can accumulate on the right plate (and positive charges on the left plate). At this time, the plate is saturated, and *the current stops*.

To further explain how the capacitor works, consider a fluid analogy with a tank divided by a semi-flexible membrane; Figure 10.6 shows such a tank. Figure 10.6a shows the tank without any

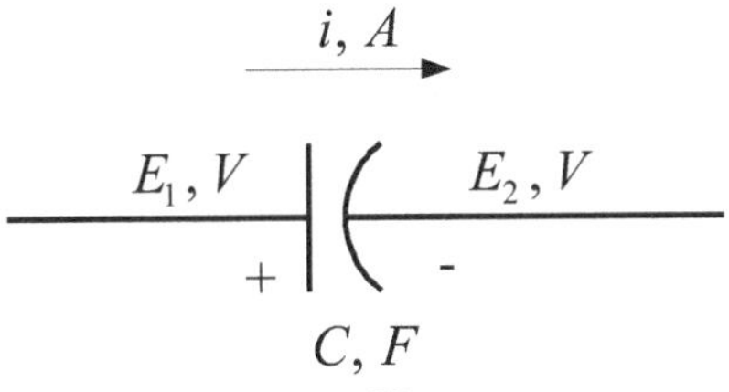

FIGURE 10.5 Capacitor.

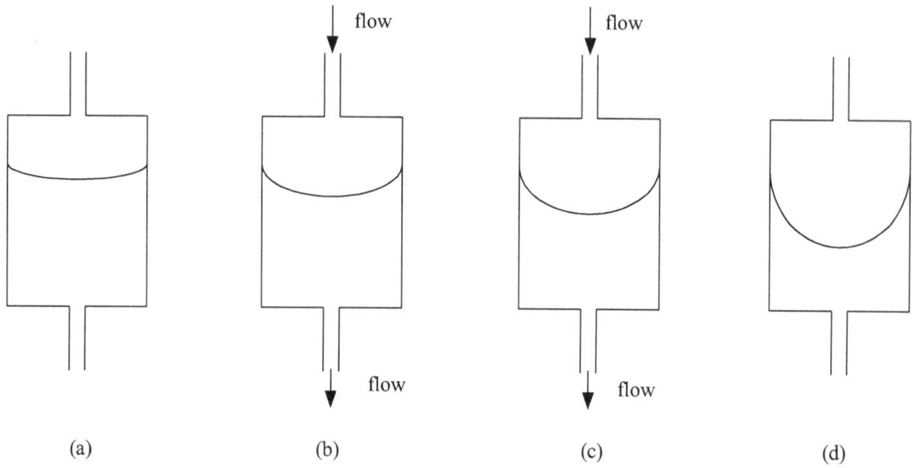

FIGURE 10.6 Tank with flexible membrane.

flow in and out. As flow enters the top chamber of the tank, Figures 10.6b and 10.6c, the flexible membrane stretches moving into the bottom chamber, building a pressure on this chamber (note that this is analogous to the accumulation of electrons on the right plate building an electric field) and forcing fluid to flow out (forcing electrons to leave the left plate). Note that no fluid crosses from the top to the bottom chamber of the tank and that this is analogous to the capacitor in that no electrons cross the dielectric. As this process continues, the fluid accumulates in the top chamber, and fluid flows out of the bottom section until the membrane cannot stretch any more. At this point, the flow stops because no more fluid can accumulate in the top chamber, and therefore, no more fluid leaves the bottom chamber (this is analogous to when no more electrons can accumulate in the right plate of the capacitor and no more electrons leave the left plate).

Thus, the capacitor stores charges on the two plates – actually, energy is stored in the electric field in the dielectric produced by the charges on the plate. The stored charges are proportional to the voltage difference between the two plates, $q \sim v_C$, or

$$q = C\, v_C \tag{10.6a}$$

or

$$v_c = \frac{1}{C} q \tag{10.6b}$$

where C is the proportionality constant, and it is called the *capacitance*. The unit of capacitance is the *farad* (F), and 1 F = 1 C/V. The value of C depends on the plate area, spacing between the plates, and on the dielectric properties of the insulating material. One *farad* is a very large capacitance, so usually *microfarads* ($\mu F = 10^{-6} F$), *nanofarads* ($nF = 10^{-9} F$), or *picofarads* ($pF = 10^{-12} F$) are used.

Differentiating Equation 10.6a with respect to time gives the current through the capacitor,

$$\frac{dq}{dt} = C\frac{dv_C}{dt} \Rightarrow i = C\frac{dv_C}{dt} \tag{10.7a}$$

or

$$\int_{v_C(0)}^{v_C} dv_C = \frac{1}{C}\int_0^t dt$$

$$v_C = v_C(0) + \frac{1}{C}\int_0^t i\,dt \tag{10.7b}$$

where $v_C(0)$ is the voltage across the capacitor at $t = 0$(initial condition).

Once the right plate is saturated with electrons, and the left plate with positive charges, no more current seems to flow, and a steady state is achieved. In this case, the capacitor becomes/behaves as an *open circuit*, which is usually represented as shown in Figure 10.7.

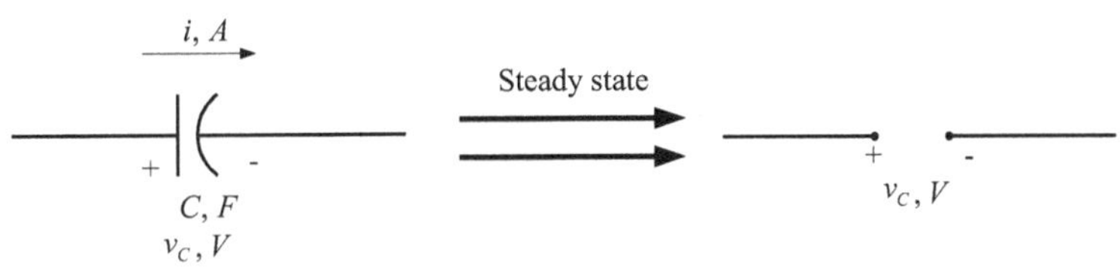

FIGURE 10.7 Capacitor becoming an open circuit at a steady state.

- Figure 10.8 shows the schematic representation of an *inductor.*

The inductor is actually a wire coiled around some type of core. As current flows through the coil, it creates a magnetic field. As the current changes, the magnetic field changes, inducing a voltage drop across the inductor.

The voltage drop across the inductor is proportional to the time rate of change of the current through the coil, $v_L \sim \frac{di}{dt}$

$$v_L = L\frac{di}{dt} \tag{10.8a}$$

or

$$\int_{i(0)}^{i} di = \frac{1}{L}\int_0^t v_L\,dt$$

$$i = i(0) + \frac{1}{L}\int_0^t v_L\,dt \tag{10.8b}$$

where L is the proportionality constant, called the *inductance*, and $i(0)$ is the current through the inductor at $t = 0$(initial condition). The unit of inductance is the *henry*, H, and $1H = 1V \cdot s / A$.

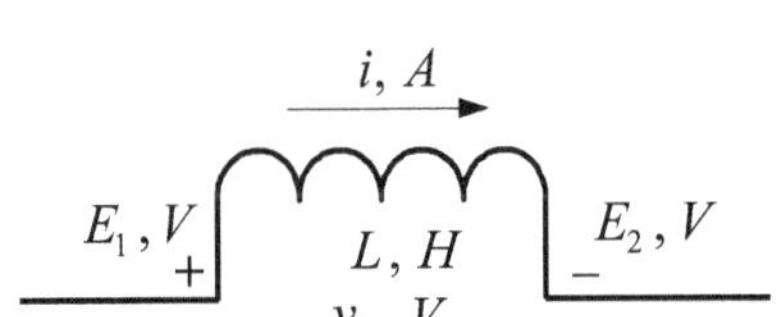

FIGURE 10.8 Inductor.

Similar to the farad, one (1) H is a large quantity, so we usually deal with *microhenry* ($\mu H = 10^{-6} H$) or *millihenry* ($\text{m} H = 10^{-3} H$).$\frac{di}{dt} = 0$.

If no current changes occur, the rate of change of the current is zero, and Equation 10.8a indicates that the voltage across the inductor becomes zero. In this case, the inductor just behaves as a simple conductor, or a "short circuit," as shown in Figure 10.9.

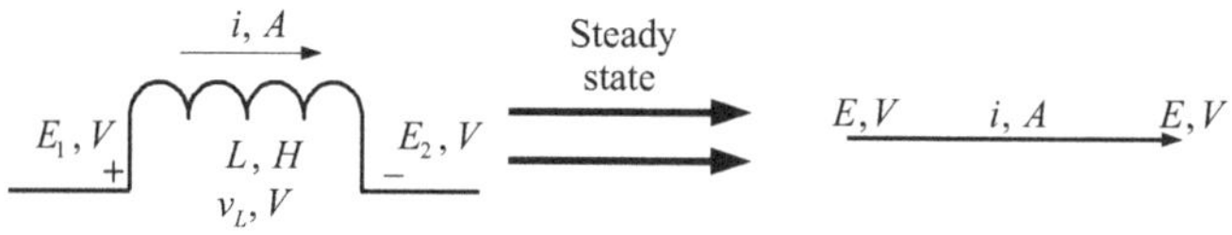

FIGURE 10.9 Inductor becoming a simple conductor at a steady state.

10.2.3 Initial Conditions

The dynamic behavior of electrical systems is much faster than the behavior of other systems we have studied in previous chapters. We briefly mentioned this difference in the last paragraph of Section 3.8 when we discussed transient and final responses. Thus, we should look at this behavior in a bit more detail before continuing.

Consider the circuit shown in Figure 10.10; this circuit is only composed of resistors and a voltage source.

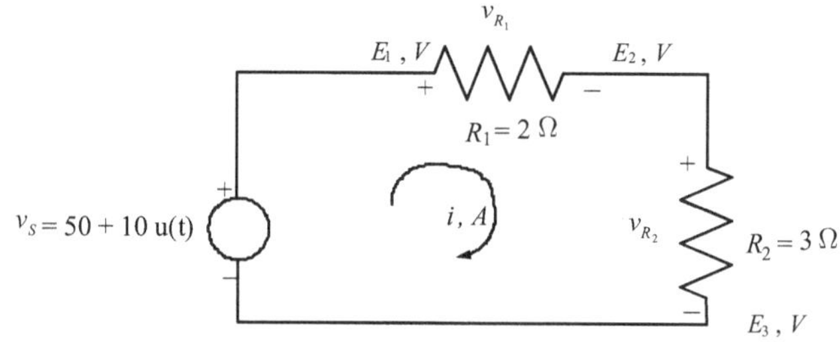

FIGURE 10.10 Electrical circuit.

The application of KVL yields

$$v_{R_1} + v_{R_2} - v_S = 0$$

or

$$R_1 i + R_2 i = v_S \tag{10.9}$$

The current at the initial state, when the voltage supply is 50 V *before* it becomes 60 V, is

$$i(0) = \frac{v_S(0)}{R_1 + R_2} = \frac{50V}{5\Omega} = 10A \tag{10.10}$$

As we have learned, the expression $u(t)$ indicates that the supply voltage at $t = 0$ becomes 60 V, or $v_S(0) = 60V$, and then $i(0) = 12A$. But Equation 10.10 shows that $i(0) = 10A$ and $v_S(0) = 50V$. So what is going on here? Let us see if we can explain this. Consider Figure 10.11.

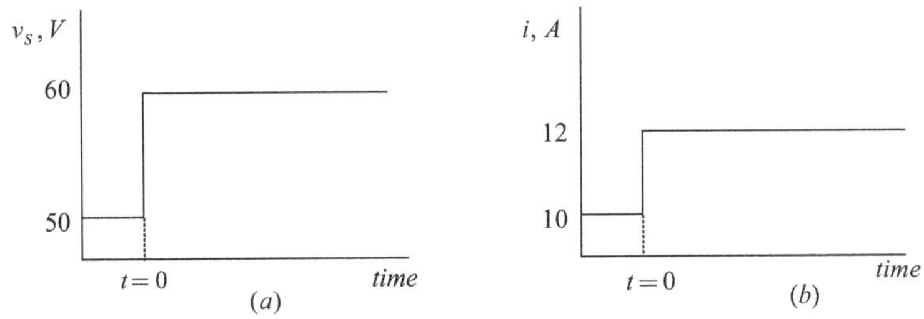

FIGURE 10.11 Voltage and current behavior.

Can you really state what is the voltage, or current, at $t = 0$? Is it 50 or 60 V? Is it 10 or 12 A? To avoid this disjunctive, electrical engineers talk about "time 0 minus," $t = 0^-$, and "time 0 plus," $t = 0^+$, and using this notation, we state

$$v_S(0^-) = 50V; i(0^-) = 10A; v_S(0^+) = 60V; i(0^+) = 12A$$

This notation is very useful when dealing with voltage drop across, or current through, resistors because the describing equation, $v_R = Ri$, is an algebraic equation. Thus, both voltage and current in a resistor change instantaneously depending on the excitation (i.e., in a step input).

Note that the current through an inductor, and the voltage across a capacitor, cannot change instantaneously. The change in the current through an inductor, Equation 10.8b, and the voltage drop across a capacitor, Equation 10.7b depends on an integration which does not provide an instantaneous change. Therefore, $i_L(0^-) = i_L(0^+)$ and $v_C(0^-) = v_C(0^+)$ for the inductor current and capacitor voltage, respectively.

The $t = 0^-$ and $t = 0^+$ notation is also useful in differentiating between the time just before and just after a switch opening or closing. In electrical circuit courses you will again be exposed to this notation.

10.3 EXAMPLES OF ELECTRICAL CIRCUITS

Example 10.1

Consider the circuit shown in Figure 10.12. Assuming that the voltage source changes from 10 V to 25 V at time $t = 0$ or, mathematically, $v_S = 10 + 15u(t)V$, find the expressions that describe the voltage drop across the resistor, across the capacitor, and the current in the loop for $t > 0$.

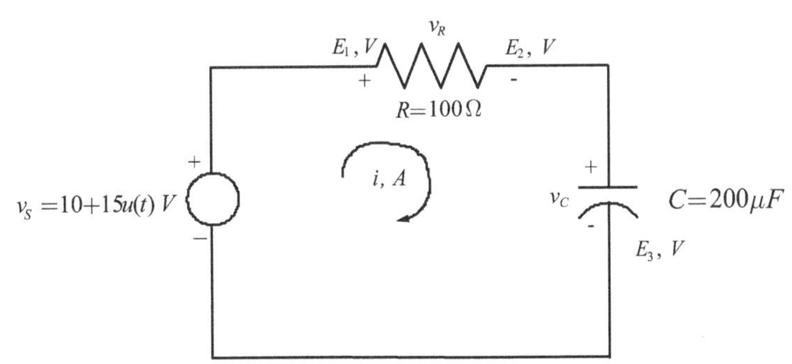

FIGURE 10.12 RC circuit.

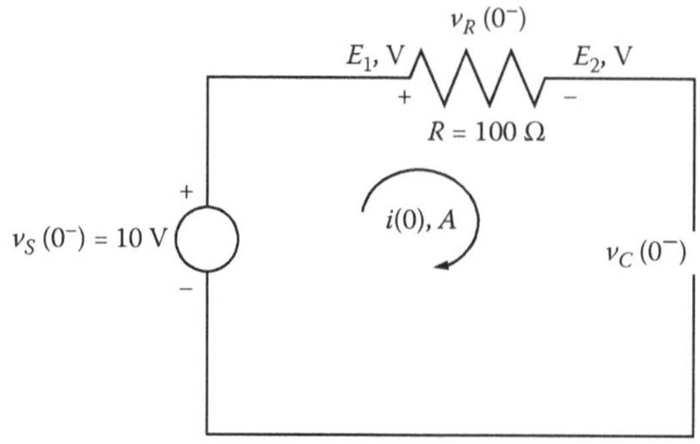

FIGURE 10.13 RC circuit at the initial steady state.

To calculate the initial current and initial voltages, note that, as learned in Section 10.2, at a steady state, the capacitor behaves as an open circuit, so we draw the circuit as shown in Figure 10.13. Because of the open circuit, there is no current in the loop, $i(0^-) = 0A$, and applying KVL for this initial condition,

$$v_R(0^-) + v_C(0^-) - v_S(0^-) = 0$$

and because $v_R(0^-) = Ri(0^-) = 0$, $v_C(0^-) = v_S(0^-) = 10V$.

Back to Figure 10.12, applying KVL around the loop,

$$v_R + v_C - v_S = 0 \tag{10.11}$$

1 equation, 2 unknowns $[v_R, v_C]$.

Note that because of the convention selected, the negative sign in front of the voltage from the voltage source.

For the resistor, from Equation 10.5,

$$v_R = iR \tag{10.5}$$

2 equations, 3 unknowns $[i]$.

From Equation 10.6b,

$$v_C = \frac{1}{C}q \tag{10.6b}$$

3 equations, 4 unknowns $[q]$.

From the definition of current,

$$i = \frac{dq}{dt} \tag{10.12}$$

4 equations, 4 unknowns.

Equations 10.11, 10.5, 10.6b, and 10.12 constitute the model for the circuit. To reduce the number of equations, substitute Equations 10.5 and 10.6b into Equation 10.11,

$$iR = \frac{1}{C}q = v_S$$

Taking the time derivative of both sides,

$$R\frac{di}{dt} + \frac{1}{C}\frac{dq}{dt} = \frac{dv_S}{dt}$$

and using Equation 10.12,

$$R\frac{di}{dt} + \frac{1}{C}i = \frac{dv_S}{dt} \tag{10.13}$$

Or

$$100\frac{di}{dt} + \frac{1}{200e-6}i = \frac{dv_S}{dt}$$

Equation 10.13 is the single equation that models the circuit. It describes how the current varies from the initial to the final steady state. It is a first-order differential equation with a bounded forcing function. Under these conditions, as presented in Chapters 3, 4, and 5, the response of the system will be stable monotonic. Using the Laplace transform of Chapter 4, the solution to Equation 10.13 is

$$i = \frac{1}{R}[v_S - v_S(0^-)]e^{-\frac{t}{RC}} \tag{10.14a}$$

or

$$i = 0.15e^{-\frac{t}{0.02}} \tag{10.14b}$$

The voltage drop across the resistance from Equation 10.5 is

$$v_R = iR = 15e^{-\frac{t}{0.02}} \tag{10.15}$$

The voltage drop across the capacitor v_C is obtained from Equation 10.11,

$$v_C = v_S - v_R = 25 - iR = 25 - 15e^{-\frac{t}{0.02}} \tag{10.16}$$

Let us show you another way of developing the model. Looking once more at Equation 10.11, which we repeat for convenience,

$$v_R = v_C - v_S \tag{10.11}$$

and substituting Equation 10.5,

$$Ri + v_C = v_S$$

Because it is the same current flowing through the resistance and through the capacitor, using Equations 10.6b and 10.12 into the previous equation and rearranging yields

$$RC\frac{dv_C}{dt} + v_C = v_S \tag{10.17}$$

Equation 10.17 is now the model that describes the voltage drop across the capacitor; Equation 10.11 is used to obtain v_R. Equation 10.17 is a simple first-order linear homogeneous differential equation with constant coefficient, so its solution can be obtained by any of the methods explained and given by Equation 5.10b,

$$v_C = v_S(0^-) + \left[v_S - v_S(0^-)\right]\left(1 - e^{-\frac{t}{RC}}\right) \tag{10.18a}$$

or

$$v_C = 10 + 15\left(1 - e^{-\frac{t}{RC}}\right) = 25 - 15e^{-\frac{t}{0.02}} \tag{10.18b}$$

which is the same as Equation 10.16. The expression for the voltage drop across the resistance is

$$v_R = v_S - v_C = 25 - 25 + 15e^{-\frac{t}{0.02}} = 15e^{-\frac{t}{0.02}} \tag{10.19}$$

which is the same as Equation 10.15, and the expression for the current in the loop is

$$i = \frac{v_R}{R} = 0.15e^{-\frac{t}{0.02}} \tag{10.20}$$

which is the same as Equation 10.14b.

Note that the solution describing the current in the circuit, Equation 1.14b or 10.20, shows that the solution is monotonic and stable, as we have noted right after the model was developed. Also note that the solution indicates that at a very long time the current becomes zero. This result makes sense because at this time, once we reach the steady state again, the capacitor becomes an open circuit and no current flows.

At this time, it is worthwhile recalling the presentation in Chapter 5 about the response of first-order systems to step changes in forcing functions; Equation 10.17 is the describing equation (model) of the voltage across the capacitor. To help us in recalling this, compare Equation 10.17, which we write again for convenience, with the general first-order equation, Equation 5.2.

$$RC\frac{dv_C}{dt} + v_C = v_S \tag{10.17}$$

or

$$0.02\frac{dv_C}{dt} + v_C = v_S$$

and

$$\tau\frac{dy(t)}{dt}+y(t)=Kx(t) \tag{5.2}$$

Thus, $\tau = 0.02\,\text{s}$ and $K = 1$. Figure 10.14 shows the response obtained from Equation 10.18b. The figure shows that

- the initial slope is the steepest slope in the response curve.
- the total change is 15 V, which is equal to the gain K times the change in v_S, $KD = 1(15)$.
- 63.2% of the total change occurs at 0.02 s which is the value of the time constant τ; the response is almost complete at $5\tau = 0.1\,\text{s}$.

As we discussed in Section 5.1, any system described by a first-order differential equation will have the same response when disturbed by a step change in forcing function. Knowing the values of time constant τ and gain K, the response curve could be drawn with some accuracy.

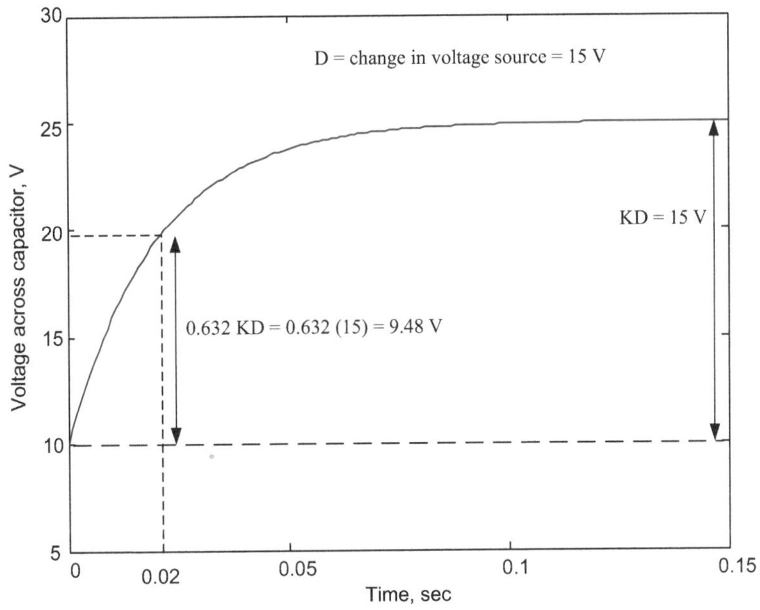

FIGURE 10.14 Response of RC circuit shown in Figure 10.12.

Most problem solutions require some number of mathematical manipulations – some more than others. For instance, in the previous example, we substituted Equations 10.5 and 10.6b into Equation 10.11, took the derivative with respect to time, and used Equation 10.12, resulting in Equation 10.13, which we repeat for convenience:

$$R\frac{di}{dt}+\frac{1}{C}i=\frac{dv_S}{dt} \tag{10.13}$$

The obvious question, Is how do we know that we did not make a mistake and that the result is correct? A good way to prove your development is by checking the units of the resulting equation. All three terms *must* have the same units. If the units are the same, then the result *may be* correct; if the units are not the same, there is *no chance* for the result to be correct. For example,

$$\frac{dv_S}{dt} = \text{has units of } \frac{V}{s}$$

$$R\frac{di}{dt} = \text{has units of } \Omega\left(\frac{A}{s}\right) = \left(\frac{V}{A}\right)\left(\frac{A}{s}\right) = \frac{V}{s}$$

$$\frac{1}{C}i = \text{has units of } \frac{A}{F} = A\left(\frac{V}{C}\right) = \frac{C}{s}\left(\frac{V}{C}\right) = \frac{V}{s}$$

Therefore, all three terms indeed have the same units, and the result may be correct. There were not many mathematical steps in arriving at Equation 10.13. It may be wise, in cases that involve many other steps, to stop after a number of steps and check the units at some intermediate result.

Another useful unit check is in Equation 10.14, which we repeat for convenience,

$$i = \frac{1}{R}[v_S - v_S(0^-)]e^{-\frac{t}{RC}}$$

A A Exponent must be dimensionless

The unit of RC is, $RC \Rightarrow \Omega F = \frac{V}{A}\frac{C}{V} = \frac{C}{A} = \frac{C}{C/s} = s$ the exponential is dimensionless.

Example 10.2

Consider the circuit shown in Figure 10.15. If the voltage source changes as $v_S = 10 + 15\sin t\, u(t), V$, find a model that describes the voltage across the resistor, across the inductor, and the current in the loop.

We first calculate the initial current and initial voltage drops in the loop. Note that, as learned in Section 10.2, at steady state the inductor behaves just as an electrical conductor, a "short circuit," and Figure 10.16 shows this state. The application of KVL at this condition indicates that

$$v_R(0^-) = v_S(0^-) = 10\ V \text{ and } i(0^-) = \frac{v_R(0^-)}{R} = 1\ A$$

Back to Figure 10.15, applying KVL around the loop,

$$v_R + v_L - v_S = 0$$

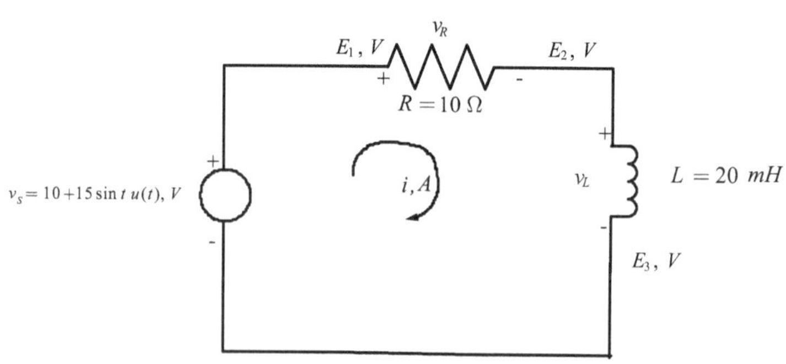

FIGURE 10.15 RL circuit.

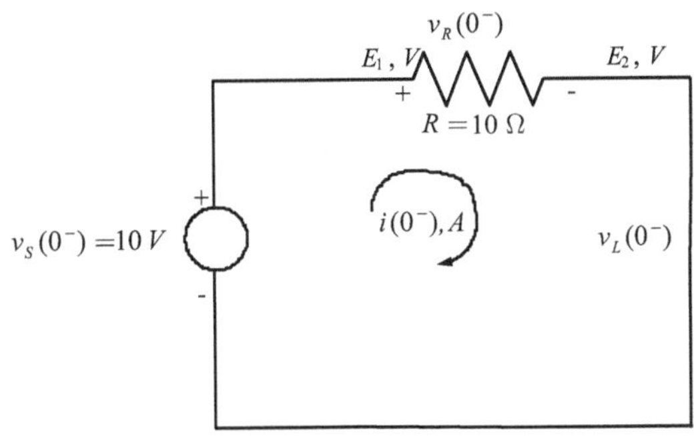

FIGURE 10.16 RL circuit at the initial steady state.

or

$$v_R + v_L = v_S \tag{10.21}$$

1 equation, 2 unknowns $[v_R, v_L]$.

For the resistor, from Equation 10.5,

$$v_R = iR \tag{10.5}$$

2 equations, 3 unknowns $[i]$,

For the inductor, from Equation 10.8a,

$$v_L = L\frac{di}{dt} \tag{10.8a}$$

3 equations, 3 unknowns.

Substituting Equations 10.5 and 10.8a into Equation 10.21 gives

$$L\frac{di}{dt} + Ri = v_S \tag{10.22}$$

This equation is the model describing the current around the loop; it is a first-order linear nonhomogeneous differential equation with constant coefficients. Because the forcing function is a bounded value (step change), the solution (response) will be stable monotonic. The solution using any of the methods we have learned is

$$i = 1 + 1.5\left[\sin t - 0.002(\cos t - e^{-t/0.002})\right] \tag{10.23}$$

The voltage drop across the resistance is

$$v_R = Ri = 10 + 15\left[\sin t - 0.002(\cos t - e^{-t/0.002})\right] \tag{10.24}$$

and across the inductor,

$$v_L = L\frac{di}{dt} = 1.5\left[\cos t - 0.002(-\sin t + 500e^{-t/0.002})\right] \tag{10.25a}$$

Or using the trigonometric identity of Equation 5.15 and some simple algebra,

$$v_L = 1.5\sin(t + 0.002) - 1.5e^{-t/0.002} \tag{10.25b}$$

Example 10.3

Consider the "RLC circuit" of Figure 10.17. Obtain an expression for the current as a function of time after $t = 0$. Initially, there is no current flowing and the capacitor is completely discharged; that is, $v_C(0^-) = 0V$ and $i(0^-) = 0A$.

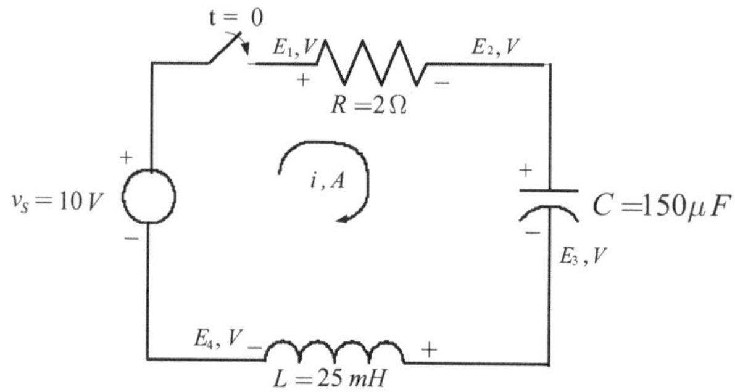

FIGURE 10.17 RLC circuit.

The combination of the voltage supply and switch can be represented mathematically as $v_S = 10u(t)$.

Applying KVL,

$$\sum_{i=1}^{n} v_i = v_R + v_C + v_L - v_S = 0 \tag{10.26}$$

1 equation, 3 unknown $[v_R, v_C, v_L]$.

For the resistor, from Equation 10.5,

$$v_R = iR \tag{10.5}$$

2 equations, 4 unknowns $[i]$.

For the inductor, from Equation 10.8a,

$$v_L = L\frac{di}{dt} \tag{10.8a}$$

3 equations, 4 unknowns.

For the capacitor, from Equation 10.7a,

$$i = C\frac{dv_C}{dt} \tag{10.7a}$$

4 equations, 4 unknowns.

The last four equations constitute the model and can be used to reduce the model to a single equation. Substituting Equations 10.5 and 10.8a into Equation 10.26 gives

$$iR + v_C + L\frac{di}{dt} = v_S$$

Because it is the same current through each element, using Equation 10.7a in the first and third terms of the equation, and rearranging the order of the terms,

$$LC\frac{d^2v_C}{dt^2} + RC\frac{dv_C}{dt} + v_C = v_S \tag{10.27a}$$

or

$$3.75\times10^{-6}\frac{d^2v_C}{dt^2} + 3x\,10^{-4}\frac{dv_C}{dt} + v_C = v_S \tag{10.27b}$$

Equation 10.27 is now the model that describes the voltage across the capacitor. Once the expression for v_C is known, it is then a simple matter obtaining the expression for the current; this is shown next. The equation is a second-order differential equation with negative real parts of complex roots. Thus, since the forcing function is bounded, the solution (response) will be stable oscillatory.

Using any of the methods to solve a linear nonhomogeneous second-order differential equation with constant coefficients presented in Chapters 3 and 4, the analytical solution of Equation 10.27b is

$$v_C = 10.0 - e^{-40t}[10\cos 515t + 0.78\sin 515t] \tag{10.28}$$

Indicating a stable (shown by the negative exponential) oscillatory (shown by the sine and cosine terms).

The expression for the current through the loop is now obtained as

$$i = C\frac{dv_C}{dt} = e^{-40t}[0.78\sin 515t - 2.55\times10^{-4}\cos 515t] \tag{10.29}$$

Example 10.4

a. The switch in Figure 10.18 is moved from position A to position B at time $t = 0$. The capacitor is initially uncharged. What is the capacitor voltage after 6 seconds?
b. At time 6 seconds, the switch is moved to position C. What is the capacitor voltage 4 seconds later?
   a. Once the switch is in position B, the circuit looks as shown in Figure 10.19. Equation 10.18a in Example 10.1 provides the voltage across the capacitor,

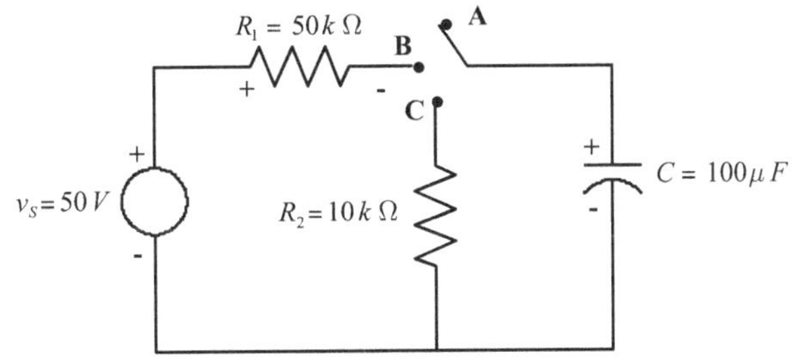

FIGURE 10.18 Circuit for Example 10.4.

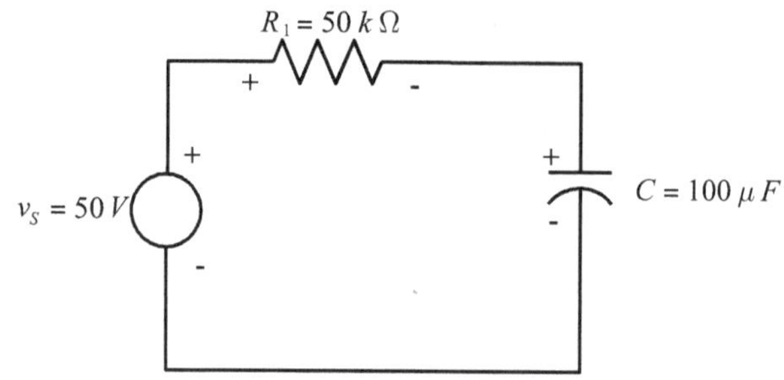

FIGURE 10.19 Circuit for Example 10.4 part (a).

$$v_C = v_C(0^-) + \left[v_S - v_S(0^-)\right]\left(1 - e^{-\frac{t}{RC}}\right) = 50\left(1 - e^{-\frac{t}{5}}\right)$$

and the voltage across the capacitor 6 seconds after the switch moves to position B,

$$v_C = 50[1 - e^{-\frac{6}{5}}] = 34.9V$$

b. Once the switch is in position C, the circuit becomes as shown in Figure 10.20.

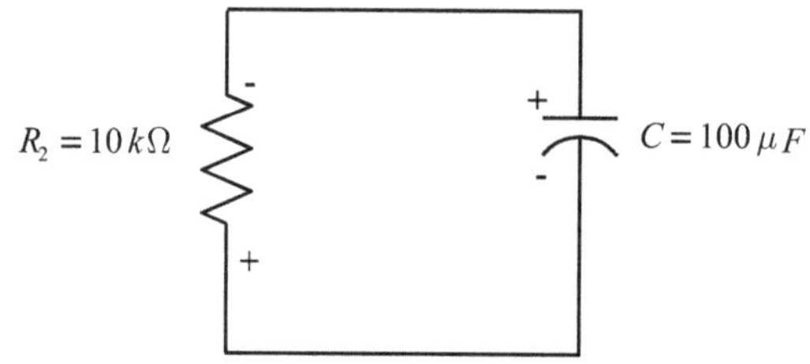

FIGURE 10.20 Circuit for Example 10.4 part (b).

In this case, $v_C(0) = 34.9V$. Applying KVL,

$$v_R + v_C = 0$$
$$Ri + v_C = 0$$

and because it is the same current i through both electrical elements, substituting Equation 10.7a for the current term,

$$RC\frac{dv_C}{dt} = +v_C = 0 \tag{10.30}$$

Equation 10.30 is the model for this part (b). This is a simple first order homogeneous differential equation with constant coefficients, which can be solved using the separation of variables as presented in Chapter 2,

$$RC\frac{dv_C}{dt} = -v_C$$
$$RC\,dv_C = -v_C\,dt$$

$$\left(10\times10^{3}\right)\left(100\times10^{-6}\right)\int_{34.9}^{v_C}\frac{dv_C}{v_C}=-\int_0^4 dt$$

Straight integration yields

$$v_C = 6.4V$$

10.3.1 Undamped Circuit (Natural Frequency and Resonance)

Example 3.11 and Sections 5.2.1.3 and 6.2.1 have presented the response of undamped systems and the significance of natural frequency and resonance. Although the examples and cases we have used so far are related to mechanical systems, these issues are also important in electrical circuits. This is the topic of this section.

Consider the circuit shown in Figure 10.21a.

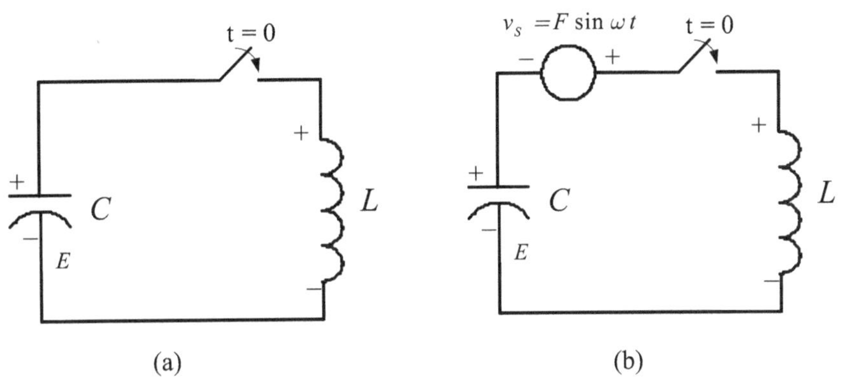

FIGURE 10.21 LC circuits.

The initial charge in the capacitor is 20 C; that is, $q(0) = 20\ C$. Let us find how the current in the circuit behaves after the circuit closes. Applying KVL,

$$v_C + v_L = 0 \tag{10.31}$$

1 equation, 2 unknowns $[v_C, v_L]$

For the capacitor,

$$v_C = \frac{1}{C}q \tag{10.6a}$$

2 equations, 3 unknowns $[q]$

For the inductor,

$$v_L = L\frac{di}{dt} \tag{10.8a}$$

3 equations, 4 unknowns $[i]$

Using the definition of current,

$$i = \frac{dq}{dt} \tag{10.32}$$

4 equations, 4 unknowns

Substituting Equations 10.6a and 10.8a into Equation 10.31 and using Equation 10.32,

$$LC\frac{d^2q}{dt^2}+q=0 \tag{10.33}$$

Equation 10.33 is the model that provides the charges in the capacitor. If the current is desired, then Equation 10.32 is used once an expression for q is obtained.

As we have learned in previous chapters, Equation 10.33 is the classical equation of an undamped system because it is a second-order differential equation without a first derivative term. For its solution, being homogeneous, we follow the characteristic method,

$$LCr^2+1=0$$

from which the roots are $r_1=i\frac{1}{\sqrt{LC}}$ and $r_2=-i\frac{1}{\sqrt{LC}}$; therefore,

$$q=C_1\cos\frac{1}{\sqrt{LC}}t+C_2\sin\frac{1}{\sqrt{LC}}t \tag{10.34}$$

Indicating that the response is continuously oscillatory with a frequency, better yet *natural frequency* $\omega_n=\sqrt{LC}$ radians/s. The C coefficients are obtained using the initial conditions; in this case, $q=10C$ and $\frac{dq}{dt}|_{t=0}=0\frac{C}{s}$, giving

$$q=10\cos\frac{1}{\sqrt{LC}}t \tag{10.35a}$$

The current in the circuit is

$$i=\frac{dq}{dt}=-\frac{10}{\sqrt{LC}}\sin\frac{1}{\sqrt{LC}}t \tag{10.35b}$$

Consider now the circuit shown in Figure 10.21b, where a power supply is added to the circuit. Following the same previous procedure, the new model is

$$LC\frac{d^2q}{dt^2}+q=Cv_S \tag{10.36}$$

Being a nonhomogeneous equation, the solution is obtained by $q=q_H+q_P$. The corresponding homogeneous solution is the same as Equation 10.34 or,

$$q_H=C_1\cos\frac{1}{\sqrt{LC}}t+C_2\sin\frac{1}{\sqrt{LC}}t \tag{10.37}$$

The form of the particular solution q_P depends on whether the frequency of the forcing function ω is equal or not to the natural frequency $\omega_n=\sqrt{LC}$ radians/s. When $\omega=\omega_n$ provides the most drastic change in the response. Following the methods we have learned, the particular solution is

$$q_P=\frac{F}{2L/\sqrt{LC}}t\cos\frac{1}{\sqrt{LC}}t \tag{10.38}$$

The addition of q_H and q_P yield the complete solution,

$$q = C_1 \cos\frac{1}{\sqrt{LC}}t + C_2\frac{1}{\sqrt{LC}}t - \frac{F}{2L/\sqrt{LC}}t\cos\frac{1}{\sqrt{LC}}t$$

Applying the same initial conditions as before,

$$q = 10\cos\frac{1}{\sqrt{LC}}t + \frac{FC}{2}\sin\frac{1}{\sqrt{LC}}t - \frac{F}{2L\sqrt{LC}}t\cos\frac{1}{\sqrt{LC}}t \tag{10.39a}$$

The equation clearly shows that, due to the third term (the forced response), the response is oscillatory with increasing amplitude. As we have shown in Chapters 3 and 6, this phenomenon is called resonance, and it happens when the frequency of the forcing function is equal to the natural frequency of the system, $\omega = \omega_n$.

The current in the circuit is

$$i = \frac{dq}{dt} = \left(\frac{F}{2L}t - \frac{10}{\sqrt{LC}}\right)\sin\frac{1}{\sqrt{LC}}t + \frac{F}{2}\left[\frac{C}{\sqrt{LC}} - \frac{\sqrt{LC}}{L}\right]\cos\frac{1}{\sqrt{LC}}t \tag{10.39b}$$

Using $C = 0.5F$ and $L = 0.5H$, Figure 10.22 shows the current responses when (a) there is no forcing function (Equation 10.35b) and (b) when $\omega = \omega_n$ (Equation 10.39b). The figure clearly shows that when there is no forcing function the system response is oscillatory at its natural frequency. Under a periodic forcing function with a frequency equal to the system's natural frequency, the response is oscillatory with increasing amplitude; this phenomenon is what we refer to as resonance.

As discussed in Section 6.2.2, there are no pure undamped systems, this is an idealized case; there is always some damping present. In the case of electrical circuits, the most common damping is the presence of resistance. There is no need for the presence of a resistor, but rather, the connectors themselves present a resistance (we have so far neglected this resistance in the chapter). Considering this resistance and lumping all of it under R, Figure 10.23 shows the circuit.

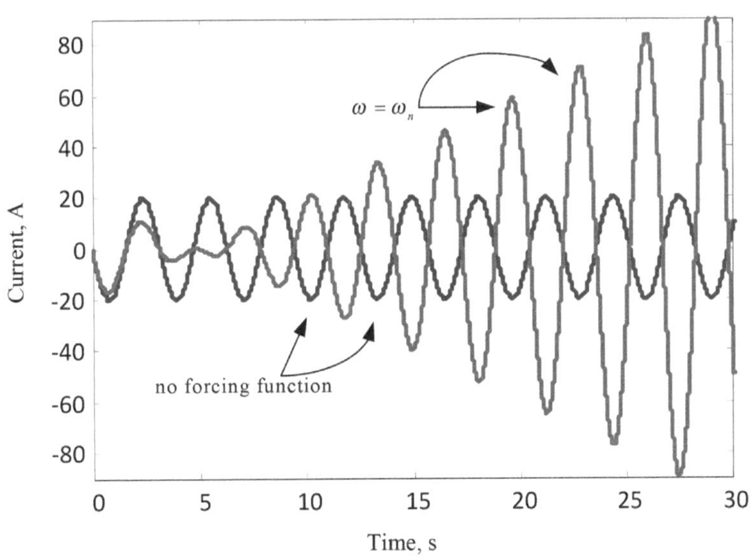

FIGURE 10.22 Responses of LC circuits.

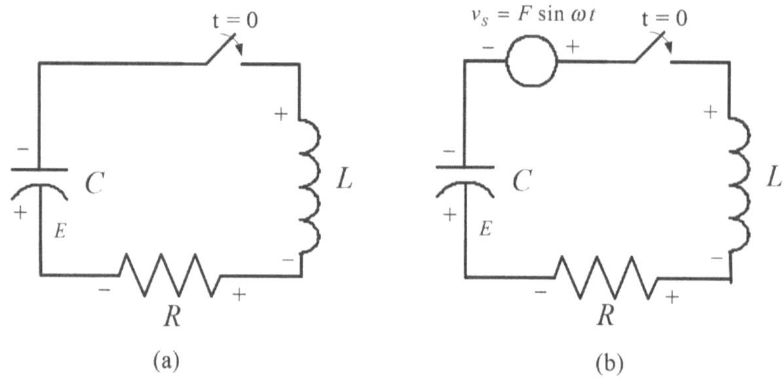

FIGURE 10.23 LC circuits with some amount of resistance.

Essentially, the figure shows two RLC circuits as we have discussed in Example 10.3. The model is

$$L\frac{d^2q}{dt^2}+R\frac{dq}{dt}+\frac{1}{c}q=v_S$$

where $v_S = 0V$ in Figure 10.23a and $v_S = F\sin\omega t V$ in Figure 10.23b.

The analytical solution for the circuit in Figure 10.23a will indicate that the current will go to zero. For the circuit in Figure 10.23b, the current will not "die out," but rather, it will reach a steady oscillation (sine wave) with a certain amplitude and phase lag and both of them are function of the forcing function ω. A *pseudo resonance*, also sometimes referred to as a *practical resonance*, occurs at the frequency that produces the largest possible final amplitude. There is a problem at the end of the chapter on this topic.

10.4 ADDITIONAL EXAMPLES

The electrical circuits discussed so far have been rather simple in that they all consist of a single closed circuit, or loop. This section presents circuits consisting of more than one loop.

There are two methods to analyze complex electrical circuits in an organized manner: the *nodal method* and the *mesh method*. In the nodal method, one of the principal nodes is selected as the reference node, and all other nodes are assigned a *voltage which is understood to be relative to the voltage of the reference node*. Because all the voltages in the diagram are relative to the voltage of the reference node, it is common practice to think of the reference node as ground, which has a potential of 0 (zero) V. We represent the reference node with the ground symbol ⏚ Once the reference node is selected, KCL is applied to each node and the resulting set of equations solved for the unknown relative voltages.

In the mesh method, currents are assigned to each loop of the system. Although the assigned direction could either be clockwise or counterclockwise, it is a common practice to assign the loop currents in the clockwise direction. If a branch is part of two loops, the current through that branch is the algebraic summation of both currents. Once these currents are assigned, KVL is applied to each loop; the resulting set of equations are solved for the unknown currents. Three examples are now presented.

Example 10.5

Consider the circuit shown in Figure 10.24. Obtain the models and their analytical solution describing the voltage drop across R_2 and the current through L_2.

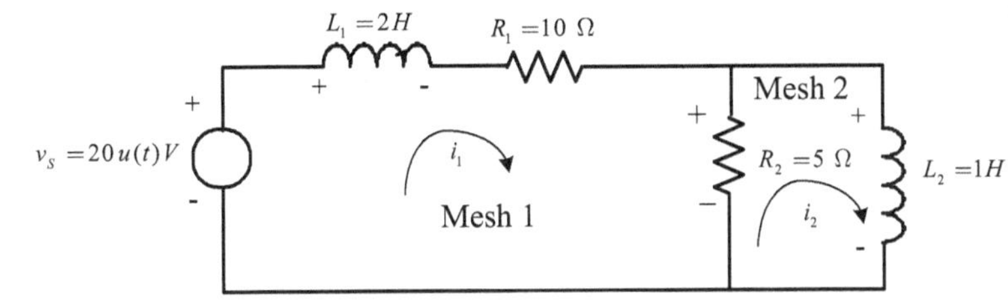

FIGURE 10.24 Electrical circuit for Example 10.5

Before the supply voltage is applied there is no current flowing and $i_1(0^-) = i_2(0^-) = 0\,A$.
We now solve this example using the mesh method as indicated in the figure. Applying KVL to Mesh 1,

$$v_{L_1} + v_{R_1} + v_{R_2} - v_S = 0 \tag{10.40}$$

1 equation, 3 unknowns $[v_{L_1}, v_{R_1}, v_{R_2}]$

Using the necessary components expressions,

$$v_{L_1} = L_1 \frac{di_1}{dt} \tag{10.41}$$

2 equations, 4 unknowns $[i_1]$

$$v_{R_1} = R_1 i_1 \tag{10.42}$$

3 equations, 4 unknowns; and

$$v_{R2} = R_2(i_1 - i_2) \tag{10.43}$$

4 equations, 5 unknowns $[i_2]$

Note this last equation. Figure 10.24 shows two currents, i_1 and i_2, flowing through resistance R_2. Current i_1 flows from the positive to the negative polarity, and by convention, this is a "positive" current. Current i_2 flows from the negative to the positive polarity, and thus, this is a "negative" current. Therefore, the total current through the resistor is the algebraic summation of the two currents or, $i_1 + (-i_2) = i_1 - i_2$.

We still have 1 degree of freedom and have completed all the equations for Mesh 1. Moving now to Mesh 2 and again applying KVL,

$$v_{L_2} - v_{R_2} = 0 \tag{10.44}$$

5 equations, 6 unknowns $[v_{L_2}]$

and

$$v_{L2} = L_2 \frac{di_2}{dt} \tag{10.45}$$

6 equations, 6 unknowns.

We now have the set of equations, Equations 10.40–10.45, with zero degrees of freedom that describes the currents in the meshes. Next, we obtain the analytical solutions, and for this, substituting Equations 10.41, 10.42, and 10.43 into Equation 10.40 yields

$$L_1 \frac{di_1}{dt} + (R_1 + R_2)i_1 = v_S + R_2 i_2 \tag{10.46a}$$

or

$$2\frac{di_1}{dt} + 15\, i_1 = 20 + 5i_2 \tag{10.46b}$$

Substituting Equations 10.43, and 10.45 into Equation 10.44 gives

$$L_2 \frac{di_2}{dt} + R_2 i_1 = R_2 i_2 \tag{10.47a}$$

or

$$\frac{di_2}{dt} + 5i_1 = 5i_2 \tag{10.47b}$$

After all the substitutions, we end up with two equations, Equations 10.46b and 10.47b, with two unknowns, i_1 and i_2. Using Laplace transform, these two equations yield the solutions

$$i_1 = 2 - 0.667e^{-10t} - 1.333e^{-2.5t} \tag{10.48}$$

and

$$i_2 = 2 + 0.667e^{-10t} - 2.667e^{-2.5t} \tag{10.49}$$

For the voltage drop across R_2,

$$v_{R_2} = R_2(i_1 - i_2) = 5(2 - 0.667e^{-10t} - 1.333e^{-2.5t} - 2 - 0.667e^{10t} + 2.667e^{-2.5t})$$

$$v_{R_2} = R_2(i_1 - i_2) = 6.67(e^{-2.5t} - e^{-10t}) \tag{10.50}$$

Interesting, this last equation indicates that as $t \to \infty$, $v_{R_2} \to 0V$! Actually, $i_1 - i_2 \to 0A$! That is, once the new steady state is reached, there will be no current flowing through R_2. You may want to think about this by analyzing the current at $t = \infty$ and obtaining $i_1(\infty)$ and $i_2(\infty)$.

Example 10.6

Consider the electrical system shown in Figure 10.25. Using the nodal and mesh methods, develop the models that describe how the voltage drop across, and current through, R_2 vary when the supply voltage v_S changes from 15 V to 20 V, $v_S = 15 + 5u(t)V$; show also the current transient through the capacitor.

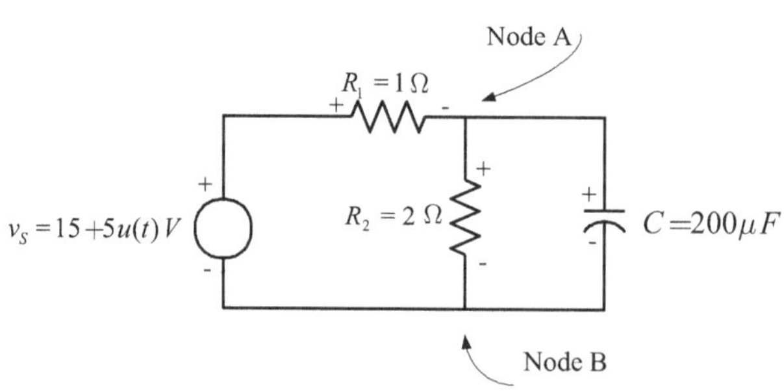

FIGURE 10.25 Electrical circuit for Example 10.6.

Let us first look at the initial steady-state conditions existing in the circuit before the supply voltage changes; these are the initial conditions necessary to solve the model. As already learned, at a steady state, the capacitor acts like an open circuit, and therefore, we can draw the circuit as shown in Figure 10.26a which is the same as the one shown in Figure 10.26b. In this case the current at the initial state is obtained applying KVL to the circuit,

$$v_{R_1}(0^-)+v_{R_2}(0^-)-v_S(0^-)=0$$

$$R_1 i(0^-)+R_2 i(0^-)-v_S(0^-)=0$$

$$i(0^-)=\frac{v_s(0^-)}{R_1+R_2}=\frac{15V}{3\Omega}=5A$$

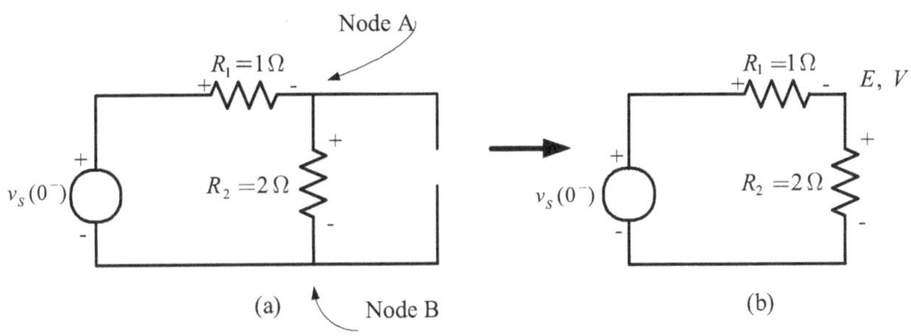

FIGURE 10.26 Electrical circuit for Example 10.6 showing open circuit.

The voltage drop across R_1 is $v_{R_1}(0^-)=R_1 i(0^-)=1\,\Omega(5A)=5V$ and consequently, the initial voltage between R_1 and R_2 is $E(0^-)=v_S(0^-)-v_{R_1}(0^-)=15-5=10V$, and $v_C(0)=10V$. We now proceed obtaining the model.

NODAL METHOD

Figure 10.27 shows that node B is selected as the reference node and shows the currents into and out of node A; the letter E denotes the voltage at node A.

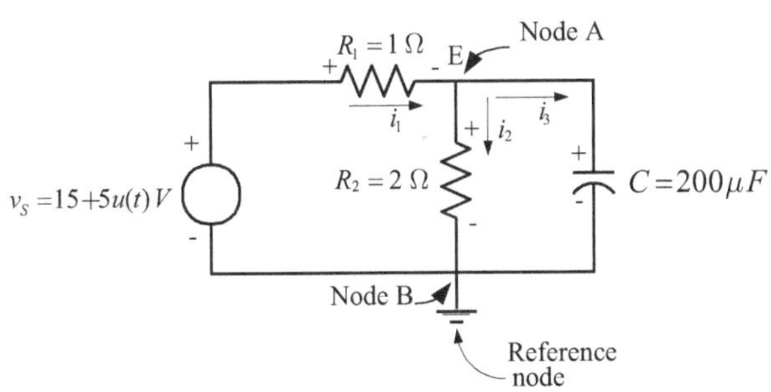

FIGURE 10.27 Electrical circuit showing information for the nodal method.

Once the reference node has been selected and the currents into and out of the other nodes defined, we proceed by applying KCL. In this case, at node A,

$$i_1-i_2-i_3=0 \tag{10.51}$$

1 equation, 3 unknowns $[i_1, i_2, i_3]$.

Using the expressions for each element; from Equation 10.5,

$$i_1 = \frac{1}{R_1}(v_S - E) = v_S - E \tag{10.52}$$

2 equations, 4 unknowns [E],

and

$$i_2 = \frac{1}{R_2}E = 0.5\,E \tag{10.53}$$

3 equations, 4 unknowns.

From Equation 10.7a,

$$i_2 = C\frac{dv_C}{dt} = C\frac{dE}{dt} = 200\times10^{-6}\frac{dE}{dt} \tag{10.54}$$

4 equations, 4 unknowns.

Equations 10.51 through 10.54 constitute the model. To reduce it to a single equation, substitute Equations10.52, 10.53, and 10.54 into Equation 10.51 and rearrange, yielding

$$200\times10^{-6}\frac{dE}{dt} + 1.5E = v_S \tag{10.55}$$

Equation 10.55 is a first-order linear nonhomogeneous ordinary differential equation with constant coefficients, and its solution provides how voltage E at node A, *relative to the reference node*, varies with time once the supply voltage v_S changes. By the way, could you say, based on only looking at Equation 10.55, by how much voltage E will eventually change once the supply – voltage changes, what will be its final value, how much time will take to reach this final value, and draw the response showing these values? (Hint: All of this can be answered using the presentation in Chapter 5).

The analytical solution of Equation 10.55 realizing that $E(0^-)$ is 10 V is

$$E = 10 + 3.33\left(1 - e^{-\frac{t}{0.000133}}\right) \tag{10.56}$$

Note that because the voltage at the exit of R_2 is the reference voltage, E is equal to the voltage drop across this resistance; that is, $v_{R_2} = E - 0$, which is one the quantities desired; because the capacitor and resistance R_2 are in parallel, E is also the voltage across the capacitor; that is, $v_C = E$. Once E is known, the current through R_2 is given by

$$i_2 = \frac{E}{R_2} = 5 + 1.667(1 - e^{-\frac{t}{0.000133}}) \tag{10.57}$$

The current transient through the capacitor is obtained applying Equation 10.7a,

$$i_3 = C\frac{dv_C}{dt} = C\frac{dE}{dt} = \frac{3.33}{1.33}C\,e^{-\frac{t}{0.000133}} = 5\,e^{-\frac{t}{0.000133}} \tag{10.58}$$

What will be i_1, i_2, and i_3 at $t = \infty$?

MESH METHOD

Figure 10.28 shows the two meshes in this circuit; a current in each circuit has been assigned in the clockwise direction.

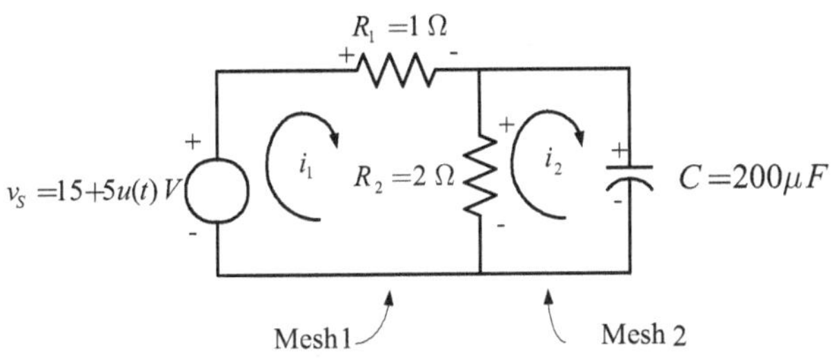

FIGURE 10.28 Electric circuit showing meshes.

The application of KVL to Mesh 1 yields

$$-v_S + v_{R1} + v_{R_2} = 0 \tag{10.59}$$

1 equation, 2 unknowns $[v_{R_1}, v_{R_2}]$.

From Equation 10.5,

$$v_{R1} = R_1 i_1 \tag{10.60}$$

2 equations, 3 unknowns $[i_1]$,

and

$$v_{R2} = R_2(i_1 - i_2) \tag{10.61}$$

3 equations, 4 unknowns $[i_2]$.

These are all the equations that can be written for Mesh 1, but we are still lacking one equation. Note that one of the unknowns is i_2, which is the current around Mesh 2. Thus, we now consider Mesh 2 and applying KVL,

$$-v_{R_2} + v_C = 0 \tag{10.62}$$

4 equations, 5 unknowns $[v_C]$.

Note the negative sign in v_{R_2}, the current i_2 flows from the negative to the positive polarity in R_2. From Equation 10.7a,

$$i_2 = C\frac{dv_C}{dt} \tag{10.63}$$

5 equations, 5 unknowns.

Equations 10.59–10.63 is the set of equations that form the model of the circuit. To reduce this set of equations, substituting Equations 10.61 and 10.63 into Equation 10.62 and rearranging gives

$$R_2C\frac{dv_C}{dt} + v_C = R_2 i_1 \tag{10.64}$$

Substituting Equations 10.60, 10.61, and 10.63 into Equation 10.59 and rearranging yields

$$i_1 = \frac{1}{(R_1 + R_2)} v_S + \frac{R_2 C}{R_1 + R_2} \frac{dv_C}{dt} \tag{10.65}$$

Substituting Equation 10.65 into Equation 10.64 and rearranging gives

$$\frac{R_1 R_2 C}{R_1 + R_2} \frac{dv_C}{dt} + v_C = \frac{R_2}{R_1 + R_2} v_S$$

or

$$0.000133 \frac{dv_C}{dt} + v_C = 0.67 v_S \tag{10.66}$$

This is another first-order linear nonhomogeneous ordinary differential equation with constant coefficients, and the solution is

$$v_C + 10 + 3.33(1 - e^{-\frac{t}{0.000133}})$$

which is the same as Equation 10.56.

From Equation 10.62,

$$v_{R_2} = v_C = 10 + 3.33(1 - e^{-\frac{t}{0.000133}})$$

Thus, the total current through R_2 is

$$i_{R_2} = i_2 - i_3 = \frac{v_{R_2}}{R_2} = 5 + 1.667(1 - e^{-\frac{t}{0.000133}})$$

which is the same as Equation 10.57.

Example 10.7

Consider the electrical circuit shown in Figure 10.29. Develop a model that describes how the voltage drop v_D varies when the supply voltage is given by $v_S = 10 \sin 5t\, u(t) V$.

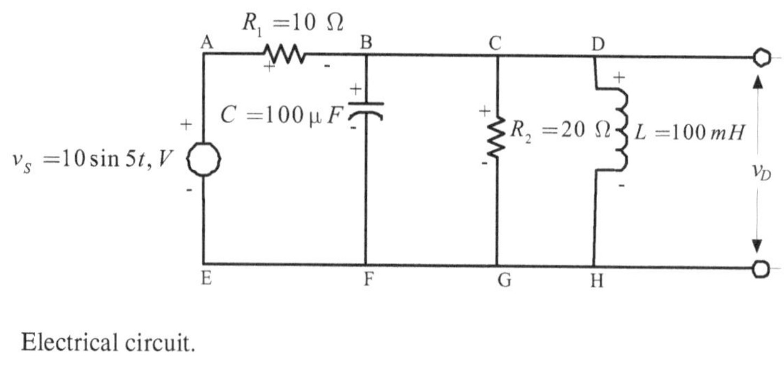

FIGURE 10.29 Electrical circuit.

Note that nodes B, C, and D have the same electrical potential, as nodes E, F, G, and H. Figure 10.30 redraws the circuit to emphasize this fact and shows that node F is selected as the reference node.

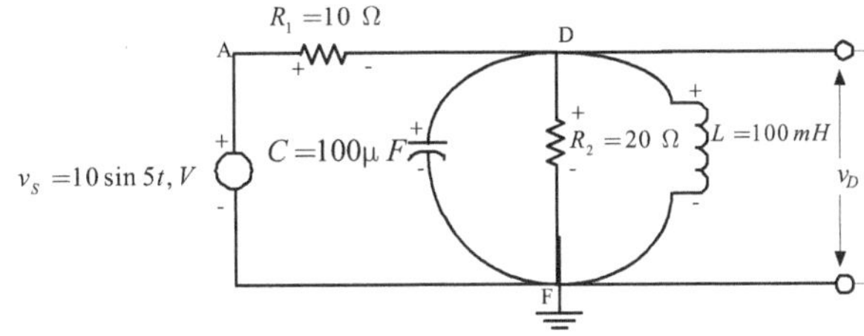

FIGURE 10.30 Redraw of Figure 10.29.

The voltage difference v_D is the variable of interest, and because node F is selected as the reference node (ground, 0 V), the voltage at node D is also v_D. In addition, because the electrical source generates a voltage v_S and because the losses in the electrical conductors are assumed negligible, the voltage at A is also v_S. Figure 10.31 shows all these assignments.

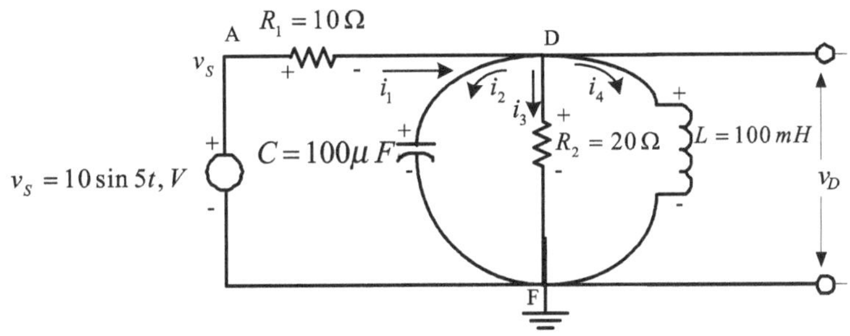

FIGURE 10.31 Redraw of Figure 10.30.

Applying KCL at node D; Figure 10.31 shows the currents entering and leaving the node.

$$i_1 - i_2 - i_3 - i_4 = 0 \tag{10.62}$$

1 equation, 4 unknowns $[i_1, i_2, i_3, i_4]$.

Using the equations for each element; for R_1 from Equation 10.5,

$$i_1 = \frac{1}{R_1}(v_S - v_D) \tag{10.63}$$

2 equations, 5 unknowns $[v_D]$.

For the capacitor, from Equation 10.7a,

$$i_2 = C\frac{dv_C}{dt} = C\frac{dv_D}{dt} \tag{10.64}$$

3 equations, 5 unknowns.

For R_2 from Equation 10.5 again,

$$i_3 = \frac{1}{R_2}v_D \tag{10.65}$$

4 equations, 5 unknowns.

For the inductor from Equation 10.8,

$$i_4 = i_4(0) + \frac{1}{L}\int_0^t v_D\,dt \tag{10.66}$$

5 equations, 5 unknowns.

Substituting Equations 10.63, 10.64, 10.65, and 10.66 into Equation 10.62,

$$\frac{1}{R_1}(v_S - v_D) - C\frac{dv_D}{dt} - \frac{1}{R_2}v_D - i_4(0) - \frac{1}{L}\int_0^t v_D\,dt = 0$$

Taking the derivative with respect to time gives

$$C\frac{d^2v_D}{dt^2} + \left(\frac{1}{R_1} + \frac{1}{R_2}\right)\frac{dv_D}{dt} + \frac{1}{L}v_D = \frac{1}{R_1}\frac{dv_S}{dt} \tag{10.67}$$

The solution of Equation 10.67 requires the initial conditions, which, based on the initial statement,

$$v_D(0) = 0V \text{ and } \left.\frac{dv_D}{dt}\right|_{t=0} = 0\,V/s$$

The analytical solution using Laplace transforms is

$$v_D = 0.0019e^{-1430t} - 0.0373e^{-70t} + 0.0373\sin 5t + 0.4973\cos 5t \tag{10.68a}$$

or

$$v_D = 0.0019e^{-1430t} - 0.0373e^{-70t} + 0.4986\sin(5t + 1.5) \tag{10.68b}$$

The following example shows the charging and discharging of a capacitor. The charging portion was already shown, but the discharging portion is new.

Example 10.8

Consider the circuit shown in Figure 10.32a.

a. The switch is originally open, and the capacitor is uncharged. The switch is moved to A until the capacitor is fully charged. Develop the model that describes the capacitor charge. What will the final capacitor charge in coulombs be?
b. Once the capacitor is charged, the switch is thrown to position B. Develop the model that describes the capacitor charge.
   a. When the switch is connected to A, the circuit looks as shown in Figure 10.32b. The application of KVL results in

$$-v_S + v_{R_1} + v_C = 0 \tag{10.69}$$

1 equation, 2 unknowns $[v_{R_1}, v_C]$

and

$$v_{R_1} = R_1 i \tag{10.70}$$

2 equations, 3 unknowns $[i]$

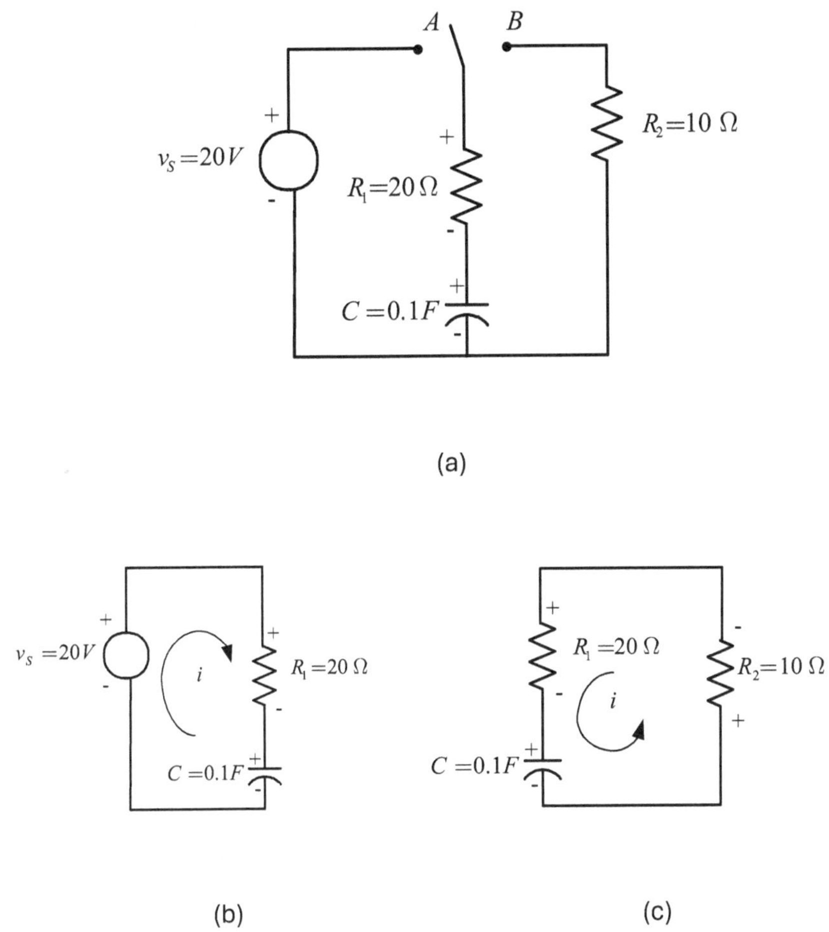

FIGURE 10.32 Circuits for Example 10.8.

$$v_C = \frac{1}{C} q \tag{10.71}$$

3 equations, 4 unknowns [q]

$$i = \frac{dq}{dt} \tag{10.72}$$

4 equations, 4 unknowns.

Using the general procedure we have followed in several examples already, from Equations 10.69–10.72, we get

$$R_1 C \frac{dq}{dt} + q = C v_S \tag{10.73a}$$

or

$$2 \frac{dq}{dt} + q = 0.1 v_S \tag{10.73b}$$

Note that in this case, $v_S = 20u(t), V$. Equation 10.73 is the model that describes the capacitor charge. The solution is

$$q = 2\left(1 - e^{-\frac{t}{2}}\right) \tag{10.74}$$

The solution indicates that $q_{final} = q(\infty) = 2\,C$.

b. When the switch is in position B, the circuit is shown in Figure 10.32c (note the polarities). This time, the current flows counterclockwise (opposite to that in part (a)). The reason is that this time the capacitor is supplying the current (charges), and the flow is from positive to negative polarity.

Applying KVL to the circuit yields

$$v_{R_1} + v_{R_2} + v_C = 0 \tag{10.75}$$

1 equation, 2 unknowns $[v_{R_2}, v_C]$

and

$$v_{R_2} = R_2 i \tag{10.76}$$

2 equations, 3 unknowns $[i]$

$$v_C = \frac{1}{C}q \tag{10.77}$$

3 equations, 4 unknowns $[q]$

$$i = \frac{dq}{dt} \tag{10.78}$$

4 equations, 4 unknowns.

Substituting Equations 10.76–10.78 and Equation 10.70 into Equation 10.75 and rearranging gives

$$(R_1 + R_2)C\frac{dq}{dt} + q = 0 \tag{10.79a}$$

or

$$3\frac{dq}{dt} + q = 0 \tag{10.79b}$$

This is the model for the circuit of Figure 10.32c; it is a first-order linear homogeneous ordinary differential equation with constant coefficients. The initial condition of Equation 10.79b is the final value of the capacitor charge of (a), or $q(0) = 2C$. The analytical solution is

$$q = 2e^{-\frac{t}{3}} \tag{10.80}$$

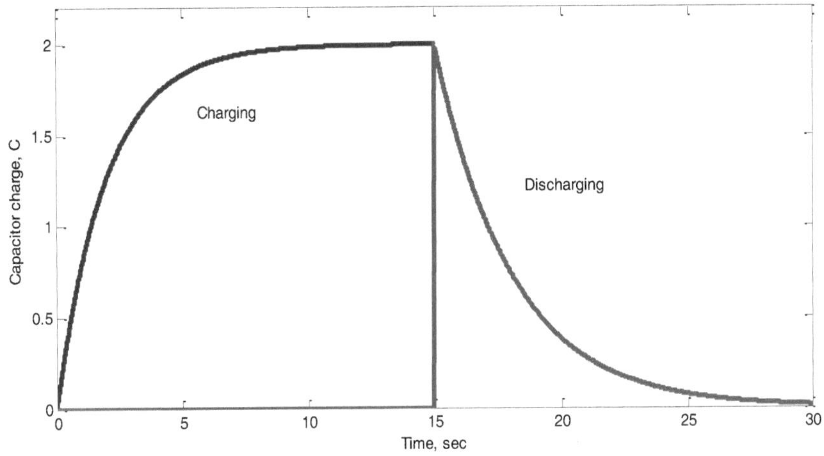

FIGURE 10.33 Charging and discharging of the capacitor in Figure 10.29a.

Figure 10.33 shows graphically the charging and discharging of the capacitor.

10.5 ENERGY AND POWER

We have been using and calculating terms related to electrical energy. The question related to what happens to this energy as electricity flows through resistors, capacitors, and inductors, is fair and important. Let us look at this question in detail for each of these electrical components.

The power (energy/time) delivered to, or absorbed by each component is given by

$$P = vi \tag{10.81}$$

Look at the units,

$$P[=]\left(\frac{Joule}{Coulomb}\right)\left(\frac{Coulombs}{\text{sec}}\right)=\frac{Joule}{\text{sec}}$$

and indeed, the multiplication of voltage drop times current has units of energy flow.

To further understand this calculation of power, consider the electrical component shown in Figure 10.34. The figure shows that $5A\left(C/s\right)$ current is delivered to the circuit and a voltage drop of $4\,V\left(J/C\right)$ are required to do so. This means that $5C/s$ go through the component and $4J$ are required to "push" each Coulomb through. Thus, an amount of energy of $20J/s(W)$ must be consumed or "absorbed" by the component. Note that this rate of energy (power) absorbed is also the result of Equation 10.81.

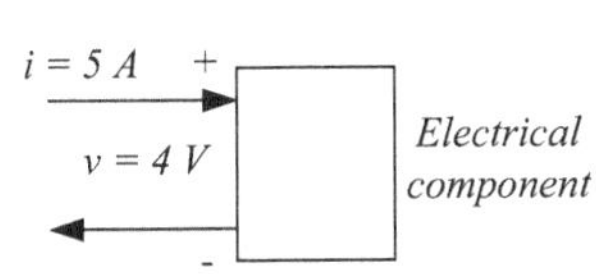

FIGURE 10.34 Electrical component.

So, what happens to the energy that is absorbed by the electrical component? As Chapter 9 presented, the law of conservation of energy (first law of thermodynamics) states that energy never disappears; therefore, something must happen. Well, in capacitors, this energy is stored in and released from the electric field between the two plates. In inductors, this energy is stored in and released from the magnetic field that is formed when current flows. Resistors do not have a place to store energy; thus, the energy entering the resistor is always released as heat to the surroundings (actually, think about an electrical heater or light bulb).

RESISTORS

As we know, resistors are devices that resist the flow of current. In doing so, some of the energy entering the resistor is dissipated as heat. Consider the resistor shown in Figure 10.35,

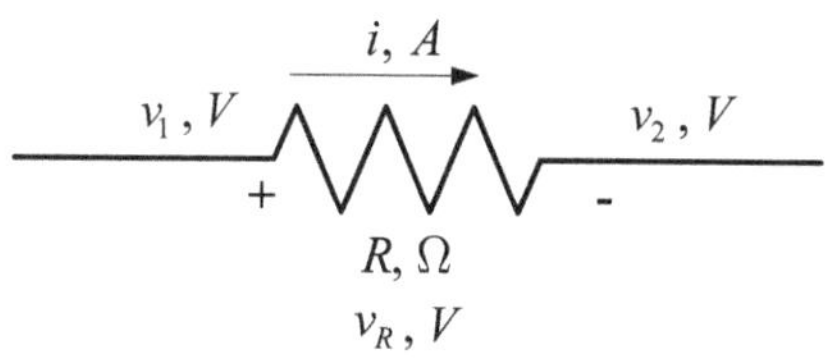

FIGURE 10.35 Resistor.

As given by Equation 10.81, the power absorbed, or better yet, in the case of resistors, the power released is

$$P_R = v_R\, i \tag{10.82}$$

Using Ohm's law the following expression is also applicable:

$$P_R = R i^2 \tag{10.83}$$

CAPACITORS

Capacitors do not dissipate energy; they store energy for later release back to the circuit. The energy is stored in the electric field that exists between the two plates. As a capacitor charges, energy is stored, and as it discharges, energy is released. Equation 10.81 is still the power delivered to the capacitor, and in this case, using Equation 10.7a, we can also write it as

$$P_C = i v_C = v_C\, C \frac{dv_C}{dt} \tag{10.84}$$

The energy stored, or released, w_C in J, during a period of time t assuming the initial charge is zero is given by

$$w_C = \int_0^t P\, dt = \int_0^t v_C\, C \frac{dv_C}{dt} dt = \int_0^{v_C} C v_C\, dv_C$$

and

$$w_C = \frac{1}{2} C v^2_C \tag{10.85}$$

INDUCTORS

Inductors, like capacitors, do not dissipate energy; they store and release the energy received in the magnetic field that is created when the current flows. The power expression is

$$P_L = iv_L = iL\frac{di}{dt} \tag{10.86}$$

The energy stored, or released, w_L in joules during a period of time t is given by

$$w_L = \int_0^t P\,dt = \int_0^t i\,L\frac{di}{dt}dt = \int_0^i L\,i\,di$$

and

$$w_L = \frac{1}{2}Li^2 \tag{10.87}$$

As the reader knows, inductors are formed by a wire coiled around some type of core. In an "ideal" inductor, the wire does not offer any resistance (actually, in this chapter, we are also assuming that no electrical conductor offers any resistance); thus, there is no energy dissipation. A real inductor would dissipate some energy in the form of heat; usually, this amount of heat is very small and negligible compared to that dissipated by the resistors in the circuit.

Example 10.9

Consider the RLC circuits shown in Figure 10.36a and b. Switch 1 is originally open; it closes at $t = 0$ and opens again at $t = 2$. Switches 2 and 3 are originally open and close at $t = 2$. Note that between $0 \le t \le 2$ both circuits are the same as shown in Figure 10.36c.

a. Model these circuits to obtain the current, the voltage drop across each element, and the power and energy absorbed/released by each element for $0 \le t \le 2$ s.
b. Model the circuits again to obtain the current, the voltage drop across, and the power and energy absorbed/released by each element in the resulting circuits for $t \ge 2$ s.

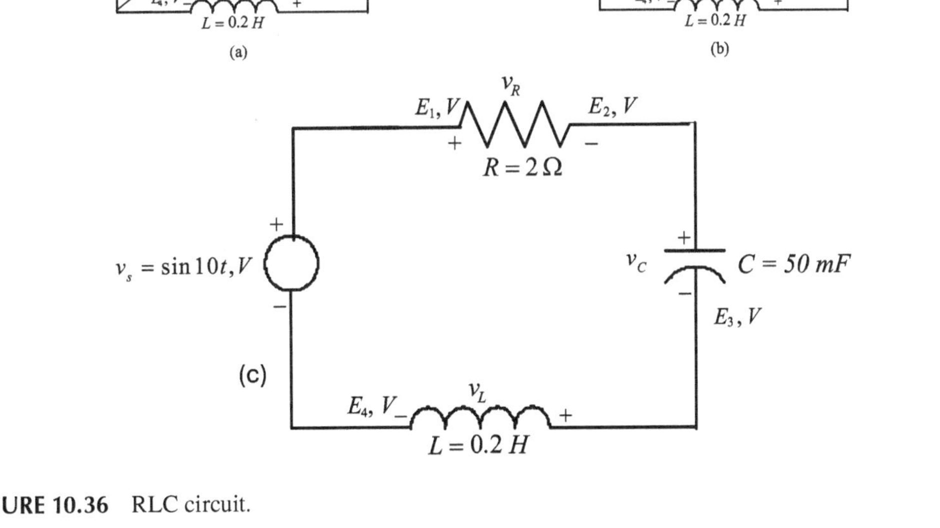

FIGURE 10.36 RLC circuit.

a. The circuit of Figure 10.36c is exactly like that of Figure 10.17 except for different supply voltage and values of each element. Therefore, for the time period $0 \le t \le 2$ s, the model that describes the voltage across the capacitor is the same as given by Equation 10.27a,

$$LC\frac{d^2v_C}{dt^2} + RC\frac{dv_C}{dt} + v_C = v_S \tag{10.27a}$$

or

$$0.01\frac{d^2v_C}{dt^2} + 0.1\frac{dv_C}{dt} + v_C = v_S \tag{10.88}$$

Following any of the methods we have previously learned, the analytical solution is

$$v_C = e^{-5t}\left[0.577 \sin 8.66t + \cos 8.66t\right] - \cos 10t \tag{10.89}$$

Using Equation 10.7a,

$$i = C\frac{dv_C}{dt} = -0.577e^{-5t} \sin 8.66t + 0.5 \sin 10t \tag{10.90}$$

Using Equation 10.8a,

$$v_L = L\frac{di}{dt} = e^{-5t}[0.578 \sin 8.66t - \cos 8.66t] + \cos 10t \tag{10.91}$$

Using Equation 10.5,

$$v_R = Ri = -1.154e^{-5t} \sin 8.66t + \sin 10t \tag{10.92}$$

The power released (J/s or W) by the resistor is

$$P_R = Ri^2 = 2\left[-0.577e^{-5t} \sin 8.66t + 0.5 \sin 10t\right]^2 \tag{10.93}$$

The power stored/released by the capacitor is

$$P_C = i\,v_C = (\text{Equation 10.90})\cdot(\text{Equation 10.89}) \tag{10.94}$$

and the energy (J) stored/released by the capacitor from time = 0 to any time t is,

$$w_C = \frac{1}{2}Cv^2{}_C = 0.025[e^{-5t}[0.577 \sin 8.66t + \cos 8.66\,t] - \cos 10t]^2 \tag{10.95}$$

The power stored/released by the inductor is

$$P_L = i\,v_L = (\text{Equation 10.90})\cdot(\text{Equation 10.91}) \tag{10.96}$$

and the energy stored/released by the inductor from time = 0 to any time t is

$$w_L = \frac{1}{2}Li^2 = 0.1[-0.577\,e^{-5t}\,\sin 8.66t + 0.5\sin 10t]^2 \tag{10.97}$$

This set of equations is what was requested by the example's statement. However, fully understanding what happens with the energy terms by themselves is not so easy. As the saying goes, a picture (or graph) "is worth a million words," so let us graph the terms for a period of 2 s; Figure 10.37 shows the graphs.

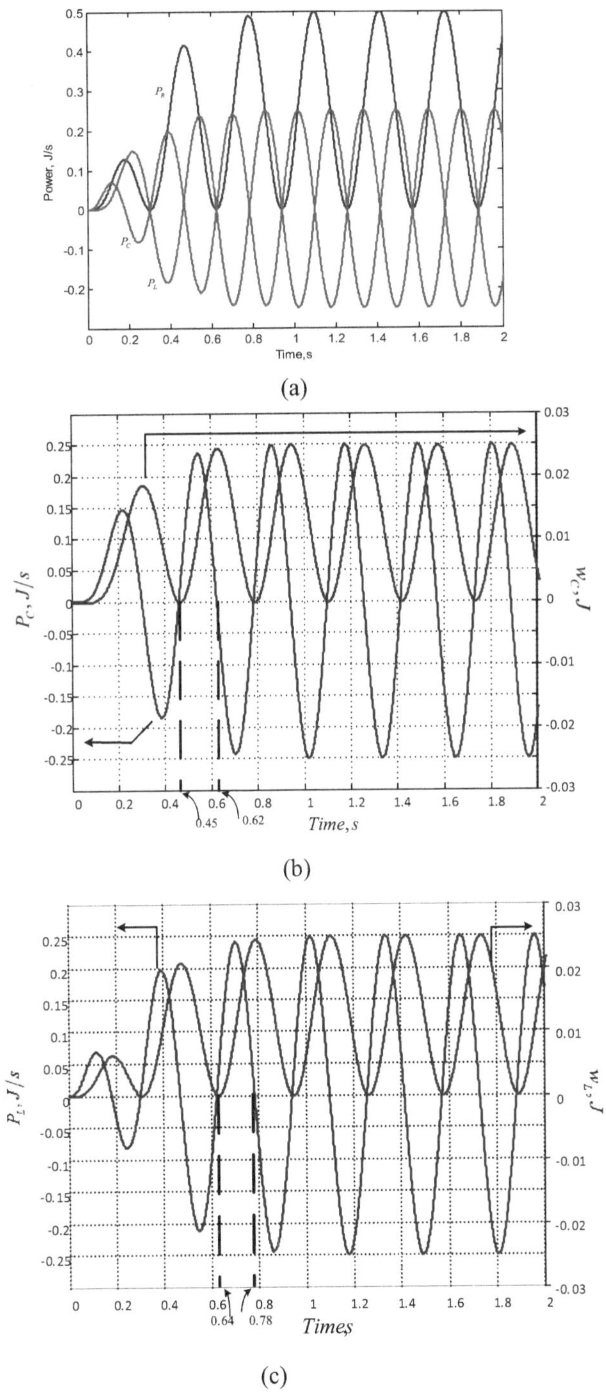

FIGURE 10.37 Power and energy responses of components in Figure 10.36.

Figure 10.37a shows the power term released by the resistance (P_R), and those stored and released by the capacitor (P_C) and inductor (P_L). Note that in both the capacitor and inductor the power term can be either positive (and this is when power is being stored by the component) or negative (and this is when power is being released back to the circuit). In the case of the resistor, power is always released to the ambient, so the curve only indicates the amount of power being released. The power curves of the capacitor and inductor show that they are "completely out of phase"; that is, one is maximum positive at the same time the other one is minimum negative. This is not always the case because it depends on the values of the capacitance and inductance. We chose those that would give this behavior; you will learn more about this in your electrical circuits course.

Figure 10.37b shows the power ($P_C, J/s$) being absorbed and released, and the energy (w_C, J) being stored and released by the capacitor. Remember that the energy being stored and released is an integral part of power. Thus, it does not change instantaneously, and it is why it "lags" the power curve. Note that at about 0.45 s, both power and energy are zero. Then from 0.45 s to about 0.62 s, the power is positive (being absorbed by the capacitor), and during that time, the energy stored increases. At about 0.62 s, the power goes from positive (being absorbed) to negative (being released), and at that moment, the energy being stored is maximum and then starts decreasing.

Figure 10.37c shows the power ($P_L, J/s$) being absorbed and released, and the energy (w_L, J) being stored and released by the inductor. Note that at about 0.64 s, both power and energy are zero. Then, from 0.64 s to about 0.78 s, the power is positive (being absorbed by the inductor) and during that time the energy stored increases. At about 0.64 s, the power goes from positive (being absorbed) to negative (being released), and at that moment, the energy being stored is maximum and then starts decreasing.

b. Now let us look at the resulting circuits for $t \geq 2$ s. Figure 10.38a shows the resulting circuit from Figure 10.36a and Figure 10.38b that from Figure 10.36b.

With respect to Figure 10.38a, applying KVL to this new circuit,

$$v_{Ca} + v_{La} = 0 \tag{10.98}$$

1 equation, 2 unknowns [v_{Ca}, v_{La}].

Note that we are using the notation v_{Ca} and v_{La} (subscript a) to differentiate these voltages from the previous ones in part (a).

For the inductor, from Equation 10.8a,

$$v_{La} = L\frac{di_a}{dt} \tag{10.8a}$$

2 equations, 3 unknowns [i_a].

For the capacitor, from Equation 10.7a,

$$i_a = C\frac{dv_{Ca}}{dt} \tag{10.7a}$$

3 equations, 3 unknowns.

From these last three equations, following the usual substitutions results in

$$LC\frac{d^2v_{Ca}}{dt^2} + v_{Ca} = 0$$

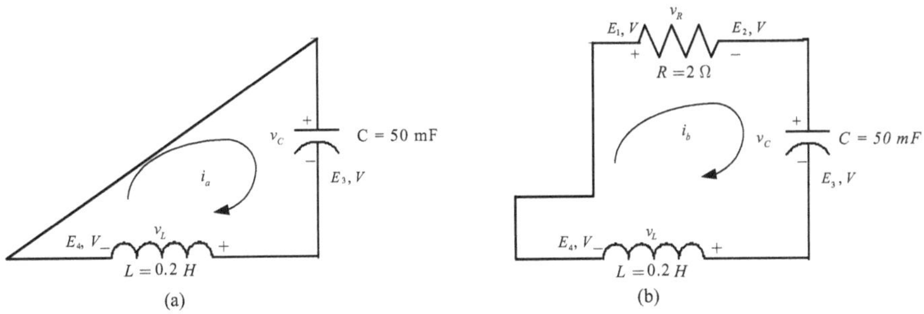

FIGURE 10.38 Circuits for $t \geq 2$ s.

or

$$0.01\frac{d^2 v_{Ca}}{dt^2} + v_{Ca} = 0 \tag{10.99}$$

Because this model has significance from $t \geq 2$, the initial conditions required, $v_{Ca}(0)$ and $\left.\frac{d^2 v_{Ca}}{dt^2}\right|_{t=0}$, for their solutions are obtained from Equation 10.89 and its derivative evaluated at $t = 2$.

In this case, for Equation 10.89,

$$v_{Ca}(0) = -0.408\,V \text{ and } \frac{d^2 v_{Ca}}{dt^2}\Big|_{t=0} = 9.13\ V/s$$

The solution of Equation 10.99 is (we encourage the reader to check this solution – don't you dislike this comment?)

$$v_{Ca} = 0.913 \sin 10t - 0.4081 \cos 10t \tag{10.100}$$

Now we obtain for $t \geq 2$,

$$i_a = C\frac{dv_{Ca}}{dt} = 0.456 \cos 10t + 0.204 \sin 10t \tag{10.101}$$

$$P_{Ca} = i_a v_{Ca} = (\text{Equation } 10.101)\cdot(\text{Equation } 10.100) \tag{10.102}$$

$$w_{Ca} = \frac{1}{2}Cv^2{}_{Ca} = 0.025[0.913 \sin 10t - 0.408 \cos t]^2 \tag{10.103}$$

$$v_{La} = L\frac{di_a}{dt} = 0.912 \sin 10t - 0.408 \cos 10t \tag{10.104}$$

and

$$P_{La} = i_a\, v_{La} = (\text{Equation } 10.101)\cdot(\text{Equation } 10.104) \tag{10.105}$$

With respect to Figure 10.38b, this is exactly the same circuit as in Figure 10.36c but without the power supply. Thus, a similar model to Equation 10.27a describes the voltage across the capacitor but, this time, without a forcing function (we use subscript b to differentiate these from part (a)):

$$LC\frac{d^2v_{Cb}}{dt^2}+RC\frac{dv_{Cb}}{dt}+v_{Cb}=0$$

or

$$0.01\frac{d^2v_{Cb}}{dt^2}+0.1\frac{dv_{Cb}}{dt}+v_{Cb}=0 \tag{10.106}$$

Using the same initial conditions as previously, the solution of this second-order linear homogeneous differential equation with constant coefficients is

$$v_{Cb}=e^{-5t}[0.819\sin 8.66t-0.408\cos 8.66t] \tag{10.107}$$

and, as we have done before, for $t\geq 2$,

$$i_b=C\frac{dv_{Cb}}{dt}=e^{-5t}[0.456\cos 8.66t-0.028\sin 8.66t] \tag{10.108}$$

$$P_{Cb}=i_b\,v_{Cb}=(\text{Equation 10.107})\cdot(\text{Equation 10.108}) \tag{10.109}$$

$$w_{Cb}=\frac{1}{2}Cv^2{}_{Cb}=0.025e^{-10t}[0.819\sin 8.66t-0.408\cos 8.66t]^2 \tag{10.110}$$

$$v_{Lb}=L\frac{di_b}{dt}-e^{-5t}[0.762\sin 8.66t+0.504\cos 8.66t] \tag{10.111}$$

and

$$P_{Lb}=i_b\,v_{Lb}=(\text{Equation 10.108})\cdot(\text{Equation 10.111}) \tag{10.112}$$

There are many equations describing the current, voltages, and energy terms in the two resulting circuits shown in Figure 10.38. As we did before, let us show some graphs to understand the difference between the two circuits. Before we do so, note that the models, Equations 10.99 and 10.106, describing the voltage drops across the capacitor, v_{Ca} and v_{Cb}, are both homogeneous differential equations meaning that there is no forcing function. Actually, you may also "see" this in Figure 10.38 in that neither resulting circuit has a power supply. So what does this say about the voltage drop and current in each circuit? Figure 10.39 shows these.

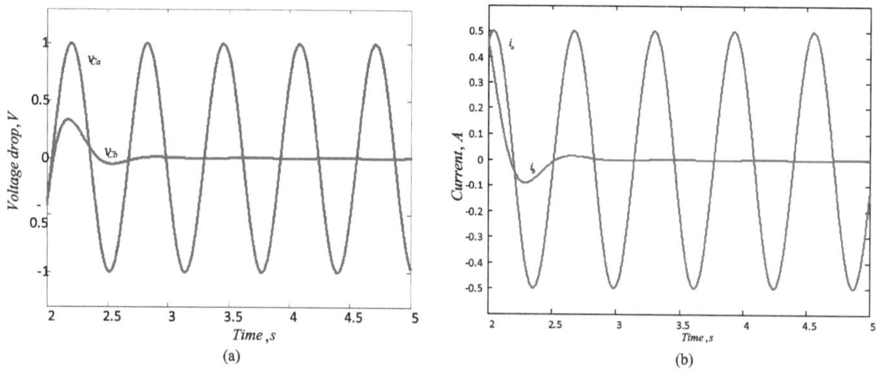

FIGURE 10.39 Voltages and currents for circuits in Figure 10.36.

The behaviors of the two circuits are obviously different. The voltage across the capacitor and the current in the circuit shown in Figure 10.38a continues oscillating as before, while that of the circuit of Figure 10.38b decay to values of zero. Note that the equations themselves (the graphs come from the equations) indicate these behaviors. Equations 10.100 and 10.101 show that the behavior of v_{Ca} and i_a follows sine and cosine with no decaying behavior; Equations 10.107 and 10.108 show that the behavior of v_{Cb} and i_b decay to zero, due to the negative exponential term (e^{-5t}), with oscillations superimposed (sine and cosine terms).

The reader may remember that in Chapters 3 and 6, and in this same chapter, we discussed the response of systems when described by second-order linear differential equations when the coefficient of the first derivative is zero. These chapters showed that the total response of a homogeneous equation, or the corresponding homogeneous response of a nonhomogeneous equation, is composed of only sine and cosine terms with no exponential term. That is, the roots of the homogeneous portion of the model are imaginary numbers (no real component).

The preceding two paragraphs analyzed the graphs and the equations that provided both graphs. But does this make physical sense for these systems? What is the difference? Recall that neither of the two circuits in Figure 10.38 have a power supply. Initially, for these circuits (at $t = 2$ for the previous circuits), there was a charge in the capacitor and a current flowing through the circuit. The circuit in Figure 10.38a has maintained the charge and the current at the same level as that the previous circuit; however, these two have disappeared in the circuit in Figure 10.39b. The *difference is that there is no resistor in the first circuit, but there is one in the second circuit.* As the reader recalls, resistors do not store energy, the energy that they absorb is dissipated as heat. Thus, as time goes on, the original total energy contained within the circuit will be dissipated as heat and the voltage drop across the elements, and the current will decay to zero. By comparison, capacitors and inductors store and release back to the circuit the energy they "absorb." Thus, in the second case, the energy contained in the circuit will remain at the same value. To help us visualize this explanation, Figure 10.40 shows the summation of the energy stored in the capacitor and inductor, $w_C + w_L$, as a function of time.

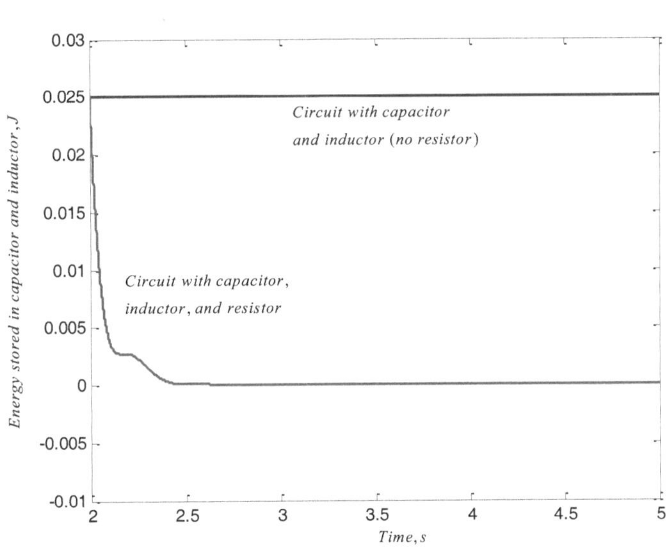

FIGURE 10.40 Energy contained in the circuits of Figure 10.38.

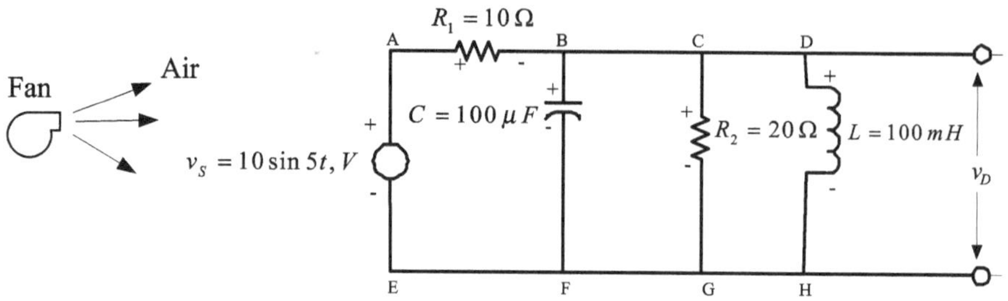

FIGURE 10.41 Electrical circuit.

Example 10.10

Consider the circuit of Example 10.7 and shown again for convenience in Figure 10.41, this time showing a variable speed fan installed next to the circuit. Let us suppose that this circuit is installed in a chip of size 3 cm × 6 cm × 2 cm. The initial temperature of the complete chip, as well as the air surrounding it is 25°C; the air surrounding the chip is still, with no movement. The mass of the entire chip is 0.5 kg, its heat capacity is $4\,\text{J}/(\text{kg}\cdot{}^{\circ}\text{C})$, the heat transfer coefficient for convection from the chip to the surrounding air is $20\,\text{J}/\text{s}\cdot\text{m}^2\cdot{}^{\circ}\text{C}$, and the chip's area in contact with the surrounding air is $0.0054\,\text{m}^2$. Initially, the fan and the circuits are off.

a. Develop a model that describes the temperature of the chip once the circuits are turned on. How long does it take for the chip temperature to reach 35°C if the ambient temperature remains at 25°C?
b. The highest temperature the chip must be maintained at is 30°C to avoid harm to the circuit. To do so, the fan blows air at different temperatures pass the chip. When the fan is off, the air is at 25°C, at low speed the fan blows air at 19°C, at medium speed the air is at 15°C, and at high speed the air temperature is at 11°C. Develop the model that describes the chip temperature and obtain the correct fan speed to avoid the chip going over 30°C. How long does it take the chip to reach its final temperature?

a. Equation 10.68a gives the voltage drop v_D:

$$v_D = 0.0019e^{-1430t} - 0.0373e^{-70t} + 0.0373 \sin 5t + 0.4973 \cos 5t \tag{10.68a}$$

To obtain the heat released by resistors R_1 and R_2 the currents i_1 and i_2 must be calculated.

$$i_1 = \frac{v_S - v_D}{R_1} = 0.00373e^{-70t} - 0.00019e^{-1430t} + 0.99627 \sin 5t - 0.04973 \cos 5t \tag{10.113}$$

$$i_2 = \frac{v_D}{R_2} = 9.5\,E - 5\,e^{-1430t} - 0.001865e^{-70t} + 0.001865 \sin 5t + 0.02486 \cos 5t \tag{10.114}$$

Thus, the power released by each resistor is

$$P_{R_1} = R_1 i_1^2 = 10\left[0.00373e^{-70t} - 0.00019e^{-1430t} + 0.99627 \sin 5t - 0.04973 \cos 5t\right]^2 \tag{10.115}$$

$$P_{R_2} = R_2 i_2^2 = 20\left[9.5E - 5e^{-1430t} - 0.001865e^{-70t} + 0.001865 \sin 5t + 0.02486 \cos 5t\right]^2 \tag{10.116}$$

and the total power is

$$P_R = P_{R_1} + P_{R_2} \tag{10.117}$$

To obtain the chip temperature as power is released by the resistors, an energy balance taking the chip as the system provides the model. As learned in Chapter 9, we write the energy balance as

$$\begin{matrix} \textit{Rate of change of} & & \textit{Rate of} & & \textit{Rate of} \\ \textit{energy accumulated} & = & \textit{energy entering} & - & \textit{energy exiting} \\ \textit{in chip} & & \textit{chip} & & \textit{chip} \end{matrix}$$

In this case the rate of energy entering the chip is the power released by the resistors, P_R, and the rate of energy exiting the chip is by heat transfer convection. Thus,

$$mC\frac{dT}{dt} = P_R - hA(T - T_A)$$

where
m = mass of chip, 0.5 kg
C = heat capacity of chip, $4\,\text{J}/(\text{kg}\cdot{}^\circ\text{C})$;
A = heat transfer area of chip, $0.0150\,\text{m}^2$;
h = heat transfer coefficient for convection, $20\ \text{J}/\text{s}\cdot\text{m}^2\cdot{}^\circ\text{C}$;
T = temperature of chip, °C; and
T_A = temperature of ambient air.

Thus,

$$2\frac{dT}{dt} = P_R - 0.108(T - T_A)$$

with $T(0) = 25^\circ\text{C}$.

The previous equation is a first-order linear nonhomogeneous ordinary differential equation. As we have previously learned, this is usually a simple equation to solve; however, what makes it difficult in this case is the form of the forcing function P_R. It is not impossible to solve but, definitely, quite cumbersome! Thus, in this case, we proceed obtaining the solution/response using simulation (Chapter 11).

Figure 10.42 shows the chip temperature. It starts at 25°C and increases due to the heat released.

a. The figure shows that when the fan is off and the ambient air temperature T_A is 25°C, the temperature reaches a maximum of 41.5°C, and it takes about 7 seconds to reach 35°C.
b. The fan needs to be at high speed to avoid more than 30°C; however, at medium speed, the temperature is not much more at about 31.5°C.

10.6 RC CIRCUITS AS FILTERS

A very common and interesting application of electrical circuits is their use as filters. Filters are used to remove, reject, or "filter out" unwanted frequencies. Sometimes, it is desired to remove low frequencies, and for those cases, we use "high-pass filters," meaning that only high frequencies exit, or "pass," the filter. Other times, removing high frequencies is desired, and for those cases we use "low-pass filters," meaning that only low frequencies exit, or pass, the filter. Figure 10.43 shows the performance of these filters. Note that the y-axis shows the ratio of the amplitude of the exiting signal to the amplitude of the input signal, which we may call the *amplitude ratio*. An "ideal" high pass filter is shown in Figure 10.43a, where the ratio below a certain frequency, called "cutoff" frequency, is zero – meaning that no signal with a frequency below the cutoff frequency passes

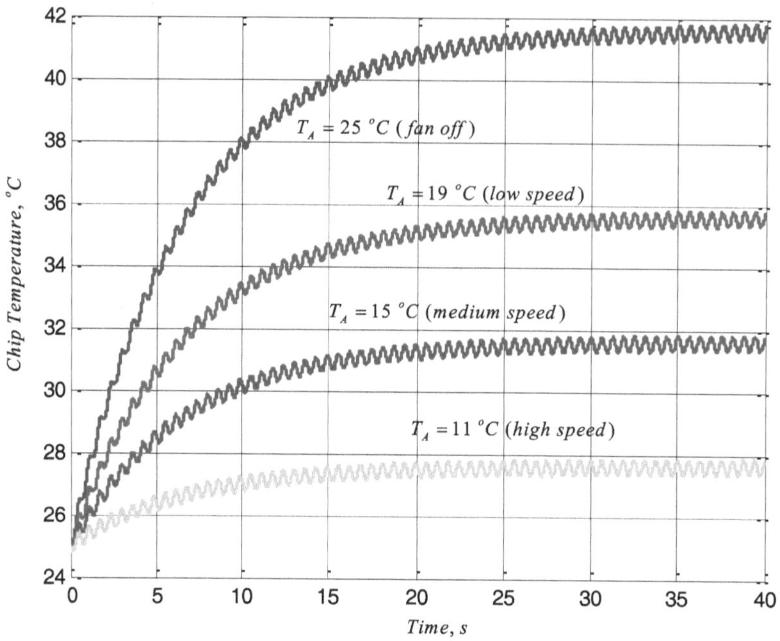

FIGURE 10.42 Chip temperature.

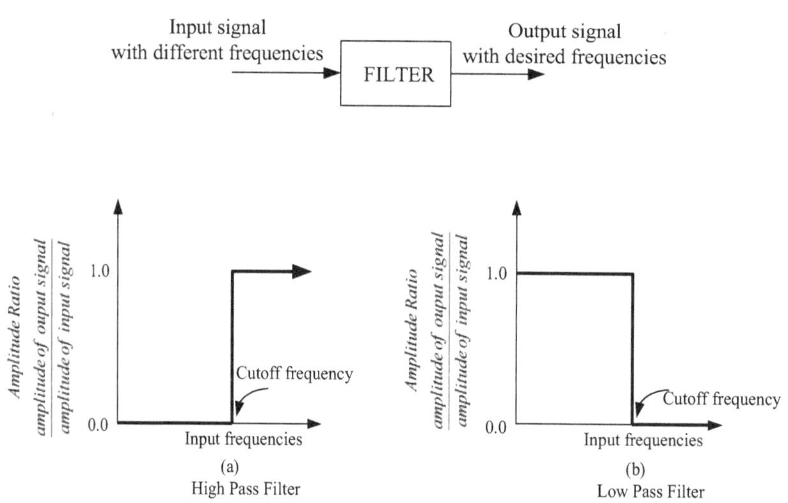

FIGURE 10.43 Ideal filters.

through. An "ideal" low-pass filter is shown in Figure 10.43b, where the ratio above the cutoff frequency is zero – meaning that no signal with frequency above the cutoff frequency passes through.

Ideal filters *are very difficult* to achieve. Fortunately, however, RC circuits provide a very simple way to *approximate* the desired performances. The cutoff frequency ω_C, mentioned in the previous paragraph is given by

$$\omega_C = \frac{1}{RC} \tag{10.118}$$

10.6.1 High-Pass Filter

Consider the RC circuit shown in Figure 10.44 with a sinusoidal voltage source, $v_S = V_o \sin \omega t$, starting at $t = 0$; for $t < 0$, there is no current flow, and the capacitor is fully discharged.

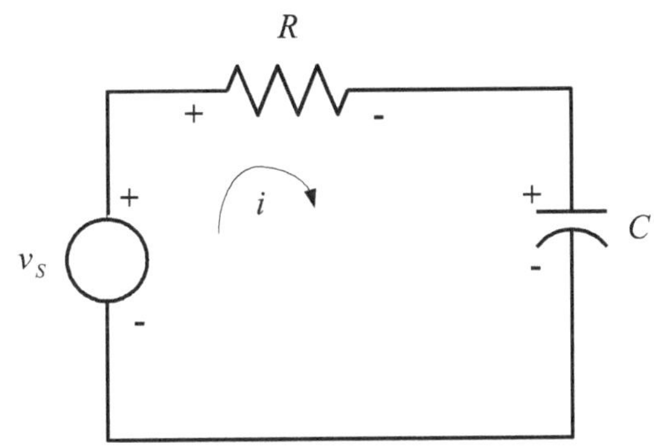

FIGURE 10.44 RC circuit.

Applying KCL,

$$v_R + v_C - v_S = 0$$

$$iR + \frac{q}{C} = v_S$$

or

$$R\frac{di}{dt} + \frac{1}{C}i = \frac{dv_S}{dt}$$

and

$$\frac{dv_S}{dt} = V_o \omega \cos \omega t$$

therefore,

$$R\frac{di}{dt} + \frac{1}{C}i = V_o \, \omega \cos \omega t \tag{10.119}$$

Equation 10.119 can be solved by different methods presented in this book; the reader is encouraged to do so (at least with one method!). The solution is

$$i = A \cos \omega t + B \sin \omega t + De^{-\frac{t}{RC}} \tag{10.120}$$

where

$$A = \frac{\dfrac{V_o \omega}{R}}{\omega^2 RC + \dfrac{1}{RC}} \quad \text{and} \quad B = \frac{\dfrac{V_o \omega}{R}}{\omega + \dfrac{1}{\omega R^2 C^2}}$$

Let us look at the solution after a very long time, that is, after the "transients" have died out; we refer to this result as the *periodic current*, $i_{periodic}$.

$$e^{-\frac{t}{RC}} \to 0 \text{ as } t \to \infty$$

then,

$$i_{periodic} = A\cos\omega t + B\sin\omega t \tag{10.121}$$

Recalling the trigonometric identity,

$$\alpha\sin(\omega t + \Phi) = \alpha\cos\Phi\sin\omega t + \alpha\sin\Phi\cos\omega t \tag{10.122}$$

Comparing Equations 10.121 and 10.122 gives

$$A = \alpha\sin\Phi \text{ and } B = \alpha\cos\Phi \tag{10.123}$$

Because the values of A and B are known, Equation 10.123 gives two equations and two unknowns. Solving these equations for α and Φ yields

$$A^2 + B^2 = \alpha^2\sin^2\Phi + \alpha^2\cos^2\Phi = \alpha^2(\sin^2\Phi + \cos^2\Phi) = \alpha^2$$

$$\alpha = \sqrt{A^2 + B^2} = \sqrt{B^2\left(1 + \frac{1}{(RC\omega)^2}\right)} \quad \text{and} \quad \Phi = \sin^{-1}\frac{A}{\alpha}$$

After substituting the expression for B and some "simple algebra,"

$$\alpha = \frac{V_o}{R}\sqrt{\frac{(RC\omega)^2}{1 + (RC\omega)^2}}$$

Finally,

$$i_{periodic} = \frac{V_o}{R}\sqrt{\frac{(RC\omega)^2}{1 + (RC\omega)^2}}\sin(\omega t + \Phi) \tag{10.124}$$

After so much algebra and a bit of trigonometry, we might as well remind the reader what this equation represents. Equation 10.124 is the current circulating in the RC circuit of Figure 10.44 after a very long time (meaning at the steady state).

Let us now obtain the expression for the voltage across the resistor.

$$v_{R\,periodic} = Ri_{periodic} = V_o\sqrt{\frac{(RC\omega)^2}{1 + (RC\omega)^2}}\sin(\omega t + \Phi) \tag{10.125}$$

The ratio of the amplitudes of the resistor voltage to the voltage source is

$$\frac{amplitude\,\{v_{R\,periodic}\}}{amplitude\,\{v_S\}} = \frac{|v_{R\,periodic}|}{|v_S|} = \frac{(RC\omega)}{\sqrt{1 + (RC\omega)^2}}$$

Note that as $\omega \to \infty$, $(\omega RC)^2 >> 1$, $\sqrt{1 + (RC\omega)^2} \to \sqrt{(RC\omega)^2} \to RC\omega$, and $\frac{|v_{R\,periodic}|}{|v_S|} \to 1.0$. Also, as $\omega \to 0$, $\frac{|v_{R\,periodic}|}{|v_S|} \to 0$. Thus, the ratio of the amplitudes is close to zero (0) for low frequencies, and close to one (1) for high frequencies. Therefore, connecting across the resistor terminals in an RC circuit as shown in Figure 10.45, gives a "high-pass filter." This term means that only high frequencies pass undisturbed, but low-pass frequencies are "filtered out," or removed, from the signal.

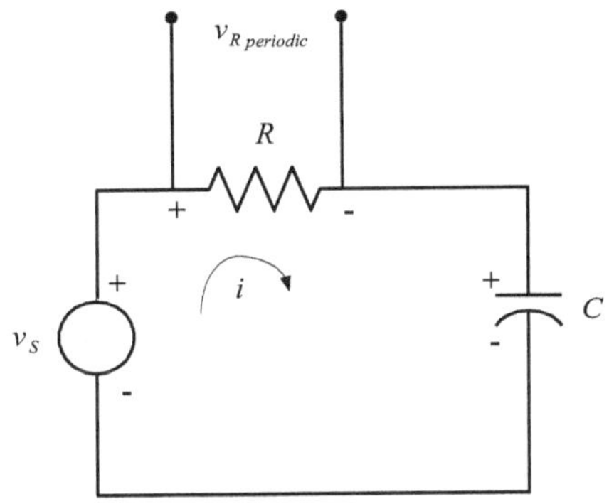

FIGURE 10.45 Voltage drop across a resistor in an RC circuit.

Assuming an RC circuit with $R = 1000\Omega$ and $C = 2.5\times10^{-6}F$, the cutoff frequency is $\omega_C = 400$ cycles / sec (*Hertz*). Figure 10.46 shows the filtering performance of the RC circuit. The figure shows the approximation given by the circuit. At the cutoff frequency, the amplitude ratio is 0.707.

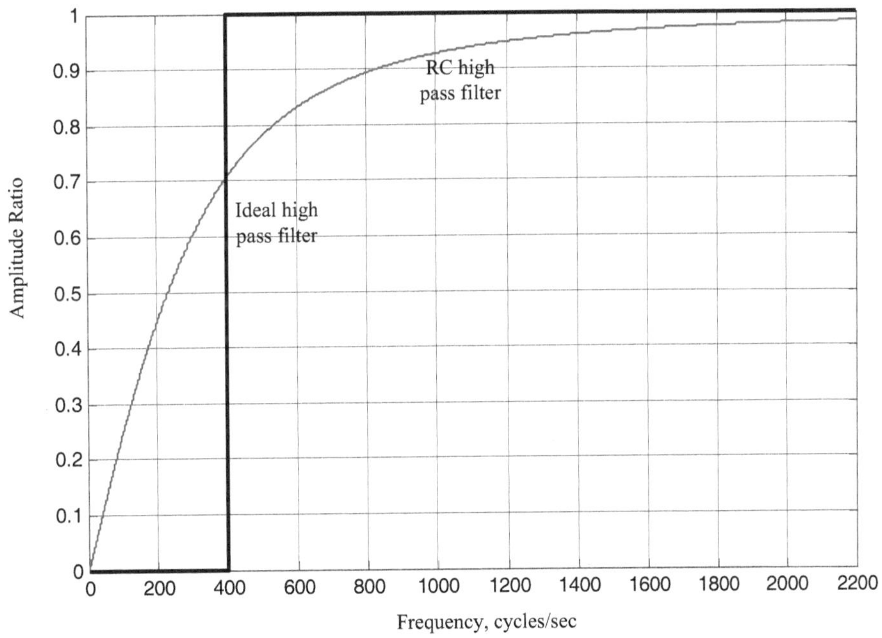

FIGURE 10.46 Performance of an RC circuit as a high-pass filter.

10.6.2 Low-Pass Filter

Let us go back to the KVL of the circuit of Figure 10.44,

$$v_R + v_C - v_S = 0$$

or considering after a very long time,

$$v_{R\,periodic} + v_{C\,periodic} - v_{S\,periodic} = 0$$

$$v_{C\,periodic} = v_{S\,periodic} - v_{R\,periodic} = V_o \sin\omega t - i_{periodic} R = V_o \sin\omega t - AR\cos\omega t - BR\sin\omega t$$

$$v_{C\,periodic} = (V_o - BR)\sin\omega t - AR\cos\omega t \tag{10.126}$$

Comparing the terms of Equation 10.126 with the terms of the trigonometric equality of Equation 10.122, gives,

$$V_o - BR = \alpha\cos\Phi \text{ and } -AR = \alpha\sin\Phi \tag{10.127}$$

Because A and B are known,

$$\alpha = \sqrt{(AR)^2 + (V_o - BR)^2}$$

Using the expressions for A and B, and some simple algebra, yields,

$$\alpha = V_o\sqrt{\frac{1}{1+(RC\omega)^2}}$$

Then,

$$v_{C\,periodic} = V_o\sqrt{\frac{1}{1+(RC\omega)^2}}\sin(\omega t + \Phi) \tag{10.128}$$

The ratio of the amplitudes of the capacitor voltage to the voltage source is

$$\frac{amplitude\,\{v_{C\,periodic}\}}{amplitude\,\{v_S\}} = \frac{|v_{C\,periodic}|}{|v_S|} = \frac{1}{\sqrt{1+(RC\omega)^2}}$$

Note that as $\omega \to \infty$, $(RC\omega)^2 >> 1$, $\sqrt{\frac{1}{1+(RC\omega)^2}} \to \sqrt{\frac{1}{(RC\omega)^2}} \to 0$, and $\frac{|v_{C\,periodic}|}{|v_S|} \to 0$. Also, as $\omega \to 0$, $\frac{|v_{R\,periodic}|}{|v_S|} \to 1.0$. Thus, the ratio of the amplitudes is close to zero (0) for high frequencies, and close to one (1) for low frequencies. Therefore, by connecting across the capacitor terminals in an RC circuit as shown in Figure 10.47, gives a low-pass filter. This term means that only low frequencies pass undisturbed, but high pass frequencies are "filtered out," or removed, from the signal.

Assuming an RC circuit with $R = 1000\,\Omega$ and $C = 2.5\times10^{-6}\,F$ the cutoff frequency is $\omega_C = 400$ cycles / s (*Hertz*). Figure 10.48 shows the filtering performance of the RC circuit. At the cutoff frequency, the amplitude ratio is 0.707.

Before concluding this section on filters, consider a signal composed of two different frequencies as given next:

$$\omega_{in} = \sin 20t + \sin 1000t \tag{10.129}$$

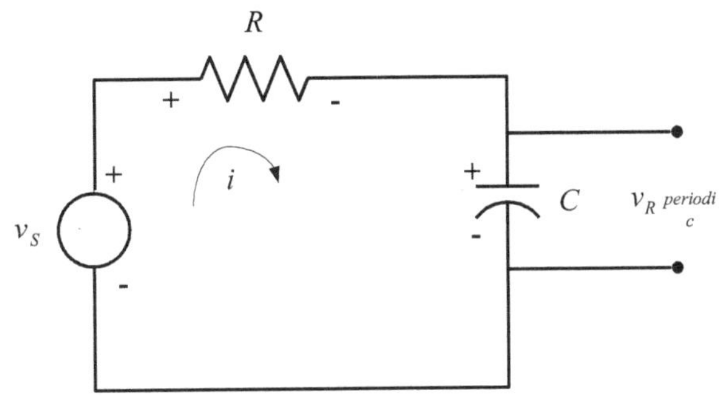

FIGURE 10.47 Voltage drop across a capacitor in an RC circuit.

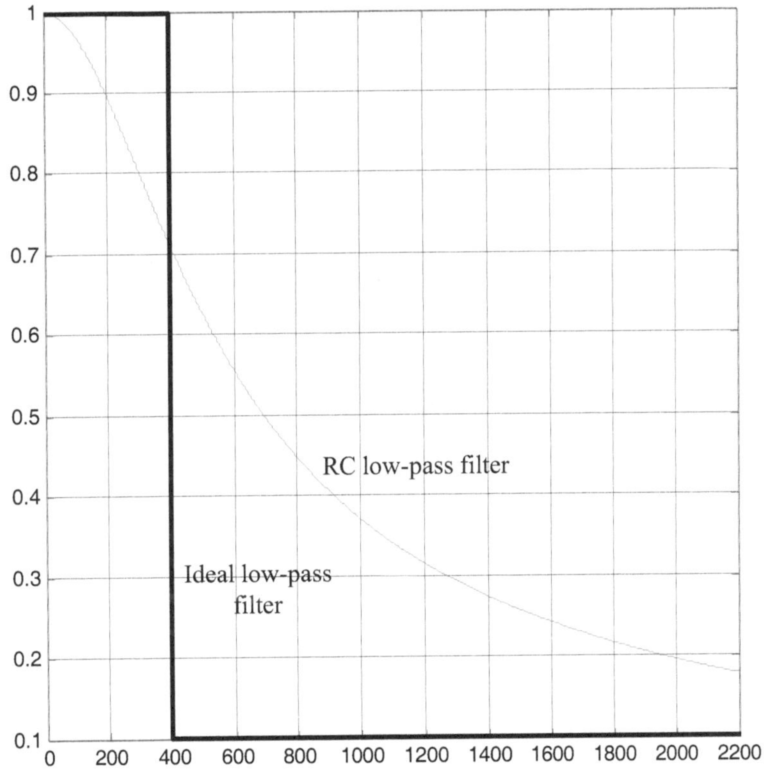

FIGURE 10.48 Performance of an RC circuit as a low-pass filter.

Using an RC circuit with $R = 1000\ \Omega$ and $C = 2.5 \times 10^{-6}\ F$, and thus $\omega_C = 400\ Hertz$, the high-pass filter, the voltage across the resistance, should pass the $\sin 1000t$ portion of the signal, while rejecting the $\sin 20t$ portion; the low-pass filter, the voltage across the capacitor, should pass the $\sin 20t$ portion of the signal, while rejecting the $\sin 1000t$ portion. Figure 10.49 shows the result.

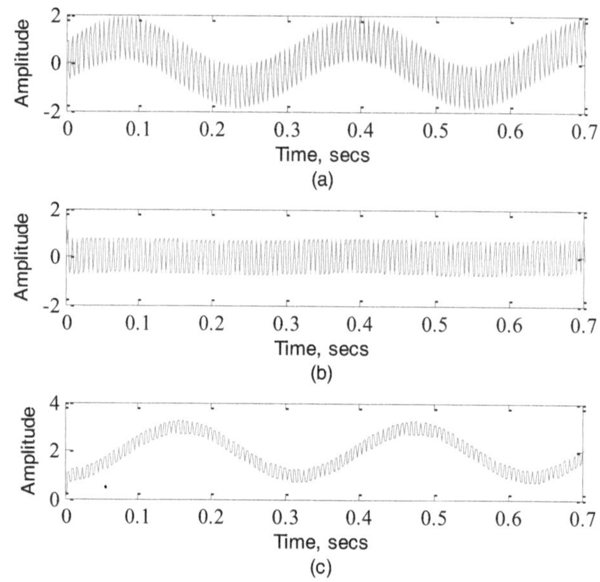

FIGURE 10.49 Performance of an RC circuit as a low-pass filter and as a high-pass filter. (a) Input signal to circuit. (b) Output of high-pass filter. (c) Output of low-pass filter.

10.7 SUMMARY

This chapter has looked into the modeling of electrical circuits. The first step in modeling is the application of KCL and KVL. The element relations are used to complete the modeling. Some moderately complex circuits were analyzed using the mesh and nodal analysis methods. The necessity, and meaning, for defining the initial conditions at "time zero minus" ($t = 0^-$) was presented and used in several examples. The effect of heat dissipation in circuits was also presented and discussed. The effect of this dissipation on the temperature of the circuitry was the topic of an example problem. Finally, a brief but complete use of an RC circuit to approximate high- and low-pass filters was presented.

Problems

Problem 10.1

Figure P10.1 shows an RC circuit. There is no initial charge in the capacitor when the switch closes. Obtain a model, and its analytical solution, that describes the charge in the capacitor and the current around the loop. How do you think the current response will look? Draw a graph with as much indication as possible.

Problem 10.2

Consider the circuit shown in Figure P10.2. The voltage source has exponential decay. The switch is closed at $t = 0$. Develop the model, and obtain its analytical solution, that describes the current in the loop after the switch closes. What is the effect of the inductance L on the response of the current, and why?

Problem 10.3

Figure P10.3 shows an RL circuit. Obtain the model, and its analytical solution, that describes the current around the loop and v_R, and v_L after the switch changes position from 1 to 2. What type of current response do you expect after the switch moves to 2? What will be the final current flow?

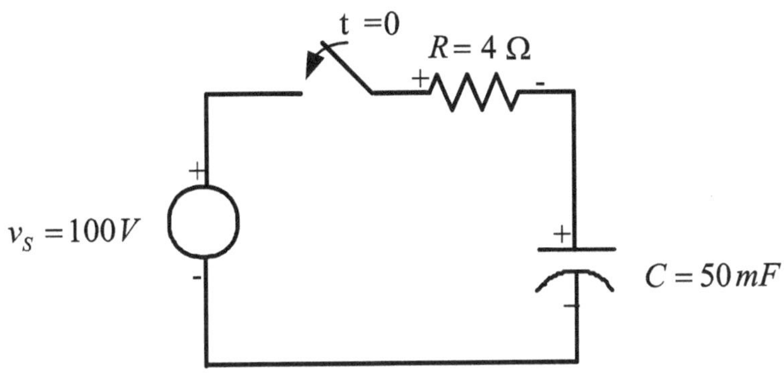

FIGURE P10.1 Circuit for Problem 10.1.

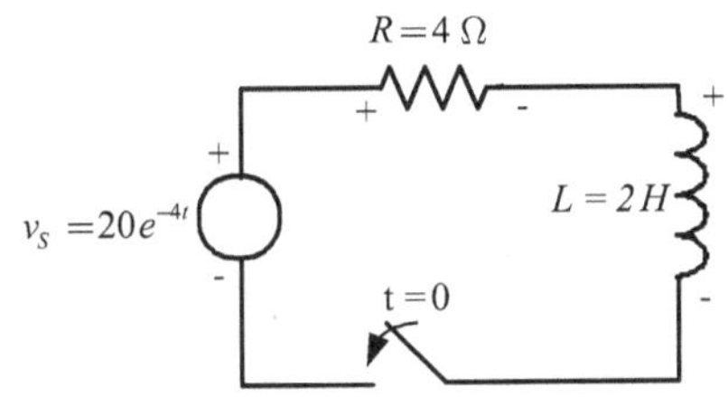

FIGURE P10.2 Circuit for Problem 10.2.

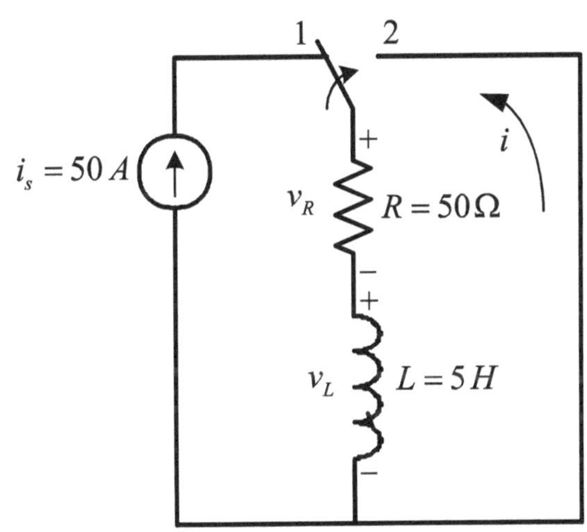

FIGURE P10.3 Circuit for Problem 10.3.

Problem 10.4

Consider the circuit shown in Figure P10.4. At time zero, the capacitor charge and current are both zero.

a. Develop a model (in terms of capacitor charge) and initial conditions that describe this circuit. Will the response be stable or not, and why? Will it be monotonic or oscillatory, and why? What is the final current in the circuit?
b. Obtain the analytical solution of the model.
c. *If you are familiar with simulation, simulate the system and compare the response of the system to that given by the analytical solution.*
d. *What is the effect of changing the capacitor in the circuit? That is, how does the capacitor affect the response of the circuit? Prove/show it.*

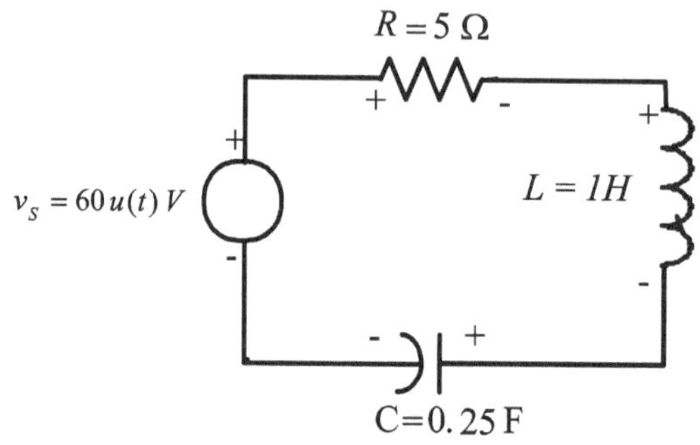

FIGURE P10.4 Circuit for Problem 10.4.

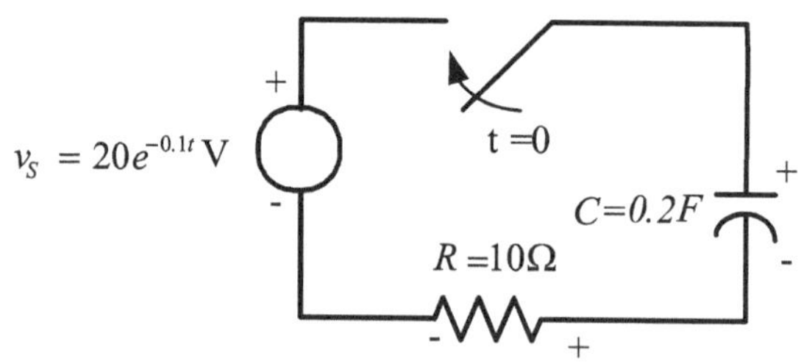

FIGURE P10.5 Circuit for Problem 10.5.

Problem 10.5

Consider the RC circuit shown in Figure P10.5. The voltage source has exponential decay. The capacitor is initially uncharged, and the switch closes at $t = 0$. Develop the model for this circuit in terms of capacitor charge. What is the steady-state final current in the circuit?

Problem 10.6

Consider the RC circuit shown in Figure P10.6. The initial capacitor charge is zero. The voltage of the source is $v_S = 20[1 - u(t-10)]V$. Develop a model for this circuit in terms of capacitor charge q.

Problem 10.7

Consider the RLC circuit shown in Figure P10.7. At time zero, when the switch closes, the current is zero and the capacitor voltage drop is 15 volts. Develop the model (in terms of capacitor charge – not current) and initial conditions that describe the behavior of this circuit. What type of response would you expect, and why?

Problem 10.8

Obtain the models for the currents in both meshes as well as the voltage drop across the capacitor in the circuit shown in Figure P10.8.

a. Without obtaining the solutions, just based on the models, answer the following three questions, Will the responses be stable or unstable, and why? Will they be monotonic or oscillatory, and why? What will be the final value of the current through the capacitor?
b. Obtain the analytical solutions.

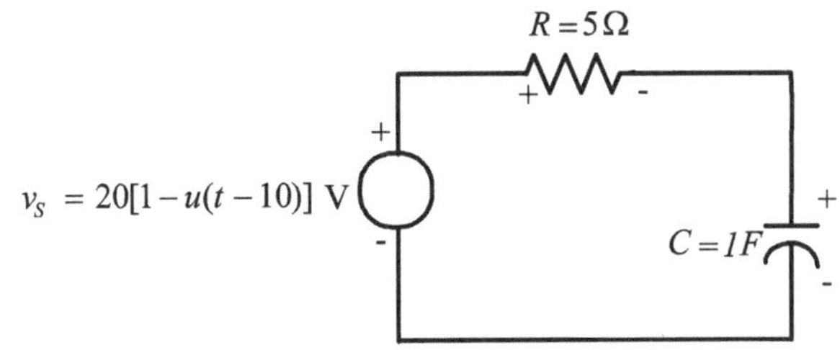

FIGURE P10.6 Circuit for Problem 10.6.

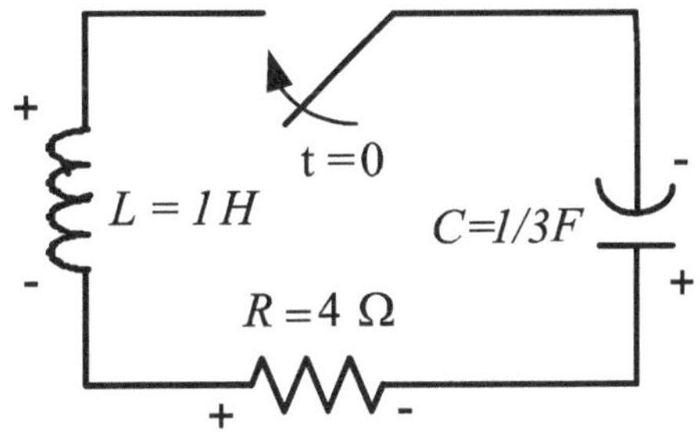

FIGURE P10.7 Circuit for Problem 10.7.

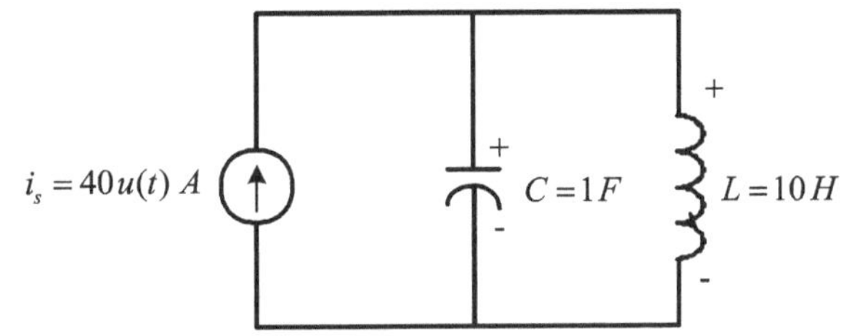

FIGURE P10.8 Circuit for Problem 10.8.

c. *If you are familiar with simulation, simulate the system and compare the response of the system to that given by the analytical solution.*
d. *Compare the final value you get from the simulation and the one you stated in part (a).*

Problem 10.9

Consider the circuit shown in Figure P10.9. Obtain the model and the analytical solution that describe currents i_1 and i_2 after the switch closes.

Problem 10.10

Consider the circuit shown in Figure P10.10. Obtain the models that describe both currents as well as the analytical solutions of both models. What are the currents after a very long time?

Problem 10.11

Consider the circuit shown in Figure P10.11. Obtain the models that describe the current i and the voltage across the inductor v_C for $t \geq 0$. What is the initial and final current in the circuit?

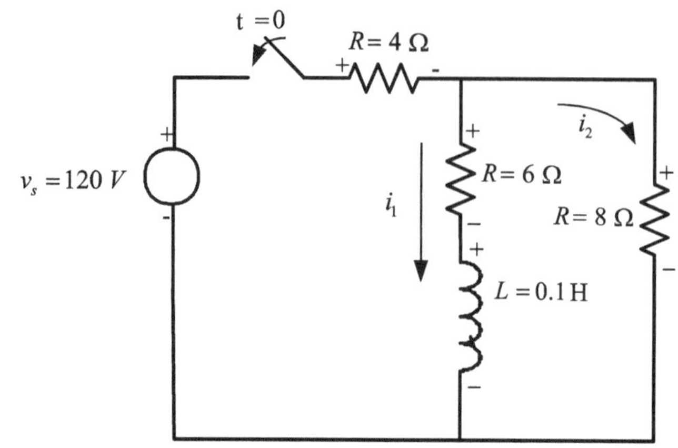

FIGURE P10.9 Circuit for Problem 10.9.

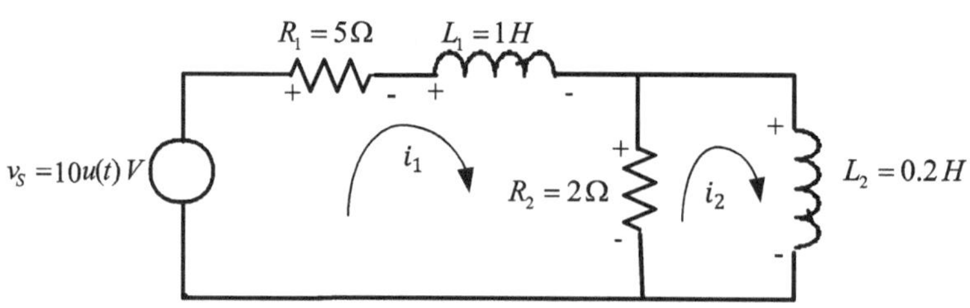

FIGURE P10.10 Circuit for Problem 10.10.

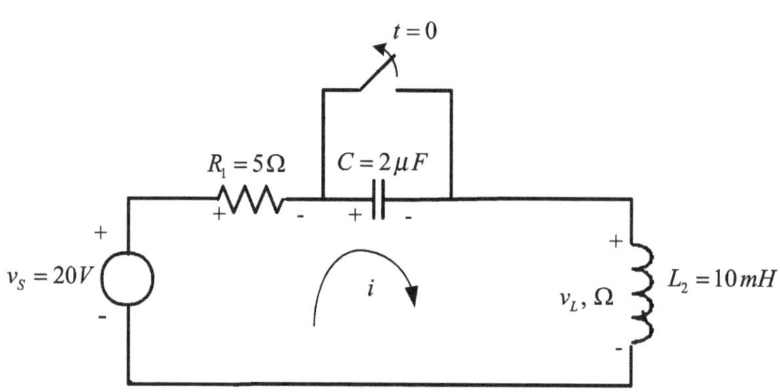

FIGURE P10.11 Circuit for Problem 10.11.

Problem 10.12

Consider the circuit shown in Figure P10.12. Obtain the models that describe the current i and the voltage across the capacitor v_C for $t \geq 0$. State the necessary initial conditions.

Problem 10.13

Consider the circuit shown in Figure P10.13. Initially, there is no charge on the capacitor and the switch is open. At time = 0 the switch is closed. Develop the model that describes the current leaving the "+" terminal of the voltage source.

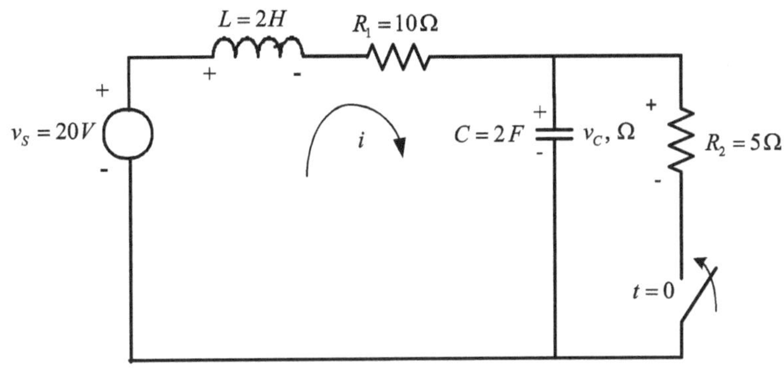

FIGURE P10.12 Circuit for Problem 10.12.

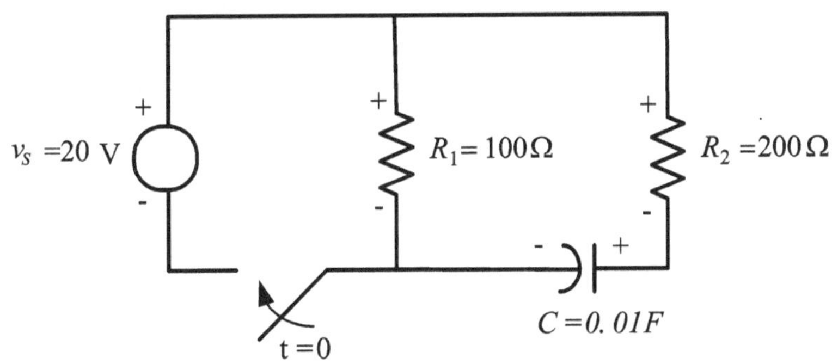

FIGURE P10.13 Circuit for Problem 10.13

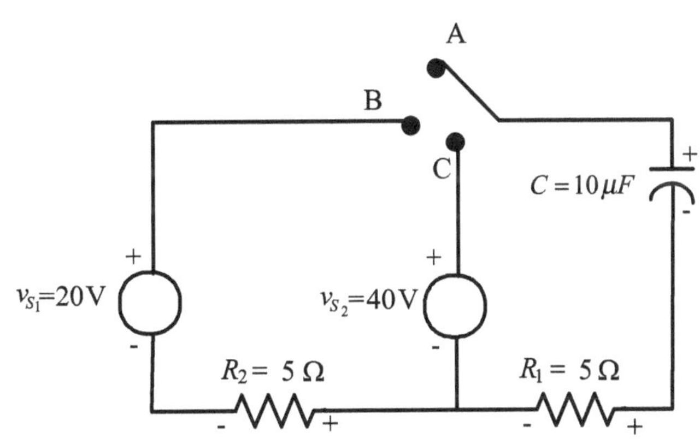

FIGURE P10.14 Circuit for Problem 10.14.

Problem 10.14

Consider the circuit shown in Figure P10.14. The capacitor is initially uncharged with the switch at position A. At $t = 0$, the switch is moved to position B for 20 seconds and then to position C. Develop models that describe the capacitor voltage at all times for $t > 0$. Provide the initial conditions for each model.

Problem 10.15

Consider the circuit shown in Figure P10.15. The switch is originally at point A, and there is no charge in any part of the circuit. The following is the desired operation of the circuit.

a. The switch is moved to point B and the system is allowed to evolve to a steady state of 20 A. Write a model that describes the current in the closed circuit (consisting of the voltage supply, R_1, and L). Decide on the required resistance R_1 to provide a fast response to a current of 20 A. How long does it take to reach the required steady value?
b. Once the current in (a) reaches the steady value, the switch is moved to position C. Up to that moment there has not been any current in the new closed circuit (including L, C, and R_2). But, as soon as the switch moves the position C, there is current equal to the current flowing through L at that moment. Write a model that describes the current, which eventually goes to zero. The operation requires that this decrease be as fast as possible without any oscillation. Decide on the capacitor needed for this requirement.
c. *If you are familiar with simulation, provide a graph showing the two currents in one graph.*

Problem 10.16

Consider the circuit shown in Figure P10.16.

a. The switch is originally open, and the capacitor is uncharged. The switch is closed, and the system is allowed to evolve to a steady state. What is the steady-state current through resistor R_1 and the steady-state voltage across the capacitor?
b. When the system reaches a steady state, the switch is opened. How many seconds will it take for the capacitor voltage to drop to half of its steady-state value?

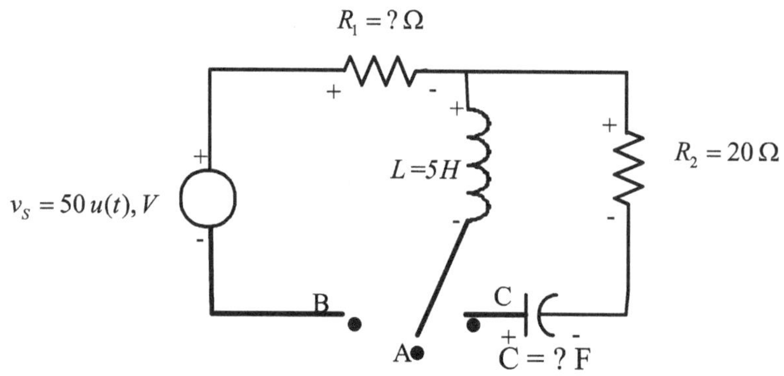

FIGURE P10.15 Circuit for Problem 10.15.

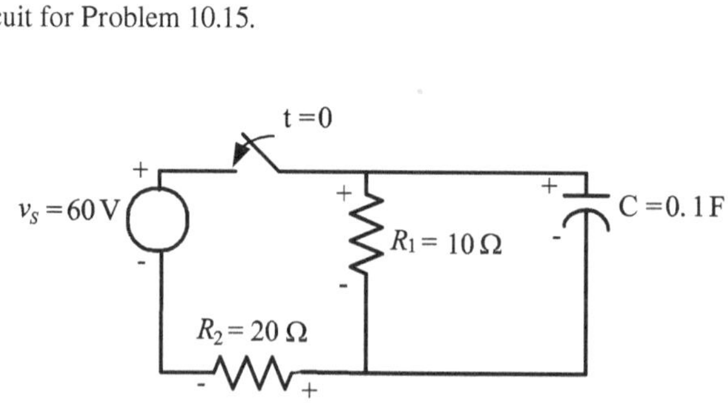

FIGURE P10.16 Circuit for Problem 10.16.

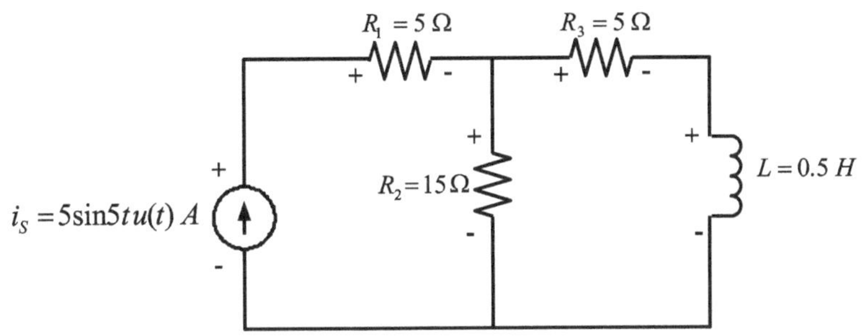

FIGURE P10.17 Circuit for Problem 10.17.

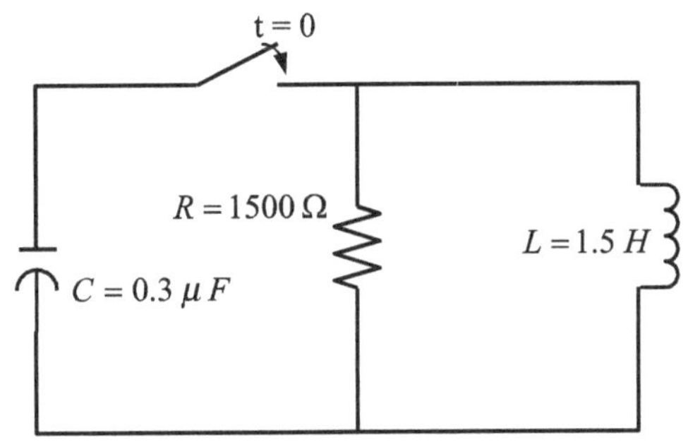

FIGURE P10.18 Circuit for Problem 10.18.

Problem 10.17

For the circuit shown in Figure P10.17, at $t = 0$ the current source goes from $i_S = 0\,A$ to $5\sin 5t\,A$. Develop the model to describe the voltage drop across the current source at $t > 0$. Provide the initial conditions.

Problem 10.18

The initial voltage across the capacitor of Figure P10.18 is 40 V.

a. Develop a model that describes the voltage drop across each branch for $t > 0$.
b. *If you are familiar with simulation, you may use it for the following. It is desired that during the response the voltage at node A does not oscillate. Obtain the resistance* R *that would provide this response. What are the final values of the currents and voltage at node A. Show the responses.*

Problem 10.19

Consider Figure P10.19, the initial current through the inductor and through the resistance is zero. Develop the model that describes the current through the resistance, through the inductor, and the voltage across the inductor for $t > 0$.

Problem 10.20

Consider the circuit shown in Figure P10.20. The voltage source changes from 10 V to 40 V at time $t = 0$. Develop the model that describes the current through each resistance, and the voltage drop across the inductor and the capacitor for $t > 0$. Provide the initial conditions. What will be the final currents through each resistor?

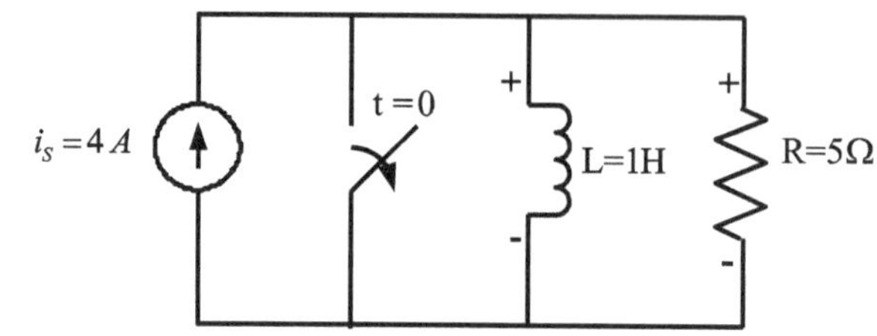

FIGURE P10.19 Circuit for Problem 10.19.

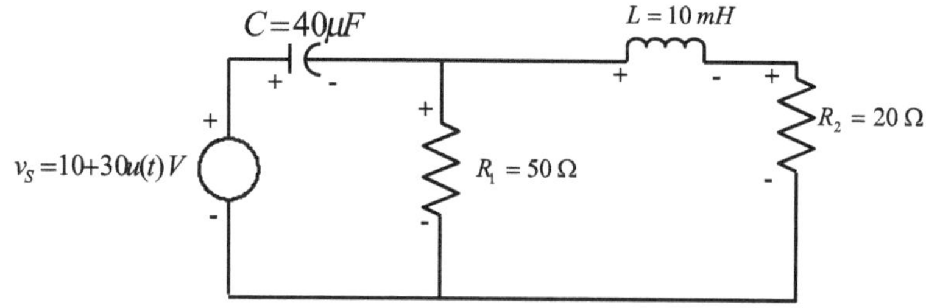

FIGURE P10.20 Circuit for Problem 10.20.

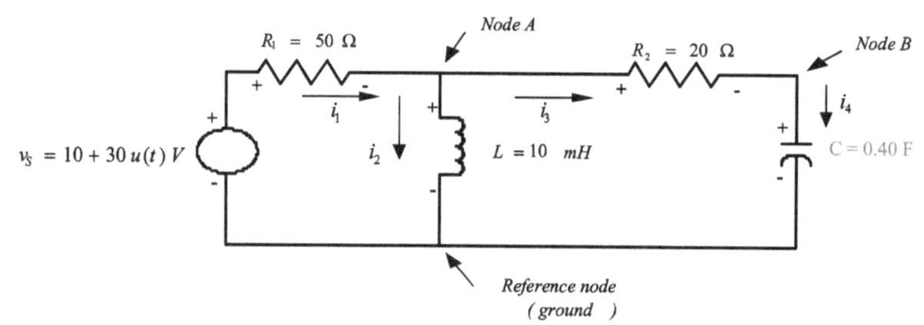

FIGURE P10.21 Circuit for Problem 10.21.

Problem 10.21

Consider the circuit shown in Figure P10.21. The voltage source changes from 10 V to 40 V at time $t = 0.02$. Develop a model that describes the current through each resistance and the voltage drop across the inductor and the capacitor for $t > 0$. Provide the initial conditions.

a. What are the initial currents in each loop?
b. Use the mesh analysis method.
c. Use the nodal analysis method.
d. *If you are familiar with simulation, determine the capacitance required to obtain a voltage drop across R equal or less than 0.1% of the voltage in. v_S This determination involves trial and error, so simulation may be the easiest way to do so.*

Problem 10.22

Find the model that describes the current through resistances R_1 and R_2, and the voltage drop across the inductor in Figure P10.22. The voltage source changes from 20 V to 40 V at $t = 0$, or $v_S = 20 + 20u(t)V$. Provide the initial conditions. What are the final steady-state currents?

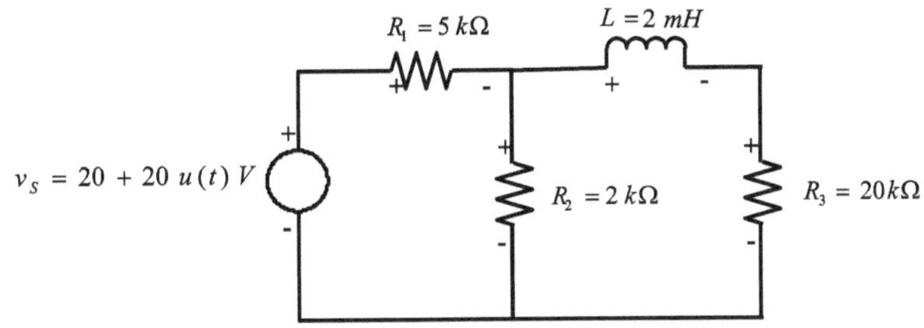

FIGURE P10.22 Circuit for Problem 10.22.

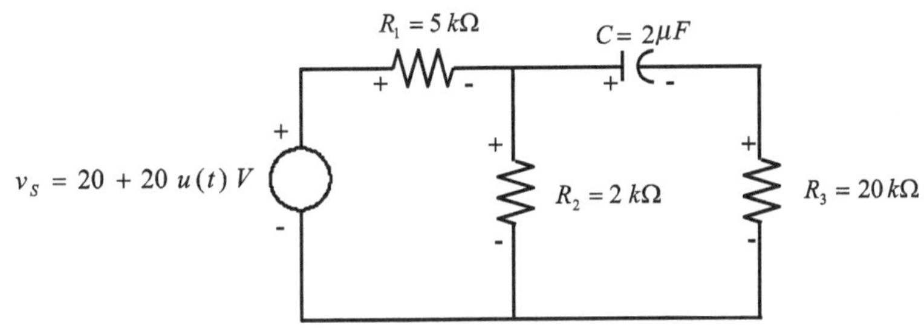

FIGURE P10.23 Circuit for Problem 10.23.

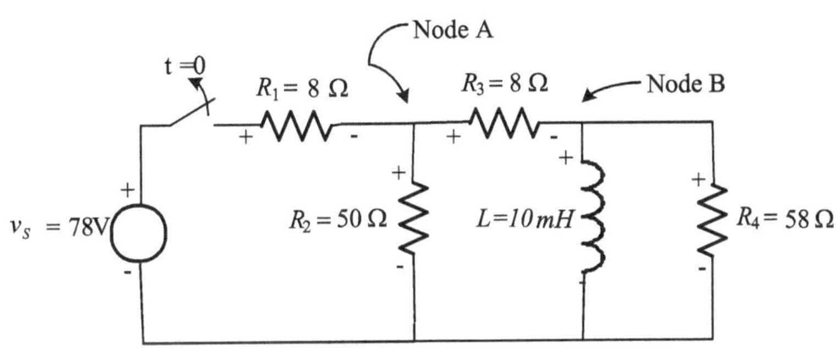

FIGURE P10.24 Circuit for Problem 10.24.

Problem 10.23

Repeat Problem 10.22 for Figure P10.23 (change the inductor to a capacitor).

a. Use the mesh analysis method.
b. Use the nodal analysis method.

Problem 10.24

Consider the circuit shown in Figure P10.24.

a. Obtain the initial currents through all the elements.
b. Obtain the model that provides the voltages at nodes A and B for $t > 0$.
c. What are the final steady state values of the voltages at nodes A and B?

Problem 10.25

Consider Figure P10.25.

a. Develop a model that describes the current through resistance R_2, the inductor, and through resistance R_3, as well as the voltage drop across the capacitor for $t > 0$.
b. What will be the final value of the current through R_1, R_2, and L? What is the final value through the capacitor?

Problem 10.26

Consider the circuit shown in Figure P10.26. Develop models that describe the currents through all the electrical components, obtain their analytical solutions, and obtain the expressions for the total power released by the resistors and the energy stored by the inductor and the capacitor.

Problem 10.27

Consider the circuit shown in Figure P10.27a in which a voltage source v_S supplies current to a load, represented by resistance R. If there is noise in the source, it will show up proportionally in the load voltage v_R. The load may not be able to resist the noise, and therefore, it will be necessary to reduce

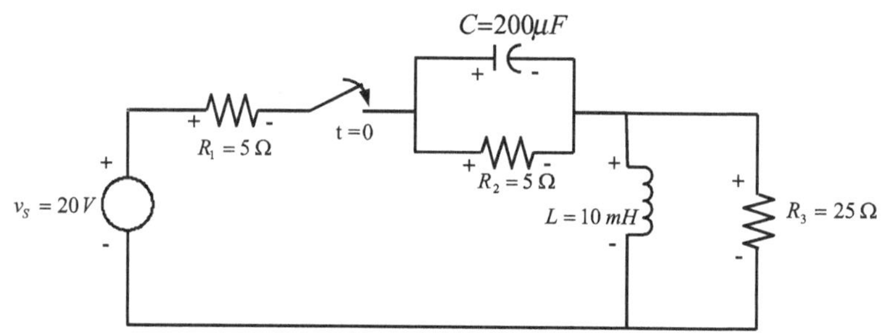

FIGURE P10.25 Circuit for Problem 10.25.

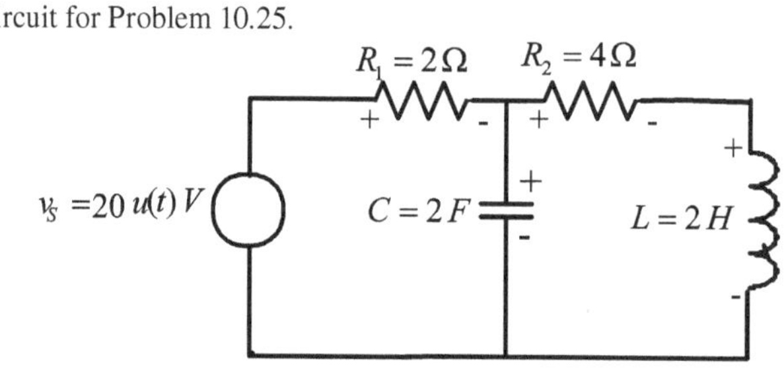

FIGURE P10.26 Circuit for Problem 10.26.

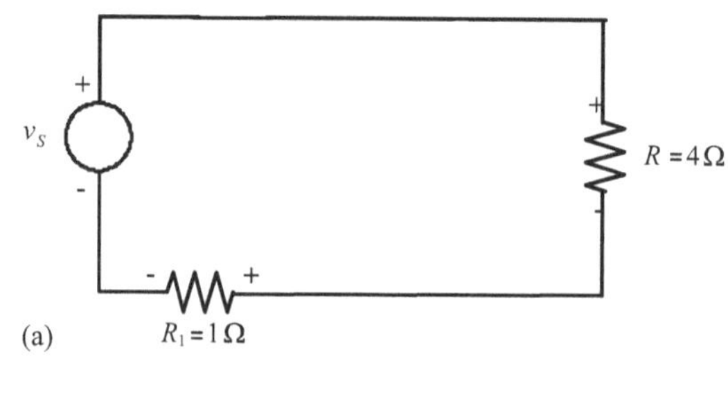

FIGURE P10.27A Electrical circuit.

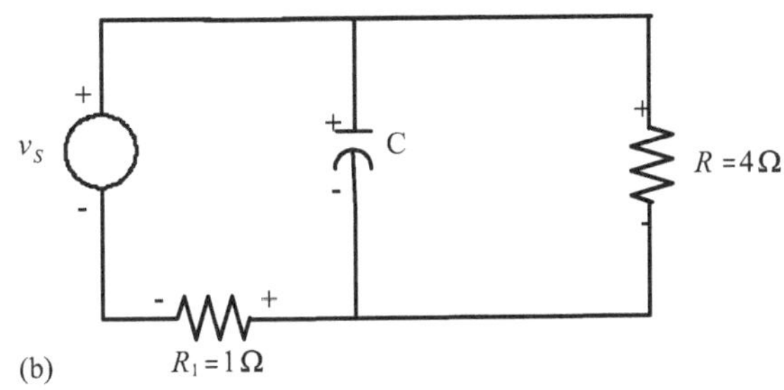

FIGURE P10.27B Circuit to damp noise.

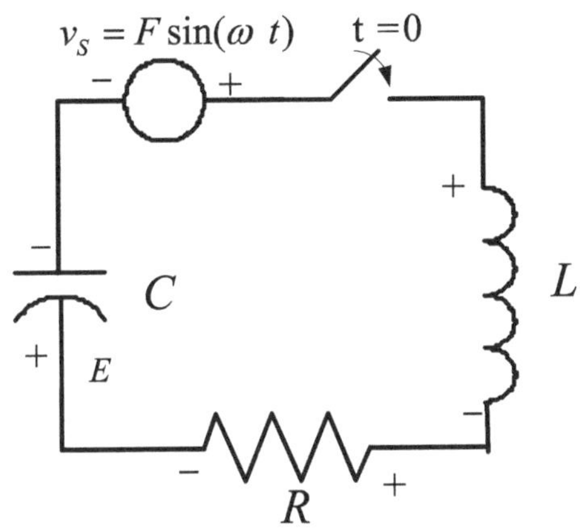

FIGURE P10.28 Electrical circuit for Problem 10.28.

the noise affecting the load. This could be done by adding a capacitor to the circuit as shown in Figure P10.27b. The purpose of this problem is to explore how the capacitor reduces noise. Assuming that the source is noisy and expressed by

$$v_S = 5 + a\sin(\omega t)$$

The noise is characterized by 50 Hz with an amplitude of 1 V; ω must be in radians/s.

a. Write the models that describe the voltage drop across resistor R in Figures P10.27a and P10.27b.
b. Obtain the analytical solution providing the voltage drop across R in both cases and compare the results.
c. *If you are familiar with simulation, determine the capacitance required to obtain a voltage drop across R equal or less than 0.1% of the voltage in v_S. This determination involves trial & error, so simulation may be the easiest way to do so.*

Problem 10.28

Consider the circuit shown in Figure P10.28.
Assuming $C = 0.5F,\ L = 0.5H, R = 1\ \Omega$, and $F = 2V$,

a. Obtain the model of the system that provides the current in the circuit for $t \geq 0$. Before the switch closes, nothing is happening in the circuit, and the capacitor has a charge of 10 C.

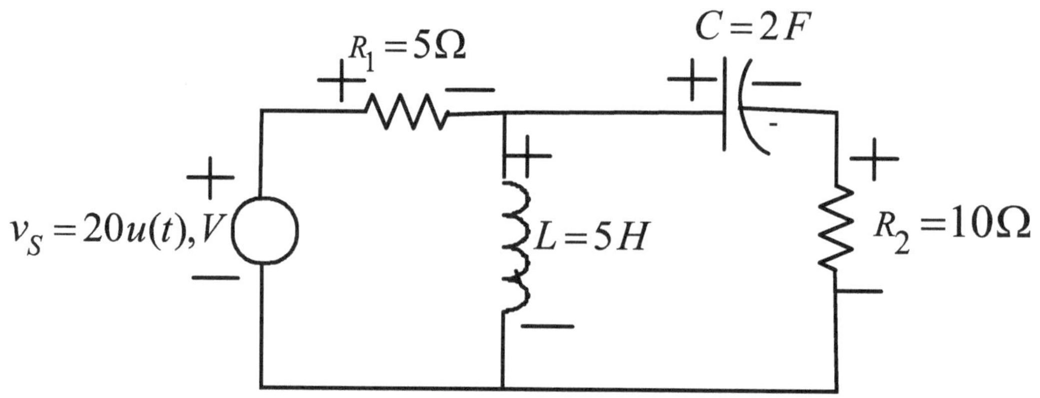

FIGURE P10.29 Electrical circuit for Problem 10.29.

b. Obtain the analytical solution of the model.
c. From the solution, how long it will take for the transient response to die out?
d. Once the response reaches its final steady oscillation (sine wave), its amplitude and phase lag are both function of the forcing function ω (the analytical solution should provide this functionality). As mentioned in Section 10.3.1, a *pseudo-resonance*, also sometimes referred to as a *practical resonance*, occurs at the frequency that produces the largest possible final amplitude. Obtain the value of this frequency.

Problem 10.29

Consider the circuit shown in Figure P10.29. All the necessary initial conditions are zero.

a. Develop a model that provides currents in each mesh.
b. Obtain the solutions of the model that provide the response of each current.
c. Based on the model, will the responses be stable or unstable, and why?
d. What will be the final currents?

11 Numerical Simulation

In this chapter, we cover the numerical simulation of ordinary first- and second-order differential equations. We begin by discussing the difference between a numerical solution to a differential equation and an analytical solution, followed by a discussion of Euler's method, which is the simplest technique for obtaining numerical solutions to differential equations. In practice, more sophisticated techniques are used but are not covered here. Instead, we introduce block diagrams, which are used as a way of entering differential equations into several computer software packages used for their numerical solution. Finally, we illustrate the use of one of these packages, Simulink®, to obtain numerical solutions to several mathematical models developed in earlier chapters.

11.1 NUMERICAL SOLUTION OF DIFFERENTIAL EQUATIONS

Before getting into Euler's method, we should first understand what a numerical solution is and why we might seek one. Until now, your solutions of differential equations have been analytical. As presented in Section 1.5, an *analytical* solution is one in which you actually find the function that satisfies the differential equation. Once you know the function, you can plug in any values of the independent variable and solve for the corresponding value of the dependent variable. For instance, in Chapter 9 we found that the solution to the differential equation

$$\frac{dT}{dt} = -\alpha(T - T_f) \tag{11.1}$$

with the initial condition $T(0) = T_o$, was

$$T = T_f + (T_0 - T_f)\exp(-\alpha t) \tag{11.2}$$

Knowing values of T_f, T_o, and α, we can use Equation 11.2 to find T for any value of t.

With a *numerical* solution, we never obtain the function that satisfies the differential equation. Rather, we obtain values of the dependent variable that correspond to values of the independent variable. In other words, our solution is a table of numbers like in Table 11.1.

These values of T versus t faithfully represent the solution to the differential equation – we just do not know what the function is.

Because the analytical solution actually provides the function, it contains more information and is desired. So why would we ever solve a differential equation numerically? Simply because we do not know how to solve it analytically – this might happen, for instance, if we develop a new form of a differential equation for which no one has worked out a solution method – or there are too many equations in the model to solve them analytically!

As mentioned at the beginning of this section, we will apply a simple technique known as Euler's method to solve ordinary differential equations numerically. The framework for the method is provided in the next section.

DOI: 10.1201/9781003619161-11

TABLE 11.1
Numerical Solution of Equation 11.1 for $T_f = 200$, $T_o = 25$, and $\alpha = 0.1$

t	*T*
0	200
0.1	198.3
0.2	196.5
0.3	194.8
0.4	193.1

11.2 EULER'S METHOD FOR FIRST-ORDER ORDINARY DIFFERENTIAL EQUATIONS

The foundation of Euler's method is the definition of the first derivative, which you should recall from calculus:

$$\left(\frac{df}{dt}\right)_{t=t_0} = f'(t_0) = \lim_{\Delta t \to 0} \frac{f(t_0 + \Delta t) - f(t_0)}{\Delta t} \tag{11.3}$$

This definition is exact in the limit as Δt approaches zero and provides the first derivative of the function *f* at $t = t_0$. If a small value of Δt is used in Equation 11.3 without taking a limit, an approximation to the derivative is obtained – and the approximation becomes better as Δt is made smaller and smaller.

Equation 11.3 in the approximate form (without the limit) is often used to estimate first derivatives of a function from tabular data of *f* versus *t*. Here, we will take a different viewpoint: Suppose we know the derivative of the function and the value of the function at $t = t_0$ and wish to estimate the value of the function at $t_1 = t_0 + \Delta t$. Rearranging Equation 11.3 and dropping the limit yields

$$f(t_1) \approx f(t_0) + \Delta t \cdot f'(t_0) \tag{11.4}$$

Once we obtain $f(t_1)$, we can imagine using a stepwise procedure in which we estimate the function value at $t_2 = t_1 + \Delta t$, at $t_3 = t_2 + \Delta t$, and so on:

$$f(t_2) \approx f(t_1) + \Delta t \cdot f'(t_1) \tag{11.5}$$

$$f(t_3) \approx f(t_2) + \Delta t \cdot f'(t_2) \tag{11.6}$$

The procedure is illustrated with an example.

Example 11.1

Solve Equation 11.1 numerically with $T_0 = 200$ °C, $T_f = 25$ °C, and $\alpha = 0.1$ s^{-1}. Use a step size Δt of 0.1 s and find *T* at 0.3 s.

The differential equation is

$$T' = -0.1(T - 25) \text{ with } T(0) = 200\ °\text{C} \tag{11.7}$$

First Step: $t_0 = 0$, $t_1 = t_0 + \Delta t = 0 + 0.1 = 0.1$, $T(t_0) = 200°C$

Using Equation (11.7) at $t = t_0$,

$$T'(t_0) = -0.1[T(t_0) - 25] = -0.1(200 - 25) = -17.5$$

Using the Euler formula, Equation 11.4,

$$T(t_1) = T(t_0) + \Delta t T'(t_0) = 200 + 0.1(-17.5) = 198.25°C$$

Second Step: $t_1 = 0.1$, $t_2 = t_1 + \Delta t = 0.1 + 0.1 = 0.2$, $T(t_1) = 198.25°C$

Using Equation 11.7 at $t = t_1$,

$$T'(t_1) = -0.1[T(t_1) - 25] = -0.1(198.25 - 25) = -17.33$$

Using the Euler formula, Equation 11.4,

$$T(t_2) = T(t_1) + \Delta t \cdot T'(t_1) = 198.25 + 0.1(-17.33) = 196.52°C$$

Third Step:

$$t_2 = 0.2, t_3 = t_2 + \Delta t = 0.2 + 0.1 = 0.3, T(t_2) = 196.52°C$$

Using Equation 11.7 at $t = t_2$:

$$T'(t_2) = -0.1[T(t_2) - 25] = -0.1(196.52 - 25) = -17.15$$

Using the Euler formula, Equation 11.4,

$$T(t_3) = T(t_2) + \Delta t\, T'(t_2) = 196.52 + 0.1(-17.15) = 194.81°C$$

As you can see, Euler's method is simply the repeated use of the differential equation to calculate derivatives at a particular time followed by use of the Euler formula to estimate the function at a later time.

11.3 EULER'S METHOD FOR SECOND-ORDER ORDINARY DIFFERENTIAL EQUATIONS

We can apply Euler's method to differential equations of higher order as well. No more theory is needed, so we illustrate with an example. We solve a second-order differential equation here. After this, you can probably work out yourself how to apply the method to third-order and higher equations.

Example 11.2

Using a time step of 0.1 s, solve the following differential equation for y at 0.3 s:

$$y'' - 2y' + y = 0 \quad y'(0) = 4 \quad y(0) = -3 \tag{11.8}$$

Right away it looks like we have a problem – if we rearrange the equation to isolate the first derivative to use in Euler's method, there is a second derivative on the right side that we cannot evaluate.

What we do in this case is to define a *new function* x that is equal to the first derivative of y:

$$x = y' \tag{11.9}$$

Noting that y'' therefore equals x', we rewrite Equation 11.8 as

$$x' - 2x + y = 0 \tag{11.10}$$

We then rearrange both Equations 11.9 and 11.10 to place the first derivatives on the left of the equations and add the initial conditions for each function:

$$y' = x \text{ with } y(0) = -3 \tag{11.11}$$

$$x' = 2x - y \text{ with } x(0) = 4 \tag{11.12}$$

Note that we have taken advantage of the fact that $x(0) = y'(0)$ by the definition of x.

In a nutshell, the definition of a new function $x = y'$ has allowed us to replace Equation 11.8, a second-order equation (with two initial conditions), with two first-order equations, each with its own single initial condition. We then apply the Euler formula to both x and y:

$$y(t_1) \approx y(t_0) + \Delta t\, y'(t_0) \tag{11.13}$$

$$x(t_1) \approx x(t_0) + \Delta t\, x'(t_0) \tag{11.14}$$

$$\text{Step 1: } t_0 = 0 \quad \Delta t = 0.1 \quad t_1 = t_0 + \Delta t = 0 + 0.1 = 0.1 \quad x(t_0) = 4 \quad y(t_0) = -3$$

Using the differential equations, Equations 11.11 and 11.12, at $t = t_0$,

$$y'(t_0) = x(t_0) = 4$$

$$x'(t_0) = 2x(t_0) - y(t_0) = 2(4) - (-3) = 11$$

Using the Euler formulas, Equations 11.13 and 11.14,

$$y(t_1) = y(t_0) + \Delta t\, y'(t_0) = -3 + 0.1(4) = -2.6$$

$$x(t_1) = x(t_0) + \Delta t\, x'(t_0) = 4 + 0.1(11) = 5.1$$

$$\textit{Step 2: } t_1 = 0.1 \;\; \Delta t = 0.1 \;\; t_2 = t_1 + \Delta t = 0.1 + 0.1 = 0.2 \;\; x(t_1) = 5.1 \;\; y(t_1) = -2.6$$

Using Equations 11.11 and 11.12 at $t = t_1$,

$$y'(t_1) = x(t_1) = 5.1$$

$$x'(t_1) = 2x(t_1) - y(t_1) = 2(5.1) - (-2.6) = 12.8$$

Using the Euler formulas, Equation 11.13 and 11.14,

$$y(t_2) = y(t_1) + \Delta t\, y'(t_1) = -2.6 + 0.1(5.1) = -2.09$$

$$x(t_2) = x(t_1) + \Delta t\, x'(t_1) = 5.1 + 0.1(12.8) = 6.38$$

$$\textit{Step 3: } t_2 = 0.2 \;\; \Delta t = 0.1 \;\; t_3 = t_2 + \Delta t = 0.2 + 0.1 = 0.3 \;\; x(t_2) = 6.38 \;\; y(t_2) = -2.09$$

Using Equations 11.11 and 11.12 at $t = t_2$,

$$y'(t_2) = x(t_2) = 6.38$$
$$x'(t_2) = 2x(t_2) - y(t_2) = 2(6.38) - (-2.09) = 14.85$$

Using the Euler formulas, Equation 11.13 and 11.14,

$$y(t_3) = y(t_2) + \Delta t\, y'(t_2) = -2.09 + 0.1(6.38) = -1.452$$

$$x(t_3) = x(t_2) + \Delta t\, x'(t_2) = 6.38 + 0.1(14.85) = 7.865$$

The last calculation for $y(t_3)$ provides the desired result.

11.4 STEP SIZE

In both of the examples, the value of the step size Δt to use in Euler's method was given. Normally, you will have to specify this value yourself. Given that the Euler formula shown in Equation 11.4 is essentially a Taylor series truncated after the linear term, you probably have the sense that if Δt is too large, the solution will not be accurate. However, if it is very small, the number of steps needed to achieve a certain value of t will be very large. Obviously, there is a trade-off between accuracy and labor. This is illustrated in Figure 11.1 for the problem of Example 11.1. The points represent the exact solution to the differential equation, given by Equation 11.2. The dashed line represents the application of Euler's method with $\Delta t = 5$ s and the solid line represents Euler's method with $\Delta t = 0.5$ s. Here, it can be seen that the smaller step size gave results that are almost identical to the exact (analytical) solution. However, it required 120 steps, as opposed to just 12 steps for the step size of 5 s.

So how does one choose a step size? Comparing the numerical solution to the analytical solution is not really an option because, if we had the analytical solution, we would not have to solve the problem numerically! Using the same number for every problem will not work either – problems where the solution is a very steep function of the independent variable will require smaller step sizes than problems for which the solution varies only mildly with the independent variable.

One reasonable approach is to assume a value for the step size and find the numerical solution. Then, solve the problem again using a step size one-tenth as large as the first time. One can continue to decrease the step size until the obtained solutions for two sequential step sizes are equal to within a desired tolerance. At this point, you can be reasonably confident that further decreases in the step size are not needed.

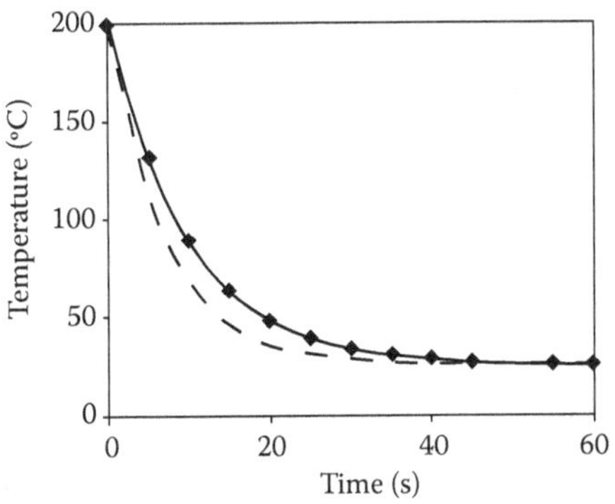

FIGURE 11.1 Illustration of the effect of step size in Euler's method using the differential equation of Example 11.1.

11.5 MORE SOPHISTICATED METHODS

Euler's method as described earlier is the simplest procedure for solving differential equations numerically. However, it is not used much in practice because the first-order Taylor series approximation requires small step sizes to obtain high accuracy. Our reason for presenting it here is to give the reader a basic appreciation of how numerical solutions are obtained.

You might imagine that constructing methods in which higher-order terms in the Taylor series expansion are used would be possible. This is essentially what more sophisticated methods do – either directly or, as in the case of *Runge–Kutta* methods, by matching the equations of the method to higher order Taylor series. As expected, these methods tend to have greater accuracy with the same step size than the first-order Euler method or the same accuracy with a larger step size.

We will not delve into the Runge–Kutta methods here. Instead, we will focus our attention on how to use software that has these methods built into them.

11.6 REPRESENTATION OF DIFFERENTIAL EQUATIONS BY BLOCK DIAGRAMS

In this chapter, we solve differential equations numerically using Simulink. There are other software packages that do the same thing, but most of them require that the differential equation be represented by a block diagram. In other words, the block diagram is how the differential equation and initial conditions are input into the computer. The focus of the rest of this chapter is on how to create these block diagrams. We will begin by showing various blocks and describing what they do. Then we address how to place them to form a block diagram.

11.6.1 Basic Blocks

Here we show the most commonly used blocks. We will learn more as the chapter proceeds. The blocks below either operate on one or more input functions, are source blocks that do not have inputs, or are display or graph blocks that do not have any outputs. Although the specific blocks shown here are used in Simulink, other software packages have comparable blocks.

11.6.1.1 The Integrator Block

The integrator block

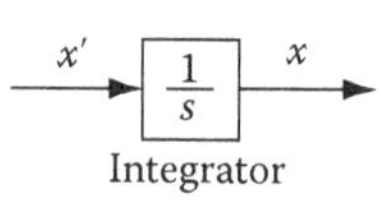

Diagram A

The integrator block is the workhorse block of any software package. The output of an integrator block is the integral of whatever function is input to it. For instance, if the input is x', the output is x. If the input is x'', the output is x'. The integrator requires the initial value of the output function so, for instance, if you input x'' to an integrator to get x', you must supply $x'(0)$. The initial condition to the integrator is specified by double-clicking on the block and inputting the value on the dialog box that comes up.

11.6.1.2 The Sum Block

The sum block

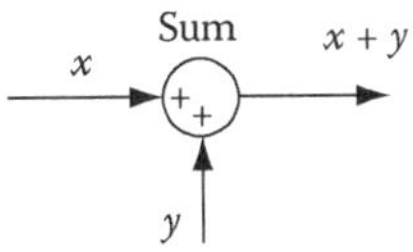

Diagram B

The sum block adds the input functions together and outputs the sum. If you want to add more than two functions at once, you can double-click on the block and add more input ports in the dialog box that appears. Also, if you wish to subtract a function from others, you can change the + sign on its input port to a minus sign using the dialog box.

11.6.1.3 The Product Block

The product block

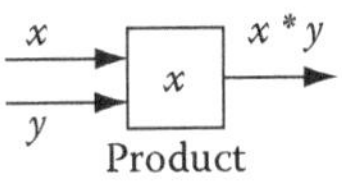

Diagram C

The product block multiplies two or more input functions together and outputs the product. Like the sum block, the number of input ports may be changed if you wish to multiply more than two functions at once.

11.6.1.4 The Gain Block

The gain block

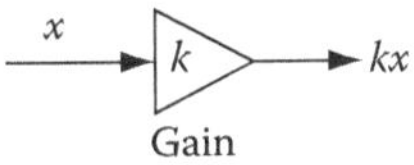

Diagram D

The gain block multiplies an input function by a constant k. This constant is also called the *gain*. You can specify a value for the gain k by double-clicking on the block and entering the value in the dialog box that appears. One source of confusion to students is the difference between the gain block and the product block. The product block is used to multiply two *functions* $x(t)$ and $y(t)$ to get $x(t) * y(t)$. The gain block is used to multiply a function $x(t)$ by a *constant* k to get $kx(t)$.

11.6.1.5 The Math Function Block

The math function block

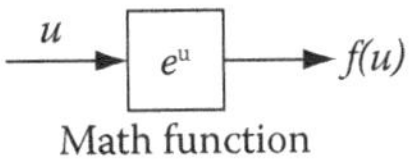

Diagram E

The math function block allows you to take logarithms, square roots, reciprocals, absolute values, and so on, of an input function and also allows you to calculate exponentials and raise the input to a power. You select which operation you want the block to perform by double-clicking on the block and selecting from a drop-down menu.

All the blocks described so far deal with *operations* on an input function. There are also blocks (called *source blocks*) that allow you to specify parameters and forcing functions or import the value of the independent variable (time). We discuss two here, and several more at the end of this section. Source blocks have no input – only an output.

11.6.1.6 The Constant Block

The constant block

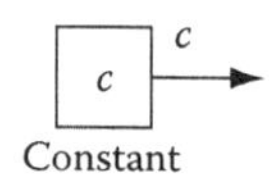

Diagram F

The constant block is used to introduce the value of a constant C. The value of the constant is specified by double clicking the block and typing the value of the constant in the dialog box. The output of the block is simply the value of that constant.

11.6.1.7 The Clock Block

The clock block

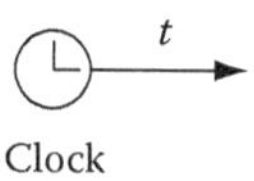

Diagram G

The output of the clock block is the instantaneous value of the simulation time. It is useful primarily when the independent variable (time) appears explicitly in the differential equation. We see its use in Example 11.4.

We need a way, or blocks, to output or graph results; Simulink refers to these as "sink" blocks. Next, we show a couple; there are others.

11.6.1.8 The Display Block

The display block

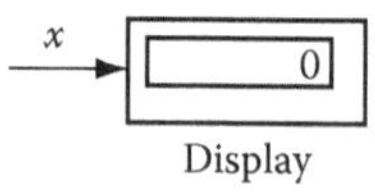

Diagram H

The display block displays the value of the input variable. There is no specification necessary, just input the desired variable.

11.6.1.9 The Scope Block

The scope block

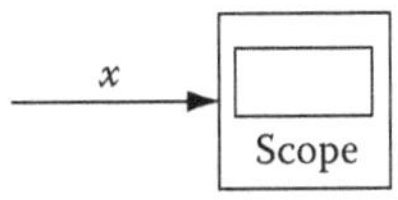

Diagram I

The scope graphs the input variable versus time. By double-clicking on the block, the user can specify the number of input variables, and the scale of each graph, or select auto scaling.

11.6.2 Guidelines for Constructing Block Diagrams

Now that when we have some blocks to work with, we need to learn how to connect them to generate a block diagram.

Constructing block diagrams is as much an art as it is a science. There are usually many different ways to construct a block diagram for a differential equation. Ten students might have 10 different block diagrams and still get the correct solution. Because you are just getting started, however, we recommend using the following guidelines. We illustrate these with an example.

Example 11.3

Construct a block diagram for $4x''+2x'+3x=0$

a. Rearrange the differential equation so that the highest-order derivative is on the left side and everything else is on the right. For instance, for the differential equation of this example write it as

$$x''=\left(-2x'-3x\right)/4$$

b Put down as many integrators as are needed. This is equal to the order of the differential equation. The differential equation is a second-order equation, so we need two integrators:

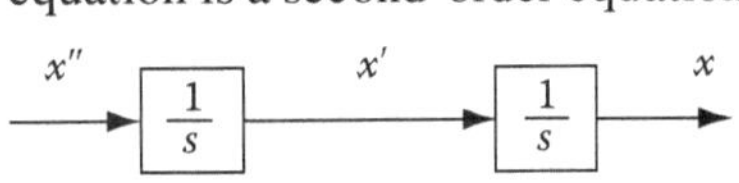

Diagram J

c. Construct the right side of the rearranged differential equation and feed it into the first integrator. Look at the rearranged differential equation, where x'' is on the left and everything else is on the right. This tells you what goes into the first integrator because the input to the first integrator is x'', and the equation tells you what x'' is. For this particular example, we begin by noting that we need $-2x'$ and $-3x$. We can get these using gain blocks:

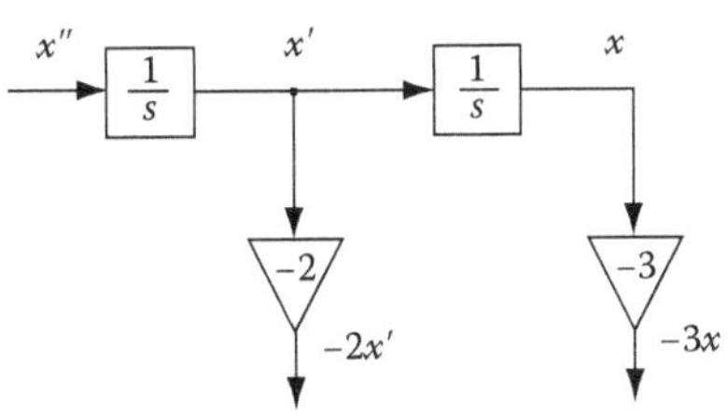

Diagram K

Now we can get $-2x' - 3x$ using a sum block.

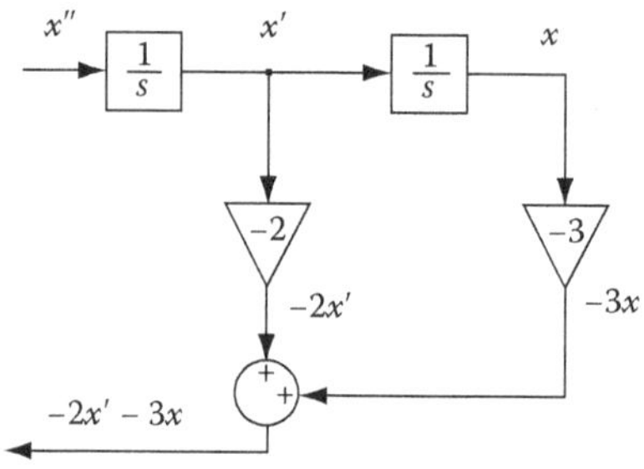

Diagram L

Note that we could have used gains of 2 and 3 instead of –2 and –3 if we had changed both + signs in the sum block to – signs.

Finally, we can get $(-2x' - 3x)/4$ by using another gain block. Instead of dividing by 4, we are multiplying by 1/4. Note from the rearranged differential equation that $(-2x' - 3x)/4$ is, in fact, equal to x''. Because x'' is the input to the first integrator, we put the output from this final gain block into the input for the first integrator:

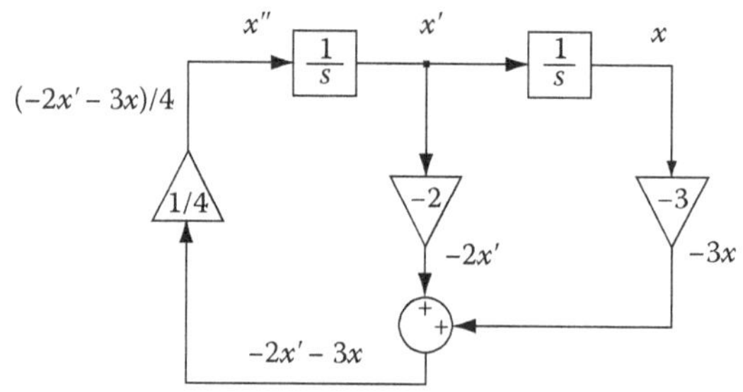

Diagram M

This is virtually a complete block diagram for this differential equation.

11.6.3 Some Additional Examples

Example 11.4

Construct a block diagram for the differential equation $y' - e^{-2t}y = 0$.

First, we rearrange the differential equation to isolate the highest derivative on the left-hand side:

$$y' - e^{-2t}y$$

Next, put down as many integrators as needed. This is a first-order equation, so we only need one:

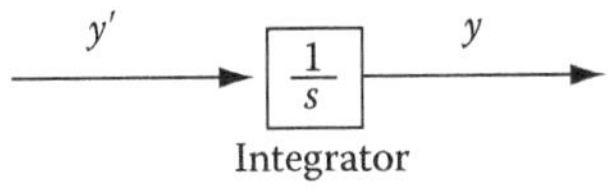

Diagram N

Now, we look at the rearranged differential equation because it tells us what gets fed into the first (in this case, only) integrator. Specifically, it tells us to construct blocks for $e^{-2t}y$ and feed that to the integrator. In the previous example, the independent variable t did not appear explicitly in the differential equation. Here, it does. When this happens, you can import it using a clock block. In this case, we can use a clock and gain its output by −2 to obtain $-2t$, followed by using a math function block (selecting exponential from the drop-down menu) to obtain e^{-2t}:

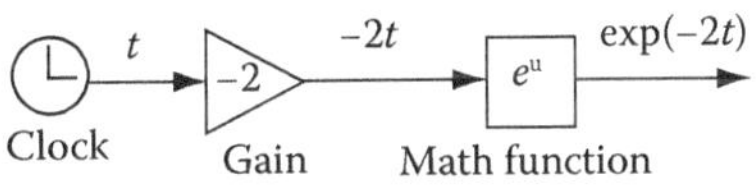

Diagram O

Now we can use a product block to multiply e^{-2t} and y to obtain ye^{-2t}. This is, in fact, y', so it gets fed into the integrator:

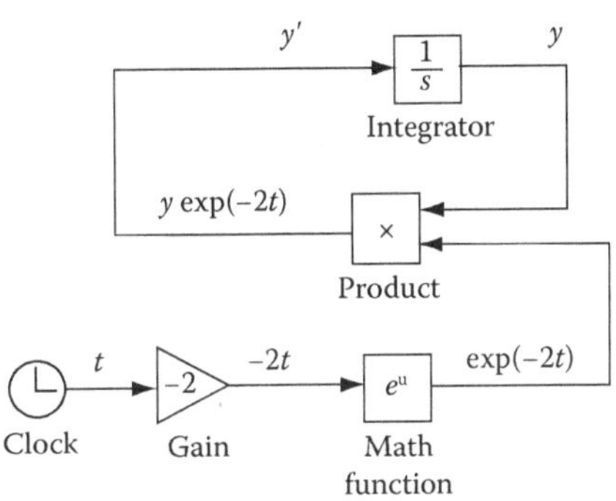

Diagram P

Example 11.5

Construct a block diagram for the coupled system:

$$x'(t)+x(t)=3y(t)$$
$$y'(t)-y(t)=2x(t)$$

Here, the functional notation is used to emphasize that there are *two* dependent variables x and y, both depending on the single independent variable t. Not only that, both dependent variables appear in both equations. As a result, these equations are *coupled*. When this happens, they cannot be solved separately, whether analytically or numerically. Because they are coupled, a *single block diagram* must be used to represent both equations.

The rules are still the same, however – rearrange the equations to get the highest derivative on the left side and put down as many integrators as needed.

$$x'(t) = 3y(t) - x(t)$$

$$y'(t) = 2x(t) + y(t)$$

Because these are first-order equations, we need one integrator for each – for a total of two:

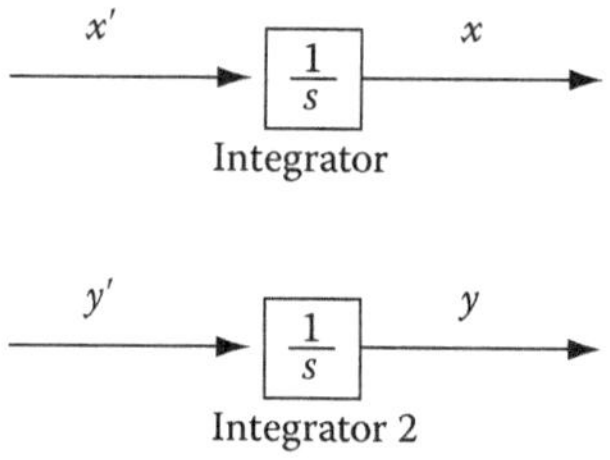

Diagram Q

Now we use blocks to construct the right-hand side of the rearranged equations and feed them into the appropriate integrators. The equations tell us we need $3y$ and $2x$, so we begin by using gain blocks:

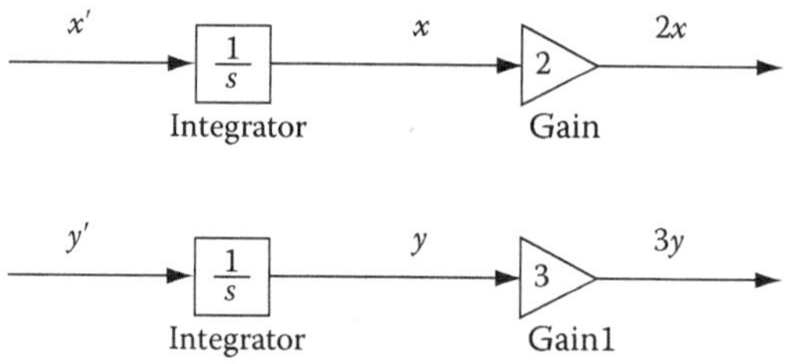

Diagram R

The first equation tells us that the feed into the x integrator is $3y - x$. So we construct that with a sum block and feed it to the x integrator. Note the use of the minus sign in the sum block:

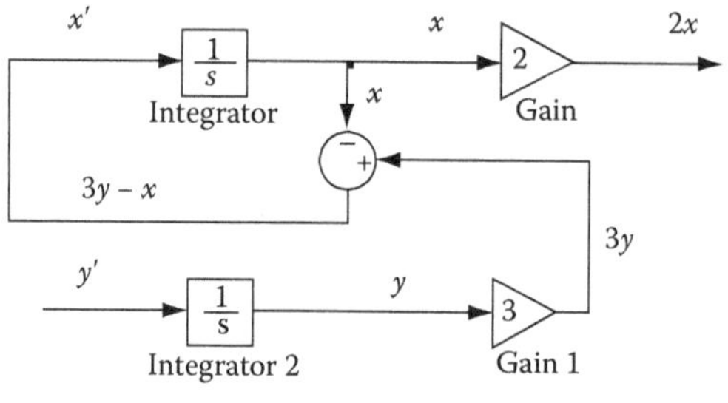

Diagram S

The second equation tells us that the feed to the y integrator is $2x + y$. So we construct that with a sum block and feed the result to the y integrator. This completes the diagram:

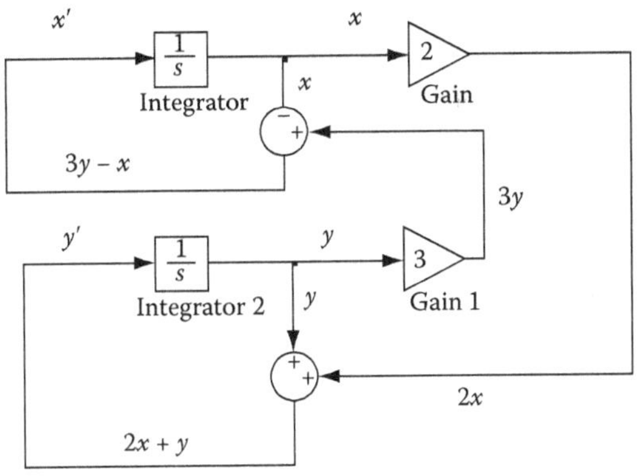

Diagram T

Let's end this section with a simple example to make a point.

Example 11.6

Construct a block diagram for the differential equation $y''(t) = -9.8$.

The equation is already arranged so that the highest derivative is on the left. The order of the equation is 2, so we need two integrators:

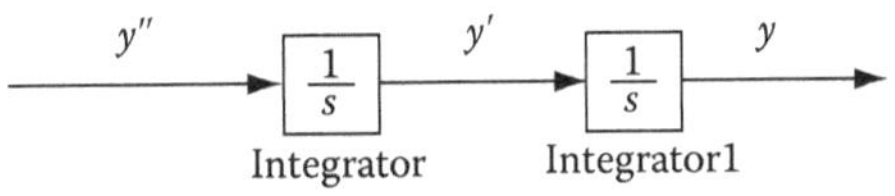

Diagram U

The right-hand side of the differential equation tells us that $y'' = -9.8$. So what goes into the first integrator? Simply a constant of −9.8. You can import a constant using a constant block:

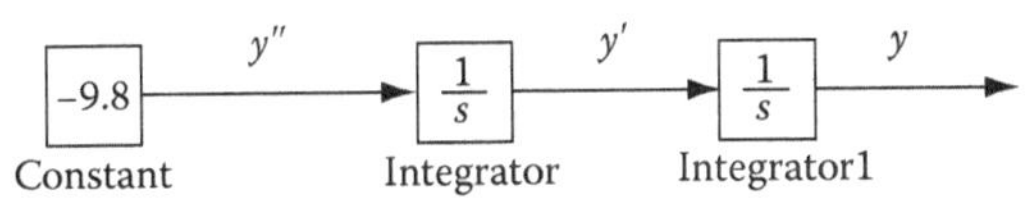

Diagram V

Notice that there is no feedback loop in this block diagram. Students often ask when a block diagram will have such a loop. The answer is really simple: Only when the rearranged equation tells you that you need one. In other words, if there are lower derivatives, or the independent variable itself, on the right side of the rearranged equation, then the input to the first integrator (the left side of the equation) is going to require information that comes out of a subsequent integrator. Hence, there will be a feedback loop in the block diagram.

11.6.4 Some Additional Source Blocks

There are a couple of additional "source" blocks worth mentioning. *Source* blocks have no inputs, and they are used to supply information to the block diagram. We have already seen two examples of source blocks earlier: The clock block (which outputs the current simulation time t) and the constant block (which outputs a constant C). Several more are given later.

Often, we wish to examine the response of a system to a forcing function. This forcing function may take different forms depending on what we wish to do. For instance, we may wish to study the temperature response of a ball bearing when its environmental temperature changes suddenly from one constant value to another. Or we may wish to study the output flow rate of a storage tank when the inlet flow to the tank undergoes random variations. Both of these forcing functions, and many others, can be modeled in Simulink using the blocks discussed in the following sections.

11.6.4.1 Step Block

The icon for the step block is shown on the left and its output on the right:

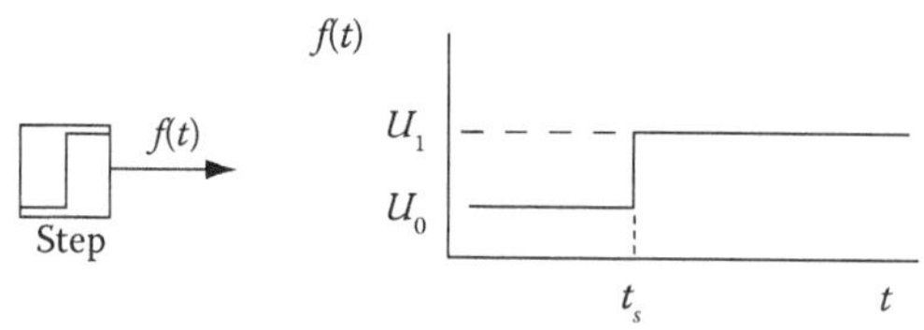

Diagram W

In a nutshell, the step block allows you to create a forcing function that changes from an initial value of U_0 to a final value of U_1 at time t_s. As you have seen, step forcing functions are commonly used by engineers to test the dynamic response of a system. Values of U_0, U_1, and t_s are specified by double-clicking on the block and typing their values into the form that comes up.

11.6.4.2 Sine Wave

Again, the icon for this block is shown on the left and its output on the right.

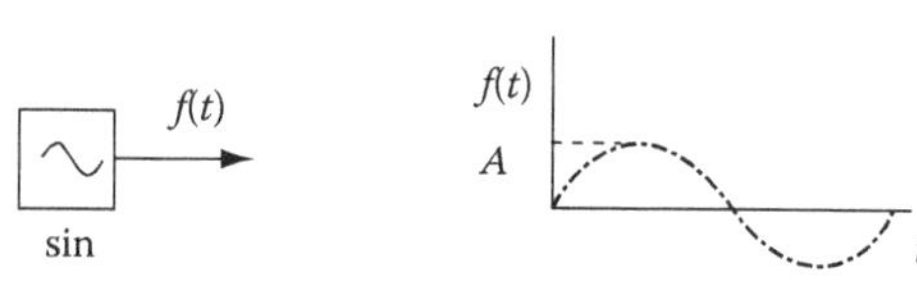

Diagram X

The amplitude A and frequency of the sine wave are specified by double-clicking on the block and typing their values into the form. If you wish to shift the sine wave, you can do that by typing a value into the *phase* box of the form. For instance, you can make the sine wave into a cosine wave by typing 1.5708 into this box. This is $\pi/2$, and it shifts the sine wave by 90°, making it a cosine function.

Two other useful source blocks, the ramp and the random number generator, are described below:

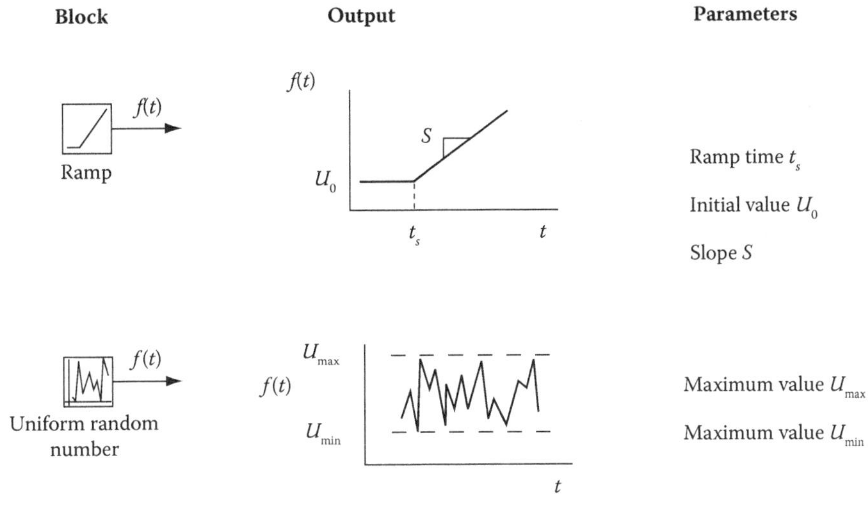

Diagram Y

11.7 ADDITIONAL EXAMPLES

The following examples present interesting aspects of physics and simulation alike. The reader is strongly encouraged to study them and the discussions following each.

Example 11.7

Let us take Example 6.1 and simulate the model. These equations are similar to those in Example 11.5 in that they are coupled, that is, they depend on each other. Section 4.5 shows the analytical solution of this model; here we simulate it.

Following the first step stated in Example 11.3, we rearrange the differential equations so that the highest-order derivative is on the left side and everything else is on the right. From Equations 6.36 and 6.37,

$$\frac{d^2x_1}{dt^2} = \frac{1}{3}\left[60x_2 + 5\frac{dx_2}{dt} - 5\frac{dx_1}{dt} - 120x_1\right] \tag{11.15}$$

$$\frac{d^2x_2}{dt^2} = \frac{1}{5}\left[f_A(t) + 60x_1 + 5\frac{dx_1}{dt} - 5\frac{dx_2}{dt} - 60x_2\right] \tag{11.16}$$

The second step calls for putting down as many integrators as are needed for each equation.

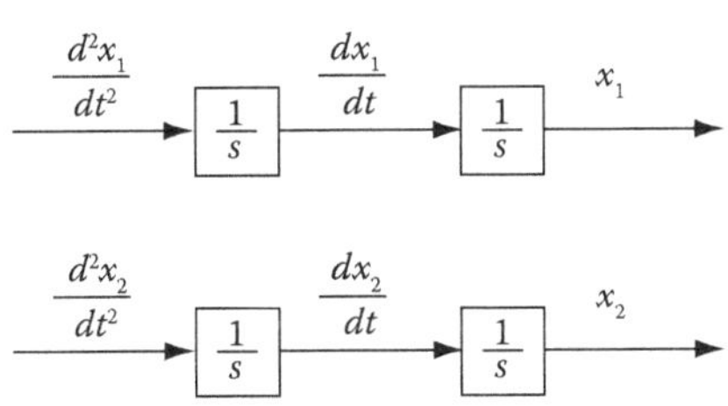

Diagram Z

Finally, the third step calls for constructing the right side of the rearranged differential equations and feeding them into the corresponding first integrator; Figure 11.2 shows the diagram. The scopes in the figure indicate that the displacements x_1 and x_2 of the carts, their first derivatives (or velocity) dx_1/dt and dx_2/dt, and the forcing function $f_A(t)$ are graphed. Figure 11.3 shows the displacement of each cart when $f_A(t)$ is applied.

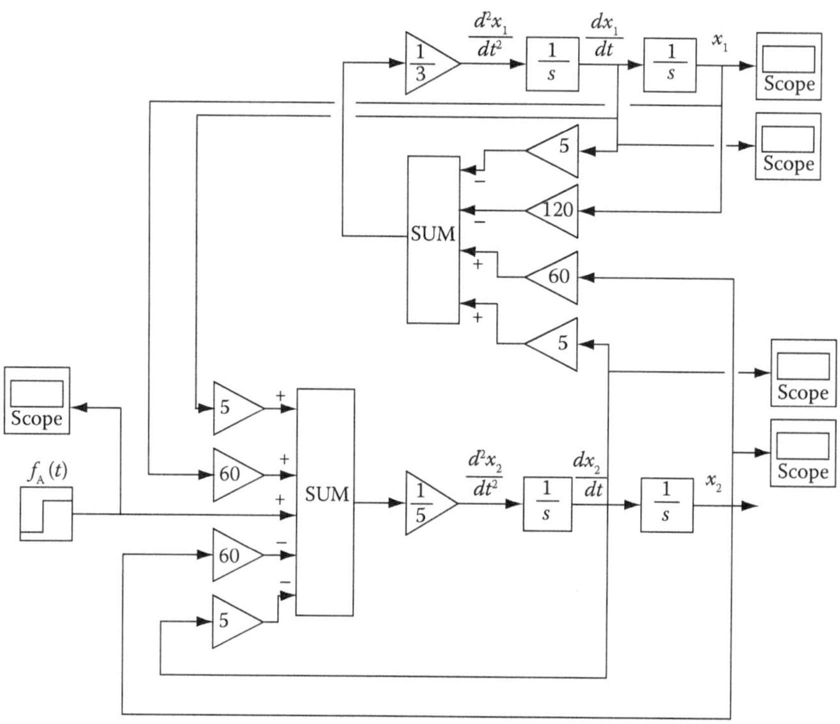

FIGURE 11.2 Diagram for the simulation of Equations 6.36 and 6.37

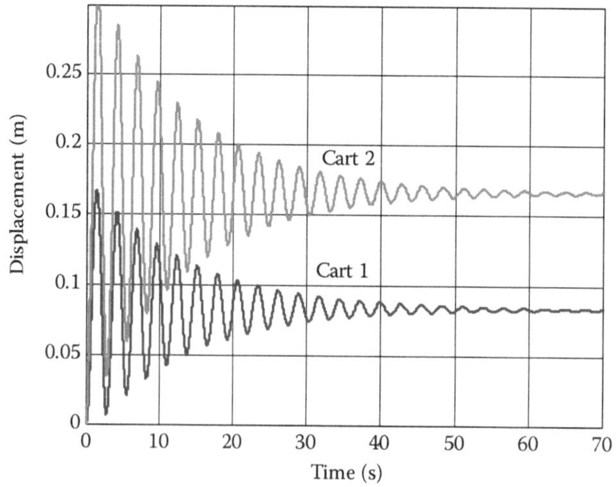

FIGURE 11.3 Response of the carts of Example 11.7

Figure 6.17 shows the response of the carts given by the analytical solution in Equations 4.52 and 4.53. It is interesting to compare the responses obtained from the simulation, Figure 11.3, with those obtained from the analytical solutions. A convenient comparison is to calculate the percentage difference between the responses; we calculate this difference as, Difference = $\left(x_{analytical} - x_{simulation}\right)\times 100 / x_{analytical}$. Figure 11.4 shows the comparison. The figure shows a minimum difference for both carts, much less than 0.2%. If the responses were graphed on top of each other, it would be impossible to differentiate between them.

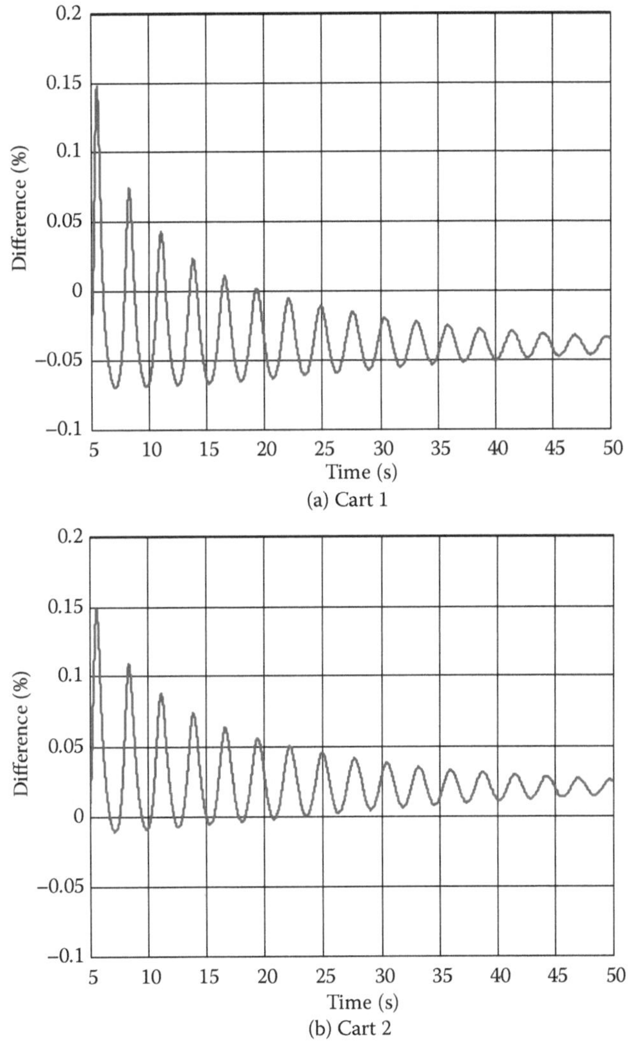

FIGURE 11.4 Difference (%) between the responses obtained from the analytical solution and simulation.

Example 11.8

Let us simulate now the tank of Example 8.1; the model is given by

$$\rho A\frac{dh}{dt} + 1.5\rho\sqrt{11.32 + 9.8h} = \rho f_1 = \rho\left[10 + 2u(t)\right] \tag{8.32}$$

Following the procedure presented to construct the block diagram, we first solve the differential equation for the highest derivative,

$$\frac{dh}{dt} = \frac{\rho[10+2u(t)] - 1.5\rho\sqrt{11.32+9.8\text{h}}}{\rho A}$$

Then we put down as many integrators as needed (one in the case),

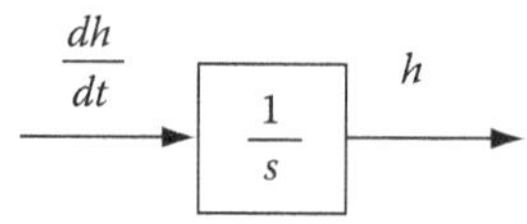

Diagram AA

And finally, we construct the right hand of the rearranged differential equation and feed it to the integrator; this is shown in Figure 11.5.

Figure 11.6 shows the response of the level in the tank. To show the change from one steady state to another more clearly, we use u(t-5) rather than u(t) in the forcing function of Equation 8.32.

Before continuing with another example, it is interesting to look in more detail at the response of the level in the tank of the previous example. Let us suppose that once the new steady state is reached, after changing the inlet flow by 2 m^3/min, the inlet flow is changed again by the same amount of 2 m^3/min, and when a new steady state is reached again, we repeat once more the change. Figure 11.7 shows the response.

Although not easily read from the figure, the first change in level was from 3.38 to 5.38 m, or a change of 2.0 m; the second change in level was of 2.36 m; and the third change was of 2.72 m. That is, the effect of a change of 2 m^3/min in inlet flow on the level is not always the same; it is different depending on the starting level or total flow through the system, or what are called the *operating conditions*. This different behavior of the process is typical of systems described by nonlinear differential equations; these systems are called *nonlinear systems*; Equation 8.32 is such an equation. If the model had been linear, all the changes in level due to the change of 2 m^3/min in inlet flow would have been the same amount.

Example 11.9

Let us now look in detail at the simulation of the model of the electrical circuit of Example 10.7. For convenience, we redraw the circuit in Figure 11.8 and present the model next.

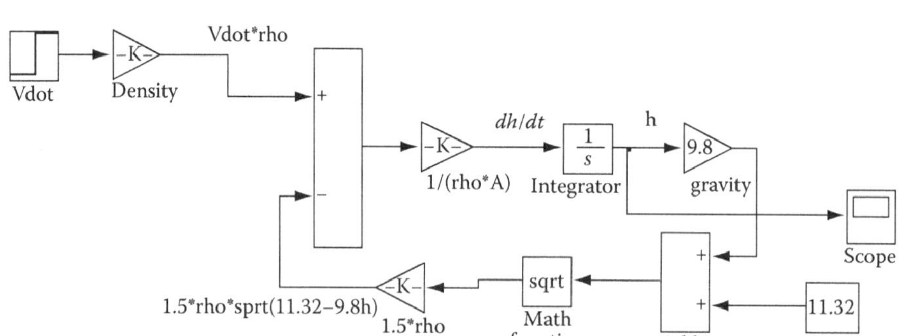

FIGURE 11.5 Block diagram for the simulation of Equation 8.32

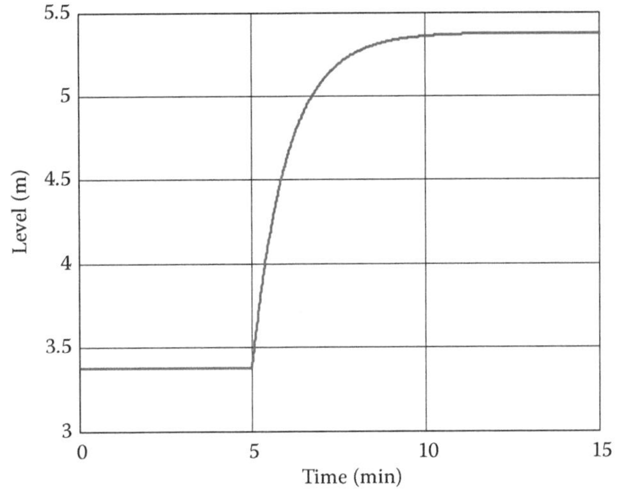

FIGURE 11.6 Response the level in the tank

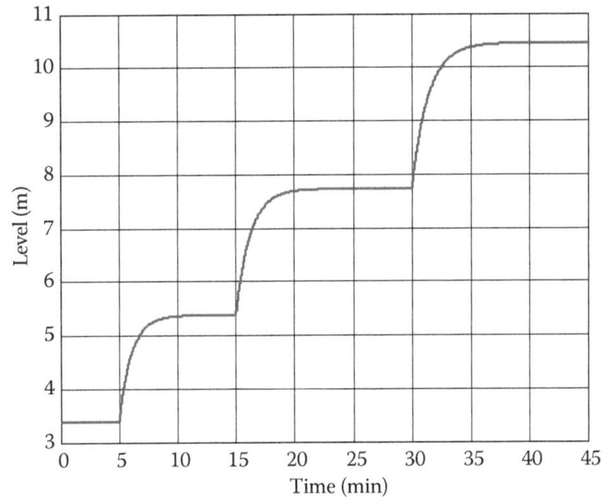

FIGURE 11.7 Response of level to three consecutive changes in inlet flow of $2\,m^3/min$

The model is given by (for a complete development go to Example 10.7),

$$i_1 - i_2 - i_3 - i_4 = 0 \tag{11.17}$$

1 eq., 4 unk. $[i_1, i_2, i_3, i_4]$

$$i_1 = \frac{1}{R_1}(v_S - v_D) \tag{11.18}$$

2 eq., 5 unk. $[v_D]$

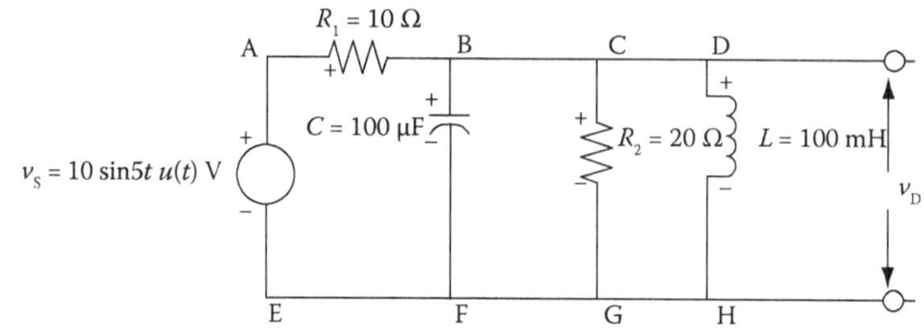

FIGURE 11.8 Electrical circuit

$$i_2 = C\frac{dv_C}{dt} = C\frac{d(v_D - 0)}{dt} = C\frac{dv_D}{dt} \tag{11.19}$$

3 eq., 5 unk.

$$i_3 = \frac{1}{R_2}(v_D - 0) = \frac{1}{R_2}v_D \tag{11.20}$$

4 eq., 5 unk.

$$i_4 = i_4(0) + \frac{1}{L}\int_0^t (v_D - 0)dt = i_4(0) + \frac{1}{L}\int_0^t v_D\, dt \tag{11.21}$$

5 eq., 5 unk.

The set of Equations 11.17 through 11.21 is the model of the circuit. If only a single equation is desired, to obtain the analytical solution or for whatever other reason, simple algebraic substitution provides the equation. Substituting Equations 11.18, 11.19, 11.20, and 11.21 into Equation 11.17 yields

$$\frac{1}{R_1}(v_S - v_D) - C\frac{dv_D}{dt} - \frac{1}{R_2}v_D - i_4(0) - \frac{1}{L}\int_0^t v_D\, dt = 0$$

Taking the derivative with respect of time gives

$$C\frac{d^2v_D}{dt^2} + \left(\frac{1}{R_1} + \frac{1}{R_2}\right)\frac{dv_D}{dt} + \frac{1}{L}v_D = \frac{1}{R_1}\frac{dv_S}{dt} \tag{11.22a}$$

or

$$0.0001\frac{d^2v_D}{dt^2} + \frac{3}{20}\frac{dv_D}{dt} + \frac{1}{0.1}v_D = \frac{1}{10}\frac{dv_S}{dt} \tag{11.22b}$$

Because $v_S = 10u(t)\sin 5t\ V$, $\frac{dv_S}{dt} = 50u(t)\cos 5t$ The solution of Eq. 11.22b requires the initial conditions which on the basis of the initial statement,

$$\left.\frac{dv_D}{dt}\right|_{t=0} = v_D(0) = 0$$

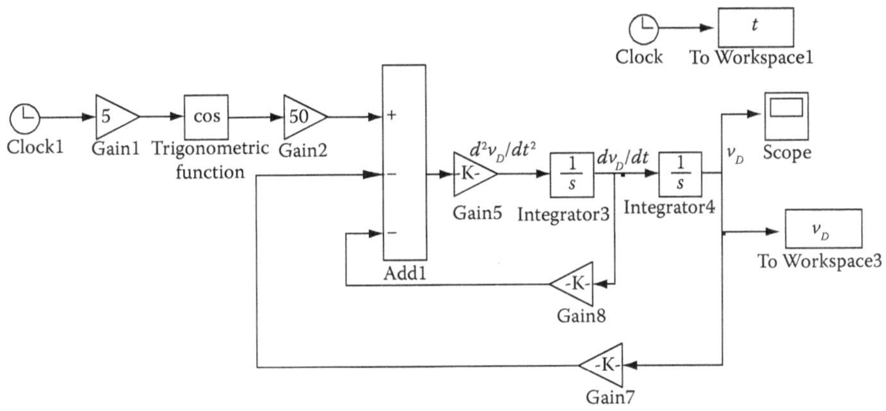

FIGURE 11.9 Simulation (block diagram) of Equation 11.22b

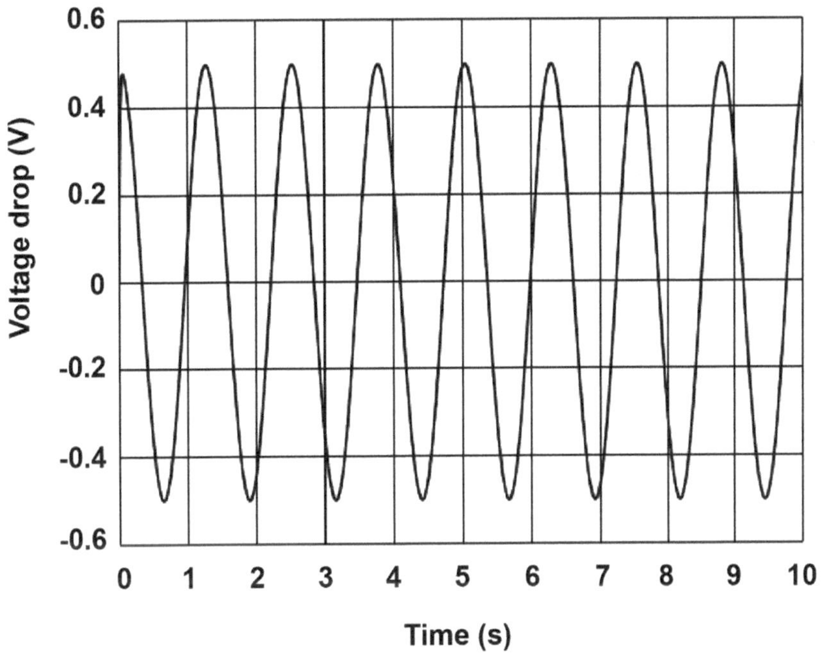

FIGURE 11.10 Response of voltage drop v_D

Figure 11.9 shows the block diagram developed following the three steps given by the guideline, and Figure 11.10 shows the response of v_D.

The model of most systems presented in this book has required more than one equation. A typical case in point is the model of the electrical circuit of Example 11.9 (or Example 10.7); this model requires five equations to fully describe the system. In all cases we have used simple algebraic substitution to reduce the set of equations to a single differential equation or maybe two. We have done this to obtain the analytical solution. However, there is no need to do so if only a simulation is used to obtain the response. In this case, each equation can be generated and connected to the others to obtain the solution. For example, from Equation 11.17,

$$i_2 = i_1 - i_3 - i_4 \tag{11.23}$$

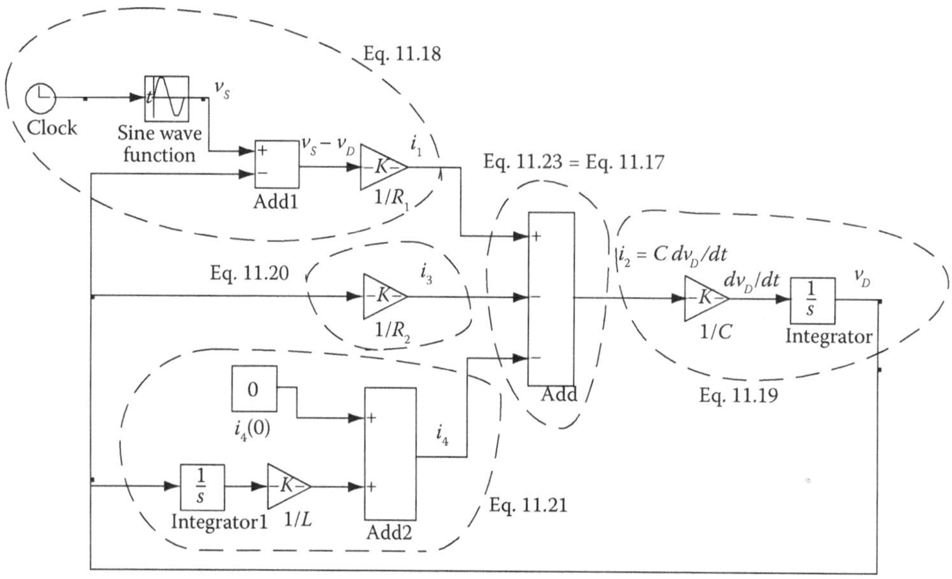

FIGURE 11.11 Block diagram for set of Equations 11.17 through 11.21

Now, we generate the block diagram as shown in Figure 11.11. Obviously, the results obtained are the same as Figure 11.10. Most industrial models are composed of many equations, and thus, instead of trying to reduce the number of equations, this is the most common method to simulate these models.

11.8 SOME ADDITIONAL TOPICS ON SIMULINK, AND MODELLING

11.8.1 Writing Equations

Consider a spherical tank like the one shown in Figure 11.12. The total volume of a sphere V_T is given by $4\pi r^3/3$ where r is the radius of the sphere. The relation between volume $V(t)$ and height $h(t)$ is given by,

$$V(t) = V_T\left[\frac{h^2(t)[3r - h(t)]}{4r^3}\right] \tag{11.24}$$

Once the radius of the sphere is known, then the volume of the tank is only a function of the height, or liquid level, $V(t) = f[h(t)]$. To program in Simulink, the volume equation would require several blocks. Simulink offers a block in which the entire equation can be written thus saving many blocks; this is shown in Figure 11.13. The figure shows the block labeled **fcn** (for MATLAB® function) which is located in the Simulink library browser. The two statements shown below the block are the statements that will perform the calculation. These statements are written inside the block which is opened by double-clicking on it. The figure shows that for a sphere of $r = 4\,\text{m}$ (which is written in the equation), and a liquid level $h(t) = 5.043\,\text{m}$, the liquid volume is $V(t) = 185.3\,\text{m}^3$.

The MATLAB function can be helpful to generate equations with minimum blocks.

11.8.2 Trial and Error (Convergence)

Let us continue using the spherical tank example. Suppose that a calculation (maybe a mass balance) provides the liquid volume existing in the tank, and the liquid level (height) is necessary to continue

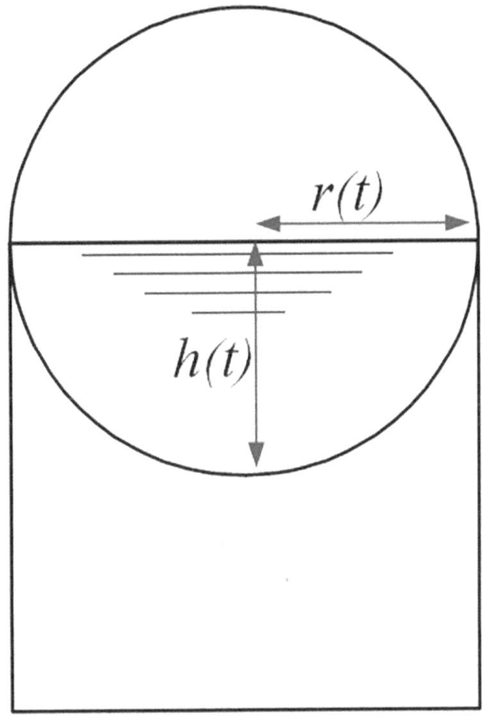

FIGURE 11.12 Spherical tank.

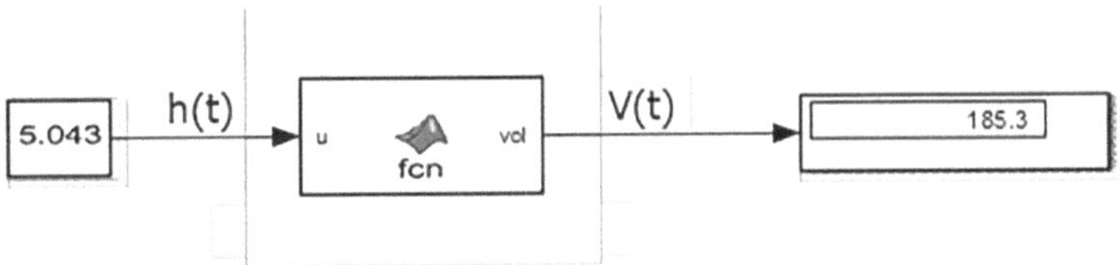


```
function vol = fcn(u)
  vol = (4*3.1416*4^3)*(u^2*((3*4)-u)/(4*(4*(4^3)))
```

FIGURE 11.13 Single block to calculate liquid volume in a spherical tank

with the calculations. Certainly, it is not simple to obtain the height from the volume equation. It would require assuming a height and checking if it yields the correct volume. If not, start again with another trial (assuming another height) until it converges to the correct volume. We refer to this procedure as "trial and error" or "convergence" and it is commonly used in engineering calculations. There are several mathematical methods to converge to a solution, and they are an important topic in numerical methods courses. Simulink offers a block that provides a convergence method. Figure 11.14 shows the program; let us explain.

The Solve block receives a signal and outputs another signal. In this example the input signal is referred to "error," and the output signal is referred to $h(t)$. The output signal is used to calculate volume. The calculated volume is compared with the actual volume in the tank. The comparison consists in subtracting the actual volume from the calculated volume to find the error. This procedure

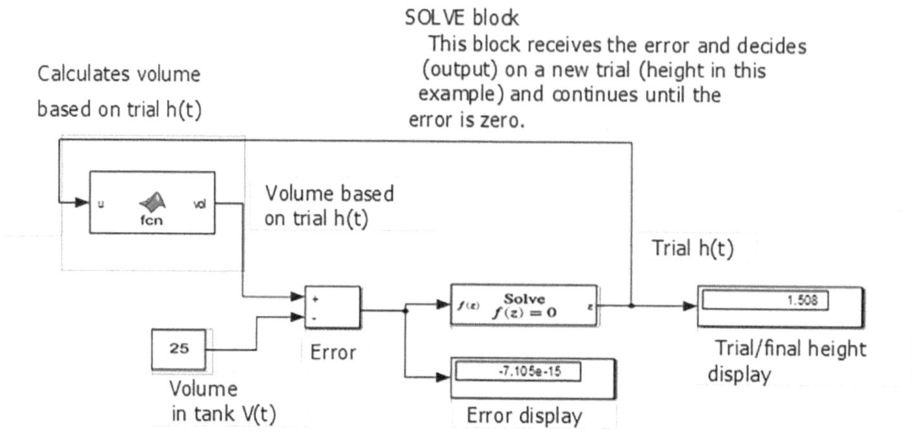

FIGURE 11.14 Simulink example for convergence

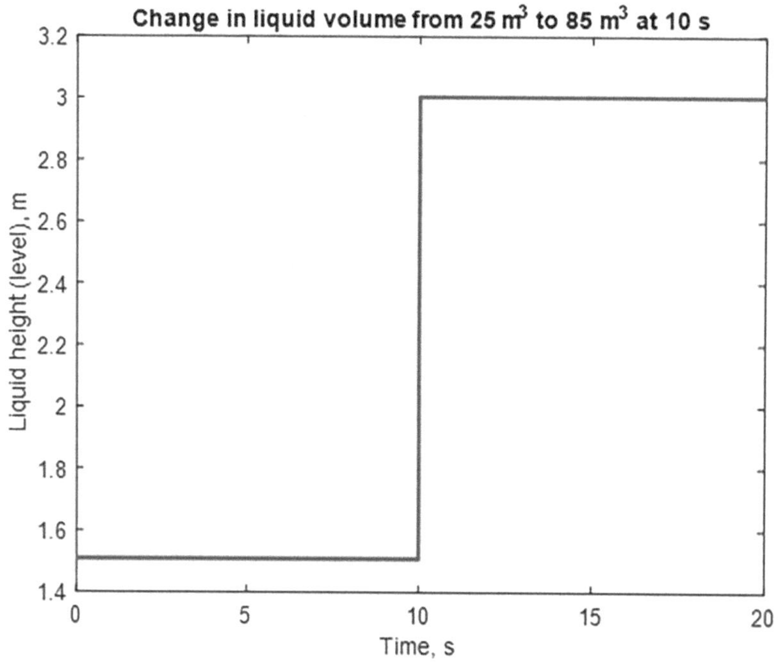

FIGURE 11.15 Change in volume and level

continues until the error becomes zero. The *h*(*t*) that provides this error is the height of liquid in the tank. So, the Solver block receives the error and using some algorithm decides on what should be the next height to drive the error to zero (volume based on *h*(*t*) equals the volume of liquid in tank). The Solve output is sent to a calculation and eventually fed back to the Solve. This "feedback path" is essential in the method.

The Solve block has three pieces of information for the output to set once you open it by double clicking on it. The third one is the initial guess for the output, in this case the first value of *h*(*t*). A simple value to choose is the height at steady state.

Figure 11.15 shows the new level when the liquid volume changes from $25\,m^3$ to $85\,m^3$. The level goes from about 1.5 *m* to a bit more than 3 *m*.

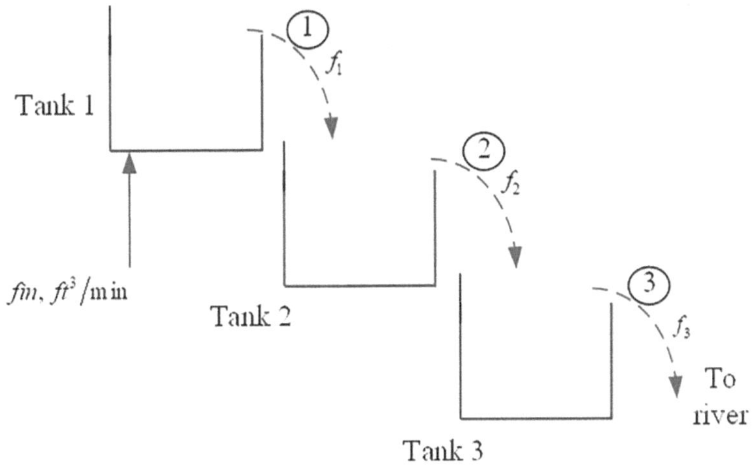

FIGURE 11.16 Three tank system

11.8.3 Another Helpful Simulation Tip

Consider the three-tank system shown in Figure 11.16. The tanks have a 2ft radius, and the right side is 8 ft tall.

The tanks are initially empty. At some time, call it $t = 0$, the inlet stream to the first tank starts flowing, $fin = 25 ft^3 / \min$. The liquid level in tank 1 starts increasing but, there is no flow out until the level reaches 8 ft at which time stream 1 starts overflowing to tank 2 by the same amount as entering the tank. This continues until eventually stream 3 flows out of tank 3.

It is desired to simulate the system and study how the liquid level behaves in the three tanks. Let us first look at tank 1. A mass balance is the appropriate first equation.

$$\rho fin - \rho f_1 = \frac{dm_1}{dt} = \rho \frac{dV_1}{dt} \tag{11.25}$$

and assuming that this is a pure component and thus all densities are equal,

$$fin - f_1 = \frac{dV_1}{dt}$$ (11.26) 1 equation, 2 unknowns $[f_1, V_1]$

For the flow out of the tank,

$$f_1 = \begin{cases} 0 & \text{if } h_1 < 8\,ft \\ f_{in} & \text{if } h_1 \geq 8\,ft \end{cases}$$ (11.27) 2 equations, 3 unknowns $[h_1]$

For the level,

$$h_1 = \frac{V_1}{A}$$ (11.28) 3 equation, 3 unknowns

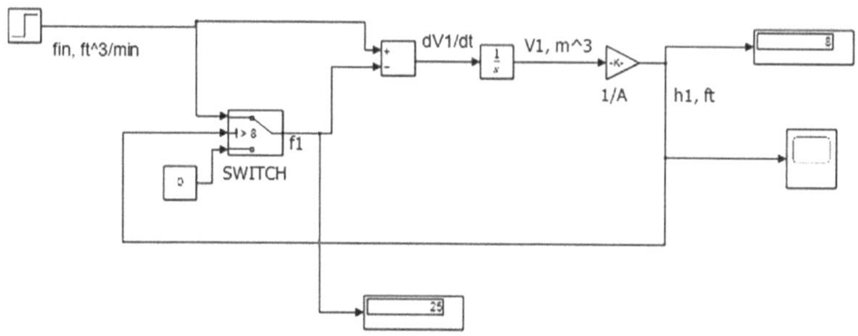

FIGURE 11.17 Simulink program

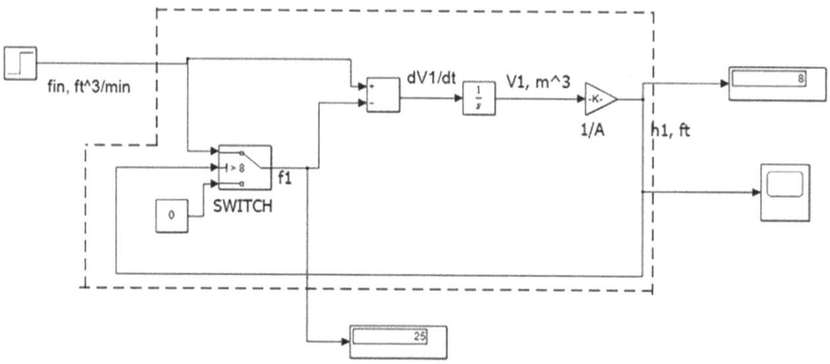

FIGURE 11.18 Blocks that will be repeated in every tank

Figure 11.17 shows the Simulink blocks for this model. The block named SWITCH generates Equation 11.27. The block is found in the Signal Routing tab in the Simulink Library Browser. It has 3 inputs. The instructions for this block state "Pass through input 1 when input 2 satisfies the selected criterion; otherwise, pass through input 3. The inputs are numbered top to bottom. The first and third input ports are data ports, and the second input port is the control port. The criteria for control port 2 are u2 >= Threshold, u2 > Threshold or u2 ~= 0." The control port in this example is the level h_1. The block shows that when the level does not satisfy the criterion (less that 8 ft) the output is $f_1 = 0$, or no flow out. When the level satisfies the criterion $f_1 = f_{in}$(the flow out is by overflow, so it is equal to the flow in).

Let us suppose now that you want to model and simulate all three tanks. Thinking about it, it is fairly simple to realize that the models for tanks 2 and 3 will be very similar to that of tank 1. The main difference will be that instead of using f_{in} for the output overflow, it will be f_1 for tank 2 and f_2 for tank 3. Therefore, the complete model will have the same blocks for each tank; Figure 11.18 shows these blocks indicated by the dashed lines.

Simulink offers an option to collapse all these blocks into a single block called a Subsystem; this is shown in Figure 11.19. The subsystem block can now be used as any other block in the library browser.

Figure 11.20 shows the program for all 3 tanks.

This section has presented a few extra topics often necessary in simulations. Obviously, it is an introduction. Remember practice is the way to learn the necessary details. Also, don't forget the extensive information on the Web.

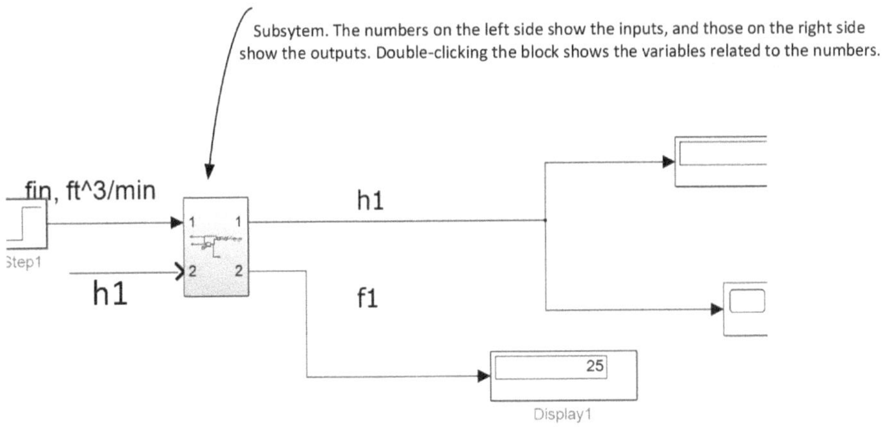

FIGURE 11.19 Subsystem

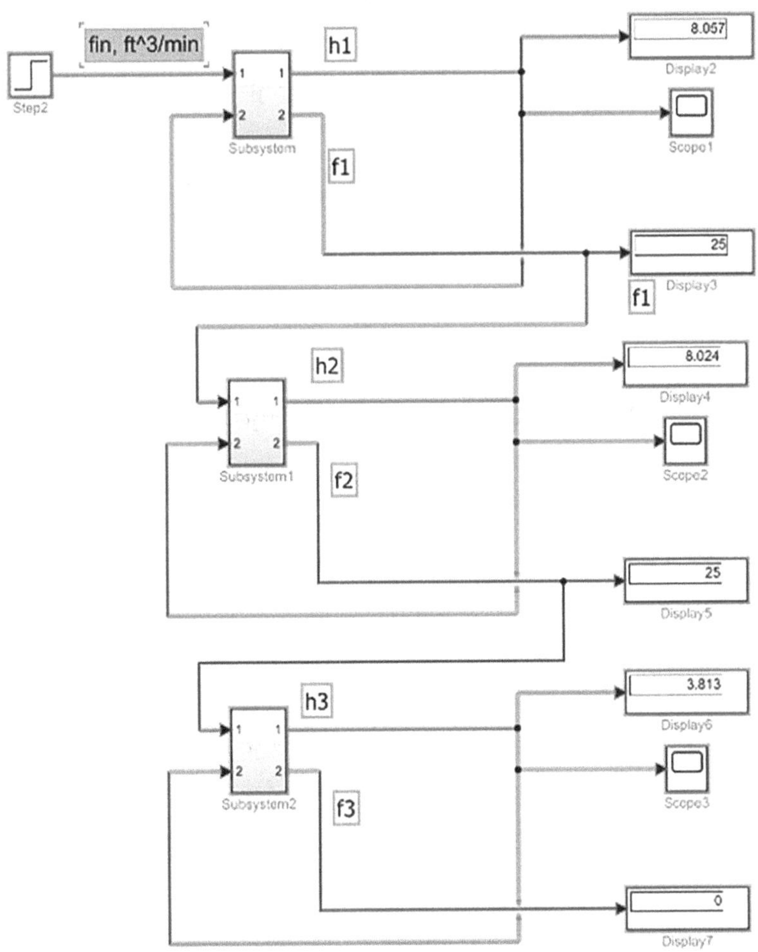

FIGURE 11.20 Simulink program for 3 tanks using the subsystems

11.9 SUMMARY

In this chapter, we described what numerical solutions to differential equations provide and why they are needed. To provide a feel for how numerical solutions are obtained, we described Euler's method, which is the simplest numerical solution technique. Rather than develop the more sophisticated methods used in practice, we showed how to construct block diagrams, which are used by computer software packages (which apply these more advanced methods) to "read" the differential equations that are to be solved numerically. We ended by using Simulink to solve several of the mathematical models that were developed earlier in the book.

REFERENCE

1. Sanjuan, M.E., D.P. Hess, Understanding Dynamics of Machinery with Friction Through Computer Simulation, Presented at ASEE Southeastern Section Conference, Clemson, South Carolina, USA, 1999.

Problems

Problem 11.1

Use Euler's method to estimate y at t = 0.3 sec. Use a step size of 0.1 sec.

a. $2\frac{dy}{dt}+4y=3e^{-t} \quad y(0)=1$

b. $\frac{dy}{dt}-y=e^{-y}e^{-t} \quad y(0)=0$

c. $\frac{dy}{dt}+2y^2=3t \quad y(0)=1$

d. $\frac{dy}{dt}=ye^{-2t} \quad y(0)=1$

Problem 11.2

Use Euler's method to estimate T at t = 3 sec. Use a step size of 1.0 sec.

$$\frac{dT}{dt}=10e^{-0.2t}-0.02(T-25) \quad T(0)=25$$

Problem 11.3

Consider the differential equation:

$$4y''-8y'+4y=0 \quad y(0)=-3 \quad y'(0)=4$$

a. Find the analytical solution to this initial value problem.
b. Solve this equation using the Euler method with a step size of Δt = 0.1 and go to t = 7. Use a spreadsheet.
c. Repeat b except use a step size of 0.01 (still going out to t=7). Plot the analytical solution and both parts a and b on the same graph. How do the numerical solutions compare to the exact solution?

Problem 11.4

In Chapter 2, we considered the drag force on falling objects to be proportional to the velocity. A more realistic model for air drag on a falling object is incorporated in the model shown below (coordinate system is positive upward):

$$dV/dt = -9.8 - (k/m)\ V\ |V| \text{ and } dY/dt = V$$

Take k/m = 0.005, V(0) = 0 and Y(0)=1000. Use the Euler method to estimate Y and V at 5 s, using a Δt of 1 s.

Problem 11.5

Variables z, y and x are all functions of t. Draw a block diagram for the coupled equations:

a. $y' = 3x - 4t$
   $x' = y - 2t$
b. $x'' + 2z' - 3x = 0$
   $z'' + 4x' - 5z = 0$

Problem 11.6

Draw a block diagram for the differential equation:

$$y''(t) = F(t) \text{ where } F(t) = \begin{cases} 3 & \text{if} \quad t < 10 \\ 1 & \text{if} \quad 10 < t < 30 \\ 2 & \text{if} \quad t > 30 \end{cases}$$

Problem 11.7

The following model, developed in Chapter 1 (Equation 1.13b), describes the liquid level in the tank shown in that chapter,

$$146\frac{dh}{dt} + 166.7\sqrt{h} = w_1(t) + w_1(t)$$

To obtain an analytical solution, Example 3.20 used Equation 3.108 to linearize the model as follows:

$$146\frac{dh}{dt} + 46.34h = w_1(t) + w_2(t) - 150.14$$

Assume h(0) = 3.24 m, w_1 = 200 kg/min, and w_2 = 100 + 50u(t) kg/min. Using simulation, obtain the response of both models and compare their results.

Problem 11.8

Consider an object dropped out of an airplane (y(0) = h, v(0) = 0). Suppose the drag force is given by

$$F_D = Pv|v|$$

a. Use Newton's second law to obtain the differential equation that represents position y as a function of t. Make the equation explicit in y (i.e. eliminate velocity and acceleration by $v = y'$ and $a = y''$.
b. Suppose the object is a parachutist ($m = 60$ kg) who drops from a plane at 2000 m and opens her chute at $t = 15$ s. Before the chute opens the drag coefficient P is 1.0 N s²/m² and after the chute opens, $P = 10$ N s²/m². Using your result from part a and the information given here, draw a block diagram that will provide the position y of the parachutist as a function of time. Absolute value can be obtained using the Function block. Label all block parameters.

Problem 11.9

Felix Baumgartner ($m = 75$ kg) recently stepped out of a balloon at an elevation of 40,000 m and parachuted to earth, setting a height record for skydivers. Here we simulate his jump from the time he steps out of the balloon until the time he opens the parachute. Assume the drag force F_D is of the form:

$$F_D = -k\left|v_y\right|v_y$$

where k is the drag coefficient and v_y is the vertical velocity. The effect of elevation on the gravitational constant is minor. However, because air density depends on elevation, and he jumped from a very high elevation, the effect of elevation on the drag coefficient k should be accounted for. A crude model for k that takes into account the variation of air density with elevation is

$$k = k_0 e^{-.00011y}$$

where k is the drag coefficient at elevation y, k_0 is the drag coefficient at sea level ($y = 0$) and where y is in meters and k and k_0 are in kg/m. Assume $k_0 = 0.16$ kg/m before the parachute opens.

a. Using a coordinate system where y is positive upward, write the differential equations that represent the velocity v_y and elevation y of Felix during the time between stepping out of the balloon and opening his chute. Include the initial conditions.
b. Simulate and graph Felix's velocity and elevation as functions of time. At what elevation will he have reached his maximum velocity? What is his maximum velocity? For comparison, measurements during the jump indicated that his maximum velocity was 1358 km/hr and that his freefall from 40,000 m to 2000 m (when his chute opened) lasted for 260 s.

Problem 11.10

For vertical flight, the equations that describe the velocity v and height y as a function of time t for a launched rocket are

$$M\frac{dv}{dt} = T - Mg - D \quad v(0) = 0$$

$$\frac{dy}{dt} = v \quad y(0) = 0$$

where T is the engine thrust in N, M is the mass of rocket plus engine and payload, v is the velocity in m/s, y is the elevation in m, and $g = 9.8$ m/s². The drag force D (in N), given by

$$D = \frac{1}{2}\rho C_D A v^2$$

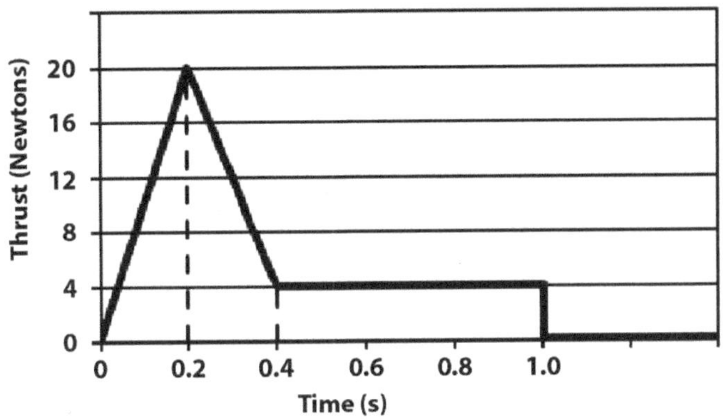

FIGURE P11.1 Thrust curve for rocket engine of Problem 11.10.

where ρ is the density of air (1.2 kg/m^3), C_D is a drag coefficient (dimensionless), and A is the frontal area (the area you would see looking straight down on the rocket) in m^2.

A model rocket has a mass of 0.03 kg including engine and propellant and a frontal area A of 0.0005m^2. The rocket has a drag coefficient of 0.65 and is to be powered with an engine that has the thrust curve shown in Figure P11.1 (the thrust fall to zero at $t = 1$ s and remains there). It is desired that the rocket carry a payload to a height of 200 m. Use simulation to determine the maximum payload mass. The change in mass of propellant is small compared to the mass of the rocket and can be neglected. Hint: Use a signal builder to represent the thrust curve.

Problem 11.11

It is widely believed that batters on a club like the Colorado Rockies have an advantage over, say, those of the Boston Red Sox, because their park is at a higher elevation. Because the air is less dense, there will be less air resistance on a batted ball, and it will therefore fly farther.

In the presence of air resistance, the equations of motion for an object moving in two dimensions (vertical and horizontal) in a gravitational field are

$$m\frac{dv_x}{dt} = -D_x \text{ and } m\frac{dv_y}{dt} = -mg - D_y$$

where v_x and v_y are the x and y components of the velocity vector, m is the mass of the object, and g is 9.8 m/s^2. The quantities D_x and D_y are the horizontal and vertical components of the drag force and are given by

$$D_x = \frac{\rho C_D A}{2} v v_x \quad D_y = \frac{\rho C_D A}{2} v v_y$$

where ρ is the density of air in kg/m^3, A is the area of the object as seen head-on (in m^2), and C_D is a (dimensionless) drag coefficient that depends on the shape of the body. The symbol v represents the magnitude of the velocity vector and is given by

$$v = \sqrt{(v_x^2 + v_y^2)}$$

Once v_x and v_y are known, the positions x and y can be found from

$$\frac{dy}{dt} = v_y \text{ and } \frac{dx}{dt} = v_x$$

Use simulation to compare the trajectories of two identically hit balls – except that one is hit in Colorado (where the air density is 0.97 kg/m^3) and one is hit in Boston (where the air density is 1.19 kg/m^3). Prepare a plot of the trajectory y versus x for both cases and determine how far in the x direction each ball flies. For a baseball, $m = 0.15$ kg, $C_D = 0.5$, and $A = 0.0043$ m^2. Assume that the initial ball speed off the bat is 45 m/s and that the ball is hit at an angle 30° from the horizontal. The initial x position is zero and the initial y position is 1 m. Is there a significant advantage to a batter in Colorado?

Problem 11.12

Equations 6.36 and 6.37 provide the model of the carts in Figure 6.16. Using $f_A(t) = 5u(t)$, simulate and graph the position of each cart, and compare it with Figure 6.17.

Problem 11.13

Equations 6.38b and 6.40b provide the model of the blocks in Figure 6.20.
Using $f_A(t) = 3u(t-5)$, simulate and graph the position and velocity of each block.

Problem 11.14

Equation 6.59 is the model of the block in Figure 6.30b. Simulate and graph the position of the block, and compare the graph with Figure 6.31. The analytical solution is given by Equation 6.60; graph this solution and compare it to the previous figure.

Problem 11.15

A simplified model for a shock absorber shown in Figure P11.2 is

$$m\frac{d^2x}{dt^2} + P\left(\frac{dx}{dt} - U'(t)\right) + k(x-U) = 0$$

where m is one-fourth of the vehicle mass, k is the spring constant, P is a damping coefficient, and $U(t)$ represents a forcing function dictated by the shape of the road surface. The dependent variable x represents the vertical position of the mass m and t is time.

Here, we will simulate the motion of the vehicle mass as it moves over a speed bump at a specified speed.

a. Determine a value for the spring constant k (N/m) from the notion that the natural frequency of the *un-damped and unforced* system should be about one cycle per second. The mass m is 300 kg.
b. Determine a value of the damping coefficient P (N•s/m) using the rule of thumb that the damping ratio $P/P_{crit} = 0.3$. Here, P_{crit} is the damping coefficient that critically damps the *unforced* system.
c. It is proposed to use a speed bump on a flat roadway that has the shape given in Figure P11.3.

The criterion used to specify the height h of the speed bump is that the maximum vertical acceleration (whether positive or negative) of the mass m be 0.5 g (g = gravitational constant) if the

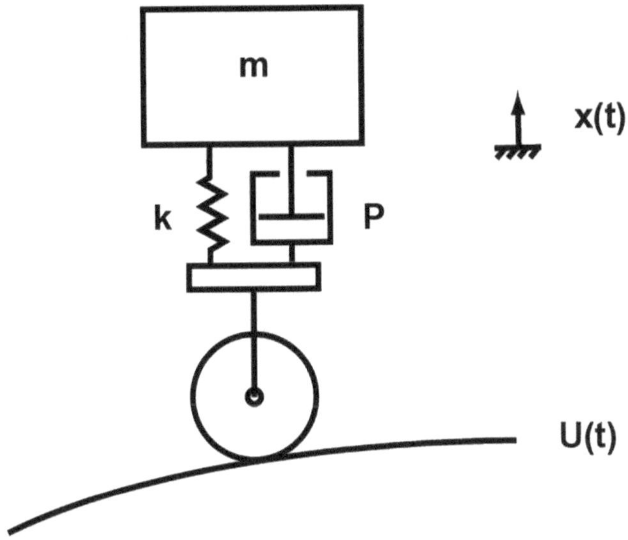

FIGURE P11.2 Shock absorber of Problem 11.15.

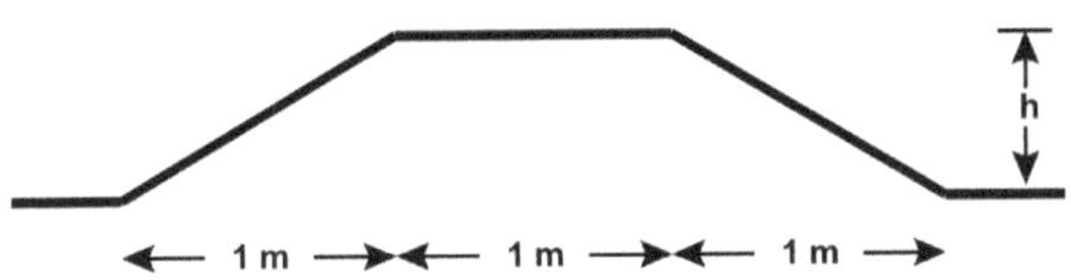

FIGURE P11.3 Speed bump of Problem 11.15.

vehicle is traveling at 5 m/s (about 11 mph). Determine the height h that satisfies this condition. Hint: The function $U(t)$ can be obtained by using the speed of the car to convert the 1 m lengths in the speed bump to time. You can use a signal builder to create the function $U(t)$ in Simulink.

Problem 11.16

The model that describes the positions of the blocks in Problem 6.8 is

$$5\frac{d^2x_1}{dt^2}+90\frac{dx_1}{dt}+100x_1=20\frac{dx_2}{dt}$$

and

$$5\frac{d^2x_2}{dt^2}+20\frac{dx_2}{dt}+100x_2=f_A(t)+20\frac{dx_1}{dt}$$

Assume that the initial conditions are $x_1(0)=\left.\frac{dx_1}{dt}\right|_{t=0}=x_2(0)=\left.\frac{dx_2}{dt}\right|_{t=0}=0$, and that the applied force is $f_A(t)=100u(t)\text{N}$. Simulate this system and graph the position and velocity of each block.

Problem 11.17

Consider the system of blocks shown in Figure P11.4. Note that block 2 is situated in the center of block 1 and that this time the reference displacement is in the center of each block. Also, note that the

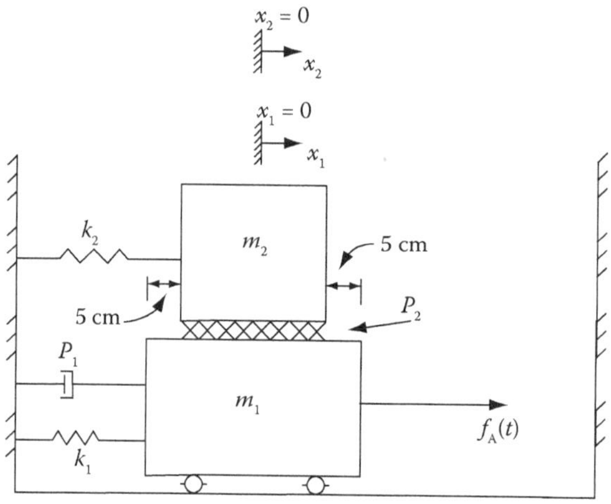

FIGURE P11.4 Mechanical system for Problem 11.17.

distance from each end of block 2 to the corresponding end of block 1 is 5 cm. The following information is known: $f_A(t) = 15u(t-5) - 15u(t-20)$ N and $m_1 = m_2 = 20$ kg; $P_1 = P_2 = 80$ N s / m; $k_2 = 5$ N / m. Find the smallest value of k_1 so that when the force is applied, block 2 stays completely on top of block 1.

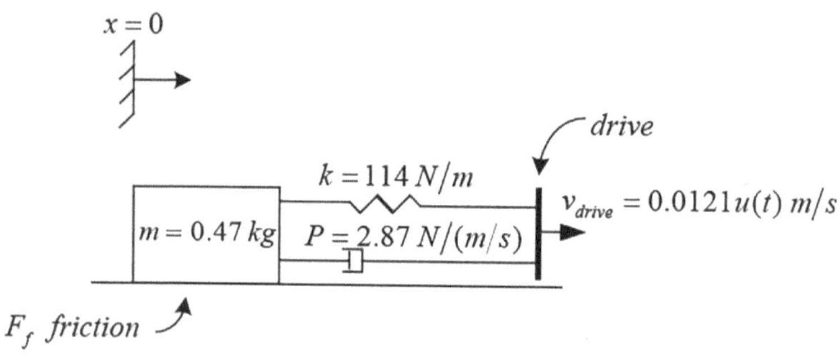

FIGURE P11.5 Mechanical system for Problem 11.18

Problem 11.18

Consider Figure P11.5.

The figure shows a block connected to a drive by a spring and a dashpot. At time equal zero, the drive starts moving to the right, which is the positive direction, with velocity 0.0121 m/s that is, $v_{drive} = 0.0121u(t)$. The following equations constitute the model that describes the displacement, x in m, and velocity, dx/dt in m/s, of the block.

$$m\frac{d^2x}{dt^2} = F_{spring} + F_{dashpot} + F_f \tag{P11.1}$$

1 equation, 4 unknowns $[x, F_{spring}, F_{dashpot}, F_f]$;

$$F_{spring} = -k\left(x - \int v_{drive}dt\right) \tag{P11.2}$$

2 equations, 4 unknowns;

$$F_{dashpot} = -P\left(\frac{dx}{dt} - v_{drive}\right) \tag{P11.3}$$

3 equations, 4 unknowns.

a. Assume for the friction with the floor the following wet friction model,

$$F_f = -4\frac{dx}{dt} \tag{P11.4}$$

4 equations, 4 unknowns.

Simulate the model and graph the response for the first 10 seconds.

b. Assume now that the friction between the block and the floor is a dry friction described by the following modified Karnopp friction model, in which case Equation P11.4 is replaced by Equation P11.5:

$$F_f = \begin{cases} \mu_D N\, sign\left(\frac{dx}{dt}\right); \text{if} \left|\frac{dx}{dt}\right| > v_{\min} \\ \text{Min}\left(\left|\sum F_{ext}\right|, \mu_s \cdot N\right) \cdot sign\left(\sum F_{ext}\right); \text{if} \left|\frac{dx}{dt}\right| \le v_{\min} \end{cases} \tag{P11.5}$$

4 equations, 4 unknowns.

This model helps explain the Slip-Stick phenomenon[(1)]. Simulate the model and graph the response for the first 10 seconds.

$\mu_D = 0.30$; $\mu_S = 0.52$; N is the normal force, in Newtons, generated by the block; $v_{\min} = 0.0005\,\text{m/s}$; the notation sign() has a value of 1 when the quantity within the parenthesis is positive, a value of −1 when the quantity is negative, and a value of 0 when the quantity is 0; $\left|\sum F_{ext}\right| = \left|F_{spring} + F_{dashpot}\right|$; *Min(,)* indicates the minimum of the quantities separated by the comma within the parenthesis.

Problem 11.19

Consider the viscous coupler of Problem 7.8. The model is

$$J_1\omega_1' = T - P\omega_1 + P\omega_2$$
$$J_2\omega_2' = P\omega_1 - (P + P_2)\omega_2$$

Solve this model by simulation and plot ω_1 and ω_2 versus time. Do the ultimate angular velocities from the simulation agree with those calculated in Problem 7.8?

Problem 11.20

Solve part (b) of Problem 7.14 (Skotch yoke) by simulation.

Problem 11.21

Consider Example 8.3. Equations 8.34 and 8.35 constitute the model of the tank shown in Figure 8.5. Simulate this tank, and graph the response when both forcing functions mentioned in the example affect the outlet concentration.

Problem 11.22

Consider the gas tank shown in Figure 8.6 and presented in Example 8.4. Equations 8.38, 8.39, 8.40, 8.41, 8.42 and 8.5 constitute the model of the tank. Obtain the response of the pressure in the tank to a change of

a. 50 kPa in the downstream pressure $p_1(t)$.
b. 3% in the signal to the fan $S_i(t)$.
c. 5% in the signal to the valve $S_o(t)$.

Problem 11.23

Example 8.5 presents a process to remove benzene (pollutant) from water. Equations 8.60, 8.61, 8.62, and 8.63 constitute the model describing the concentration of benzene in both basins. Simulate this process, and graph the response of the concentration of benzene as a function of time in each basin for the period from 0 to 1 hour when the entering concentration of benzene spikes to 5 gr/m^3 and remains there for 0.5 hours, at which point it drops back to zero. Compare your result to Figure 8.12. Example 8.5 provides all the necessary information.

Problem 11.24

Consider part (b) of Example 8.7 describing a reactor where the reaction $S \rightarrow P$ takes place and P is the desired product. Equations 8.73 through 8.75 and 8.79 through 8.80 provide the model. Simulate this process, and obtain the times required to reach 90% and 95% conversions. If the membrane surface area is changed to 1.25 m^2, what are the new times?

Problem 11.25

Consider Example 8.8 which describes a water filter. Using simulation, and the parameter values given in the example, obtain the time required to filter the water and reach 10% of the initial height. Now, assume that a new clay with a value $C = 1.5 \times 10^{-7} \text{cm}^2 \cdot \text{s} / \text{gr}$ is used. How does this change affect the design and operation?

Problem 11.26

Consider Example 8.9. This example presents the conversion of carbohydrates to glucose and then to glycogen; the example also presents some actual information from an individual. Equations 8.97 and 8.99 provide the model for the concentration of carbohydrate and glucose; the example provides the necessary information. Simulate this system, and compare the results with the individual's information.

Problem 11.27

Example 8.10 discusses the amount of antibiotic in the blood volume of a patient and how often the antibiotic must be injected to maintain a minimum required level. Equations 8.102, 8.103, 8.92, and 8.104 provide the model. Simulate this system and compare your results with Figure 8.20.

Problem 11.28

Example 8.12 considers the fermentation of glucose $(C_6H_{12}O_6)$ producing ethanol (C_2H_5OH) and carbon dioxide (CO_2). Equations 8.117, 8.118, 8.119, and 8.120 provide the model yielding the concentration of ethanol once the reaction starts. Using the initial concentrations given in the example, simulate this reaction system and graph the concentrations of ethanol and biomass versus time.

Problem 11.29

Example 8.13 considers the production of gluconic acid by fermentation of glucose. Equations 8.121, 8.122, and 8.123 provide the model of the reaction process. Using the initial concentrations given in the example, simulate this reaction system and graph the concentrations of glucose, and biomass, versus time.

Problem 11.30

The following equation describes the concentration of a component in a fermentation reaction (Chapter 8).

$$\frac{dX}{dt} = \frac{X^2(-0.5+X)}{X+1} \quad \text{with } X(0) = 0.6\,\text{mmole/L}$$

a. Using simulation obtain the time required to go from the initial concentration of 0.6 mmole/L to a concentration of 1.0 mmole/L; the time unit is hour.
b. Example 2.2 presented the analytical solution using Separation of Variables. Compare the simulation and analytical solution responses.

Problem 11.31

Problem 8.2 shows that the equation that describes the pressure *P*(*t*) in a tank containing a gas when it is filled is

$$6.915C_V\sqrt{1000-P(t)} = 1.08\frac{dP(t)}{dt}$$

The gas enters the tank through a valve that must be sized by choosing the numerical value of C_V. The initial pressure in the tank is 100 *kPa*. The filling procedure calls for raising the gas pressure from its initial value of 100 *kPa* to a final value of 500 *kPa* in 4 minutes. Using simulation, obtain C_V that would satisfy this requirement.

Problem 11.32

Solve the problem of Example 7.4 by simulation. The model is given by Equations 7.32 and 7.33. Verify your results by comparing them to Figure 7.13.

Problem 11.33

This problem was created by engineers at DuPont Company in the early 1960s. The waste disposal tank shown in Figure P11.6 is fed by three identical pumps and discharges through a very long pipe. Because of the inertia of the liquid in the pipe, the tank level response is underdamped. This causes a problem because, although the tank is tall enough at steady state, the overshoot in level can cause the tank to overflow when all three pumps are turned on.

A material balance on the tank, assuming constant liquid density, gives,

$$\frac{dh(t)}{dt} = \frac{4}{\pi D_t^2}\left(\dot{V}_1+\dot{V}_2+\dot{V}_3-\dot{V}_{out}(t)\right) \tag{P11.6}$$

where
$h(t)$= liquid in the tank, *ft*;
D_t= diameter of tank, *ft*; and
$\dot{V}_1, \dot{V}_2, \dot{V}_3$= volumetric flows from pumps, $\frac{ft^3}{s}$.

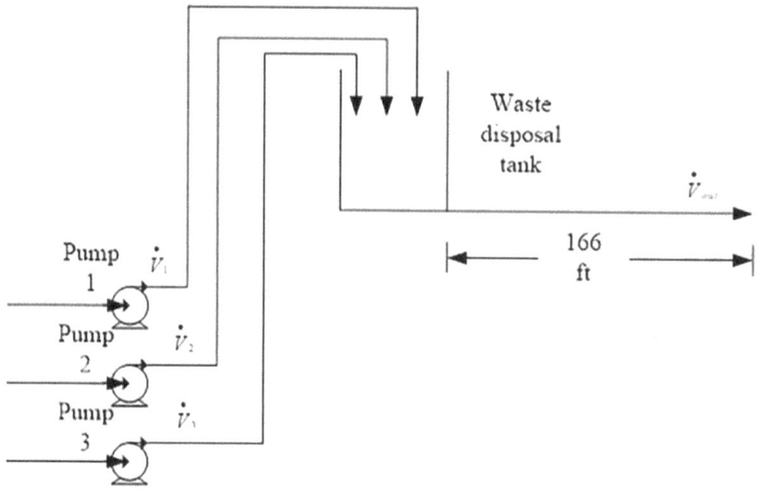

FIGURE P11.6 Waste disposal tank with three pumps.

The flow out of the tank is given by a momentum balance on the exit pipe,

$$\frac{d\dot{V}_{out}(t)}{dt}=\frac{A_p g_c}{L}\left(\frac{g}{g_c}h(t)-0.0185\dot{V}_{out}(t)-3.6\times10^{-5}\dot{V}_{out}^{2}(t)\right) \tag{P11.7}$$

where

A_p= area of exit pipe, ft^2

$g_c = 32.2\dfrac{lbm\cdot ft}{lbf\cdot s^2}$

L= length of exit pipe, ft; and

g= acceleration due to gravity $=32.2\dfrac{ft}{s^2}$

The term in parentheses contains a correlation for the friction losses in the pipe in feet of head. The tank is 25 ft (7.6 m) in diameter and 9 ft (2.7 m) tall. The exit pipe is 2 ft (0.61 m) in diameter and 166 ft (50.6 m) long. Each pump delivers 90 ft³/s.

Equations P11.6 and P11.7 constitute the model for the process. Simulate the process, and determine the minimum time sequence for starting the pumps that prevents the level in the tank from exceeding the height of the tank (overflow).

Problem 11.34

Solve part c of problem 7.9 by simulation rather than by analytical solution.

Problem 11.35

Section 9.9 develops the model of an adiabatic batch reactor; the model is given by the following two nonlinear differential equations,

$$\frac{dC^A}{dt} = -k_0 e^{-a/T} C^A$$

and

$$\frac{dT}{dt} = \frac{-Lk_0}{\rho C} e^{-a/T} C^A$$

where $L = -209{,}000$ J/mol, $\rho = 1050$ kg/m^3, $C = 3800$J/kg·K, $T(0) = 293$K, $C^A(0) =$ 500 mol/m^3, $a = 5627$ K, and $k_0 = 24.15 \times 10^6 \text{min}^{-1}$. Simulate this reactor and show that the responses of the concentration and temperature are the ones shown in Figure 9.9.

Problem 11.36

Consider Problem 9.16. The mathematical model is

$$mC\frac{dT}{dt} = 0 - hA(T - T_f)$$

$$m_f C_f \frac{dT_f}{dt} = hA(T - T_f) - 0$$

Using the parameters and initial conditions given in Problem 9.16, simulate this system and prepare graphs of T and T_f as functions of time.

Problem 11.37

Equation 10.73b describes the charging of the capacitor in Example 10.8 (Figure 10.32), and Equation 10.79b describes the discharging of the same capacitor. Simulate each case and compare the results with Figure 10.33.

Problem 11.38

Equation 10.67 describes the voltage v_D in Figure 10.29 (Example 10.7). Simulate and graph the response of this voltage for the specification given in the example.

Problem 11.39

Consider the RC circuit examined in Section 10.6 ($R = 1000\Omega$ and $C = 2.5 \times 10^{-6}$F). Suppose the supply voltage consists of a direct and alternating part: $v_S = 10 + 2\sin(2000t)$. Simulate this circuit and plot the resistor voltage v_R and the capacitor voltage v_C. What do you observe?

Problem 11.40

Consider problem 10.27. The model representing Figure P10.27B is

$$\frac{dv_R}{dt} = \frac{v_S}{R_1 C} - \frac{v_R}{\alpha R_1 C}$$

where $\alpha = R/(R + R_1)$. Use this model to solve part c of problem 10.27 by simulation.

Answers to Selected Problems

CHAPTER 1

1.1

a. $f(t)=1+\frac{3}{2}(t-2)u(t-2)-\frac{3}{2}(t-4)u(t-4)-3u(t-6)$

b. $f(t)=4-3u(t+16)+\frac{3}{2}(t+16)u(t+16)-\frac{3}{2}(t+14)u(t+14)-3u(t+12)$

1.2

a. $w_1(t)=200+200u(t-10)$

b. $146\frac{dh}{dt}+166.7\sqrt{h}=w_1+w_2$

$h=9\text{m}$

1.3

Approximately

$$\begin{aligned} Thrust = &\frac{20}{0.2}t\,u(t)-\frac{36}{0.2}(t-0.2)u(t-0.2) \\ &-\left[\frac{20}{0.2}t\,u(t)-\frac{36}{0.2}(t-0.2)u(t-0.2)-4\right]u(t-0.4)-4u(t-1) \end{aligned}$$

CHAPTER 2

2.1

a. $y=\sqrt{e^{2t}-2t-1}$

b. $y=2\exp[\sin(t)+t]$

c. $y=\sqrt{t^2+4t+4}$

d. $y=\frac{t^2}{2}+2\ln(t)+\frac{1}{2}$

e. $y=8t+2e^{-4t}-1$

f. $y=e^{\frac{t^3}{3}}$

g. $y=\frac{2}{e^{2t}+1}$

h. $y=2e^{(t^2+t)}$

i. $y=\frac{1}{1+2t^2}$

j. $y=4+t-4\cos\left(\frac{t}{2}\right)$

k. $y = -2t + e^{-2t} - 1 + 2\sin(2t)$

l. $y = -\frac{1}{2}t + \frac{1}{4}e^{2t} - \frac{1}{4}$

2.2

a. $u = \frac{1}{3}y^3 - y + C$

b $T = 5 + Ce^{-0.0002t}$

d. $u = \dfrac{-1}{\frac{1}{\pi}\sin(\pi t) + 2}$

e. $\frac{1}{6}\ln(y-1) + \frac{5}{6}\ln(y+5) = \frac{1}{2}t^2 + C$

f. $\frac{1}{2}\ln(u+1) - \ln(u+2) + \frac{1}{2}\ln(u+3) = t + C$

g. $y = 4\left(e^{-\frac{t^2}{2} - 3t} - 1\right)$

h. $y = e^{\sin(t) - 2t}$

i. $y = -e^t$

2.3

$$t = 4.44 - 1.2\ \text{in}C^S - 0.1C^S$$

2.5

$$v = 4\left(1 - e^{-2t}\right)$$

2.6

$$i_1 = 57.96 \times 10^{-4} + 0.76 \times 10^{-4}\left(1 - e^{-\frac{t}{9.3\times10^{-8}}}\right)$$

$$i_2 = 2.66 \times 10^{-4} + 0.000267\left(1 - e^{-\frac{t}{9.3\times10^{-8}}}\right)$$

2.7

$$X = \frac{18.91e^{0.377t}}{0.986 + e^{0.377t}}$$

2.8

$$C_V = 0.596$$

2.9

$$\frac{dC^A}{dt} = -85.28\left(C^A\right)^2\left(0.132 + 0.5C^A\right)$$

$$t = 0.18\,\text{h}$$

2.12

a. $v_y(t) = 151(1 - e^{-0.2t}) \rightarrow v_y(2) = 151(1 - e^{-0.4}) = 49.8\,\text{m}/\text{s}$

b. $t = 2 - 5\ln\left(\dfrac{-0.49}{-0.49 - 0.01(49.8)}\right) = 5.5\text{s}$

Thus, after the thrusts stop, the rocket continues upward for 3.5 seconds more.

2.13

a. $m\dfrac{dv}{dt} = -kv$ with $v(0) = 50\,\text{m/s}$

$\dfrac{dx}{dt} = v$ with $x(0) = 0\,\text{m}$

b. $t = 19.5\,\text{s}$

c. $x = 245\,\text{m}$

2.14

Going Up

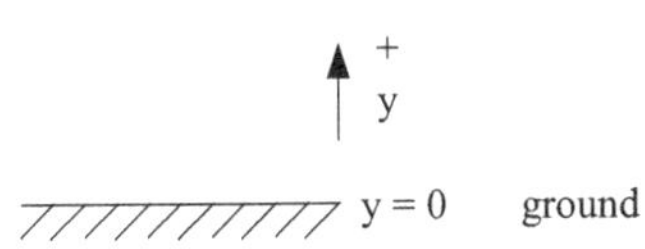

$$y_{max} = 62.5\,\text{m}$$

Going Down

$$t = 3.57\,\text{s}$$

2.15

$$\sum F_y = ma_y = m\frac{dv_y}{dt} \qquad \text{(1) 1 } eq.\text{, 2 } unk\left[\sum F_y, v_y\right]$$

Considering air resistance (drag force due to air)

$$\sum F_y = F_g + F_d = -\text{mg} - 1.0\left|v_y\right|v_y \qquad \text{(2) 2 eq., 2 unk.}$$

Eqs (1) and (2) constitute the model for the velocity. For the position,

$$\frac{dy}{dt} = v_y \qquad (3)3eq3unk[y]$$

Units of 1 are $\frac{\text{N}}{\text{m}^2/\text{s}^2}$

2.16

$$\text{m} \approx 2.3\,\text{kg}$$

2.17

$$\rho A\frac{dh}{dt} + C_v\sqrt{h} = w_1 + w_2$$

$$t = 3.15\,\text{min}$$

2.20

1.39 and 5.095 sec

CHAPTER 3

3.1

a. $y = e^{-t}\left[\frac{t^2}{2} + 1\right]$

b. $y = 1 - e^{-\sin(t)}$

c. $y = e^{-t^2/2}\left[t + 1\right]$

d. $y = 1 - e^{-t^3}$

e. $y = te^{-0.1t}$

f. $y = 10e^{-0.1t} + t - 10$

g. $y = e^{-2t}$

h. $y = e^{-2t}\left[\frac{t^3}{3} + 1\right]$

i. $y = 10 - 20e^{-0.1t} + 10e^{-0.2t}$

j. $y = \frac{1}{2}e^{-t^2}\left[1 - e^{-2t}\right]$

k. $y = e^{-2t}\left[1 - \cos(t)\right]$

l. $x = t^2 - \frac{1}{t^2}$

3.2

a. $y = 2e^{3t} - 2e^{-4t}$

b. $y = e^{-3t}\left[2\cos(t) + 6\sin(t)\right]$

c. $y = e^{-t} - e^{-2t}$
d. $y = \sin(2t)$
e. $y = 4e^{-2t}\sin(3t)$
f. $y = 3te^{3t} - e^{3t} = e^{3t}(3t - 1)$

3.3

a. $y = e^{-0.5t}\left[0.75\sin(2t) - \cos(2t)\right]$
b. $y = 1 - 2e^{-t}$
c. $y = 0.00224e^{5t} - 0.90603e^{-t}$
d. $y = e^{-t}\left[2\cos(2t) + \sin(2t)\right]$
e. $y = e^{-2t}\left[1 + 6t\right]$

3.4

a. $y = \frac{71}{54}e^{-3x} + \frac{67}{54}e^{3x} + \frac{2}{9}x - \frac{14}{9}$
b. $y = \frac{9}{36}e^{-3x} + \frac{7}{36}e^{3x} + \frac{1}{6}xe^{-3x} - \frac{4}{9}$
c. $y = -1.1795e^{-2t} + 0.2745e^{-5t} - 0.095\cos(3t) + 0.0045\sin(3t)$
d. $y = 3.5e^{-t} - 0.02e^{-5t} + 0.4t - 0.48$

3.5

a. $y = e^{-t} + (2t + 3)e^{-3t}$
b. $y = te^{-t} + 3t^2e^{-t}$
c. $y = (2t + 1)e^{-t} - e^{t}\cos 2t$
d. $y = 2\,e^{-t}\sin 3t + e^{-2t}$
e. $y = 3(1 + e^{-2t}) - 6e^{-3t} - 12te^{-3t}$
f. $y = -1.25e^{-2t}(1 + t) + 0.75(t - 1) + 2e^{-t}$
g. $y = 2(\cos 3t - \sin 3t) + 2\sin 2t$
h. $y = 6e^{-t} + 0.5778e^{1.618t} - 6.58\mathrm{e}^{-0.618t}$
i. $y = 2 + (2t^2 - 12t - 2)e^{-6t}$
j. $y = -\frac{1}{2} + 2t + \left(\frac{1}{2} - t\right)e^{-2t}$
k. $y = 3 + 2e^{-3t} + (2t - 2)e^{-2t}$

3.6

$$y = C_1e^{-5t} + e^{4t}(C_2\cos 3t + C_3\sin 3t)$$

3.9

$$C^G = 90 + 240.74\left(e^{-0.0224t} - e^{-0.0453t}\right)$$

3.10

a. $T(t) = \frac{5}{32} e^{-2t} \cos 2t + \frac{5}{16} t + \frac{315}{32}$

b. The roots $r_1 = -2 + i2$ and $r_2 = -2 - i2$ indicate a stable (bounded) and oscillatory response. The response given by the complete analytical solution is not bounded because the forcing function is not bounded.

3.11

a. $x = -e^{-\frac{t}{3}} [0.3 \cos 2.56t + 0.039 \sin 2.56t]$

b. $x = 0.09e^{-5.44t} - 0.39e^{-1.23t}$

3.16

$$x = 0.05 - e^{-3.5t}[0.05 \cos 2.78t + 0.063 \sin 2.78t]$$

3.17

$$v = 4\left(1 - e^{-2t}\right)$$

3.18

$$x = 0.1e^{-2t} + 0.2te^{-2t}$$

3.19

$$x = 0.1 - 0.1 \cos 3t$$

3.21

$$v_c = 10 - e^{-40t}\left[10 \cos(515t) + 0.78 \sin(515t)\right]$$

3.22

$$i_2 = 5.34 \times 10^{-4} - 2.68 \times 10^{-4} e^{-1.071 \times 10^7 t}$$

$$i_1 = 0.143 \times 10^{-3} v_s - 0.77 \times 10^{-4} e^{-1.071 \times 10^7 t} + 1.53 \times 10^{-4}$$

3.23

$$E = e^{-1111.11t}\left[40 \cos(993.81t) + 44.72 \sin(993.81t)\right]$$

3.24

$$i = 5.1338e^{-1.025t} - 0.135e^{-38.97t}$$

3.25

$$q = 5e^{-4t} - 20e^{-t} + 15$$

3.26

$$i = 5\left(e^{-2t} - e^{-4t}\right)$$

3.27

$$q = 7.45e^{-t} - 2.46e^{-3.03t}$$

3.28

a. $y = 3/4e^{-2t}(1+t) - 3/4\left(1 - 2t + 2t^2\right) + 3/4t(2t-1)$

b. $y = -3e^{-3t} - 3te^{-3t} - 3e^{-2t}(t-1) + 3te^{-2t}$

CHAPTER 4

4.1

a. $F(s) = \dfrac{1}{s^2}$

c. $F(s) = \dfrac{s}{s^2 + \omega^2}$

d. $F(s) = \dfrac{s+a}{(s+a)^2 + \omega^2}$

4.2

a. $F(s) = \dfrac{1}{s} + \dfrac{2}{s^2} + \dfrac{6}{s^3}$

b. $F(s) = \dfrac{1}{s+2} + \dfrac{2}{(s+2)^2} + \dfrac{6}{(s+2)^3}$

c. $F(s) = \dfrac{1}{s} + \dfrac{1}{s+2} - \dfrac{2}{s+1}$

d. $F(s) = e^{-2s}\left[\dfrac{1}{s} - \dfrac{1}{(s+2)^2 + 1}\right]$

4.3

a. $f(t) = u(t) + 2t + 3t^2 \Rightarrow F(s) = \dfrac{1}{s} + \dfrac{1}{s^2} + \dfrac{6}{s^3}$

Initial Value

$$\lim_{t\to 0}\left(u(t) + 2t + 3t^2\right) = 1 \qquad \lim_{s\to\infty} s\cdot\left(\frac{1}{s} + \frac{2}{s^2} + \frac{6}{s^3}\right) = \lim_{s\to\infty}\left(1 + \frac{2}{s} + \frac{6}{s^2}\right) = 1$$

Final Value

$$\lim_{t\to\infty}\left(u(t) + 2t + 3t^2\right) = \infty \qquad \lim_{s\to 0}\left(1 + \frac{2}{s} + \frac{6}{s^2}\right) = \infty$$

b. $f(t)=e^{-2t}\left(u(t)+2t+3t^2\right) \Rightarrow F(s)=\dfrac{1}{s+2}+\dfrac{2}{(s+2)^2}+\dfrac{6}{(s+2)^3}$

Initial Value

$$\lim_{t\to 0} e^{-2t}\left(u(t)+2t+3t^2\right)=1 \quad \lim_{s\to\infty} s\left[\frac{1}{s+2}+\frac{2}{(s+2)^2}+\frac{6}{(s+2)^3}\right]=\frac{\infty}{\infty}$$

Must apply L'Hopital's rule

$$\lim_{s\to\infty}\left[\frac{1}{1}+\frac{2}{2(s+2)}+\frac{6}{3(s+2)^2}\right]=1$$

Final Value

$$\lim_{t\to\infty} e^{-2t}\left(u(t)+2t+3t^2\right)=0\cdot\infty \qquad \lim_{s\to 0} s\left[\frac{1}{s+2}+\frac{1}{(s+2)^2}+\frac{6}{(s+3)^2}\right]=0$$

Apply L'Hopital's rule

$$\lim_{t\to\infty}\left(\frac{0}{2e^{2t}}+\frac{2}{2e^{2t}}+\frac{6t}{2e^{2t}}\right)=0 \text{ Check!}$$

4.6

a. $y=2e^{3t}+3e^{2t}$

b. $y=e^{-4t}\left[2\cos 3t+\sin 3t\right]$

c. $y=e^{-4t}\left[2\cos 2t+\sin 2t\right]$

d. $y=e^{-2t}\left[4\cos 2t-3\sin 2t\right]$

e. $y=e^{-2t}\left[\cos 2t+4\sin 2t\right]$

f. $y=2e^{-5t}\cos 4t$

g. $y=3e^{-3t}+4e^{-t}$

h. $y=-\dfrac{1}{5}e^{-t}+\dfrac{1}{5}\cos 2t+\dfrac{2}{5}\sin 2t$

i. $y=1-3e^{-2t}+2e^{-3t}$

j. $y=-1+3t+e^{-3t}$

k. $y=1+3e^{-4t}-4e^{-3t}$

l. $y=-2e^{-4t}+2\cos 3t+\sin 3t$

4.7

$$T=20+0.16\left[e^{-0.1t}-e^{-2t}\right]$$

4.12

$$x=0.05-e^{-3.5t}\left[0.05\cos 2.78t+0.064\sin 2.78t\right]$$

4.14

$$C^S(t) = 1.136 + 3.86e^{-0.11t}$$

$$C^P(t) = 3.86\left(1 - e^{-0.11t}\right)$$

4.16

$$\frac{X_1(s)}{F_A(s)} = \frac{\frac{20}{25}s}{(s+17.95)(s+1.11)(s+1.47-i4.22)(s+1.47+i4.22)}$$

On the basis of the roots the response will be stable and oscillatory

4.18

$$x = 0.9875 - e^{-0.5t}\left[0.3756\cos(3.97t) + 0.0474\sin(3.97t)\right]$$

4.19

$$x = 0.5 - e^{-2t}\left[0.5\cos(4t) + 0.25\sin(4t)\right]$$

4.20

$$v = 4 - 4e^{-2t}$$

4.21

$$x = 0.1e^{-2t} + 0.2te^{-2t}$$

4.22

$$x(t) = 0.1t - 0.0334\sin(3t)$$

4.29

$$C_A(t) = e^{-k_1 t} \quad C_B(t) = \frac{k_1}{k_1 - k_2}\left(e^{-k_2 t} - e^{-k_1 t}\right)$$

For $k_1 = k_2$, $C_B(t) = k_1 t e^{-k_1 t}$

CHAPTER 5

5.1

a. Yes. The time to steady state behavior does not depend on the nature of the forcing function.
b. No. While it does not depend on the forcing function, it will now depend on whether the 2nd order ODE is underdamped, critically damped, or overdamped.
c. Both first and second order ODES have the same steady state response, which can be expressed in one of two ways: $y_{steady-state} = KB$ or $y_{steady-state} = K(B-A) + y(t=0)$, assuming $y(t=0)$ is the steady state value when the forcing function has a value of A.

d. The natural response of a first order ODE will never oscillate. The same is not true for a second order ODE which will oscillate if underdamped.
e. First order ODES never display resonance.
f. Underdamped second order ODEs will display resonance if the driving frequency is near the natural frequency.

5.2

a. $\tau = \frac{V}{\dot{V}} = 500\,\text{s}$, $K = \frac{1}{\dot{V}} = 50\,\text{s}/\text{m}^3$

b. The steady state concentration is $c_{dye} = 0.0083 \frac{\text{kg}}{\text{m}^3}$, which will take $41.67\,\text{min}$ to reach based on 5τ. These answers are independent of the initial conditions as steady state only does not depend on initial conditions.

$$\frac{Km_{injected}}{\tau} = 50\frac{\text{s}}{\text{m}^3} \times \frac{0.01\,\text{kg}}{500\,\text{s}} = 0.001\frac{\text{kg}}{\text{m}^3}.$$

c. No, as the injection time is 600 seconds and $\tau = 500\,\text{s}$, this will have to be modeled as a rectangular pulse.

5.3

$$\tau = \frac{1}{k}$$

$$m_a(0) = 4\,\text{mg}$$

a. From the graph, $\tau =$ 30 s, hence $k = \frac{1}{30\,\text{s}} = 0.033\,\text{s}^{-1}$

b. Steady state is achieved after 5τ, or $150\,\text{s}$

5.4

a. $\tau = 12\,\text{min}$, hence $T_{water} = 68.96^\circ\text{F}$ after 12 minutes
b. $60\,\text{min}$
c. As 1 minute is much less than τ, the solution can be approximated as a response to an impulse where $T_{water}(t) = (100-70)^\circ\text{F} \times \left(\frac{1\,\text{min}}{12\,\text{min}}\right) e^{-\frac{t}{12\,\text{min}}} + 70^\circ\text{F}$

d. As opposed to the answer, we now need to model as a rectangular pulse

$$T_{water} = \begin{cases} (100-70)^\circ\text{F}\left(1 - e^{-\frac{t}{1\,\text{min}}}\right) + 70^\circ\text{F} & 0 \le t < 1\,\text{min} \\ (100-70)^\circ\text{F}\left(e^{1} - 1\right)e^{-\frac{t}{1\,\text{min}}} + 70^\circ\text{F} & t \ge 1\,\text{min} \end{cases}$$

5.5

a. $\tau = 202.5\,\text{s}$
b. $\dot{q}_{in} = 4{,}000\,\text{J}/\text{s}$

c. 5τ is $16.9\,\text{min}$

d. As τ is much less than 40 minutes and $K=\frac{1}{UA}=\frac{1}{80\,\text{J/s}\cdot{}^{\circ}\text{C}}=0.0125\frac{\text{s}\cdot{}^{\circ}\text{C}}{\text{J}}$, the temperature will increase approximately as $T(t)=0.0125\frac{\text{s}\cdot{}^{\circ}\text{C}}{\text{J}}\times\dot{q}_{in}+25\,{}^{\circ}\text{C}$,

5.6

a. $T=1000\,\text{s}$

b. No. The transient response never depends on the forcing function. The transient response is determined only by the complementary solution in which the forcing function is set to zero.

c. τ is 1 min and the period of the forcing function is $0.1\,\text{s}$. Hence, the forcing function changes much faster than v_c can respond, hence v_c will not reach $5\,\text{V}$. In other words, $\tau\omega \gg 1$.

d. τ is 1 min and the period of the forcing function is $69.4\,\text{days}$. Hence, v_c responds instantaneously to changes in the forcing function and therefore v_c will reach $5\,\text{V}$. In other words, $\tau\omega \ll 1$.

5.7

a. $\tau=\frac{L}{R}=0.001\,\text{s}$, $K=\frac{1}{R}=0.05\,\text{A/V}$

b. $5\tau=0.005\,\text{s}$

c. $i=\frac{Kv_{in}}{\sqrt{1+(\omega\tau)^2}}=\frac{0.05\,\text{A/V}\times 15\,\text{V}}{\sqrt{1+(200\times 0.001)^2}}=0.74\,\text{A}$

$$\phi=\tan^{-1}(-\tau\omega)=-0.197\,\text{rad}$$

$$\Delta t_{lag}=\frac{T\phi}{2\pi}=-\frac{0.197\,\text{rad}}{200\,\text{rad/s}}=-0.000985\,\text{s}$$

5.8

a. Here τ is much less than the pulse width (5 s), hence the steady state velocity is achieved in the duration of the pulse. The gain is $K=\frac{2\,\text{m/s}}{1\,\text{N}}$, which is defined as $K=1/\mu_f$. Hence the friction is $\mu_f=0.5\,\text{N}\cdot\text{s/m}$. Note that $\tau=\text{m}/\mu_f=\frac{0.1}{0.5}=0.2\,\text{s}$

b. Here τ is much greater than the pulse width (5 s) and the transient dies off after about 50 s. τ can be estimated as 10 s, and $\mu_f=\text{m}/\tau=0.5\,\text{N}\cdot\text{s/m}$.

5.9

a. $\tau=787.5\,\text{s}$

b. $5\tau=62.63\,\text{min}$

c. The forcing functions are $\dot{q}_{in}$ and T_{∞}. The gains associated with each are $K=\frac{1}{hA}=1\frac{\text{s}\cdot{}^{\circ}\text{C}}{\text{J}}$ and $K=1$ [dimensionless], respectively.

d. $\dot{q}_{in} = 75\,\mathrm{J/s}$

e. $\Delta\dot{q}_{in} = 75\,\mathrm{J/s}$

5.10

a. $\tau = \sqrt{\frac{\mathrm{m}}{\mathrm{k}}}$, $\zeta = \frac{P}{2\sqrt{\mathrm{mk}}}$, $K = \frac{1}{k}$

b. $K = \frac{2\,\mathrm{m}}{10\,\mathrm{N}}$ =0.2 m/N

c. Decay Ratio $\approx \frac{0.1}{0.6} = 1/6$

d. From the definition of the Decay ratio, $\zeta = 0.292$. $k = \frac{1}{K} = 5\,\mathrm{N/m}$ and $P = 4.13\,\mathrm{N \cdot s/m}$

e. Here, when $\zeta = 1.0$ the system will reach steady state displacement as quickly as possible without oscillations (critically damped). $P = 14.14\,\mathrm{N} \cdot \frac{\mathrm{s}}{\mathrm{m}}$.

5.11

a. $\tau = \sqrt{\frac{a}{1+2G}}$, $K = \frac{2G}{(1+2G)}$, and noting that $\frac{b}{1+2G} = 2\tau\zeta$, $\zeta = \frac{1}{2\tau}\left(\frac{b}{1+2G}\right)$ or $\zeta = \frac{b}{2\sqrt{a(1+2G)}}$

b. $G = 9.4$

c. $K = 0.95$, hence $C = 0.95 \times 5 = 4.75$

d. $\frac{5\tau}{\zeta} = \frac{5 \times 1.12}{0.22} = 25$ seconds.

5.12

a. $\tau = \sqrt{LC}$, $\zeta = \frac{R}{2}\sqrt{\frac{C}{L}}$, $K = C$

b. The derivative of a step function is a delta function, $\frac{dv_s}{dt} = 5V \cdot \delta(t)$

c. If there is no resistance, $\zeta = 0$. Hence, the undamped frequency is $\omega_0 = \frac{1}{\tau} = \frac{1}{\sqrt{LC}} = \frac{1}{\sqrt{1.25\,\mathrm{H} \times 20\,\mathrm{F}}} = 0.2\frac{\mathrm{rad}}{\mathrm{s}}$.

d. The period of the response is 72.525 s – 39.585 s = 32.94 s. Hence the frequency of the response is $\omega = \frac{2\pi}{32.94\,\mathrm{s}} = 0.1907\,\mathrm{rad/s}$.

e. The damping ratio is $\zeta = \sqrt{1 - \left(\frac{\omega}{\omega_0}\right)^2} = 0.3$. The resistance is $R = 2\zeta\sqrt{\frac{L}{C}} = 2 \times 0.3\sqrt{\frac{1.25\,\mathrm{H}}{20\,\mathrm{F}}} = 0.15\,\Omega$

CHAPTER 6

6.3

$$m\frac{d^2x}{dt^2}+P\frac{dx}{dt}=f_A(t)$$

with $\frac{dx}{dt}|_{t=0}=x(0)=0$. At $t\to\infty$ the velocity reaches a constant rate meaning that the acceleration equals zero. Thus, $\frac{dx}{dt}|_{t\to\infty}=4$

6.4

$$\mathrm{m}\frac{d^2x}{dt^2}+kx=ky \text{ with } \frac{dx}{dt}|_{t=0}=x(0)=0$$

The analytical solution is $x=-0.0333\sin 3t+0.1t$

6.5

a. $10\frac{d^2x}{dt^2}+10\frac{dx}{dt}+40x=f_A(t)$ with $\frac{dx}{dt}|_{t=0}=x(0)=0$ and $f_A(t)=20u(t)N$

b. $x=0.5-e^{-0.5t}\left[0.5\cos 1.94t+0.13\sin 1.94t\right]$

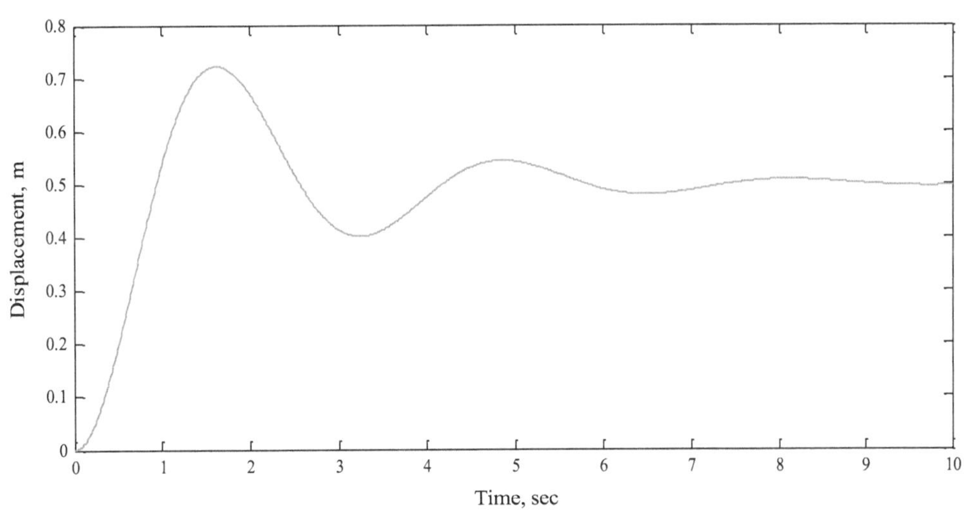

6.6

a. $10\frac{d^2x}{dt^2}+70\frac{dx}{dt}+200x=f_A(t)$ with $\frac{dx}{dt}|_{t=0}=x(0)=0$ and $f_A(t)=10u(t)N$

b. $x=0.05-e^{-3.5t}\left[0.05\cos(2.78t)+0.06\sin(2.78t)\right]$

6.7

a. $10\frac{d^2x_1}{dt^2}+40\frac{dx_1}{dt}+20x_1=f_A(t)+10x_2+20\frac{dx_2}{dt}$

$10\frac{d^2x_2}{dt^2}+40\frac{dx_2}{dt}+20x_2=10x_1+20\frac{dx_1}{dt}$

b. $x_1=1.333-0.371e^{-0.551t}-1.0te^{-t}-1.0e^{-t}+0.037e^{-5.449t}$

$x_2=0.667+0.371e^{-0.551t}-1.0te^{-t}-1.0e^{-t}-0.037e^{-5.449t}$

6.9

a. $10\frac{d^2x_1}{dt^2}+35\frac{dx_1}{dt}+70x_1=15\frac{dx_2}{dt}$

$10\frac{d^2x_2}{dt^2}+15\frac{dx_2}{dt}+150x_2=f_A(t)+15\frac{dx_1}{dt}$

b. Ultimately, $x_1 = 0$ and $x_2 = 4/15$ m

6.13

$$2\frac{d^2x}{dt^2}+8\frac{dx}{dt}+40x=8(2.5)u(t)$$

$$x(\infty)=0.5\,\text{m}$$

6.15

a. $5\frac{d^2y}{dt^2}+40\frac{dy}{dt}+200y=(m_1g+30)u(t)$ with $x(0)=\frac{dy}{dt}|_{t=0}=0$

b. $y=0.395-e^{-4t}[0.395\cos 4.9t+0.32\sin 4.9t]$

c.

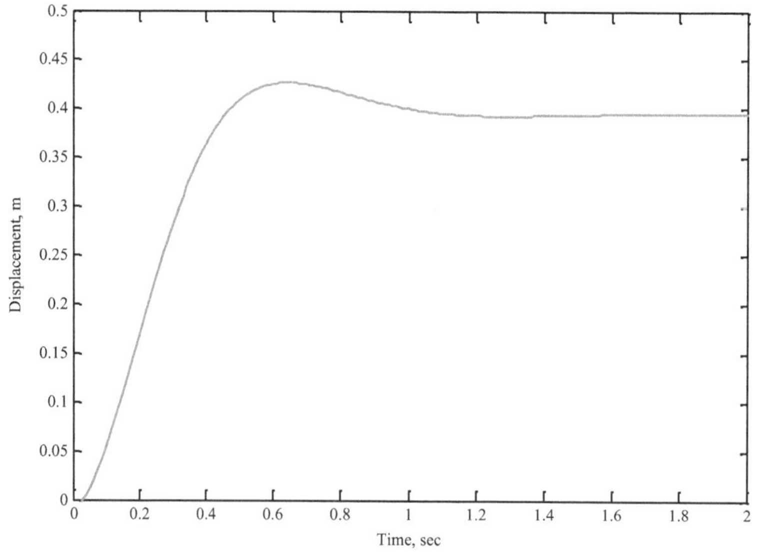

6.19

Block 1

$$m_1\frac{d^2x_1}{dt^2}=\sum F_1$$

$$\sum F_1=-k_1x_1-P_1\frac{dx_1}{dt}-P_2\frac{dx_1}{dt}-k_2(x_1-x_2)$$

Block 2

$$m_2\frac{d^2x_2}{dt^2}=\sum F_2$$

$$\sum F_2=-k_2(x_2-x_1)-P_3\frac{dx_2}{dt}+f_A(t)$$

6.20

$$0.47\frac{d^2x}{dt^2}+6.87\frac{dx}{dt}+114x=2.87v_{drive}+114x_{drive}$$

$$\text{or }0.47\frac{d^2x}{dt^2}+6.87\frac{dx}{dt}+114x=0.0347+1.379t$$

or

$$x=e^{-7.31t}\left[-0.00067\sin 13.75t+0.0004\cos 13.75t\right]+0.0121t-0.0004$$

CHAPTER 7

7.1

a. $J = mR^2/2$.
b. $J = mR^2$.

7.2

b. $\theta=\theta_0e^{-3.25t}\left[\cos(6.2t)+\frac{3.25}{6.2}\sin(6.2t)\right]$

7.3

b. $\theta=e^{-3t}\left(\frac{\pi}{2}\cos t+\frac{3\pi}{2}\sin t\right)$

7.4

b. $\theta_1(t)=\frac{1}{16}(-1+\cos 4t)$

$\theta_2(t)=\frac{1}{16}(1-\cos 4t)$

7.5

b. T = 2.09 N·m, 5τ = 7.5 s

7.6

b. $\omega_{1,t} = 110$(rad)/s $\omega_{2,t} = 100$(rad)/s

c. 5τ = 15s

d. $t = 5.4$s

7.7

a. t_{eng} = 1.875s

b. t_{lock} = 7.5 sec. $\omega_1 = \omega_2$ = 249.9 (rad)/s.

7.8

b. ω_1 = 55 and ω_2 =15 (rad)/s.

c. $\omega_1(t) = 55 - 54e^{-0.2t} - e^{-1.2t}$ $\omega_2(t) = 15 - 18e^{-0.2t} + 3e^{-1.2t}$

7.9

b. $\theta_1 = 1/15 = 0.06\overline{6}$ rad

$\theta_2 = 1/(29/3) + 0.06\overline{6} = 0.170$ rad

c. $\theta_1(t) = 13.\overline{3}\left(0.005 + e^{-2t}\{0.001\cos 4t + 0.0055\sin 4t\} - e^{-t}\{0.006\cos 3t + 0.008\overline{6}\sin 3t\}\right)$

$\theta_2(t) = \dfrac{1}{0.725}\left(0.12\overline{3} + e^{-2t}\{0.044\overline{6}\cos 4t - 0.004\overline{3}\sin 4t\} - e^{-t}\{0.168\cos 3t + 0.020\overline{4}\sin 3t\}\right)$

7.10

c. $v_t = 0.98$ m/s

d. $v = 0.98\left(1 - e^{-2t}\right)$

7.11

b. θ_{ult} = 0.5 rad.

c. $\theta = 0.5 - 0.5e^{-0.2t} - 0.1te^{-0.2t}$

7.12

b. x_{ult} = 0.05 m

c. $x = -0.1e^{-t} + 0.05e^{-2t} + 0.05$

7.13

b. Ultimate values

$v_t = \dfrac{f_p R_p}{kR_w\left(R_1 / R_s\right)}$

$$\omega_{p,t} = \frac{f_p R_p}{k R_w^2 \left(R_1 / R_s\right)^2}$$

c. $R_1/R_S = 3.06$, $v_t = 21.5$ mile/hr

7.14

c. $\omega = \sqrt{\frac{fR}{J}\left[\theta - \sin\theta\cos\theta - \frac{\pi}{2}\right]}$

7.15

b. $\omega_1 = \left(\frac{FR_H}{mR_1^2} - \frac{g}{R_1}\right)t$

$$\theta_1 = \left(\frac{FR_H}{mR_1^2} - \frac{g}{R_1}\right)\frac{t^2}{2}$$

c. F = 98 N
d. F = 99.57 N

7.16

b. T = 0.6 N·m
c. $\theta_2 = \frac{3}{2}\left(t + \frac{81.25}{2}e^{-2t/81.25} - \frac{81.25}{2}\right)$
d. t = 173.4 sec.

7.17

b. $\theta_{eq} = 1.225$ (rad).
c. $\theta = 0.775\cos 4t + 1.225$

7.18

b. $\omega_{3t} = 0.5\text{(rad)/s}$

$$\theta_{4t} = 0.2\text{(rad)}$$

7.19

$$\theta = 0.08e^{-0.2t} + 0.04te^{-0.2t} - 0.06\sin(0.4t) - 0.08\cos(0.4t)$$

7.20

$$\theta_1 = \frac{1.5}{16}\cos(4t) - \frac{1.5}{16} + 0.25t^2 \qquad \theta_2 = \frac{-1.5}{16}\cos(4t) + \frac{1.5}{16} + 0.25t^2$$

CHAPTER 8

8.1

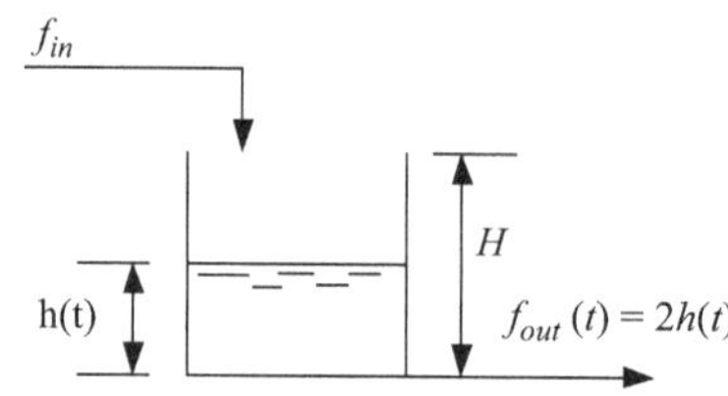

$$D = 6.15\,\text{m};\ H = 3.05\,\text{m};\ A = \frac{\pi D^2}{4} = 29.7\,\text{m}^2$$

a. $2 \Rightarrow \dfrac{\text{m}^3}{\text{min} * \text{m}} = \dfrac{\text{m}^2}{\text{min}}$

b. $h = 1.4\,\text{m}$

c. $29.7\dfrac{dh}{dt} + 2h = f_{in}$

d. $f_{in} = 2.8 + 4.3u(t)$ $h = 3.55 - 2.15e^{-\frac{t}{14.85}}$

e. Maximum height will occur as t approaches ∞, at which time $h(\infty) = 3.55\,\text{m}$.

This indicates that the tank will overflow. To obtain the time when the tank will overflow set the height equal to 3.05 meters

$$3.05 = 3.55 - 2.15e^{-\frac{t}{14.85}}$$

and t = 21.66 minutes

8.3

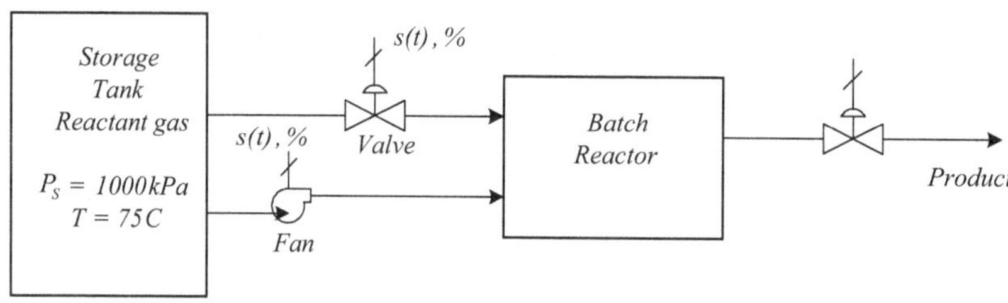

a. $k_{fan} = 4.167\dfrac{\text{m}^3}{\text{min}}$

b. $C_V = 0.16\dfrac{\text{m}^3/\text{min}}{kPa^{1/2}}$

8.4

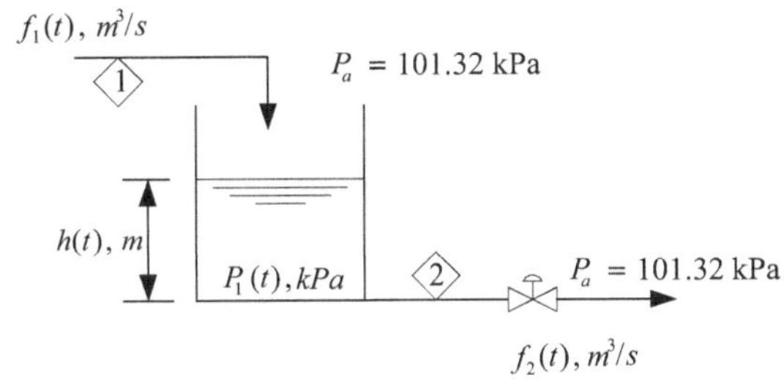

a. Height at steady state = 1.5 m

b. $A\dfrac{dh(t)}{dt}+1.565\times10^{-3}\sqrt{\rho gh(t)}=f_1(t)$

$h(0)=1.5\text{m}$

8.8

Pond is the System

$$\frac{X^p_{out}(s)}{X^p_{in}(s)}=\frac{1}{10S+1}$$

$$X^p_{in}(s)=\frac{5\times10^{-4}}{s}$$

$$t = 12.04 \text{ h}$$

8.9

Hole is 0.2 cm in diameter; $A=\dfrac{\pi D^2}{4}=0.0314\text{cm}^2$

Exit velocity through hole $= v=\sqrt{2gh(t)}\text{ in}\dfrac{\text{cm}}{\text{s}}$

$t=160\text{s}$

8.10

a. $vp_1=\dfrac{w_1}{500C_{v1}\sqrt{G_f\Delta P_1}}=\dfrac{14{,}000}{500(7.3)\sqrt{1.12(103)}}=0.357$

$vp_2=\dfrac{W_2}{500C_{v2}\sqrt{G_f\Delta P_2}}=\dfrac{14000}{(500)(7.28)\sqrt{(1.12)(52.88)}}=0.5$

b.

Mass Balance

$w_1(t)-w_2(t)=\dfrac{dm(t)}{dt}=\rho\dfrac{dV(t)}{dt}$ (1) 1 equation, 3 unknowns $[w_1(t),w_2(t),V(t)]$

Valve 1

$w_1(t) = 500C_{v1}vp_1(t)\sqrt{G_f \Delta P_1}$ (2) 2 equations, 3 unknowns

Valve 2

$w_2(t) = 500C_{v2}vp_2(t)\sqrt{G_f \Delta P_2(t)}$ (3) 3 equations, 4 unknowns [$\Delta P_2(t)$]

$\Delta P_2 = P_2 + \frac{\rho g h(t)}{1000} - P_3$ (4) 4 equations, 5 unknowns [$h(t)$]

$V(t) = V_T \left[\frac{h^2(t)[3r - h(t)]}{4r^3}\right]$ (5) 5 equations, 5 unknowns

c.

Mass Balance

$w_1(t) - w_2(t) = \frac{dm(t)}{dt} = \rho \frac{dV(t)}{dt}$ (6) 1 equation, 3 unknowns [$w_1(t), w_2(t), V(t)$]

$w_1 = 500C_{v1}vp_1\sqrt{G_f \Delta P_1(t)}$ (7) 2 equations, 4 unknowns [$\Delta P_1(t)$]

$\Delta P_1(t) = 448 - 395 - \frac{\rho g h_1(t)}{1000}$ (8) 3 equations, 5 unknowns [$h_1(t)$]

$h_1(t)$ is the liquid level height about the entrance point. It is a positive value or zero (cannot be negative)

$h_1(t) = h(t) - H_1$ if $h_1(t)$ is negative then make it 0 (9) 4 equations, 6 unknowns [$h(t)$]

H_1 is the height of the entrance port.

Valve 2

$w_2(t) = 500C_{v2}vp_2(t)\sqrt{G_f \Delta P_2(t)}$ (10) 5 equations, 7 unknowns [$\Delta P_2(t)$]

$\Delta P_2 = P_2 + \frac{\rho g h(t)}{1000} - P_3$ (11) 6 equations, 7 unknowns

$V(t) = V_T \left[\frac{h^2(t)[3r - h(t)]}{4r^3}\right]$ (12) 7 equations, 8 unknowns

8.11

a.

i. $\rho f_i x_i(t) - \rho f_o x_o(t) = \frac{dm\, x_o(t)}{dt}$

ii. $t = 0.1686\,\text{h}$

b.

i. $f_i(t) - 2500 = \frac{dV(t)}{dt}$

ii. $V(t) = 4000 - 1500t$

t = 2.67 h

8.12

a. Tank 1

Total Mass Balance

$$\rho f_5(t) + \rho f_1(t) - \rho f_3(t) = \frac{dm_1(t)}{dt} = \rho \frac{dV_1(t)}{dt}$$

If $V_1(t) \geq 7000\,\text{gal}$ then $f_3(t) = f_5(t) + f_1(t)$ if not $f_3(t) = 0$

Mass Balance - Component A

$$\rho f_5(t)x_5(t) - \rho f_3(t)x_3(t) = \rho \frac{dV_1(t)x_3(t)}{dt}$$

b. Tank 2

Total Mass Balance

$$\rho f_3(t) + \rho f_2(t) - \rho f_4(t) = \frac{dm_2(t)}{dt} = \frac{\rho dV_2(t)}{dt}$$

If $V_2(t) \geq 7000\,\text{gal}$ then $f_4(t) = f_3(t) + f_2(t)$ if not $f_4(t) = 0$

Mass Balance - Component A

$$\rho f_3(t)x_3(t) + \rho f_2(t)x_2(t) - \rho f_4(t)x_4(t) = \rho \frac{dV_2(t)x_4(t)}{dt}$$

c. Tank 3

Total Mass Balance

$$\rho f_4(t) + \rho f_7(t) - \rho f_6(t) = \frac{dm_3(t)}{dt} = \rho \frac{dV_3}{dt}$$

If $V_3(t) \geq 7000\,\text{gal}$ then $f_6(t) = f_4(t) + f_7(t)$ if not $f_6(t) = 0$

$$\rho f_4(t)x_4(t) + \rho f_7(t)x_7(t) - \rho f_6(t)x_6(t) = \rho \frac{dV_3(t)x_6(t)}{dt}$$ 9 equations, 9 unknowns

8.13

a. Mass Balance – $Mg(OH)_2$

$$-25x = 300\frac{dx}{dt}$$

b. Time for reducing mass fraction to 0.15:

$0.15 = 0.45e^{-0.0833t}$

t = 13.19 min

Time for reducing mass fraction to 0.03

$0.03 = 0.45e^{-0.0833t}$

t = 32.51 min

c.

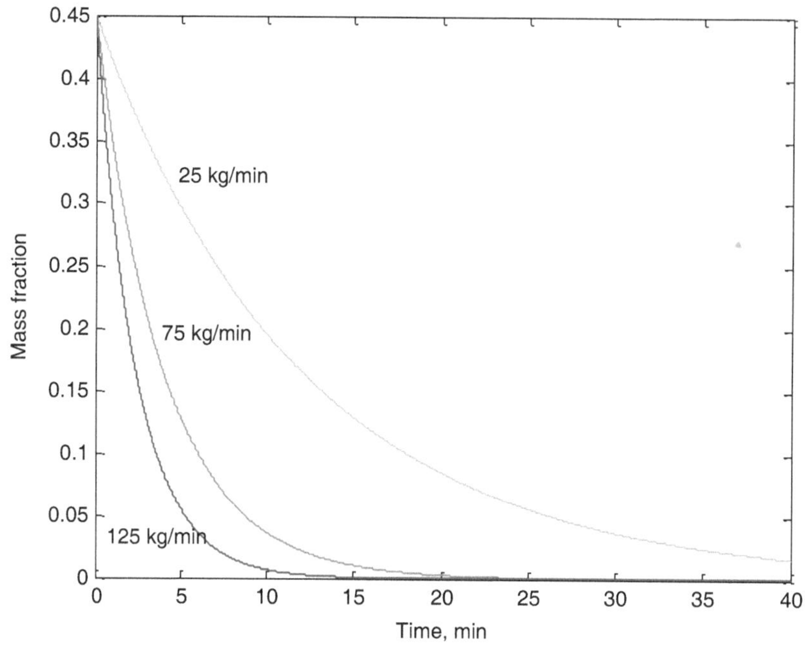

We can write (1) as $-wx = 300\dfrac{dx}{dt}$

where w is the flow in/out. The analytical solution is then,

$x = 0.45e^{-\frac{w}{300}t}$

As w increases the negative exponent also increases making the exponential term decrease faster.

8.15

a. Mass Balance

$-\rho f(t) = \dfrac{dm(t)}{dt} = \rho\dfrac{dV(t)}{dt}$ (1) 1 equation, 2 unknowns [$f(t), V(t)$]

Valve

$f(t) = C_V(s)\sqrt{h(t)}$ (2) 2 equations, 3 unknowns [$h(t)$]

Volume of a vertical cylindrical tank,

$V(t) = \pi r^2 h(t)$ (3) 3 equations, 3 unknowns

Note that the question to obtain the tank height H that delivers in 4 hours (14400 sec) 11.0 m^3 of liquid. H is really another unknown. We can write this specification as,

$V(14400) = V(0) - 11.0 = \pi r^2 H - 11.0$ (4) 4 equations, 4 unknowns [H]

b. The model for this part is very similar to that of part (a) with some changes on unknowns.

Mass Balance

$-\rho f(t) = \dfrac{dm(t)}{dt} = \rho\dfrac{dV(t)}{dt}$ (5) 1 equation, 2 unknowns [$f(t), V(t)$]

Valve

$f(t) = C_V(s)\sqrt{h(t)}$ (6) 2 equations, 4 unknowns [$h(t), s$]

Volume of a vertical cylindrical tank,

$V(t) = \pi r^2 h(t)$ (7) 3 equations, 4 unknowns

The final specification is related to the requirement that in 4 hours (14400 sec) 3.5 m³ of liquid be delivered,

$$V(14400) = V(0) - 7.5 = \pi r^2 H - 7.5 \quad (4) \text{ 4 equations, 4 unknowns}$$

8.16

a. $11.7\frac{dh(t)}{dt} + 599.5h^{1.5} = f_i(t)$

with $h(0) = 0.202$ ft and $f_i(t) = 30 + \Delta f_i u(t)$.

b. The differential equation is nonlinear. Therefore, the methods of Chapters 3 and 4 do not work except if we linearize the nonlinear term as shown in Chapter 3. It is probably better to use simulation.

8.17

a. $f_i(t) - 1.25s(t) = 7.07\frac{dh(t)}{dt}$

b. $t = 9.9\,\text{min}$

c. $h(5) = 8.84\,ft$

8.18

a. $\frac{dC}{dt} + 0.2C = 1.6 - 1.4e^{-0.04t}$

$C = 8 - 8.75\left(e^{-0.04t} - e^{-0.2t}\right)$

b. From the above expression, it takes 5.158 h or 18568.8 sec for C to drop to $4\,\text{mg}/\text{liter}$. Note that as time t increases the exponentials will drop and eventually become zero and C will become 8 again. This happens because once the microorganism has broken down the organic material no more oxygen will be used, and its concentration will rise because it continues being transferred from the atmosphere. At t = 18.2 h or 65520 sec C will rise above $4\,\text{mg}/\text{liter}$ again. At 1 m/s speed, for the first 18.6 km the oxygen will be at an acceptable level, then after 65.52 km it will be again at an acceptable level.

8.19

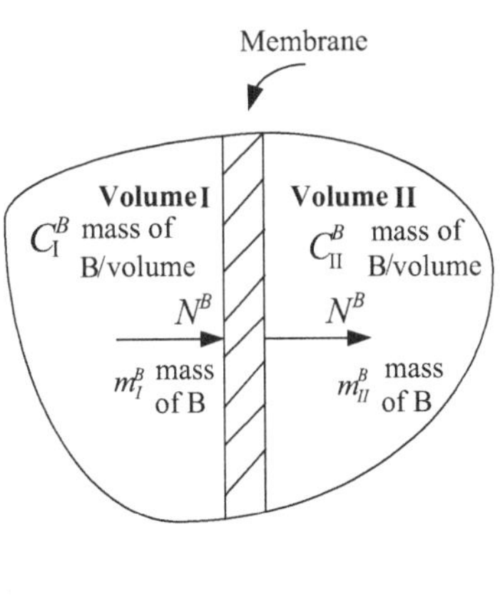

New model

$$V_I \frac{dC_I^B}{dt} = -kA(C_1^B - C_{II}^B) \text{ and } V_{II} \frac{dC_{II}^B}{dt} = kA(C_1^B - C_{II}^B)$$

Because the two differential equations are coupled, use Laplace transform to obtain the analytical solution, and then use Equation 6 to obtain the necessary time.

The batch time should be longer because the mass transfer rate will be slower due to the effect of the B concentration in volume II.

8.20

$$m^{C1}(t) = 2.5 - 0.214e^{-4t} - 2.286e^{-0.5t}$$

and

$$m^{C2}(t) = 2 + 0.286e^{-4t} - 2.286e^{-0.5t}$$

8.21

$$m^{theophylline}(t) = 7500(1 - e^{-0.0013t}) \text{ for } 0 < t \le 30\,\text{min}$$

$$m^{theophylline}(t) = 286.87e^{-0.0013(t-30)} \text{ for } t > 30\,\text{min}$$

8.23

$$\frac{dC^A}{dt} = -85.28\left(C^A\right)^2\left(0.132 + 0.5C^A\right)$$

$$t = 0.14\,\text{h} = 8.4\,\text{min}$$

8.25

$$R = 12.64\,\text{cm} \quad H = 79.69\,\text{cm}$$

$$t = 16185.2\,\text{s}$$

8.26

a. $0.1 - 0.00002336(0.9) f_{air\ steady\ state} = 0 \Rightarrow f_{air\ steady\ state} = 4756.47\, ft^3/\text{min}$

b. $f_{air\ steady\ state} = 100\,s(t) \Rightarrow s_{steady\ state} = 47.56\%$

c. Mass Balance – Toluene

$$0.1 - f_{air} C^{toluene} = \frac{dm^{toluene}}{dt} = V\frac{dC^{toluene}}{dt} = 10\frac{dC^{toluene}}{dt} \quad (1)$$

1 equation, 2 unknowns [f_{air}, $s(t)$]

Fan

$$0.1\frac{df_{air}}{dt}+f_{air}=100\,s(t) \qquad (2)$$

1 equation, 2 unknowns

We first obtain the analytical solution of Equation 2 as $s(t)$ goes from 0% to 47.56 %. Equation 2 is a first order differential equation, and Equation 5–1.10b gives the solution,

$$f_{air}=f_{air}(0)+KD(1-e^{-\frac{t}{\tau}})=4756(1-e^{-\frac{t}{0.1}})$$

Using this solution in Equation 1,

$$\frac{dC^{toluene}}{dt}=0.01-475.6(1-e^{-\frac{t}{0.1}})C^{toluene}$$

Using simulation, the figure below shows the response.

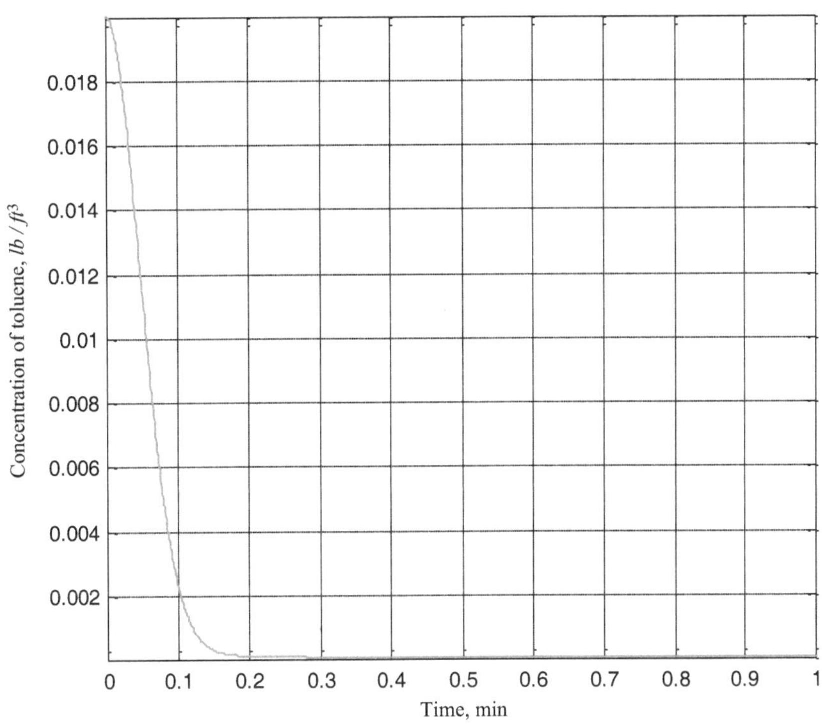

CHAPTER 9

9.1

b. t = 2678 s

9.2

b. $\tau = 10.7\,\text{s}$

c. Time to reach 39.9 °C is 53.6 s

9.3

b. $t = 722.8\,\text{s}$
c. $T = 763^{\circ}\text{C}$

9.4

a. t = 2844 s
b. t = 786 s

9.5

b. $T_{ss} = 37.5^{\circ}\text{C}$, $5\tau = 94$ sec
c. $t = 9.6\,\text{s}$

9.6

b. $T = T_0 + \dfrac{A}{mCb}\left(1 - e^{-bt}\right)$

c. $T_{\infty} = 33^{\circ}\text{C}$, $5\,\tau = 10$ s

9.7

T = 45.8 °C

9.8

b. $T_{ss} = \dfrac{UAT_A + U_C A_C T_C}{UA + U_C A_C}$, $A_C = 0.02667\ \text{m}^2$.

c. t = 2071 s

9.9

a. $mC\dfrac{dT}{dt} = hA_E(T_E - T) - UA(T - T_A)$ $T(0) = T_A$

$m_E C_E \dfrac{dT_E}{dt} = \dot{q}_{in} - hA_E(T_E - T)$ $T_E(0) = T_A$

b. $T_{ss} = 75^{\circ}\text{C}$, $T_{Ess} = 115^{\circ}\text{C}$

9.10

$T(t) = 25 + 500e^{-0.125t}$

The maximum temperature of 525 °C occurs at t = 0.

9.11

$$t = 6.93\,\text{min}$$

9.12

$$t = 2.04\,\text{min}$$

9.13

$$T = 319.2\,\text{K}$$

9.14

a. $t = 41.8\,\text{min}$

b. $T = 312.0\,\text{K}$

9.15

$$T = 9.9 + 0.01t + 0.1e^{-0.1t}$$

9.16

$$T_f(t) = 26.47 - 1.47e^{-0.0255t}$$

$$T(t) = 26.47 + 73.53e^{-0.0255t}$$

9.17

a. $T = T_A + \dfrac{(T_0 - T_A)}{e^{mL} - e^{-mL}}\left[\left(e^{mL} - 1\right)e^{-mz} + \left(1 - e^{-mL}\right)e^{mz}\right]$ where $m^2 = 2h/kR$

b. 27.4 J/s.

c. $T_{\min} = 54.6^{\circ}\text{C}$

9.18

a. $T = T_A + \left(\dfrac{T_o - T_A}{e^{mL} + e^{-mL}}\right)\left(e^{m(L-z)} + e^{-m(L-z)}\right)$ where $m^2 = 4h/kW$

b. 18.9 J/s

c. 0.3 J/s, a factor of 63 lower

9.19

a. $T = T_A + \dfrac{\dot{s}R}{2h} + \dfrac{\dot{s}}{4k}\left(R^2 - r^2\right)$

b. 1150 °C

9.20

a. $T = T_f + (T_f - T_s)\dfrac{\left(\dfrac{1}{r} - \dfrac{1}{R}\right)}{\left(\dfrac{1}{R} - \dfrac{1}{R_0}\right)}$

b. $T_f = T_A + \dfrac{4\pi R^3 \dot{s}/3}{U_o A_o}$

$$T_s = T_A + \frac{4\pi R^3 \dot{s} / 3}{hA_o}$$

c. $R_0 = 0.0258\,\text{m} \quad T_f = 50.4\,^{\circ}\text{C}$

9.21

As one example, $T_A = -3.5\ ^{\circ}C$ for a velocity of 6 m/s.

CHAPTER 10

10.1

$$RC\frac{dq}{dt} + q = Cv_s \Longrightarrow \tau\frac{dq}{dt} + q = K100\,u(t)$$

where $\tau = RC = 200\,\text{ms} = 0.2\,\text{s}$, and $K = C = 50\frac{mC}{V} = 0.05\frac{C}{V}$

$$q = q(0) + KD\left(1 - e^{-\frac{t}{\tau}}\right) = 5\left(1 - e^{-\frac{t}{0.2}}\right), C$$

and $i = \frac{dq}{dt} = 25e^{-\frac{t}{0.2}},\ A$

10.2

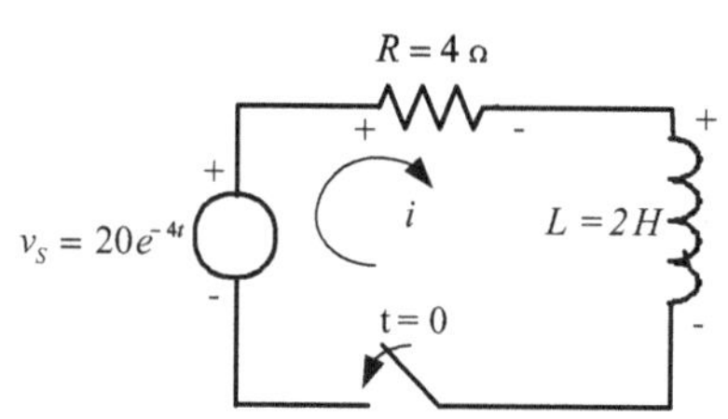

$$0.5\frac{di}{dt} + i = 0.25v_s$$

with $i(0) = 0$

$$i = 5\left(e^{-2t} - e^{-4t}\right)$$

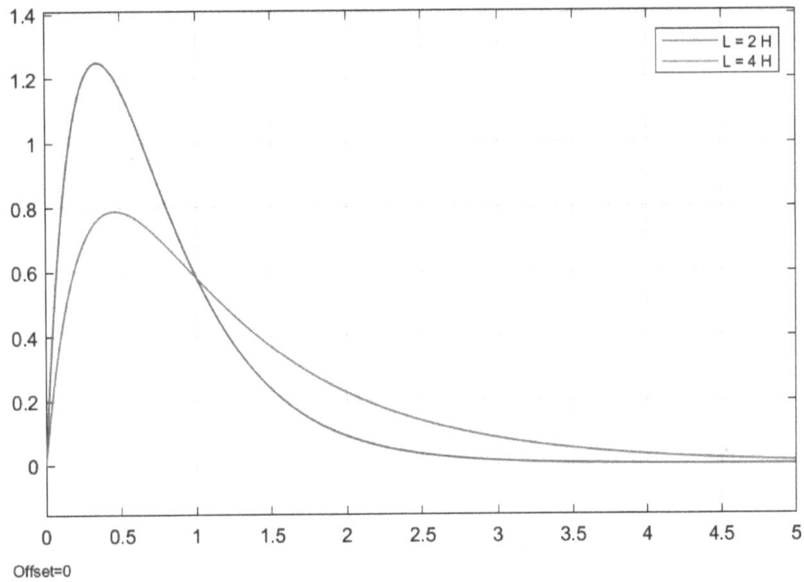

10.3

Obviously, the current flowing through the circuit before the switch moves to position 2 is 50 A. Once the switch changes, the circuit looks as

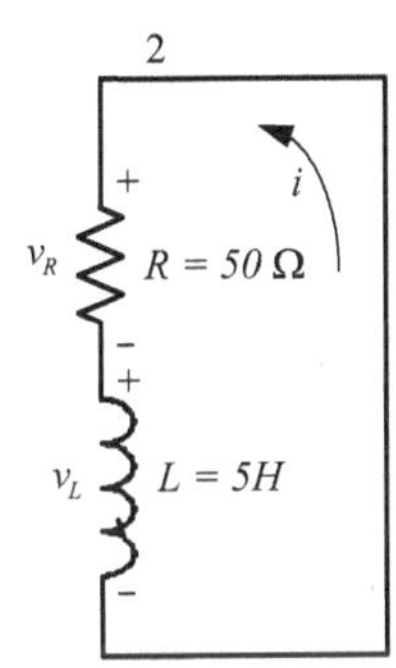

$$Ri + L\frac{di}{dt} = 0 \quad \text{or} \quad \frac{L}{R}\frac{di}{dt} + i = 0 \text{ with } i(0) = 50\,A$$

$$i = 50e^{-10t},\ v_R = Ri = 2500e^{-10t}, \text{and } v_L = L\frac{di}{dt} = -2500e^{-10t}$$

10.4

$$0.25\frac{d^2q}{dt^2} + 1.25\frac{dq}{dt} + q = 0.25v_s$$

with q(0)=0 and $\frac{dq}{dt}|_{t=0} = 0$

After a very long time, $i(\infty) = 0$ A, because the capacitor becomes an open circuit again.

10.5

$$RC\frac{dq}{dt}+q=Cv_s$$

with $q(0)=0$

After a very long time, $i(\infty)=0$ A, because the capacitor becomes an open-circuit again.

10.6

$$R\frac{dq}{dt}+\frac{1}{C}q=v_s \text{ or } RC\frac{dq}{dt}+q=Cv_s$$

with q(0) = 0 C

10.7

$$LC\frac{d^2q}{dt^2}+RC\frac{dq}{dt}+q=0 \text{ or } 0.33\frac{d^2q}{dt^2}+1.33\frac{dq}{dt}+q=0$$

with $q(0)=Cv_c(0)=5$

10.8

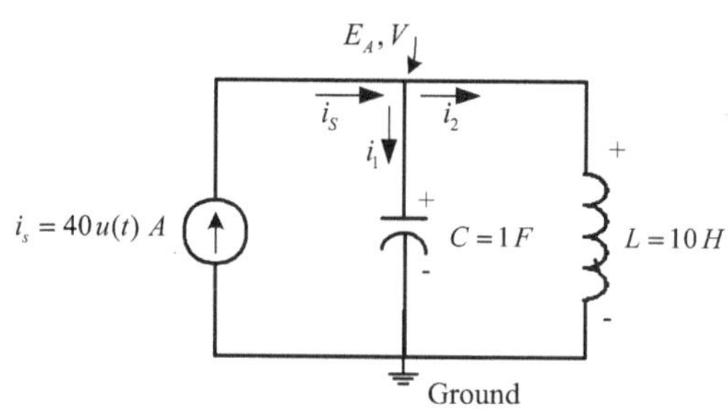

$$\frac{d^2E_A}{dt^2}+0.1E_A=\frac{di_S}{dt}$$

$$i_1=\frac{dE_A}{dt}=40\cos 0.316t \text{ and } i_2=0.1\int E_A dt=-40\cos 0.316t\,|_0^t=40-40\cos 0.316t$$

10.10

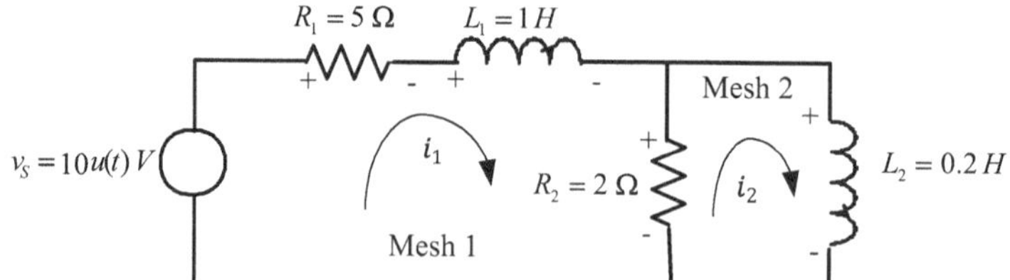

$$\frac{di_1}{dt} + 7i_1 = v_S + 2i_2$$

$$0.2\frac{di_2}{dt} + 2i_2 = 2i_1$$

Both differential equations are of first order therefore, the responses will be stable and monotonic.

After a very long time, $\left.\frac{di_1}{dt}\right|_{t\to\infty} = \left.\frac{di_2}{dt}\right|_{t\to\infty} = 0$ so $7i_1(\infty) = 10 + 2i_2(\infty)$,

$2i_2(\infty) = 2i_1(\infty)$, and from these two equations, $i_1(\infty) = i_2(\infty) = 2A$

10.12

Before the switch closes there is no current flowing anywhere; the capacitor acts as an open circuit, and the inductor as a short circuit. Once the switch closes the circuit is as shown below.

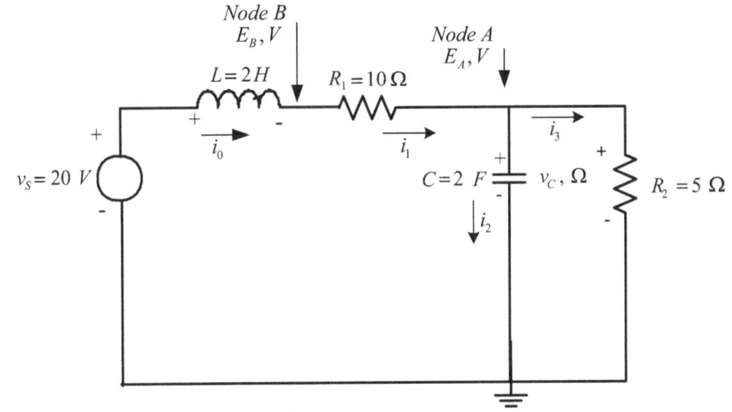

We will use the nodal analysis method to develop the model; the figure shows the necessary information.

Node A

KCL $i_1 - i_2 - i_3 = 0$ (1) 1 equation, 3 unknowns [i_1, i_2, i_3]

For R_2 $\quad i_3 = \frac{E_A}{R_2} = \frac{E_A}{5}$ (2) 2 equations, 4 unknowns [E_A]

For the capacitor $\quad i_2 = C\frac{dE_A}{dt} = 2\frac{dE_A}{dt}$ (3) 3 equations, 4 unknowns

For R_1 $\quad i_1 = \frac{E_B - E_A}{R_1} = \frac{E_B - E_A}{10}$ (4) 4 equations, 5 unknowns [E_B]

Node B

KCL $i_0 - i_1 = 0$ (5) 5 equations, 6 unknowns [i_0]

For the inductor $i_0 = i_0(0) + \frac{1}{L}\int (20 - E_B)dt = \frac{1}{2}\int (20 - E_B)dt$ (6) 6 equations, 6 unknowns

These 6 equations constitute the model.

10.13

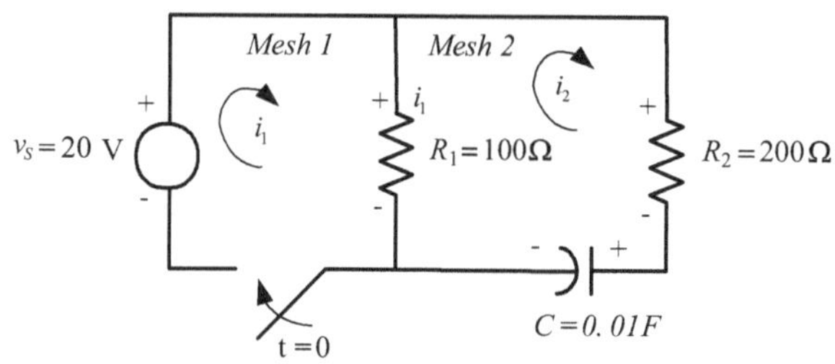

$$i_1 = \frac{1}{R_1} v_s + i_2$$

$$R_2 \frac{di_2}{dt} + \frac{1}{C} i_2 = \frac{dv_s}{dt}$$

with $i_2(0) = 0$

10.14

Switch in position B

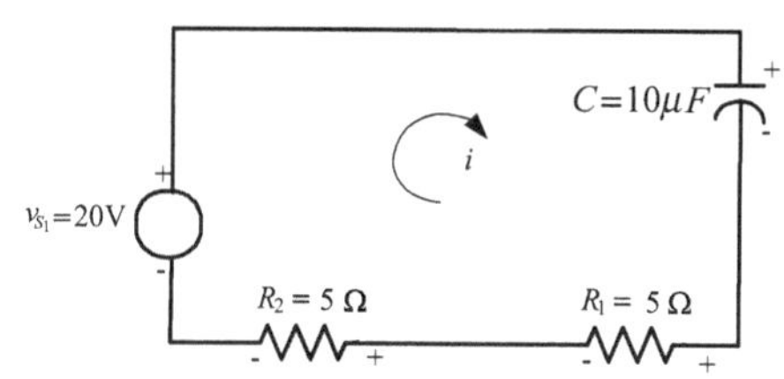

$$v_{R1} = R_1 C \frac{dv_c}{dt} \text{ and } v_{R2} = R_2 C \frac{dv_c}{dt} \text{ and } (R_1 C + R_2 C)\frac{dv_c}{dt} + v_c = v_{s1}$$

with $v_c(0) = 0V$

The solution is $v_c = 20(1 - e^{-10000t})$

and $v_c(20) = 20V$

Switch in position C

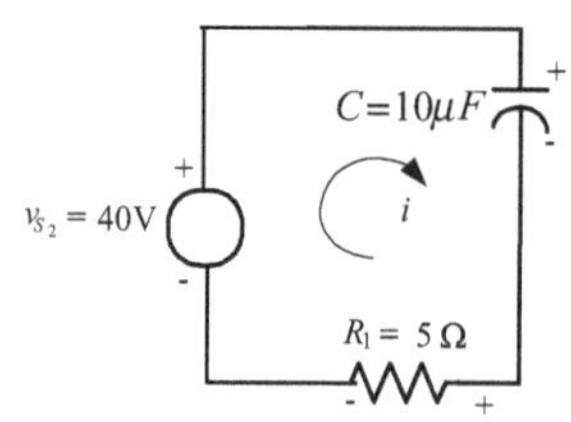

$$R_1 C \frac{dv_c}{dt} + v_c = 40$$

with $v_c(0) = v_c(20)$

10.15

a. Switch at point B

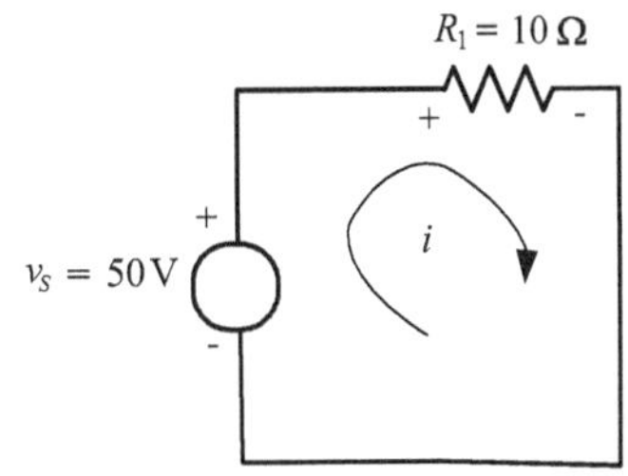

At steady-state the inductor becomes a "short circuit". Then,

$$i(\infty)=\frac{v_s}{R_1}=\frac{50}{10}=5A$$

b. Switch at point C

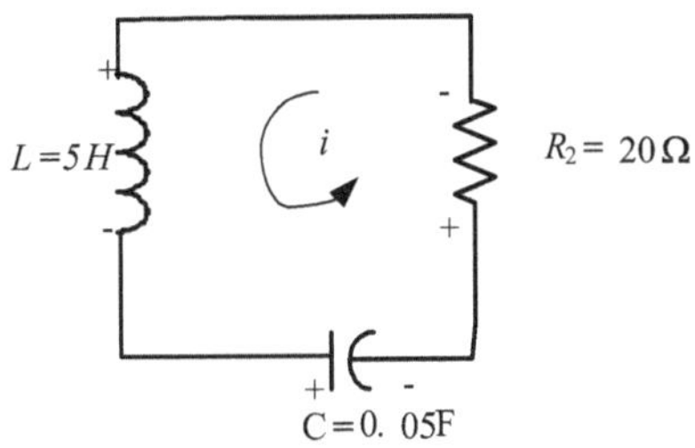

$$0.25\frac{d^2i}{dt^2}+\frac{di}{dt}+i=0$$

with $i(0)=5A;\ \frac{di}{dt}|_{t=0}=0$

10.16

a. Switch is closed

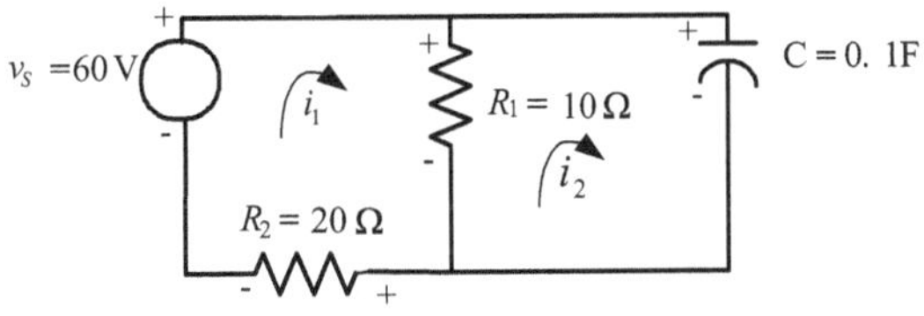

$i_1(\infty)=\frac{1}{30}[v_s]=2A$

$v_c(\infty)$ is equal to $v_{R1}(\infty)$, so

$v_c(\infty)=v_{R1}(\infty)=R_1(i_1(\infty)-i_2(\infty))$

$v_c(\infty)=v_{R1}(\infty)=20V$

Note that $i_2(\infty)=0A$ makes sense because of steady state, the equation acts as an open circuit

b. Switch is opened

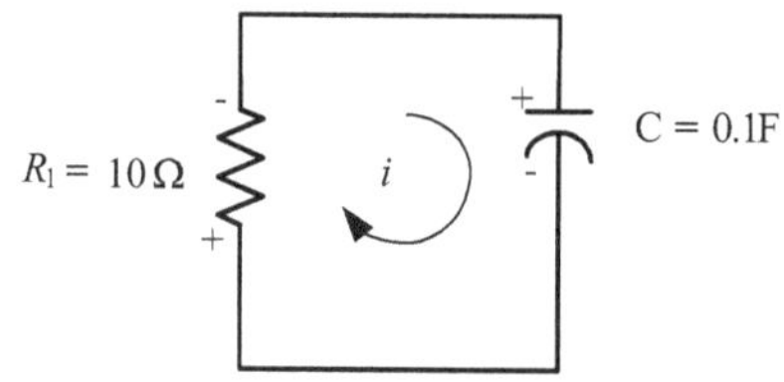

$t = 0.693\,\text{s}$

10.17

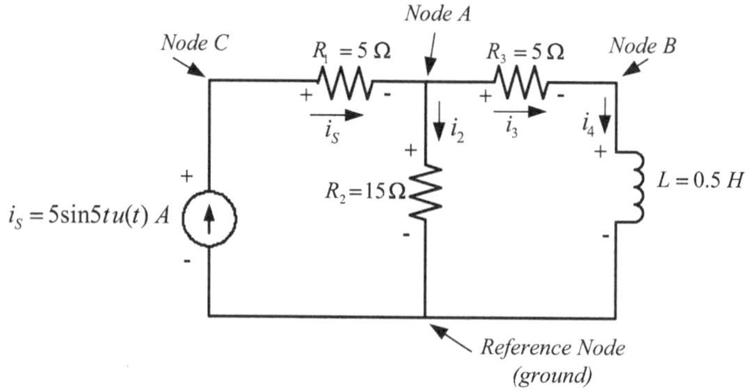

$$\frac{dE_A}{dt} = \frac{R_2R_3}{R_3+R_2}\frac{di_s}{dt} + \frac{R_2}{R_3+R_2}\frac{dE_B}{dt}$$

$$\frac{1}{R_2+R_3}\frac{dE_B}{dt} + \frac{1}{L}E_B = \frac{R_2}{R_2+R_3}\frac{di_s}{dt}$$

We now have 2 equations with 2 unknowns, E_A and E_B. Knowing E_A and i_S, E_C is calculated and the voltage drop across the voltage source is equal to - E_C.

10.19

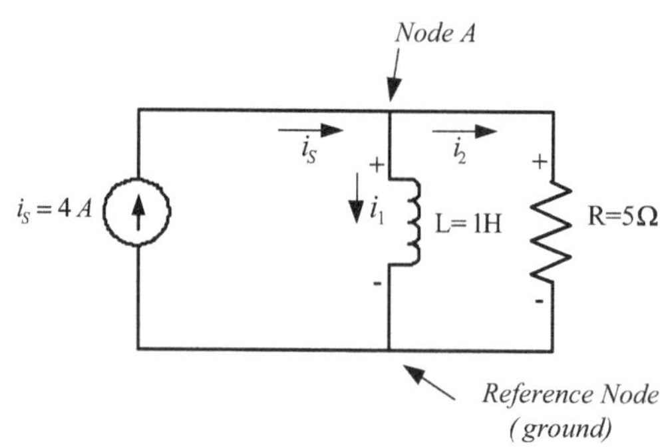

$$\frac{1}{R}\frac{dE_A}{dt} + LE_A = \frac{di_S}{dt}$$

10.20

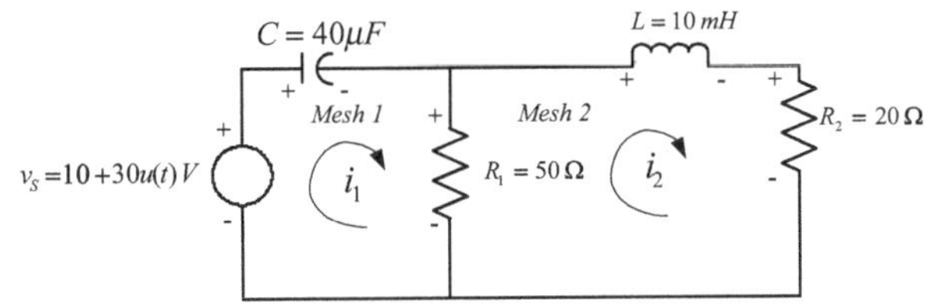

$$R_1\frac{di_1}{dt}+\frac{1}{C}i_1=\frac{dv_S}{dt}+R_1\frac{di_2}{dt}$$

$$L\frac{di_2}{dt}+(R_1+R_2)i_2=R_1i_1$$

These two equations form the model that provides the currents through each resistor - i_2 through R_2, and i_1-i_2 through R_1. Knowing i_1, v_C is calculated.

Note that at steady state the capacitor is an open circuit and thus, $i_2(0)=i_1(0)=0\,A$.

10.21

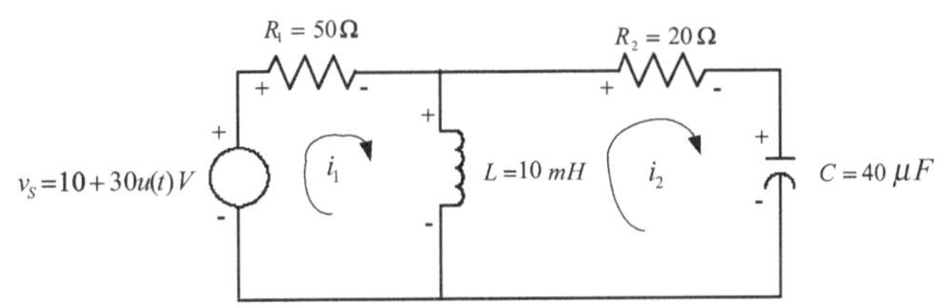

a. Mesh Analysis

$$L\frac{di_1}{dt}+R_1i_1=v_s+L\frac{di_2}{dt}$$

$$L\frac{d^2i_2}{dt^2}+R_2\frac{di_2}{dt}+\frac{1}{C}i_2=L\frac{d^2i_1}{dt^2}$$

To find the initial conditions consider the diagram at the initial steady state.

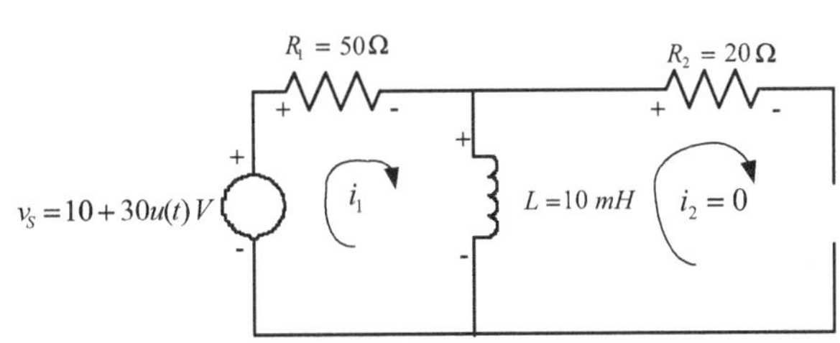

Because, the capacitor behaves as an open circuit, $i_2(0)=0A$.
Mesh 1 $v_{R1}-v_S=0$

$$R_1i_1(0)=v_S(0)$$

$$i_1(0)=\frac{10}{50}=0.2\,A$$

b. Nodal Analysis

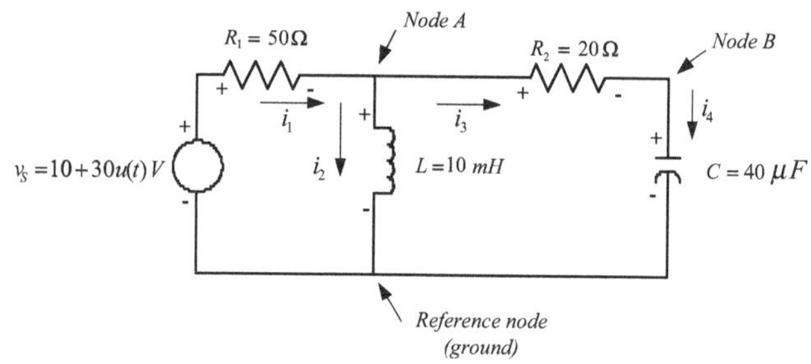

$$\left(\frac{1}{R_1}+\frac{1}{R_2}\right)\frac{dE_A}{dt}+\frac{1}{L}E_A=\frac{1}{R_1}\frac{dV_s}{dt}+\frac{1}{R_2}\frac{dE_B}{dt}$$

$$C\frac{dE_B}{dt}+\frac{1}{R_2}E_B=\frac{1}{R_2}E_A$$

with $E_A(0)=E_B(0)=0V$.

10.22

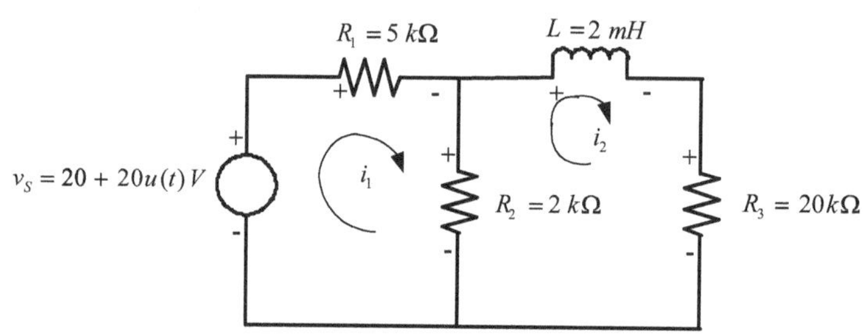

$$i_1 = 0.143\times10^{-3}\ v_S + 0.286 i_2$$

$$2\times10^{-3}\frac{di_2}{dt}+21428 i_2 = 0.286\,V_s$$

and $i_1(0^-)=0.00293A$; $i_2(0^-)=0.000267A$; $i_3(0^-)=0.00266A$

10.23

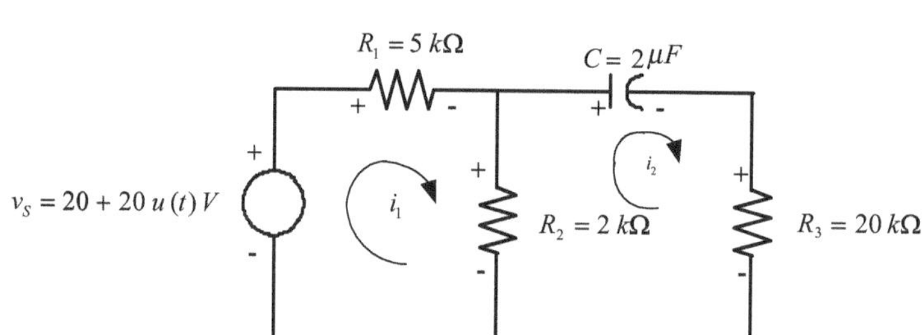

a. Mesh Analysis

$$i_1 = \frac{1}{R_1 + R_2}(v_s + R_2 i_2) \qquad \left(R_2 + R_3 - \frac{R_2^2}{R_1 + R_2}\right)\frac{di_2}{dt} + \frac{1}{C}i_2 = \frac{R_2}{R_1 + R_2}\frac{dv_s}{dt}$$

We now have 2 equations and 2 unknowns, i_1 and i_2.: $\quad v_C = v_C(0) + \frac{1}{C}\int i_2 dt$

To obtain the initial conditions, describe the circuit at the initial steady state.

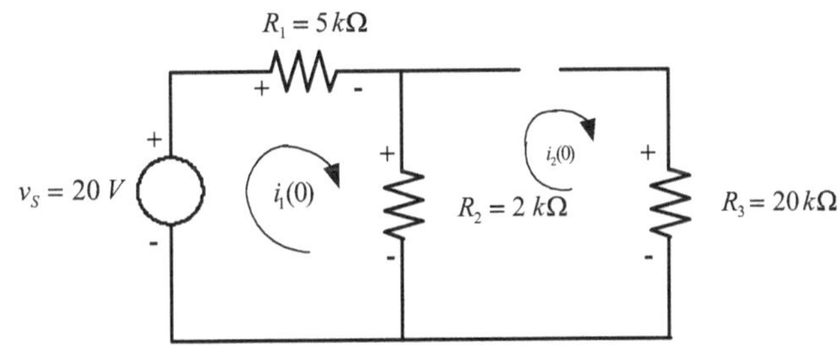

The capacitor acts like an "open circuit." So, $i_2(0) = 0A$
Then, $v_{R1}(0) + v_{R2}(0) - v_s(0) = 0$

$$(R_1 + R_2)i_1(0) = v_s(0)$$

$$i_1(0) = \frac{v_s(0)}{(R_1 + R_2)} = \frac{20V}{7000\Omega} = 0.00285A$$

b. Nodal Analysis

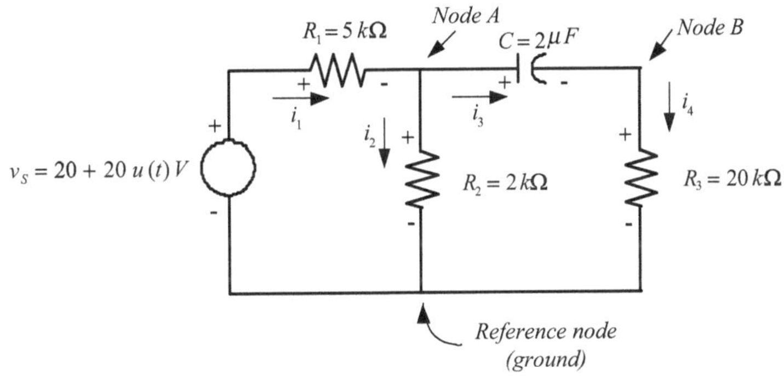

$$C\frac{dE_A}{dt} + \frac{R_2 + R_1}{R_1 R_2}E_A = \frac{1}{R_1}v_s + C\frac{dE_B}{dt}$$

$$C\frac{dE_B}{dt} + \frac{1}{R_3}E_B = C\frac{dE_A}{dt}$$

We now have 2 equations with 2 unknowns, E_A and E_B.
To obtain the initial conditions, consider again the drawing of the circuit at the initial steady state.

$$E_A(0) = \frac{R_2}{R_1 + R_2}v_s(0) = \frac{2000}{7000} \times 20 = 5.71V$$

$$E_B(0) = 0$$

10.24

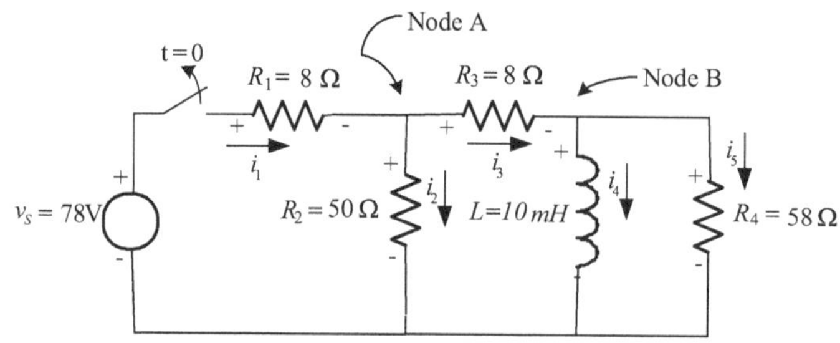

a. Initially, before the switch is opened.

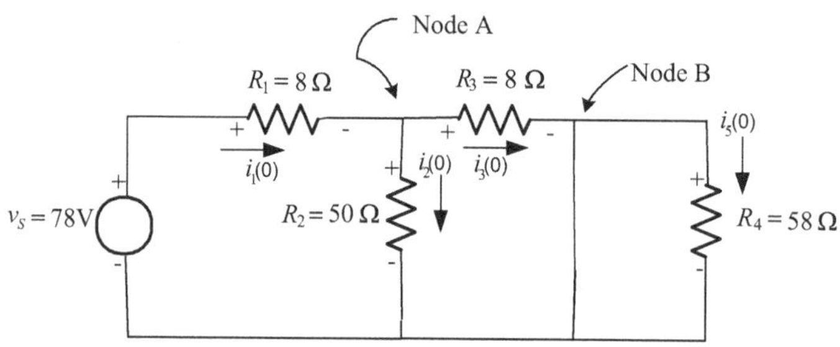

At steady state the inductor becomes a short circuit. In this case, the potential at Node B is equal to ground, and there is no current through R_4.

$E_A(0) = 36.11V;\ E_B(0) = 0V$

And $i_1(0^-) = 5.236\,A;\ i_2(0^-) = 0.722\,A;\ i_3(0^-) = 4.514\,A$

b. $E_A = 0.463 v_s + 0.463 E_B;\ 0.084\dfrac{dE_B}{dt} + 100\mathrm{E_B} = 0.058\dfrac{dv_s}{dt}$

10.25

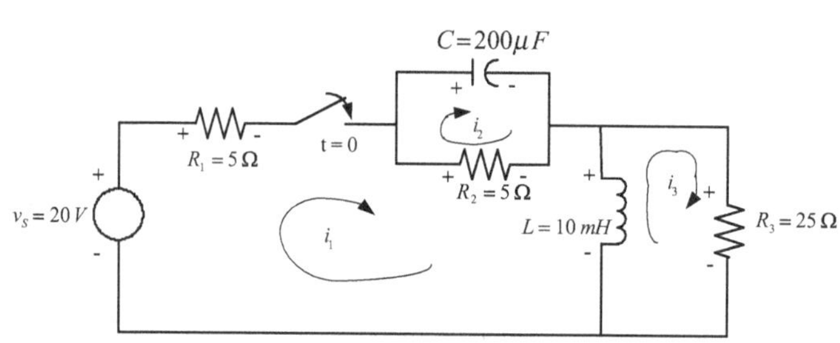

$$L\frac{di_1}{dt} + (R_1 + R_2)i_1 = v_s + R_2 i_2 + L\frac{di_3}{dt};\ R_2\frac{di_2}{dt} + \frac{1}{C}i_2 = R_2\frac{di_1}{dt};$$

$$L\frac{di_3}{dt} + R_3 i_3 = L\frac{di_1}{dt};\ v_c = v_C(0) + \frac{1}{C}\int i_2 dt$$

The solution of the set of equations provides the three currents, i_1, i_2, i_3, and v_C. All initial conditions are equal to zero. Use Laplace transform to obtain the analytical solution.

10.26

We will use the nodal method of analysis; the following figure shows the necessary information.

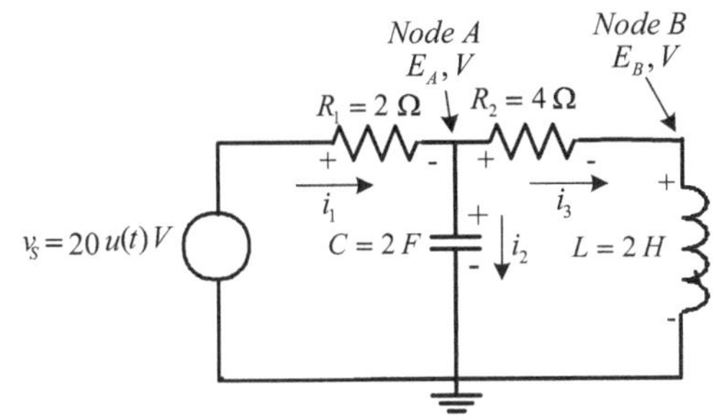

$$2\frac{dE_A}{dt} + 0.75E_A = 10 + 0.25E_B\ ;\ 0.25\frac{dE_B}{dt} + 0.5E_B = 0.25\frac{dE_A}{dt}$$

$$E_A = 13.33 - 13.63e^{-0.41t} + 0.30e^{-1.84t}\ ;\ E_B = 3.48\left(e^{-0.41t} - e^{-1.84t}\right)$$

The currents are,

$$i_1 = \frac{20 - E_A}{2} = 3.33 + 6.82e^{-0.41t} - 0.15e^{-1.84t}\ ;\ i_2 = 2\frac{dE_A}{dt} = 11.09e^{-0.41t} - 1.09e^{-1.84t}$$

$$i_3 = \frac{E_A - E_B}{4} = 3.33 - 4.28e^{-0.41t} + 0.95e^{-1.84t}$$

Then, $P_{R_1} = R_1 i_1^2 = 2\left[3.33 + 6.82e^{-0.41t} - 0.15e^{-1.84t}\right]^2$

$$P_C = i_2 v_C = i_2 E_A = \left[13.33 - 13.63e^{-0.41t} + 0.30e^{-1.84t}\right]\left[11.09e^{-0.41t} - 1.09e^{-1.84t}\right]$$

$$P_L = i_3 v_L = i_3 E_B = \left[3.33 - 4.28e^{-0.41t} + 0.95e^{-1.84t}\right]\left[3.48\left(e^{-0.41t} - e^{-1.84t}\right)\right]$$

10.27

a. For Figure P10.27b

<u>Mesh 1</u>

KVL	$v_C + v_{R_1} - v_S = 0$	(1) 1 equation, 2 unknowns [v_C, v_{R_1}]
For the capacitor	$v_C = v_C(0) + \frac{1}{C}\int (i_1 - i_2)dt$	(2) 2 equations. 4 unknowns [i_1, i_2]
For R_1	$v_{R_1} = R_1 i_1$	(3) 3 equations, 4 unknowns

<u>Mesh 2</u>

KVL	$v_R - v_C = 0$	(4) 4 equations, 5 unknowns [v_R]
For R	$v_R = R i_2$	(5) 5 equations, 5 unknowns

b. These equations can be combined into a single equation for v_R:

$$R_1 C\frac{dv_R}{dt} + \left(1 + \frac{R_1}{R}\right)v_R = v_S$$

One way to proceed is to find the particular solution for v_R, which represents the resistor voltage after the transients die out. Express the resulting sin and cos solution as a single sin term and find C by matching its amplitude to 0.004 (0.1% of the signal).
C = 0.8 F

10.28

a. $0.5\frac{d^2q}{dt^2}+\frac{dq}{dt}+\frac{1}{C}q=2\sin\omega t$

b. $q=e^{-t}\left[(10-B)\cos\sqrt{3}t+\left(\frac{10-B-A\omega}{\sqrt{3}}\right)\sin\sqrt{3}t\right]+A\sin\omega t+B\cos\omega t$

where $A=\frac{2(2-0.5\omega^2)}{(2-0.5\omega^2)^2+\omega^2}$ and $B=\frac{-2\omega}{(2-0.5\omega^2)^2+\omega^2}$

c. It takes roughly 5 seconds for the transients to "die-out."
d. The frequency producing the practical resonance is about 2.5 rad/s.

CHAPTER 11

11.1

a. 0.839
b. 0.279
c. 0.663
d. 1.270

11.2

49.33

11.4

V = –37.504, y = 910.8

11.9

Maximum velocity of 341 m/s at 50 sec.

11.10

0.052 kg

11.11

94 m in Boston, 102 m in Colorado

11.15

a. $k=11844\,\text{kg}/\text{s}^2$
b. p = 1131 kg/s
c. h = 0.11 m

11.17

$$k_1 \sim 180\,\mathrm{N/m}$$

11.30

a. 5.25 h

11.31

$$C_V = 0.6$$

11.33

0 s, 48 s, and 112 s

11.40

$$C = 0.8\ \mathrm{F}$$

Index

Note: Page numbers in *italics* indicate a figure and page numbers in **bold** indicate a table on the corresponding page.

A

accumulated energy, thermal systems, 331–332
acetic anhydride, hydrolysis of
  in adiabatic batch reactor, 343–345, *345*
adiabatic batch reactor, 85–87
  hydrolysis of acetic anhydride in, 343–345, *345*
  overview, 343
  thermal effects in, 343–345, *345*
air resistance, 16, 20, 199, 204–205
  drag force due to, 19–22, 200, *201*
algebraic equations, 2, 5, 101, *see also* differential equations (DEs)
algebraic manipulations, using Laplace transform, 134–136
ampere, 366
amplitude ratio, 407
analytical solution, of differential equation, 13
angular acceleration, 238
angular displacements, 249
angular velocity, 238, 246
antidifferentiation technique, 17–19
approximate, 81–82
Arrhenius, 343

B

batch processes, 287
  batch reactor, 289–290, *289*
  batch separation, 287–289, *287*
  batch time, 287
  closed system, 287, *287*
  mass balances, 287–290, *287*
  open system, 287
batch reactor, 289–290, *289*, 343, *see also* adiabatic batch reactor
  example, 293–296, *294*
batch separation, 287–289
batch time, 287
beats, 178–180
bioengineering
  example, 40
  overview, 298–299
bioengineering processes
  mass balance applications, 298–311, *299*
    biological reactions, 301–307
    compartmental modeling, 298–301, *300*
    fermentation, 307–311
biological reactions
  examples, 301–307
  mass balance application, 301–307
  product, 301
  substrate, 301
  substrate concentration *vs.* time, 307, *308*
biomass, 307
biomedical processes, 283, 298, *see also* bioengineering processes
biotechnology processes, 283, 305, 307, *see also* bioengineering processes
Biot number, 334, 335, 346, 350
block diagrams, DEs representation by
  basic blocks, 432–435
  clock block, 434
  constant block, 434
  construction, guidelines for, 435–436
  for coupled system, 437–439
  display block, 434
  examples, 436–439, 441–448
  gain block, 433
  integrator block, 432
  math function block, 433–434
  product block, 433
  scope block, 435
  sine wave, 440–441
  source blocks, 439–441
  step block, 440
  sum block, 433
Bode plot 170, *170*
boundary conditions, differential equation, 6–8, 345
boundary value problems
  defined, 345
  in heat transfer, 345–358
    examples, 345–358
  steady-state heat transfer from extended surface (example), 354–358, *354*
  steady-state temperature profile (example), 350–353, *350*, *354*
  temperature profile and heat flow in wire insulation (example), 345–350, *346*

C

capacitance, 370
capacitors, 369–371, *370*, *371*
  energy/power, 398
cart system (example), 134–135
  free body diagrams for, *211*
  with friction and external force, 214–216, *215–216*
  frictionless carts with external force, 211–214
characteristic equation, 48
  complex roots, 51–52
  real roots, 51
  repeated real roots, 51
  roots of, 51–52, *53*, 133
  using Euler's identity, 52
characteristic equation method, 47–50
  constant coefficients, 49
  homogeneous differential equation, 47–50

characteristic time, second-order systems, 150
chemical reactions
expressions for, 285–287, *285*
half-life, 287
reaction rate, 285
reaction rate constant, 286
chemical reactor, *285*
clock block, 434
closed-loop electrical circuit, *367*
closed system, 287, *287*, *see also* batch processes
coefficient of kinetic friction, 200
compartmental modelling, mass balance application, 298–301, *300*
complementary solution, nonhomogeneous differential equations, 55
complete solution, 47, 55
completing the square method, 116, 127–129
complex differentiation theorem, of Laplace transform, 107–108
complex roots
characteristic equation, 51–52
inversion of Laplace expressions, 114–117, 119–121
first-order expansion, 114–115
second-order expansion, 115–117
complex translation theorem, of Laplace transform, 107–108
concentration gradient, 283–284
concentrations, mass fraction, 273–275
conduction, 327–328, *328*
and convection in series, 329–331, *330*
defined, 327
through plate, 327–328, *328*
conservation of energy, *see* energy, conservation of
conservation of mass, *see* mass, conservation of
multicomponent systems, 270–272
types of processes, 272
constant block, 434
constant coefficients, 49
first-order linear differential equation with, 150
linear differential equation, 43–44
contact force, 250
control valve, 275, *275*
convection, 328–329, *329*
and conduction in series, 329–331, *330*
defined, 327
convective heat transfer
examples, 332–335
with internal generation, 336–339, *337*
conventional current, 366
coulomb, 366
Coulomb's friction law, 200
coupled first-order ODEs, 171–173, *171*
coupled system, block diagram for, 437–439
coupled undamped second-order ODEs, 173
damping ratio, *173*
mass-spring-damper, 174–176, *175*
oscillation mode, 177
responses of, *178*
weakly coupled oscillators and beats, 178–180
Cramer's rule, 76–77, 79
critically damped systems, 162
current
conventional, 366
defined, 366
cutoff frequency, 407–408, 411, 412
cycles per second/hertz, frequency in, 203

D

D'Alembert's principle, 207–209
damped system, 204–206
defined, 204
final response, 206
practical resonance, 206
pseudo-resonance, 206
response, 206
dampening in rotational system, 246–249, *247*
dampers, 198–199, *198*, *199*
in parallel, 220–222, *220*
in series, 220–222, *221*
damping coefficient, 199
damping forces, mechanical systems (translational), 199–200, *201*
damping of oscillations, 173
damping ratio
example, 180–181, 183
second-order systems, 161, **162**, 163
Darcy's equation, 298
Darcy's law, 297
dashpots (pistons or dampers), 198–199, *198*, *199*
decaying exponential, 333
decay ratio
examples, 182–183
second-order systems, 164
definite integrals, 18
degrees of freedom (DoF)
defined, 24
objects in gravitational field, 22–25
dependent variable, 5, 9
at steady state, 8
DEs, *see* differential equations (DEs)
deviation variables, Laplace transform, 136–140
dielectric, 369
differential equations (DEs), 4, 5–9
analytical solution of, 13
bioengineering, 40
boundary condition, 6–8
defined, 5
dependent variable, 5, 9
electrical circuit, 42–43, *43*
examples, 40–43
first-order, 5–6
fluid system, 41–42, *41*
forcing function, 5, 10–13
homogeneous, *see* homogeneous differential equation
independent variable, 5, 8, 9
initial condition, 6–8
initial value problem, 7
integrating factor method, 40
linear, *see* linear differential equation
mechanical translational model, 41, *41*
nonhomogeneous, *see* nonhomogeneous differential equations
*n*th-order, 7
numerical simulation of, 427, **428**
order, 5–6

ordinary (ODEs), *see* ordinary differential equations (ODEs)
partial (PDEs), *see* partial differential equations (PDEs)
representation by block diagrams
basic blocks, 432–435
clock block, 434
constant block, 434
construction, guidelines for, 435–436
display block, 434
examples, 436–439, 441–448
gain block, 433
integrator block, 432
math function block, 433–434
product block, 433
scope block, 435
sine wave, 440–441
source blocks, 439–441
step block, 440
sum block, 438
second-order, 7–8, 48, 51
separation of variables, 20–22
thermal system, 42, *42*
third-order, 5
differential equations (DEs), solution using Laplace transform, 109, 110–132
inversion by partial fraction expansion, 112–129
completing the square method, 116, 127–129
complex roots, 114–117, 119–121
examples, 117–129
repeated real roots, 118–119
unrepeated real roots, 117–118
step response, 111
time delays, handling, 130–132, *131*
Dirac delta function, 103
display block, 434
DoF, *see* degrees of freedom (DoF)
drag forces, 199, *200*
due to air resistance, 19–22, 200, *201*
drum and mass system (example), *252*, 253
dry friction forces, 199, *200*
dynamic equilibrium, 207
dynamic models, 10
dynamic response, 440

E

Einstein's relation, 268
electrical charge, 366
electrical circuits
examples, 42, *43*, 139–140, *139*, 182–183, *183*, 373–386
LC circuits, 383–386, *383*, *385*
mesh method, 386–389, 391–392
nodal method, 386, 389–390
RC circuit, 373–378, *373–374*
RLC circuit, 380–381, *380*
RL circuit, 373–378, *373–374*
undamped circuit (natural frequency and resonance), 383–386
electrical conductors, 367
electrical laws, 367–368
electrical systems, 366–414
capacitance, 370
capacitor, 369–371, *370*, *371*
closed-loop electrical circuit, *367*
conventions, 366
current, 366
definitions, 366
electrical charge, 366
electrical components, 369–372
electrical conductors, 367
electrical laws, 367–368
energy and power, 397–407
capacitors, 398
examples, 399–407
inductors, 398–399
resistors, 398, *398*
examples, 387–389, 392–397, 399–407
electrical circuits, 373–386, *see also* electrical circuits
ideal current source, 366, *366*
ideal voltage source, 366, *366*
inductance, 371
inductor, 371–372, *372*
initial conditions, 372–373, *372–373*
Kirchhoff's current law (KCL), 367, *367*, 368
Kirchhoff's voltage law (KVL), 367–368, *367*
mesh method, 391–392
nodal method, 389–391
overview, 366
RC circuits as filters, 407–413
high-pass filter, 409–411, *411*
low-pass filter, 412–413, *413*, *414*
resistance, 369
resistor, 369, *369*
voltage drop, 366, 367
energy, conservation of, 326
energy balance, 335, 337, 341, 343–344
energy/power, electrical systems, 397–407
capacitors, 398
examples, 399–407
inductors, 398–399
resistors, 398, *398*
enthalpy, 340–341
Euler formulas, 430–431
Euler's identity, 52, 115
using, characteristic equation, 52
Euler's method
for first-order ordinary differential equations, 428–429
for second-order ordinary differential equations, 429–431
step size, 431, *431*
experimental facts and elements, mass balances
flow element, 275, *275*
gas service, 277
liquid service, 275–276, *275*
extended surface, steady-state heat transfer from (example), 354–358, *354*

F

FBD, *see* free body diagram (FBD)
fermentation
examples, 309–311
mass balance application, 307–311
yield coefficient, 309

filters, RC circuits as, 407–413, *408*
 high-pass filter, 409–411, *411*
 low-pass filter, 412–413, *413*, *414*
final response
 damped system, 206
 linear differential equations, 87–90, *89*
final value theorem, of Laplace transform, 107
first-order differential equation, 141–142, 148
first-order expansion, Laplace expressions, 114–115, 120
first-order linear differential equation
 with constant coefficients, 149–150
 integrating factor method, 44–47
 standard form of, 187–188
 system gain, 150
 time constant, 150
first-order ordinary differential equations
 antidifferentiation, 17–19
 Euler's method for, 428–429
 separation of variables, 20–22
first-order systems, 171–173, *171*
 first-order linear differential equation
  with constant coefficients, 149–150
  standard form of, 187–188
  system gain, 150
  time constant, 150
 impulses, 153–157, **154**, *155*, *156*
 problems, 187–195
 rectangular pulse, 153–157, **154**, *155*, *156*
 response of, 148–160
 sinusoidal signal, 157–160, *159*
 step changes, 150–151, **150**, *151*, *153*
  gain for, 152–153
  step change in, 151–152
  time constant for, 153
 transfer function, 160
flow element, mass balance, 275, *275*
flow rate(s), 339
 defined, 273
 mass, 273
 mass balances, 273–275
 mole, 273
 volumetric, 273
flow system
 heating of vessel with inflows and outflows (example), 341–343
 heat transfer in, 339–343, *340*
fluid friction forces, 200, *200*
fluid friction model, 200, *200*
fluid system, example, 41–42, *41*
forced response
 nonhomogeneous differential equations, 55, 56–57
forces
 damping forces, 199–200, *201*
 gravity force, 199
 mechanical systems (translational), 199–200
forcing functions, 5, 10–13
 examples, 12
 Laplace transforms, 108–109
 linear differential equation, 43
 nonhomogeneous differential equations, 58
  multiple forcing functions, 68–69, *69*
Fourier's law, 347, 351, 352, 355, 357, 365
free body diagram (FBD), 207–209, *207–208*, 238
frequency
 in cycles per second/hertz, 203
 in radians per second, 203
friction, 199

G

gain block, 433
gas service, mass balance, 277
gas tank system (example), 281–283, *281*
gears, 249–252, *250*
general solution, 47, 280, 356
governing equation, 9
gravitational field, objects in
 antidifferentiation, 17–19
 complex roots, 32
 degrees of freedom, 22–25
 drag force due to air resistance (example), 19–22
 equations, 22–25
 example, 16–20
 first-order expansion, 32–34
 Newton's second law, 16
 overview, 16
 partial fraction expansion, 25–34
 problems, 34–39
 repeated real roots
  algebra, 28–30
  limit, 31–32
 separation of variables, 20–22
 unknowns, 22–25
 unrepeated real roots
  algebra, 27–28
  limit, 30–31
gravity, 16
 vertical systems and, 222–223
gravity force, mechanical systems (translational), 199

H

half-life, 287
head of liquid, 2
heat, 326
 capacity, 42, 331, 341
heat flow, 326
 temperature profile and, in wire insulation (example), 345–350, *346*
heat of reaction, 344
heat transfer, 326
 boundary value problems in, 345–358
 by convection and conduction, 329–331, *330*
 in flow system, 339–343, *340*
 modes of, 326–327
 steady-state, from extended surface (example), 354–358, *354*
heat transfer coefficient, 329
higher-degree polynomials, 26
higher-order systems, variation of parameters method, 77–80
high-pass filter, RC circuits as, 409–411, *411*
homogeneous differential equation, 44
 characteristic equation method, 47–50
 corresponding, 56–57
 natural response, 56–57

qualitative system response, 52–55
roots of characteristic equation, 51–52
solution of, 47–55
Hooke's law, 197, 209, 244
hydrolysis, of acetic anhydride
in adiabatic batch reactor, 343–345, *345*
hydrostatic pressure, 276

I

ideal current source, 366, *366*
ideal gas law, 277
ideal pulley, 199, *199*
ideal spring, 198, *198*
ideal voltage source, 366, *366*
indefinite integrals, 18
independent position variable, 345
independent variable, 5, 8, 9
inductance, 371
inductors, 371–372, *372*
energy/power, 398–399
inertial force, 207
inertial torque, rotational mechanical systems, 239
initial conditions
differential equation, 6–8, 345
electrical systems, 372–373, *372–373*
second-order differential equation, 7–8
initial value problems, 345
differential equation, 7
initial value theorem, of Laplace transform, 108
insulated wire, temperature profile and heat flow in (example), 345–350, *346*
integrating factor method, 40, 44–47
development of, 46–47
integrator block, 432
internal energy, 326, 331–332, 344
internal heat generation, 336, 350–353
inverse Laplace, 101
isothermal systems, 355

J

jacketed, stirred tank, liquid heating in (example), 335–336, *335*

K

Kirchhoff's current law (KCL), 367, *367*, 368, 386
Kirchhoff's voltage law (KVL), 367–368, *367*, 372, 386, 409, 411–412

L

Laplace domain, 101
Laplace transforms/Laplace transform method, 25, 80, 101–140, 212, 295, 375
algebraic manipulations using, 134–136
background, 101
of common functions, **105**
complex differentiation theorem, 107
complex translation theorem, 108
definition, 102–104
pulse, 103, *104*
sine wave, *102*, 104
unit constant, 103
unit impulse function, *102*, 103–104
unit step function, *102*, 103
deviation variables, 136–140
differential equations solution using, *see* differential equations (DEs), solution using Laplace transform
final value theorem, 107
forcing functions, 108–109
initial value theorem, 108
linearity property, 104
overview, 101
partial fraction expansion, 34
properties, 104–110
real differentiation theorem, 105–106
real integration theorem, 106
real translation theorem, 106–107, *107*
theorems of, 104–110
transfer function, 112, 132–134
law of conservation of mass, *see* mass, conservation of
LC circuits (example), 383–386, *383*, *385*
linear approximation, 81–87, *82*
linear differential equation
constant coefficients, 44
defined, 43–44
example, 44
forcing function, 44–45
integrating factor method, 44–47
ordinary, *see* ordinary linear differential equations
time-invariant coefficients, 44
linearity property, of Laplace transform, 104
linearization, of nonlinear differential equations, 81–87
linear systems, forcing functions, 43
line of action, 241
liquid heating, in stirred tank (example), 335–336, *335*
liquid service, mass balance, 275–276, *275*
logarithms, development of, 101
low-pass filter, RC circuits as, 412–413, *413*, *414*
Luedeking–Piret model, 309 310
lumped parameter, 332, 333

M

Maclaurin series, 81
mass, conservation of, 268–272, *268*
law of, 9, 268
multicomponent systems, 270–272
types of processes, 272
mass acceleration diagram (MAD), 207
mass balance(s), 268–311
application of, 268–269
batch processes, 287–290, *287*
bioengineering processes application of, 298–311, *299*
biological reactions, 301–307
compartmental modeling, 298–301, *300*
fermentation, 307–311
chemical reactions, 285–287, *285*
concentrations, 273–275
conservation of mass, 268–272, *268*
multicomponent systems, 270–272
types of processes, 272
elements and experimental facts

flow element, 275, *275*
gas service, 277
liquid service, 275–276, *275*
examples, 277–283, 290–298
batch reactor, 293–296, *294*
gas tank system, 281–283, *281*
mixing tank system, 279–280, *279*
tank system, 277–279
water treatment basins, 290–293, *290*, *292*
expression for, 268
flow rates, 273–275
generic system, 268–269, *268*
mass transport, 283–285, *284*
overview, 268
steady-state assumption, 269
mass density, defined, 273
mass flow rate, 273
mass fraction, defined, 273–274
mass moment of inertia, rotational mechanical systems, 239–241
mass-spring-dashpot system (example), *41*, 137–138, 205, *205*, *207*
mass-spring system (example), 180–182, *182*
mass transfer rate, 283
mass transport
concentration gradient, 283–284
expressions for, 283–285, *284*
mass transfer rate, 283
Mathcad™, 26
mathematical model(ing), 2, 9, 13
development, 9
governing equation, 9
math function block, 433–434
MATLAB™, 26
mechanical law, 196
rotational systems, 238–239
translational systems, 196
mechanical systems (rotational), *see* rotational mechanical systems
mechanical systems (translational), *see* translational mechanical systems
mechanical translational model, example, 41, *41*, 180, *181*
mesh method, electrical circuits, 386–389, 391–392
Michaelis–Menten equation, 301
mixing tank process (example), 1–4
mixing tank system (example), 279–280, *279*
model, defined, 13
modeling, 9–10, *see also* mathematical model(ing)
development, 10
governing equation, 10
molar density, defined, 273
mole balance, *see also* mass balance(s)
expression for, 270
steady-state systems, 270
unsteady-state systems, 270
molecular mass, 269
mole flow rate, 273
mole fraction, defined, 274
mole percentage, 274
mole(s), defined, 269
moment arm, 241
Monod model, 307, 308
monotonic response, 60–61
multicomponent systems, mass balance, 272–274
multiple forcing functions, nonhomogeneous differential equations, 68–69, *69*

N

natural frequency, 62, 203, 170, 203
undamped circuit, 383–386
natural response
homogeneous differential equation, 56–57
Newton's law of cooling, 329, 349, 355
Newton's second law, 16, 19, 22, 196, 201, 205, 207, 211, 238–239
Newton's third law, 196
nodal method, electrical circuits, 386, 389–390
nonhomogeneous differential equations, 44
complementary solution, 55
forced response, 55–57
forcing functions, 58
multiple forcing functions, 68–69, *69*
particular solution, 55, 57, **57**
solution of, 55–69
undetermined coefficients, 57–68
nonlinear differential equations
adiabatic batch reactor (example), 85–87
linearization of, 81–87
nonlinear systems, 444
*n*th-order constant-coefficient (time-invariant) linear homogeneous ordinary differential equation, 47
particular solution of, 57, **57**
*n*th-order differential equation, 7, 51
numerical simulation, 427–454
block diagrams, DEs representation by
basic blocks, 432–435
clock block, 434
constant block, 434
construction, guidelines for, 435–436
display block, 434
examples, 436–448
gain block, 433
integrator block, 432
math function block, 433–434
product block, 433
scope block, 435
sine wave, 440–441
source blocks, 439–441
step block, 440
sum block, 433
of differential equations, 427, **428**
Euler's method
for first-order ordinary differential equations, 428–429
for second-order ordinary differential equations, 429–431
examples, 441–448
overview, 427
Runge–Kutta methods, 432
Simulink and modelling
Simulation tip, 451–454
trial and error (convergence), 448-451, *449–451*
writing equations, 448

step size, 431, *431*
system and study, 451–452, *451*, *452*
Taylor series, 432

O

objects in a gravitational field, *see* gravitational field, objects in
ODEs, *see* ordinary differential equations (ODEs)
Ohm's law, 369
open system, 287, 340
operating conditions, 444
order, differential equation, 5
ordinary differential equations (ODEs), 8–9
first-order, technique for solving, 17–19
antidifferentiation, 17–19
separation of variables, 20–22
ordinary linear differential equations
classical solutions, 40–90
defined, 43–44
examples, 40–43
final response, 87–90
homogeneous
characteristic equation method, 47–50
natural response, 56–57
qualitative system response, 52–55
roots of characteristic equation, 51–52
solution of, 47–55
integrating factor method, 44–47
development of, 46–47
nonhomogeneous
forced response, 56–57
multiple forcing functions, 68–69, *69*
solution of, 55–69
undetermined coefficients, 57–68
nonlinearities and variable coefficients, 80–87
linearization, 81–87
Taylor series, 80–81
transient response, 87–90
variation of parameters
higher-order systems, 77–80
second-order systems, 70–77
oscillatory response, 53
overall heat transfer coefficient, 330
overshoot, second-order systems, 165

P

partial differential equations (PDEs), 8–9
partial fraction expansion, 25–27, 34
completing the square method, 116, 127–129
complex roots, limit concept, 32
first-order expansion, 32
inversion of Laplace expressions, 112–129
complex roots, 114–117, 119–121
examples, 117–129
repeated real roots, 118–119
unrepeated real roots, 117–118
objects in gravitational field, 25–34
repeated real roots, 28–30
limit concept, 31–32
unrepeated real roots, 27–28
limit concept, 30–31
particular solution, nonhomogeneous differential equations, 55, 57, **57**
PDEs, *see* partial differential equations (PDEs)
pendulum (example), 241–244, *241*, *242*
periodic current, 410
physical law for systems, 268
pistons, 198–199, *198*, *199*
polynomials, 25
higher-degree, 26
power, electrical systems, *see* energy/power, electrical systems
practical resonance, 386
damped system, 206
principle of superposition, 44, 56
process(es)
components, 9
mass balances, 272
types of, 272
valve, 275
product, biological reactions, 301
product block, 433
pseudo force, 207
pseudo-resonance, 426
damped system, 206
pulleys, 199, *199*
pulse, Laplace transform, 103, *104*

Q

qualitative system response, homogeneous differential equation, 52–55
monotonic, 53
oscillatory, 53
stable, 53
unstable, 53

R

radians per second, frequency in, 203
radiation, defined, 327
rate of mass, 2–3
RC circuits
examples, 373–378, *373–374*
as filters, 407–413, *408*
high-pass filter, 409–411, *411*
low-pass filter, 412–413, *413*, *414*
voltage drop across resistor in, 411
reactive system, thermal effects in, 343–345, *345*
real differentiation theorem, of Laplace transform, 105–106
real gases, 277
real integration theorem, of Laplace transform, 106
real roots, characteristic equation, 51
real spring, 198, *198*
real translation theorem of Laplace transform, 106–107, *107*
relative displacement, 210
relative velocities, 210

repeated real roots
  characteristic equation, 51
  partial fraction expansion, 28–32
    inversion of Laplace expressions, 118–119
resistance, 369
resistors, 369, *369*
  energy/power, 397–398, *398*
resonance, 64
  damped system, 203
  undamped circuit, 383–386
  undamped system, 203–204
resonant frequency, 62, 170, 203
response, system
  damped system, 206
  FBD method, 209, *209*
  first-order systems, 148–160
  forced, 55
  monotonic, 53
  oscillatory, 53
  qualitative, 52–55
  second-order systems, 160–171
  stable, 52, 53
  for undamped system, 203–204, *204*
  unstable, 52, 53
rise time, second-order systems, 165
RLC circuit (example), *7*, 185–186, 380–381, *380*, 399–406, *399*
RL circuit (example), 373–378, *373–374*
rotational damping systems, 246–249, *247*
rotational mechanical systems, 238–254
  angular acceleration, 238
  angular position, 238, *239*
  angular velocity, 238
  gears, 249–252, *250*
  inertial torque, 239
  mass moment of inertia, 239–241
  mechanical law, 238–239
  moment, 238–239
  overview, 238
  rotational damping, 246–249, *247*
  torque, 239, 241–244
  torsion springs, 244–246, *245*
  with translational elements, 252–254, *252*
  and translational systems
    relations between, 238, **239**
Runge–Kutta methods, 432

S

scope block, 435
second-order constant-coefficient (time invariant) linear nonhomogeneous ordinary differential equation, 56
second-order differential equation, 48, 51, 161–162
  initial conditions, 7–8
second-order expansion, Laplace expressions, 115–117, 120–121
second-order ordinary differential equations, Euler's method for, 429–431
second-order systems, 160–162, **161**, **162**, 186–187
  bode plots, 170, 187
  characteristic time, 163
  coupled undamped second-order ODEs, *see* coupled undamped second-order ODEs
  damping ratio, 161, *162*
  decay ratio, 165
  examples, 180–186
  impulses, 166–168, **166**, *167*, *168*
  overshoot, 165
  period of oscillation, 165
  problems, 187–195
  rectangular pulse, 166–168, **166**, *167*, *168*
  response of, 160–171, *163*
  rise time, 165
  second-order differential equation, 161–162
  settling time, 165
  sinusoidal signal, 168–171
    bode plots, 170
    magnitude plot, 171
    phase plot, 171
  step changes, 162–164, **163**, *163*
    critically damped response, 165
    overdamped response, 165–166
    undamped response, 164
    underdamped response, 164–165, *164*, **165**
    unstable responses, 166
  system gain, 184
  transfer function, 171
  variation of parameters method, 70–77
settling time, second-order systems, 165
simulation, 14, *see also* numerical simulation
Simulink®, 427, 432, 434, 440, 448, *450*, 452, *452*, *453*
sine wave
  block diagram, 440
  Laplace transform, *102*, 104
single block diagram, 437
  second-order systems, 168–171
  sinusoidal signal response first-order systems, 157–160, *159*
slender bar (example), 240–241, *240*
source blocks, 439–441
  sine wave, 440–441
  step block, 440
specific enthalpy, 340
spring constant, 198
spring–mass system, *205*
springs, 196–198, *197*, *198*
  examples, 222–225
  forces generated by, 196–198, *197*
  ideal, 198, *198*
  in parallel, 218–219, *218*
  real, 198, *198*
  reference position, 197
  in series, 219–220, *219*
  torsion springs, 244–246, *245*
stable response, 53
steady state, 1–2, *3*
  dependent variable at, 8
steady-state assumption, mass balance, 269
steady-state heat transfer from extended surface (example), 354–358, *354*
steady-state models, 10
steady-state response, 206
steady-state systems, 270

steady-state temperature profile (example), 350–353, *350*, *354*
step block, 440
step function response
first-order systems, 150–153, **150**, *151*, *153*
second-order systems, 162–164
undamped response, *163*, 164
underdamped response, 164–165, *164*, **165**
step size, in Euler's method, 431, *431*
stored energy, thermal systems, 331–332
substrate, biological reactions, 301
sum block, 433
system gain
first-order linear differential equation, 150
second-order systems, 184
system identification examples, 183–185

T

tank system (example), 277–279
Taylor series, 80–81, 432
temperature profile and heat flow in wire insulation (example), 345–350, *346*
terminal velocity, 22
thermal conductivity, 328
thermal systems, 326–358
accumulated/stored energy, 331–332
boundary value problems in heat transfer, 345–358
conduction, 327–328, *328*
conduction and convection in series, 329–331, *330*
conservation of energy, 326
convection, 328–329, *329*
effects in reactive system, 343–345, *345*
examples, 42, *42*
convective heat transfer, 332–335
convective heat transfer with internal generation, 336–339, *337*
liquid heating in stirred tank, 335–336, *335*
heat transfer in flow system, 339–343, *340*
heat transfer modes, 327
overview, 326
third-order differential equation, 5–6
time constant, 153
first-order linear differential equation, 153
time delays, handling, Laplace transforms and, 130–132, *131*
time-dependent differential equations, 345
time-invariant coefficients, linear differential equation, 44
torque, rotational mechanical systems, 239, 241–244
torsion springs, rotational mechanical systems, 244–246, *245*
transfer function
concept of, 132
defined, 132
first-order systems, 160
Laplace transform, 112, 132–134
second-order systems, 171
transient response, ordinary linear differential equations, 87–90, *89*
translational mechanical systems, 196–225
D'Alembert's principle and free body diagrams, 207–209
dampers
in parallel, 220–222, *220*
in series, 220–222, *221*
dashpots (pistons or dampers), 198–199, *198*, *199*
examples, 209–218
forces, 199–200
damping forces, 199–200, *201*
gravity force, 199
ideal pulley, 199, *199*
mechanical law, 196
overview, 196
with rotational elements, 252–254, *252*
and rotational systems, relations between, 238, **239**
springs
in parallel, 218–219, *218*
in series, 219–220, *219*
system components
dashpots, 198–199
ideal pulley, 199
springs, 196–198, *197*, *198*
types of, 200–206
undamped system, 201
classic equation of, 202
mass–spring system, 202
natural/resonant frequency, 203
with and without resonance, *204*
vertical systems, 222–225
trial solution, 57

U

ultimate velocity, 254
undamped circuit (example), 383–386
undamped step response, second-order systems, *163*, 164
undamped systems, 63, 201–204, *202*
classic equation of, 202
defined, 201
resonance, 203–204
response for, 203–204, *204*
underdamped response, 164–165
underdamped step response, second-order systems, 164–165, *164*, **165**
undetermined coefficients method, nonhomogeneous differential equations, 57–68
unit constant, Laplace transforms, 103
unit impulse function, Laplace transforms, *102*, 103–104
unit step function, Laplace transforms, *102*, 103
unknowns, 22–25
unrepeated real roots, partial fraction expansion, 27–28
inversion of Laplace expressions, 117–118
unstable response, 53, 166
unsteady-state systems, 270

V

valve coefficient, 2, 275
variable coefficients, 80–87
approximate analytical solution, 81–87
Taylor series, 80–81
variable(s)
dependent, 5
independent, 5
separation of, 20–22

variation of parameters method
linear differential equations
higher-order systems, 77–80
second-order systems, 70–77
velocity
model of drag force due to air resistance, 19, 20, 22
terminal, 22
vertical mechanical system, 222–225, *223*
examples, 222–225
gravity and, 222–223
viscous friction model, 200
volt, 366, 367
voltage drop, 366, 367
across resistor in RC circuits, *411*
volumetric flow rate, 273

W

water treatment basins (example), 290–293, *290*, *292*
weakly coupled oscillators, 178–180, *180*
Wronskian of the *k*th column, 76, 273

FSC
www.fsc.org
MIX
Papier | Fördert
gute Waldnutzung
FSC® C083411

Kolibri 360 GmbH
Ferdinand-Jühlke-Straße 7
99095 Erfurt, Deutschland
produktsicherheit@kolibri360.de